A First Course in Computational Fluid Dynamics

Fluid mechanics is a branch of classical physics that has a rich tradition in applied mathematics and numerical methods. It is at work virtually everywhere you look, from nature to technology.

This broad and fundamental coverage of computational fluid dynamics (CFD) begins with a presentation of basic numerical methods, and proceeds with a rigorous introduction to the subject. A heavy emphasis is placed on the exploration of fluid mechanical physics through CFD, making this book an ideal text for any new course that simultaneously covers intermediate fluid mechanics and computation. Ample examples, problems and computer exercises are provided to allow students to test their understanding of a variety of numerical methods for solving flow physics problems, including the point-vortex method, numerical methods for hydrodynamic stability analysis, spectral methods and traditional CFD topics.

A First Course in Computational Fluid Dynamics

H. AREF
Virginia Polytechnic Institute and State University

S. BALACHANDAR
University of Florida

CAMBRIDGE
UNIVERSITY PRESS

University Printing House, Cambridge CB2 8BS, United Kingdom
One Liberty Plaza, 20th Floor, New York, NY 10006, USA
477 Williamstown Road, Port Melbourne, VIC 3207, Australia
4843/24, 2nd Floor, Ansari Road, Daryaganj, Delhi – 110002, India
79 Anson Road, #06-04/06, Singapore 079906

Cambridge University Press is part of the University of Cambridge.

It furthers the University's mission by disseminating knowledge in the pursuit of education, learning and research at the highest international levels of excellence.

www.cambridge.org
Information on this title: www.cambridge.org/9781107178519
DOI: 10.1017/9781316823736

First published 2018

Printed in the United Kingdom by TJ International Ltd. Padstow Cornwall

A catalog record for this publication is available from the British Library

Library of Congress Cataloging-in-Publication Data
Names: Aref, Hassan, 1950–2011 author. | Balachandar, S., author.
Title: A first course in computational fluid dynamics / H. Aref (Virginia Polytechnic Institute and State University), S. Balachandar (University of Florida).
Description: Cambridge, United Kingdom ; New York, NY : Cambridge University Press, 2017. | Includes bibliographical references and index.
Identifiers: LCCN 2016052764 | ISBN 9781107178519 (hardback ; alk. paper) | ISBN 1107178517 (hardback ; alk. paper) | ISBN 9781316630969 (pbk. ; alk. paper) | ISBN 131663096X (pbk. ; alk. paper)
Subjects: LCSH: Computational fluid dynamics–Textbooks.
Classification: LCC TA357.5.D37 A74 2017 | DDC 532/.0501515–dc23
LC record available at https://lccn.loc.gov/2016052764

ISBN 978-1-107-17851-9 Hardback
ISBN 978-1-316-63096-9 Paperback

Contents

Preface

This book has been in preparation for over a decade. Hassan Aref and I had been making substantial additions and revisions each year, in our desire to reach the perfect book for a first course in Computational Fluid Dynamics (CFD). I sincerely wish that we had completed the book a few years ago, so that Hassan was there when the book was published. Unfortunately this was not the case. September 9th 2011 was a tragic day for fluid mechanics. We lost an intellectual leader, a fearless pioneer and for me an inspirational mentor. It is quite amazing how my academic career intersected with Hassan's over the years. He was my Masters Thesis advisor at Brown University. But when Hassan left for the University of California at San Diego, I decided to stay and finish my PhD at Brown. A few years later, when I was an assistant professor at the Theoretical and Applied Mechanics Department at the University of Illinois at Urbana–Champaign, he was appointed as the head of the department. This is when he asked me to join him in this project of writing a non-traditional introductory book on CFD. The project was delayed when Hassan moved to Virginia Tech as the Dean and I moved to the University of Florida as the Chair of the Mechanical and Aerospace Engineering Department. His passing away a few years ago made me all the more determined to finish the book as a way to honor his memory. I am very glad that the book is now finished and can be a monument to his far-sighted vision. Decades ago when CFD was in its infancy he foresaw how powerful computers and numerical methods would dominate the field of fluid mechanics.

Hassan and I shared a vision for this book. Our objective was to write something that would introduce CFD from the perspective of exploring and understanding the fascinating aspects of fluid flows. We wanted to target senior undergraduate students and beginning graduate students. We envisioned the student to have already taken a first level course in fluid mechanics and to be familiar with the mathematical foundations of differential equations. For such a student this book would serve as an ideal source for learning intermediate fluid mechanics, numerical methods and basic CFD – all three topics in an integrated manner. There are many excellent books on numerical methods that cover different techniques of temporal and spatial discretization of ordinary and partial differential equations. They are often written from the perspective of applied mathematics and their focus is on establishing accuracy, stability and convergence properties. In these books CFD is viewed as numerical methodology for a system of nonlinear partial differential equations. On the other hand, there are outstanding books that are dedicated to the topic of CFD. These generally dive straight in the deep-end, with descriptions of complex numerical implementations and associated code listings for solving Navier–Stokes equations in complex three-dimensional settings. From this engineering perspective CFD is viewed as a design and development tool and therefore a considerable fraction of these books is spent on training students to be developers and expert users of large-scale CFD codes.

Our book is different. We start at a considerably lower level by presenting a sequence of basic numerical methods for time and space discretization and build our way towards CFD. Admittedly, as a result, this book will only be a first course in CFD, but the philosophy of CFD as a computational approach for exploring the science of fluid mechanics will be firmly established at an early stage. We take the view that CFD is not limited to time-dependent numerical solutions of the three-dimensional Navier–Stokes equations. On the way to the Navier–Stokes equations we will address numerical techniques for solving point-vortex equations, Kelvin–Kirchhoff equations of motion of a solid in ideal fluid, and Orr–Sommerfeld equations of hydrodynamic stability, which are simpler mathematical models derived from application of appropriate fluid mechanical principles. It is not always necessary to solve the full-blown Navier–Stokes equations in 3D – it may be far more insightful to first obtain a simpler mathematical model and to seek its numerical solution. Furthermore, even if the final answer being sought means solving the Navier–Stokes equations in 3D, it may be advantageous to first start with a simpler model problem. For example, by starting with the Euler equations and progressively increasing viscosity one could pin-point the role of viscosity. Or, before solving the full-blown Navier–Stokes equations in 3D, one could (or must) validate the code by testing it on a simpler limiting problem, whose solution is either known analytically or obtained to very high accuracy with a simpler code.

In essence, the philosophy followed in the book is best articulated by Hamming (1973): the purpose of CFD is physical insight and not just numerical data. As a corollary, the following guiding principle is adopted in this book: always first reduce the problem to its simplest possible mathematical form before seeking its numerical solution. Or in the language of this book, first formulate your (mathematical) *model* before starting to work on your (numerical) *method*. Thus, this book strives to establish a solid foundation where CFD is viewed as a tool for exploring and explaining the varying facets of fluid mechanics. With the foundation laid in this book you will be ready to tackle any in-depth treatment of CFD found in more advanced textbooks.

Pedagogically we feel it is important to learn numerical methods and CFD techniques in context. We will present numerical algorithms and methods in conjunction with a rich collection of fluid mechanical examples, so that the book will also serve as a second course in fluid mechanics. For example, what better way to learn time-integration techniques for a system of initial-value ODEs than by solving point-vortex equations. Regular and chaotic motion, roll-up of a vortex sheet, conservation constraint in the form of a Hamiltonian, and the extreme sensitivity of the vortex sheet to round-off error are prime examples of characteristics exhibited by numerical solutions of many physical systems that can be best studied in the context of point-vortex systems. Similarly, numerical techniques for two-point eigenvalue problems are best illustrated in the context of Orr–Sommerfeld or Rayleigh stability equations. This way the reader gets to learn not only appropriate numerical techniques for two-point eigenvalue equations, but also gets excellent exposure to hydrodynamic stability theory and important related fluid-mechanical concepts such as inviscid and viscous instability, neutral curve, Squires theorem and inflection-point theorem.

Another motivation for this book stems from our desire to impact the typical undergraduate engineering curriculum in the United States and perhaps in many other countries as well. With decreasing credit hours needed for graduation and increasing amount of information needed to be at the cutting-edge, a typical mechanical, aero, chemical, civil or environmental engineering undergraduate curriculum does not include a CFD course or a second course in fluid mechanics, even as a technical elective. This book will enable a single three-credit course that combines three topics that must otherwise be taught separately: (i) introduction to numerical methods; (ii) elementary CFD; and (iii) intermediate fluid mechanics. The content of this book can also be used as the material for an introductory graduate level course on CFD. Such undergraduate and graduate courses based on versions of this book have been offered both at the University of Illinois, Urbana–Champaign campus and at the University of Florida.

I want to thank Susanne Aref for her support and encouragement in seeing this book to completion as one of the lasting contributions of Hassan to the field of fluid mechanics. I want to thank David Tranah and the Cambridge University Press for their immense help with the preparation and publication of the book. I want to thank B.J. Balakumar for the MATLAB codes that produced some of the figures in the book. I also want to thank the students who took the Computational Mechanics course at the University of Illinois and the CFD course at the University of Florida for spotting numerous typos and for allowing me to reproduce some of their homework figures. Finally, I want to thank my family for their support and understanding, as I spent countless hours working on the book.

S. Balachandar
Gainesville, Florida
December 2016

1 CFD in Perspective

Computational fluid dynamics, usually abbreviated CFD, is the subject in which the calculation of solutions to the basic equations of fluid mechanics is achieved using a computer. In this sense, CFD is a subfield of fluid mechanics, but one that relies heavily on developments in fields outside of fluid mechanics proper, in particular *numerical analysis* and *computer science*. CFD is sometimes called numerical fluid mechanics.

It is a prevailing view that the computer revolution has produced a "third mode" of scientific investigation, often called *scientific computing* or *computational science*, that today enters on a par with the more established modes of (theoretical) analysis and (laboratory) experimentation. CFD is this new "third mode" of doing fluid mechanics.

1.1 The Nature of CFD

Our objective is to explore CFD from a basic science perspective. We take as a given the power and elegance of analytical reasoning in fluid mechanics, and the precision and persuasiveness of well-executed experiments and observations. We wish to demonstrate that CFD has ingredients from both these traditional ways of doing fluid mechanics, and, furthermore, injects a set of unique viewpoints, challenges and perspectives. The subject has many very new aspects, due in large part to the rapid and recent development of computer technology. Yet it also has considerable historical and intellectual depth with many contributions of great beauty and profundity that derive directly from cornerstone developments in mathematical analysis and physical theory.

There is often a very special flavor to a CFD investigation. Sometimes this arises from the unique combination of trying to describe a continuum using a finite set of states (and, thus, ideas from discrete mathematics). On other occasions one has the challenge to address a question that is still too difficult for conventional analysis and is largely inaccessible to laboratory experiments. An example of this kind is whether the three-dimensional Euler equations for incompressible flow generate singularities in a finite time starting from smooth initial data. In a computation we can often impose a constraint or symmetry that is very difficult to achieve or maintain experimentally, such as exactly two-dimensional flow. In a computation we can increase the nonlinearity in the problem gradually, from a regime conforming to a linearized description, that is usually amenable to analysis, to a weakly nonlinear regime (where analysis typically is much harder), and finally, to a fully nonlinear range (where analytical understanding may be altogether lacking). We usually have control over initial and boundary conditions to a degree of precision unimaginable in the laboratory. In a computation we can realize with ease experimentally dubious yet theoretically useful approximations, such as potential flow or periodic boundary conditions. Constitutive relations for the fluid concerned, e.g., non-Newtonian fluids, can be stated exactly and varied more or less at will. One might say that computational studies of fluid motion are not constrained by reality! For many computational methods flow geometries can be changed quickly and with great flexibility. Theoretically motivated experiments such as turning certain terms on or off in the equations may be done. And with the detail available in most CFD investigations stunning images of flow quantities can be computed and displayed for later scrutiny and analysis. Indeed, the uncritical use of color graphics in CFD investigations has led some cynics to state that the "C" in CFD stands for "colorized" or "colorful."

Adopting a computational outlook on problems often reveals totally new and sometimes unexpected aspects. We shall see that the process of conducting numerical experiments on a computer has a lot in common with its laboratory counterpart, while the tools employed in setting up the calculation are those of theory. The laboratory experimentalist may be confident of manipulating the right quantities. The numerical experimentalist, on the other hand, is equally assured of manipulating the right equations! The results of laboratory and CFD experiments often complement one another in remarkable ways, as we shall have occasion to illustrate. This complementarity is part of the richness and allure of our subject.

Done correctly, CFD involves more than taking a code package off the shelf and running it to obtain a numerical answer to a pressing engineering or environmental problem. The capabilities of CFD in this respect are, of course, responsible for the leverage that the subject commands in the allocation of societal resources, and should in no way be underestimated. The true intellectual fascination of the subject lies, however, in the interplay between the physics and mathematics of fluid motion, and the numerical analysis and computer science to which these give rise. This multi-faceted nature of CFD ultimately is what gives the subject its attraction and from which future developments of the subject will spring. Hamming (1973) concludes with a philosophical chapter on such matters, the main thesis of which is: "The purpose of computing is insight, not numbers" – a very interesting notion for a monograph on numerical methods.

Because fluid mechanics is ruled by well-known partial differential equations, viz. the Navier–Stokes equations and specializations thereof, much more pragmatic visions of what CFD is abound. An applied mathematician or numerical analyst might see CFD simply as part of numerical partial differential equations (PDEs). An engineer working with one of the several available comprehensive, commercial fluid dynamics packages might look upon CFD as a part of computer aided design. A meteorologist might classify CFD as simply another tool for weather and climate prediction. There are powerful arguments to support all these points of view. For example, in engineering the economics of conducting CFD-based design calculations versus full-scale construction and testing needs little in the way of general epistemology to promote the field! Similar arguments pertain to aspects of weather prediction, such as storm warnings. More fundamentally, numerical studies of climate dynamics, large-scale oceanography, convective and magnetohydrodynamic processes in Earth's interior, flow problems in astrophysics, etc., allow researchers to conduct *virtual experiments* in those fields, as they often are the only possible experiments. With the recent interest in global warming and the observation of the ozone hole, computer experiments on atmospheric dynamics are taking on an entirely new and societally urgent dimension.

Because so many different scientists with different outlooks and different objectives view CFD as everything from a tool to a basic discipline in its own right, texts on CFD vary widely. Most texts assume that the reader is already convinced of the merits of the CFD approach, and basically wants to improve technique. Hence, detailed discussions of computational methodologies are presented, often with code listings and comparisons in terms of accuracy and efficiency. The present volume starts at a considerably more elementary level, and attempts to build up a general philosophy of CFD as various technical issues are introduced. This philosophy is, we believe, important in establishing CFD solidly within the field as an intellectual equal to many of the profound physical and mathematical subjects that fluid mechanics is so well known for.

1.2 Overview of the Book

Before embarking on any detailed discussions it may be useful to give a brief outline of what follows. In this chapter we define some of the *basic concepts* of CFD and delineate the *intrinsic limitations* that arise when trying to simulate fluid motion with a large number of degrees of freedom on a finite machine. The evolution of computer hardware is of interest to this discussion and is briefly traced.

In Chapter 2 we discuss the general idea of a discrete mapping. Any computer representation of fluid motion with discrete steps in time can be viewed as such a mapping, and our study of mappings may be seen as a paradigm of CFD. Furthermore, the study of mappings, usually computer assisted, has revealed many fascinating features, some of which have a direct bearing on fluid flow. The coding required to compute mappings is usually very simple, and the topic thus allows us to get started on computer exercises with elementary code that yields interesting output. We encounter the wide-ranging topics of chaos and fractals, which arise so naturally in fluid mechanics and have firm roots in the subject.

Chapters 3 and 5 are devoted to ordinary differential equations (ODEs). An ODE can be classified as *initial*, *boundary* and *eigenvalue problem*. The mathematical nature of these problems are so different that they warrant different numerical techniques. As we briefly review at the beginning of these chapters, techniques for solving these different types of ODEs are well in hand. Chapter 3 will cover initial value ODEs, where we will consider classical Adams–Bashforth, Adams–Moulton and Runge–Kutta family of time integration techniques. Chapter 5 will cover boundary and eigenvalue ODEs, where we will consider shooting and relaxation methods to these problems. As a preliminary, in Chapter 4 we will consider spatial discretization schemes. We will cover finite difference and compact difference schemes for spatial differentiation, and Simpson and quadrature rules for spatial integration. The concept of discrete derivative operators in the form of matrices will be introduced in this chapter.

A surprising number of interesting fluid mechanics problems present themselves in the format of a system of ODEs. We discuss by way of example the shape of capillary surfaces (which historically turns out to be the problem that motivated the development of what we today call the Adams–Bashforth method), the Blasius and Falkner–Skan boundary layer profiles, the Kármán solution for a rotating disk, the Graetz problem of thermal development in a pipe flow, onset of Rayleigh–Bénard convection from linear instability, the Orr–Sommerfeld equation for the stability of parallel shear flows, the Rayleigh–Plesset equation for collapse of a gas-filled cavity in a liquid, the Kelvin–Tait–Kirchhoff theory of motion of a solid body in irrotational, incompressible flow, the equations for interacting point vortices in two-dimensional flow, and the topic of chaotic advection. Meaningful computer exercises on all these can be completed with rather brief coding, and should help in developing an appreciation for the mode of work in CFD.

While finite difference methods work with spatially localized approximations for derivatives and field values, methods based on functional expansions use global representations in which the unknown field is expanded in a series of known modal shapes with variable coefficients. Methods based on such representations include finite element methods and, in particular, spectral methods. Chapter 6 is mainly concerned with spectral methods, although the theory of functional expansions is presented in rather general terms. The fast Fourier transform (FFT) algorithm is explained. The theory and utility of Chebyshev polynomials is discussed.

In Chapter 7 aspects of the theory of numerical solutions to partial differential equations are given. This chapter, in a sense, is the starting point of many advanced books on CFD. Thus, much more elaborate treatments of this material may be found in any of several CFD texts. Again, the intent is to provide an introduction that includes enough of the details to be interesting and to give you, the reader, an appreciation for some of the issues, yet avoid the comprehensive treatment that the expert user of these techniques will need to acquire. The issue of stability, first mentioned in the context of mappings in Chapter 2, is re-examined, and plays an important role in structuring the discussion. Computer exercises for the simpler methods can be carried out with a nominal programming effort. In Chapter 8 we consider multi-dimensional PDEs. In this chapter we finally address direct and iterative numerical methods for the Navier–Stokes equations. In this chapter we introduce the powerful time-splitting or operator-splitting methodology, which reduces the Navier–Stokes equation into a set of Helmholtz and Poisson equations. We also briefly cover spectral methodology for PDEs. The Navier–Stokes equations are written out in spectral form. In principle, the material covered here should allow the reader to construct a spectral code for the three-dimensional Navier–Stokes equation in a periodic box. Finally, in Chapter 8 we also briefly cover CFD methodologies for solving fluid flow in complex geometries. The first approach considered is solving Navier–Stokes equations in a curvilinear body-fitted grid. As an alternative we also consider two Cartesian grid methods: sharp interface methods and immersed boundary techniques.

1.3 Algorithm, Numerical Method, Implementation and Simulation

We start by introducing some basic concepts. The first is **algorithm**. The word has the curious etymology of being derived from the name of the Arabic mathematician *Muhammed ibn-Musa al-Khowarizmi*, who flourished around 800 AD. Initially algorithm simply meant the art of calculating using Hindu–Arabic numerals. The modern meaning of the word is: *a systematic procedure for performing a calculation*. An algorithm is thus an independent construct – a set of rules – that gives the principle or strategy for carrying out a calculation. Algorithms exist quite independently of any implementation on a computer.

A well-known example of an algorithm is the sieve of Eratosthenes, dating from the third century BC, for finding all prime numbers between 1 and N. The algorithm consists of a series of steps. First we construct a table of all integers from 2 to N. Then:

(1) Cross out all multiples of 2 (except 2 itself).
(2) Find the first number after 2 that has not been crossed out (i.e., 3).
(3) Cross out all multiples of 3 (except 3 itself).
(4) Find the first number after 3 that has not been crossed out (i.e., 5).
(5) Cross out all multiples of 5 (except 5 itself).

And so on until the table is exhausted. We will be left with the primes $2, 3, 5, 7, \ldots$. Clearly the steps can be grouped together in a loop:

(n) Cross out all multiples of p (except p itself).
($n+1$) Find the first number after p that has not been crossed out, call it p'. Repeat the above with p' instead of p until $p \geq N$.

The algorithm can now be assessed with regard to efficiency, robustness, etc. For example, in this case an immediate improvement in speed is realized by stopping when the integer that is having its multiples eliminated exceeds $\sqrt{N}$, rather than N, since further testing accomplishes nothing.

On a modern computer system the lowest level algorithms for addition and multiplication are implemented in hardware. Some of the algorithms for often-required tasks, such as the computation of common functions (cos, sin, tan, log, etc.) or vector–matrix operations, are coded in highly optimized form and collected in libraries that are called as the default with the compiler/loader. Yet other algorithms are collected in libraries, for which the source code in a high-level language, such as FORTRAN or C, is often accessible. A good example is BLAS – a library of *Basic Linear Algebra Subprograms* (Dongarra *et al.*, 1990), which has become the *de facto* standard for elementary vector and matrix operations. Highly optimized BLAS modules that take full advantage of the machine architecture can be found on most parallel computers. Finally, you, the CFD programmer, will have to provide those algorithms that are particular to the problem you wish to compute.

The algorithms that perform standard calculus operations, such as differentiating or integrating a function, are treated in texts on numerical analysis. Many of them have keyword names, such as *trapezoidal rule*, *predictor–corrector method*, and *fast Fourier transform*. Others are named for their originators, such as *Gauss–Legendre quadrature*, *Simpson's rule*, and the *Newton–Raphson method*. From the above names it is quite clear that the terms *algorithms* and *methods* have traditionally been used interchangeably. CFD presumes and utilizes much of this material, and we shall only pause to treat some of it in detail. In most cases we shall refer to treatments in the numerical analysis literature.

It may be useful to point out that successful algorithms and elegant mathematics do not always go hand in hand. An elegant mathematical formulation may not lead to an efficient calculation procedure, or it may not have suitable stability properties, or it may require excessive storage. For example, in the relatively simple problem of solving systems of linear equations an elegant mathematical solution is given by *Cramer's algorithm*, where the solution is written in terms of ratios of certain determinants. However, the direct evaluation of a determinant of an $N \times N$ system requires some $N!$ operations. It is not difficult to see that this operation count can be beaten handily by the elementary process of successive elimination that one learns well before encountering determinants. In numerical analysis this method is called Gaussian elimination.

A collection of algorithms and basic numerical methods are often needed to perform more complex mathematical operations. For example, common problems in linear algebra, such as solving systems of equations, linear least square problem, singular value decomposition, and eigenvalue and eigenvector computation, require more basic vector–vector and matrix–vector operations. Transportable libraries in linear algebra, such as LAPACK, are built on top of BLAS routines. There have been a number of powerful numerical libraries, such as Portable

Extensible Toolkit for Scientific Computation (PETSc) that have been widely accepted and used in the CFD community. They provide ready-to-use algorithms and methods that are well designed for parallel computing.

We shall then think of a **numerical method** for CFD as a collection of many elementary algorithms and methods. This viewpoint has many potential advantages. First of all this provides a modular approach which greatly simplifies the programming and debugging of the numerical method. This approach also allows for easy replacement of an individual algorithm by another equivalent algorithm as and when necessary. Furthermore by using standard packages from libraries, such as PETSc and LAPACK, the user is relieved of having to code standard algorithms in common use. Full attention can therefore be focused on developing algorithms specific to the problem at hand.

What determines the merit of a given CFD method are typically the algorithms that transcribe the continuum PDEs describing fluid motion into a set of algebraic equations and solve them. This may be accomplished in a variety of ways, all of which, in general, incur some kind of approximation error. The most common approach is for the fluid domain to be fitted with a grid and the amplitudes of the fields of pressure, velocity, etc., at the nodes of the grid chosen as basic variables. This procedure is used in *finite difference methods*. Alternatively, one may transform the pressure and velocity fields to a set of amplitudes that enter in a series expansion in terms of modes of fixed shape. Such an expansion, of which the Fourier series may be the best-known example, is called a *Galerkin expansion*. Truncating the expansion and working with only a finite number of amplitudes leads to *spectral methods* (when the modes that we expand in are global orthogonal functions) and to *finite element methods* (when the modes are local low-order polynomials).

A less common but often very powerful approach is to abandon the Eulerian view altogether and use the computer to follow (many) individual fluid particles. In this way one arrives at any one of a variety of *particle methods*. Particularly useful are representations where the particles traced carry some important attribute of the fluid motion, such as vorticity, or delineate a physically significant boundary in the fluid.

In general, a set of ODEs involving grid values or Fourier amplitudes or particle positions results. Except for a few special cases, these provide only an approximate representation of the PDEs from which they came. We speak of having performed a **discretization** of the continuum equations in space. As seen above there are usually several choices here. Some may be arguably better than others, and considerations of the relative capabilities and merits of different methods make up much of CFD. The numerical method must further contain algorithms for handling the integration in time of the coupled set of ODEs. Again there are a variety of choices, and again the continuous variable time must be discretized in some fashion. For example, integration of the discretized equations can be done either *explicitly* or *implicitly*, and the time and space discretizations can be either coupled or decoupled. The success of a particular method usually depends on a balance between purely numerical considerations and considerations arising from the underlying physics of the flow. The interplay of physics and numerics is often very important.

A numerical method turned into a computer program that will compile and execute on some machine and produce suitable output we shall call an **implementation** of the method. Implementations can vary considerably depending on the skill of the programmer. Implementations may be written in different computer languages. Implementations might depend on the machine architecture of the computer as well. For optimal performance, the choice of algorithms and methods needs to depend on the vector or parallel nature of the computer. The speed of the computation critically depends on these choices. The implementations may have many diagnostics built in that continually check the calculation for consistency and reliability. For example, if some flow quantity is known to be conserved in time by the PDE and/or the ODEs resulting from the discretization, it would be prudent of an implementation to monitor it, and a good implementation will have such a monitoring capability. There are many options available when it comes to the questions of how to interact with the program, i.e., how the program handles the inputting of variables and parameters and how the program presents output in specific numbers, tables and diagrams. Given an implementation of some numerical method we speak of an actual run of the program as a **numerical simulation** of the flow.

Most texts on CFD discuss many different numerical methods that have been used to solve the equations of fluid mechanics. Sometimes these methods are compared. More often they are not. Frequently, direct experimentation with implementations of different methods is the only real test of what works and what does not for a given type of flow. One can classify numerical codes broadly into general purpose and special purpose implementations. As the names indicate, implementations in the former category attempt to be applicable to a very wide variety of flow situations. Implementations in the latter category treat a more limited class of flows, hopefully and typically in greater detail than a general purpose implementation.

Many general purpose implementations are available commercially as codes. They are usually designed to be used as black boxes, i.e., the user is not supposed to change major elements of the source code (which may not be accessible). Such programs can have immense importance in engineering design situations, and many of them are used routinely in that way. However, for our present purposes they are less useful, since we want to start at a more primitive stage and trace the development from theory (PDEs) to executable code. Many human-years have been invested in commercial packages. Here we shall focus on smaller problems where the entire process of producing a numerical simulation is achievable in a limited time, and where one person can retain an overview of what has gone on in developing the program from start to finish.

Awareness of and experience with commercially available packages is certainly a marketable skill. However, neither an exhaustive discussion of techniques nor the detailing of commercially available software will be our main goal in this book. Some general description of what is available and what has been and can be done is inevitable, but there are many other issues that we must explore, which involve the physics of fluid motions and illustrate the basic philosophy of computation. These include such topics as the computability of random processes (in our case a chaotic or turbulent flow), the adaptability of numerical methods to the flow physics (e.g., how to treat shocks or sharp interfaces, or how to limit diffusion), the manifestations of numerical errors as physical effects (e.g., as diffusion or surface tension), the retention in a discretization of integrals of the motion of a PDE, and so on. The general purpose codes invariably come with many switches and options. Good understanding of the anticipated general physics of the flow and the property of the numerical methods and algorithms employed is important to turn on the right switches in the general purpose code. Such understanding of the interplay between flow physics and numerical methods is the focus of this book. Issues of this kind will occupy us extensively.

The construction of numerical methods is an integral part of the field of CFD, and the attendant questions of stability, efficiency, robustness, etc., are all very important and interesting. Some of these come up in the context of very special problems, where the ability of an algorithm and attendant computer program to elucidate complicated analytical behavior of the flow solutions is in focus. General purpose methods typically have little to offer in addressing such questions. (A general purpose package will deal with the issues we raise in some way, but if you do not like the solution strategy adopted, you generally have little recourse, and you typically cannot experiment with methodological changes.) Our point of view is that although some of these issues might on occasion be dismissed as academic, they do, in general, provide a valuable avenue for looking behind the scenes and ultimately for assessing the quality of a given program. They constitute, in fact, the science of CFD.

1.4 Models and Methods

Issues of principle are often best elucidated by model problems where the mathematical mechanism under discussion is clearly displayed. It is probably not unfair to say that some of the most profound discoveries in CFD have come from the study of models using (quite obviously) special purpose numerical methods. In fact, most of these discoveries have arisen from programs that just barely qualify as implementations of numerical methods.

For example, calculations on a model called the *Lorenz equations* first showed evidence of the structure we now call a *strange attractor* (Figure 1.1). These computations changed our thinking about the transition from laminar to turbulent flow in profound ways. However, a program to solve the Lorenz equations and display graphs of the attractor certainly does not qualify as general purpose software for turbulent thermal convection – Lorenz equations are the simplest

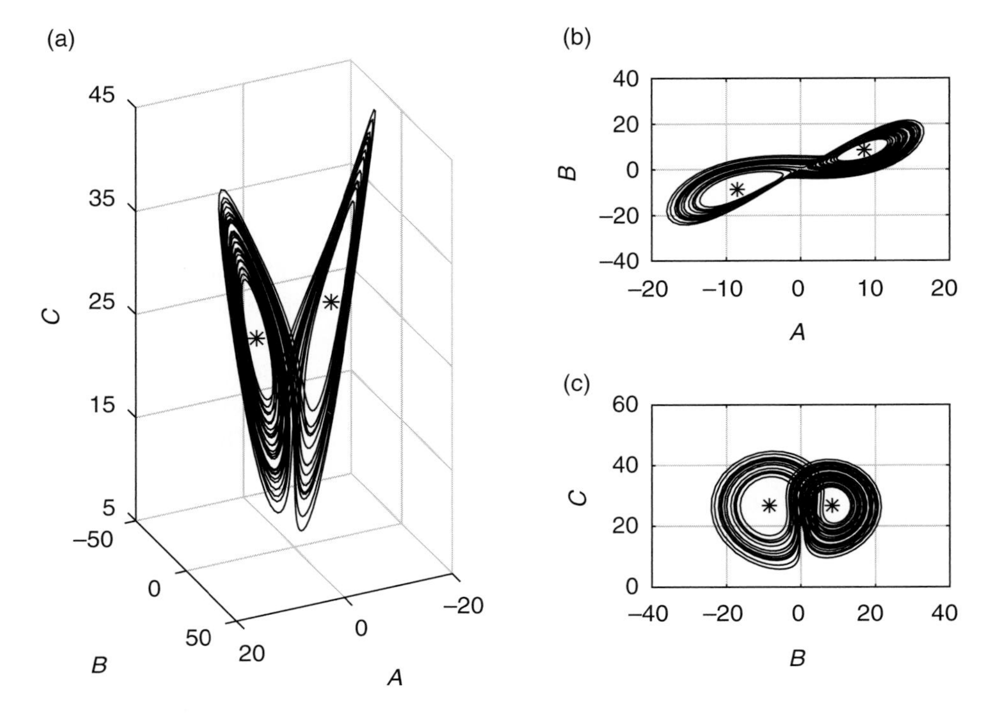

Figure 1.1 The initial computer generated evidence of a strange attractor (Lorenz, 1963). The equations that have been integrated numerically to produce these illustrations are $\frac{dA}{dt} = \mathrm{Pr}(B - A)$; $\frac{dB}{dt} = -AC + rA - B$; $\frac{dC}{dt} = AB - bC$. These equations result from expanding the equations governing thermal convection in a Fourier series and retaining the three lowest modes. In this expansion A is the amplitude of the stream-function mode, while B and C are modal amplitudes of the temperature field. The parameter Pr is the Prandtl number, which was chosen to be 10. The parameter b depends on the most unstable wavelength. To make the above Lorenz equations agree with the full linear stability calculation for the onset of thermal convection one needs to set $b = 8/3$. The final parameter, r, is the ratio of Rayleigh number to the critical Rayleigh number for the conduction-to-convection transition. Using the model equations the steady convection state that ensues for $r > 1$ becomes unstable for $r = 470/19 = 24.74$. The calculations were performed for $r = 28$. Panel (a) shows the trajectory over 30 time units. Panels (b) and (c) show the trajectory projected onto the AB- and BC-planes. The stars correspond to fixed points or steady convection.

model of thermal convection. Or think of the discovery of *solitons* by the numerical integration of certain PDEs (Figure 1.2), features that are not in fact all that easy to realize exactly in real flows.[1]

A system of equations abstracted from the general laws of fluid motion will be referred to as a **model**. On the other hand, the tradition in CFD has been to develop **methods**, and, indeed, to vary the resolution parameters to gain ever

[1] Closely related solitary waves are quite abundant, but solitons proper require a very special balance of steepening and dispersion in the governing equations.

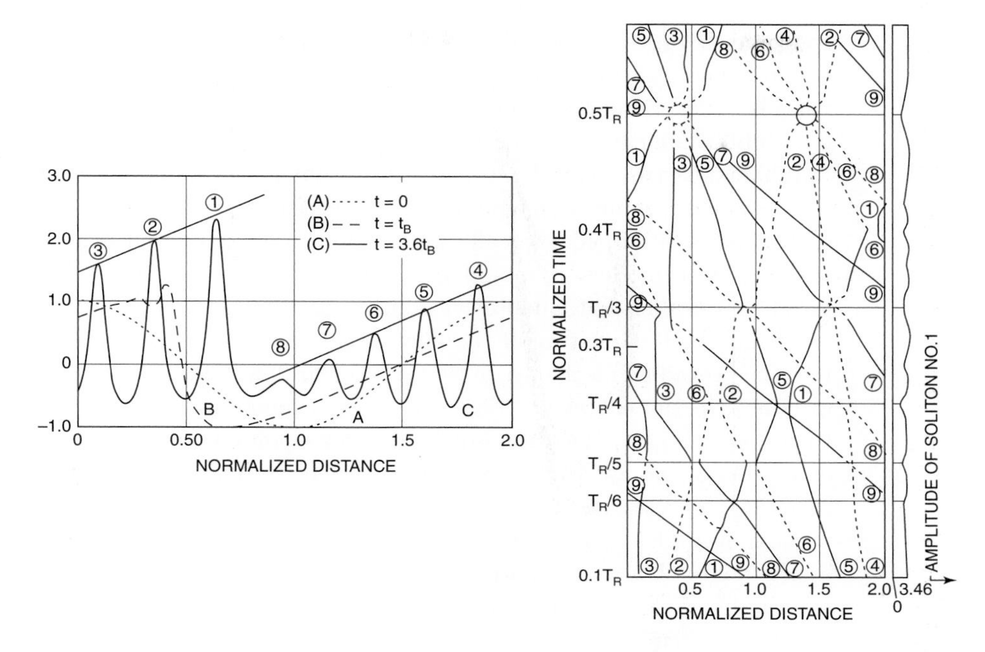

Figure 1.2 The equation being integrated is the Korteweg–de Vries equation: $u_t + uu_x + \delta^2 u_{xxx} = 0$. This is a PDE for $u(x,t)$ with one parameter δ (which is set here to 0.022). The diagram on the left shows the waveform u at three different times labeled A, B, and C, starting from a sine wave at $t = 0$ (A). The remarkable discovery was that the coherent wavepackets labeled 1–8 appeared, and that they underwent collisions but re-emerged essentially unscathed except for shifts in position relative to where they would have been without collisions. The right panel traces the location of the various peaks as a function of time. Figure reprinted with permission from Zabusky and Kruskal (1965), © 1965, American Physical Society.

more accurate representations and solutions of the governing PDEs or numerical simulations of a given flow. With this very broad definition the discretization used in any numerical method qualifies as a model. However, we want to maintain a distinction between a model derived from the governing PDEs and a numerical method constructed for simulating them, and we may do this by focusing on the important issue of **convergence**. In a numerical method we start from the PDEs, discretize them in both space and time in some way, and then compute with the large but finite system of algebraic equations that results. We get some solution depending on the fineness of the spatial grid, the size of the time step, the number of particles used, etc. If everything is done intelligently, we expect that the solution will converge to the solution of the PDEs as the grid is made finer, the time step smaller, and/or the number of particles increased. Certainly every prudent calculation of a flow field for which quantitative accuracy is claimed should be checked for convergence. In fact, most fluid mechanical journals now have an explicit requirement that a manuscript must demonstrate convergence of its numerical methodology. Interestingly, in common implementations of large eddy simulations of turbulent flows, the model, and as result what is being computed, itself varies with the level of spatial discretization. This certainly complicates the question of convergence and how to establish it.

While subscribing to this practice of establishing convergence for numerical methods, the computational pursuit of models can be equally challenging in many cases, and is often pedagogically useful. Hence, several of our examples in what follows will be models. In a model, we give ourselves more leeway. Now the main goal is physical insight, qualitative and/or quantitative. Even if we have to create what may be a caricature of the flow, we are willing to do so if it gives insight into the prevailing mechanics of that particular flow. Thus, the Lorenz equations, discussed in the context of Figure 1.1, are a model of thermal convection. On the other hand, the derivation of the Lorenz equations, as a truncated Galerkin expansion of the full PDEs describing 2D thermal convection, is the basis for a systematic spectral method of solving such equations.[2] As the number of retained mode amplitudes is increased, we are presented with a sequence of models that approximate convection more and more accurately. Following this sequence to convergence is a *numerical method* route. Opting for and staying with the three-equation system of Lorenz means exploring a *model* of the flow. Both approaches have their merits, and both require computation.

Often even our model is very difficult to study. Vladimir Arnold, the distinguished Russian mechanician and mathematician, writes provokingly in his text on classical mechanics (Arnold, 1978) that analyzing systems with two degrees of freedom (described typically by four coupled, nonlinear ODEs) is "beyond the capability of modern science." That doesn't leave much hope for fluids in general, governed as they are by PDEs!

Although models result in equations much simpler than the original PDEs, they usually still require numerical algorithms and methods for their solution. Of course, these should be subjected to the same rigors and scrutiny of numerical analysis and computer science as the numerical methods employed for the PDEs. The models we are willing to consider often will not converge to a solution of the full equations of motion, nor may they make much sense if viewed from that perspective. Usually this can be seen analytically, since some drastic assumption will have been made in the full Navier–Stokes equations to arrive at the model in the first place. But once the model has been decided upon, we demand

[2] Indeed, one point of controversy in the literature has been to what degree the chaos arising in the Lorenz equations corresponds to the chaos observed in convection experiments. Curry *et al.* (1984) write: "It has been noticed previously ... that too little spatial resolution in the solution of partial differential equations can induce chaos in the computed solution which disappears when adequate resolution is used. This effect may perhaps be considered surprising as one might expect that, as partial differential equations involve an infinite number of degrees of freedom, their solutions would be richer and more chaotic than solutions to finite-mode dynamical systems. On the contrary, the sea of weakly excited small-scale modes in a dissipative dynamical field seems to act as an effective damping, or eddy viscosity, which tends to reduce the tendency of modes to exchange energy in a chaotic way."

that its numerical investigation be subjected to all the rigors of numerical and computational methodology that we require of any numerical method.

The distinction between models and methods is particularly clear in turbulence research. Turbulent flow is an immensely complicated state of fluid motion, which is only barely possible to capture computationally. Representations of turbulent flow using effective equations for the mean motion, that differ from the Navier–Stokes equations by parametrizing the action of all turbulent fluctuations back on the mean flow, have been pursued and developed for many years. This approach to computing turbulent flow generally goes by the name *Reynolds-averaged Navier–Stokes* (RANS) simulations. The RANS simulations vary widely in their parametrization of the effect of the turbulent fluctuations on the mean flow. The literature cites many turbulence closure models by name or abbreviation, e.g., *eddy viscosity*, *k–ϵ*, *algebraic-stress*, *EDQNM* and *DIA*. Such models are used in computational studies of turbulent flow. Another approach increasing in popularity is the *large eddy simulation* (LES). Here the turbulent flow is separated into large-scale and small-scale motion. Only the dynamics of the large scales are computed, while the effect of the neglected small scales on the dynamics of the computed large scales is accounted for through subgrid scale models, such as the Smagorinsky model, scale-similarity model, and dynamic subgrid models.

There is (or should be) no pretense that it is theoretically possible to compute the real turbulent flow without any models. All one has to do is to revert to the fundamental governing Navier–Stokes equations and increase the spatial and/or temporal resolution to resolve all the important length and time scales of turbulent motion. This is often stressed in the engineering literature by referring to such direct discretization and computation of the Navier–Stokes equations as *direct numerical simulation* (DNS). With sufficient resolution in both space and time the DNS is guaranteed to yield the right answer. Unfortunately, for computational limitations this approach is limited to low Reynolds numbers and becomes prohibitively expensive with increasing Reynolds number. On the other hand, by increasing the resolution in a computer simulation of a LES or RANS simulation of turbulence one gets (at best) a numerically accurate solution to the large eddy or Reynolds-averaged equations. In our terminology an approach based directly on the Navier–Stokes equations, implemented according to the tenets of numerical analysis, leads to a direct numerical method. An implementation of a turbulence model, however large and however well done, is still numerical solution to a model.[3]

Several of the examples and exercises treat established and successful models, since these tend to be useful testbeds for ideas, and valuable for gaining computational experience. Among the models that we shall describe and analyze are *Burgers' equation*, the Orr–Sommerfeld equation, the equations of motion for

[3] One may rightfully argue that the Navier–Stokes equation is itself a model. In this book we take the position that the Navier–Stokes equations are exact enough at the continuum scale to describe all the flow phenomena discussed here.

interacting *point vortices*. Computer trials, or numerical experiments as they are often called, with such models can provide considerable insight.

The insight gained from a model can often augment a method in various ways. Often to assess a given method it is tried out on some simpler set of equations that are more analytically tractable, creating in effect a model of the flow and a testbed for the numerical analysis features of the method. For example, when we come to discuss finite difference methods, we shall see that the problem of advection of a scalar field by a constant velocity is essential for an understanding of the features of this important class of numerical methods. The problem is analytically solvable and physically intuitive. Nevertheless, it yields a wealth of insight, and is part of the foundation of the theory of finite difference methods.

For models and methods alike the basis is, of course, some set of *bona fide* fluid mechanical equations. These may well be the full Navier–Stokes equations, but a surprising amount of what goes on in the realm of both CFD and computer modeling deals with other equations. These are, to press a point, subsets of the Navier–Stokes equation, but often are so distinct that appeal to the full version of the equations is not helpful. For example, many computational studies have been concerned with the Euler equations, the limiting form of the Navier–Stokes equations when viscous effects are ignored from the outset and the no-slip boundary condition is relinquished. One favorite topic in this realm, on which we shall have more to say later, has been the *two-dimensional Euler equation with embedded regions of vortical fluid.* For example, Pozrikidis and Higdon (1985) considered an idealized flow in which a band of uniform vorticity of a certain width is perturbed by a sinusoidal wave. This is a highly idealized model of a shear layer. Owing to Kelvin–Helmholtz instability, the band rolls up into discrete vortices. Using this flow model the authors discovered an intriguing amplitude dependence of the rolled-up state on the amplitude of the initial perturbation. The numerical method used in their work depended on a representation of the flow in which all the computational degrees of freedom are concentrated on the boundary of the vorticity carrying region. Such methods belong to a large class known collectively as *boundary integral methods.*

The final point we would like to stress before we leave this topic is that while the Navier–Stokes equations are PDEs, the models often result in simpler ODEs and sometimes even algebraic equations. Traditional CFD books are generally devoted to techniques needed for solving the governing Navier–Stokes equations in three spatial dimensions. Whereas introductory texts in numerical methods abound, and many of them provide excellent coverage of methods for ODEs and linear and nonlinear algebraic systems, this is in a general context without special attention to the needs and constraints posed by flow physics.

1.5 Round-off Error

Round-off error is a manifestation of the way floating point numbers are stored in a computer. In the computer any given real number can only be retained with so many digits. To map the number r onto a sequence of 0s and 1s of some given length we write it as

$$r = (\text{sign}) \times M \times 2^{\rho-\rho_0}, \tag{1.5.1}$$

where (sign) is $\pm$, and where $\rho - \rho_0$ is the integral part of $\log_2 |r|$. With a 32-bit word, for example, we can use 1 bit for the sign, 8 bits for the exponent ρ, and the remaining 23 bits for the mantissa M. The bias in the exponent, ρ_0, is usually chosen to be 127 and therefore the exponent part ranges from 2^{-127} to 2^{128}. With the exponent chosen as described, the mantissa is in the interval $1 \leq M < 2$, and it is expressed as $1+x$, and the 23 bits are used for representing the fractional field x. We can think of x as a binary number

$$x = \sum_{n=1}^{\infty} a_n \, 2^{-n}, \tag{1.5.2}$$

where the coefficients a_n are all 0 or 1. We can in turn write this as the binary decimal $.a_1\, a_2\, a_3 \ldots$. In this notation we have $1 = .111\ldots$, $2^{-1} = .1000\ldots = .0111\ldots$, etc.[4] For example, the 32-bit word

1 10000011 11000000000000000000000

represents the floating point number -28.0. The exponent field (10000011) is 131, the fractional part is $.110\ldots = 0.75$, and together they represent $-1.75 \times 2^4 = -28.0$.

The range of numbers that can be represented on a computer depends on how many bits are available for each number. On a 32-bit machine using IEEE standard this number will be of order $10^{\pm 38}$. However, when two numbers are multiplied or added, it is the number of bits available for the fractional field that determines when significance is lost. Rounding of the fractional part of the result can produce errors, particularly when working with numbers that are of very different orders of magnitude. Similarly, if two numbers the fractional part of which agree in almost all bits are subtracted, the result will have poor accuracy because only one or two significant bits survive. This effect, labeled **round-off**, will be detectable at the accuracy level of 1 part in 2^{23} (*not* 1 part in 2^{32}) or, equivalently, 1 part in 10^7 (using the example of the 32-bit word with 23-bit mantissa from before). We shall see shortly that the operations we perform in a numerical computation may act to exacerbate the role of round-off errors. The **unit round-off error** is the smallest number U such that $1.0+U$ is found to be greater than 1.0 by the computer. Some code packages, e.g., an integrator with

[4] We agree to use the terminating form of the representation in cases where two representations give the same number. For example, $3/4 = .11$ instead of $3/4 = .1011\ldots$.

built-in logic to reduce the integration step, may require determination of this quantity for the computer system being used.

Problem 1.5.1 Determine (i.e., write a computer program to determine) the unit round-off error on a computer of your choice.

The round-off error can often be reduced by special programming, e.g., by grouping terms together in such a way that the effects of round-off are ameliorated. Most modern computing machines have double precision (64-bit arithmetic) as the standard precision. Most programming languages also have an option, known as quadruple precision (128-bit arithmetic) for increasing the amount of bits allocated to each number, thus further reducing effects of round-off in this way. Quadruple precision is not recommended, however, since it can be substantially slower than a machine's standard precision arithmetic.

Round-off error, in principle, destroys many of the familiar laws of arithmetic when done on a computer. For example, the associative law of multiplication, $a \times (b \times c) = (a \times b) \times c$, is no longer valid in computer arithmetic with round-off. The seriousness of such violations depends sensitively on the relative magnitudes of the numbers involved.

A familiar manifestation of round-off arises from simple numbers in the decimal representation, such as 0.1, not having a terminating representation in the computer's binary representation. Round-off then leads to a loss of the tail of this representation so that 10×0.1 may be reported as $0.999\ldots$ rather than 1.00000. This, in turn, amplifies the warning never to use real numbers for logical tests, i.e., incrementing by 0.1 ten times may not result in an increment by 1.0 because of round-off. A test based on equating ten increments of 0.1 to 1.0 would then fail. It is always better to use integer counters for such purposes.

Problem 1.5.2 Write the decimal fraction 0.1 as a binary decimal and verify that it does not terminate.

For many applications round-off error can be reduced in significance to the point of being ignorable, but that is certainly not always the case. It is always something to be on the lookout for. In the following example a particularly pernicious effect of round-off is traced.

Example 1.5.3 Consider the function $f(x)$ defined on the interval $0 \le x \le 1$ by

$$f(x) = (2x \bmod 1)\,. \tag{1.5.3}$$

If we represent x in binary decimal, then the mapping $f(x)$ corresponds to a shift of the binary decimal point one place to the right and the integer part is lost due to the *mod*.

$$f(.\,a_1 a_2 a_3 \ldots) = .\,a_2 a_3 \ldots . \tag{1.5.4}$$

Let us now look at these results from the point of view of the round-off error. In terms of binary decimals round-off means that two numbers that agree in the first N places are indistinguishable as far as this particular computer is concerned. The number of places that are distinguished, N, depends on the design of the computer (and the program that is running). Suppose now that our calculation calls for iterating f a number of times. Each iteration moves the binary decimal one place to the right. Eventually, after just N iterations, the entry a_N will have moved up just to the left of the binary decimal point. At this point all the significant digits in the representation of x depend on how the machine (and/or program) handles round-off!

This example may seem a bit extreme in various respects. First, the function f when iterated clearly must have some rather vicious properties. For more details in this direction see the paper by Kac (1938). The function f appears quite regular, but, as we shall see later, iteration of f will generate random numbers. It stands to reason, then, that results purely dependent on round-off will ensue after a while.

The above simple example beautifully illustrates sensitivity to initial condition. If x is initially known to 23-bit accuracy, this initial knowledge is good for only 23 iterations of f, after which x entirely depends on what we did not know at the beginning. Improving the initial knowledge to more bits only delays this process. The impact of such extreme sensitivity to initial condition can be seen in weather prediction. Information on today's weather is sufficient to make only reasonably accurate short-range forecast (although four-day forecast and hurricane warnings are becoming increasingly accurate, everyone has probably seen or experienced the wrath of their unpredictability). In this case, part of the problem may be in the mathematical models and numerical methodologies used in medium and long-range weather prediction, but certainly the incomplete knowledge of the initial condition needs to bear substantial part of the blame.

It is premature, yet irresistible to ask: What happens when one in CFD asks the computer to simulate a turbulent flow? From the statistical theory of turbulence we expect turbulent flows to be so complicated that the velocity at a point sampled at sufficiently well-spaced instants in time provides a sequence of random numbers. Thus, is it not meaningless to simulate turbulence on a computer for any appreciable length of time? Will the entire result not be saying more about the round-off errors on the computer system used than the statistics of complex solutions to the Navier–Stokes equation? Or, conversely, since the value of the velocity on a finite machine can only assume a finite number of values, will we not get spurious recurrences in a simulation that would not occur when measuring in a real fluid? We leave these as questions for now. Partial answers will be attempted later. Suffice it to say that significant issues arise long before we reach systems as complicated as the 3D Navier–Stokes equation.

1.6 A Hierarchy of Computation

When one looks at the computational tasks faced by a discipline of physical science such as fluid mechanics, they fall quite clearly into a hierarchy, with relatively simple and well-defined tasks at one end, and with a progression through ever more composite and complex tasks. In the case of fluid mechanics, ultimately, at the pinnacle of the hierarchy one arrives at the full capability of simulating a general class of flows by solving a set of three-dimensional time-dependent PDEs represented by Navier–Stokes equations with a robust and reliable method. Often with reasonable physical assumptions (such as steady state and two-dimensionality) and appropriate theoretical/analytical development and modeling the mathematical problem to be solved can be substantially simplified, leading to a hierarchy of models from the simplest one to the most complex. Let us look at some of the steps in this hierarchy.

1.6.1 Function Evaluation

At the lowest level – although any connotations of a value judgment are unintended – we find the calculation of numbers from a specific formula. We shall refer to this kind of computation as **function evaluation**. This is what you need to do when you have arrived by analysis at some formula, and you want to know specific numerical values. For example, you might have performed a linear stability analysis for some problem, and arrived at a dispersion relation. The formula is all there, yet it may be presented to you such that the physical content is not obvious to the naked eye. You need to generate tables and/or plots in order to understand the formula in more detail. What we are calling function evaluation here is the type of task that has already been performed for many of the more standard formulae, such as those that describe special functions in mathematics. You can find detailed descriptions of techniques for many of these in texts on numerical analysis. What will frequently be missing in your particular fluid mechanical application is the variety of formulae that are available for special function evaluations. For the better known and named special functions, years of analysis often have singled out some formulae best suited for numerical computation, e.g., a series that is more rapidly convergent than other more obvious series. In any particular fluid mechanics problem you might be stuck with a representation that for numerical purposes is less than optimal. Often you will just have to work around such obstacles. By and large techniques exist for extracting numbers from any well-defined mathematical formula. The more explicit the formula, the easier it is going to be to get reliable numerical information.

Example 1.6.1 In the linear stability analysis of the *Rayleigh–Taylor instability* the following dispersion relation, due originally to Harrison (1908), arises:

$$\begin{aligned}
&-\left\{\frac{gk}{n^2}\left[\alpha_1-\alpha_2+\frac{k^2T}{g(\rho_1+\rho_2)}\right]+1\right\}(\alpha_2 q_1+\alpha_1 q_2)-4k\alpha_1\alpha_2 \\
&+\frac{4k^2}{n}(\alpha_1\nu_1-\alpha_2\nu_2)\left\{(\alpha_2 q_1-\alpha_1 q_2)+k(\alpha_1-\alpha_2)\right\} \\
&+\frac{4k^3}{n^2}(\alpha_1\nu_1-\alpha_2\nu_2)^2(q_1-k)(q_2-k)=0\,.
\end{aligned} \tag{1.6.1}$$

The physical situation is that we have two fluids, one above the other, each filling a half-space, in a homogeneous gravitational field of acceleration g. Subscripts 1 and 2 refer to the upper and lower fluids, respectively. The fluids individually are uniform with densities ρ_1 and ρ_2, and kinematic viscosities ν_1 and ν_2, respectively. The interfacial tension is denoted by T.

Such a configuration is a possible equilibrium state. It will not be stable if the density of the upper fluid is larger than that of the lower fluid. The dispersion relation expresses the amplification rate, n, of a disturbance of a given two-dimensional wave vector, $\mathbf{k}$, in the interfacial plane separating the two fluids. The process is isotropic in $\mathbf{k}$, so that n only depends on the wavenumber $k=|\mathbf{k}|$. The dependence of n on k is obtained from a linear stability analysis of the equilibrium state. The result is (1.6.1). The parameters α and q are derived quantities defined as follows:

$$\alpha_i=\frac{\rho_i}{\rho_1+\rho_2}\quad;\qquad q_i=\sqrt{k^2+\frac{n}{\nu_i}}\,. \tag{1.6.2}$$

It is clear that a functional relation of this complexity can benefit from computer elucidation. Indeed, to make analytical progress with (1.6.1) it has been necessary to set the two kinematic viscosities equal, and even then several diagrams and pages of explanation are required (see Chandrasekhar, 1961, §94).

Example 1.6.2 Another instance where the computer can be very useful as a tool in a function evaluation mode is in the drawing of streamlines and other contours on the basis of an analytically given flow solution. Consider as an example the two-dimensional flow between two steadily rotating cylinders. If the cylinders are concentric, this problem is trivially solved in elementary fluid mechanics texts. It is the Couette flow problem. However, if the cylinders are *not* concentric the solution is not known analytically for the full Navier–Stokes equations, but the solution is known for Stokes flow. It is not a simple matter to write it down! In the article by Ballal and Rivlin (1976) the history of this flow solution is traced from Reynolds' considerations a century ago in the context of lubrication theory to more recent formulations. A transformation to *bipolar coordinates* is required, and even after that Ballal and Rivlin need about 25 lines to write out the full solution for the streamfunction due to the many parameters that are involved, such as cylinder radii, angular speeds, and the offset of centers. The angular speeds of rotation enter the expression for the streamfunction *linearly*. Thus, even with the full analytical formula in hand there is still a substantial project left to actually interpret that solution. There are several flow regimes depending on the values of the parameters, and these have to be sorted out, a task where a reliable computer representation of the results is extremely valuable. Figure 1.3 shows some representative flow configurations.

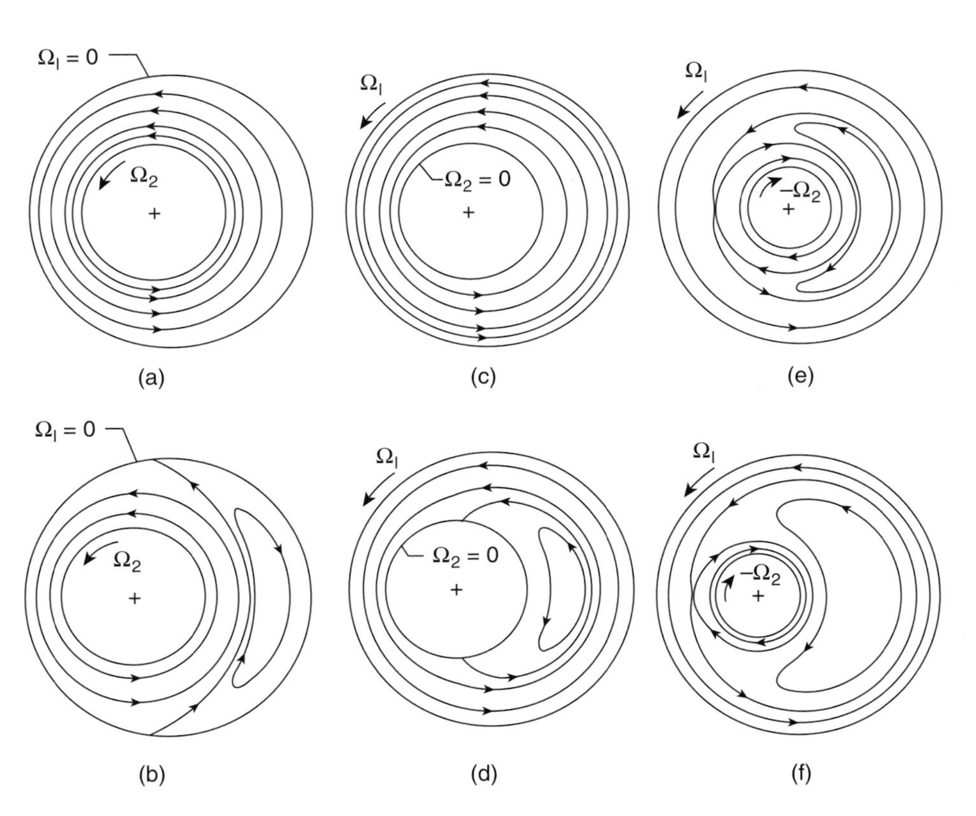

Figure 1.3 Computer generated streamlines for Stokes flow between eccentric rotating cylinders, from Ballal and Rivlin (1976), reprinted with permission of Springer-Verlag. The parameter R is the ratio of the inner to outer cylinder radius. The offset of the centers scaled by the difference in radii is called ϵ. (a) Outer cylinder stationary, $R = 0.25$, $\epsilon = 0.5$; (b) same but $\epsilon = 0.5$; (c) inner cylinder stationary, $R = 0.5$, $\epsilon = 0.25$; (d) same but $\epsilon = 0.5$; (e) counter-rotating cylinders, $\Omega_2/\Omega_1 = -4.0$, $R = 0.3$, $\epsilon = 0.1$; (f) same but $\epsilon = 0.4$.

Example 1.6.3 Another application of function evaluation, which we shall consider in great detail in Chapter 2, is in the context of *mappings*. For example, a simple quadratic mapping connecting the current position (x, y) with the position at a finite time later (x', y') in an elliptic stagnation point flow has been analyzed extensively by Hénon (1969). This mapping can be written as

$$x' = a\,x + b\,y + c\,x^2 + d\,xy + e\,y^2\,, \tag{1.6.3}$$

$$y' = A\,x + B\,y + C\,x^2 + D\,xy + E\,y^2\,, \tag{1.6.4}$$

where a, b, c, d, e, A, B, C, D, and E are constants that describe the elliptic stagnation. The above mapping can be viewed as evaluation of quadratic algebraic expressions for the position (x', y'). Further discussion of this mapping will be presented in Section 2.7 under *area-preserving maps.*

1.6.2 Ordinary Differential Equations

Function evaluation involves finding numerical values of a function for a series of values of an independent variable. The next step in the hierarchy requires the determination of an entire function or system of functions from functional equations, typically a system of ordinary differential equations. This kind of problem arises often in steady flows or in flows allowing similarity solutions. Reduction to a system of ordinary differential equations also requires that the flow varies in only one geometric direction.

Example 1.6.4 In boundary-layer theory one of the basic equations that must be tackled is the *Blasius equation* (Batchelor, 1967, §5.8),

$$f''' + \frac{1}{2} f\, f'' = 0 \tag{1.6.5}$$

with boundary conditions

$$f(0) = f'(0) = 0 \quad \text{and} \quad f'(\eta \to \infty) \to 1\,. \tag{1.6.6}$$

This equation arises by making the similarity *ansatz*

$$\frac{u}{U} = f'(\eta) \quad ; \quad \eta = y\sqrt{\frac{U}{\nu x}}\,, \tag{1.6.7}$$

where U is the free-stream velocity about a semi-infinite flat plate aligned with the positive x-axis, u is the local x-velocity, y is the coordinate perpendicular to the plate, and ν is the kinematic viscosity of the fluid. Substitution of (1.6.7) into the boundary layer equations (with no imposed pressure gradient) yields (1.6.5).

The nonlinear ordinary differential equation (1.6.5) along with the boundary conditions (1.6.6) forms a **two-point boundary value problem** that has no known solution in terms of elementary functions. We would like to obtain the solution numerically.

Other examples where the flow equations collapse to a boundary value problem consisting of a few nonlinear ODEs include the *Falkner–Skan boundary layer solutions* (Batchelor, 1967, §5.9), the equations for *Jeffery–Hamel flow* (ibid., §5.6) and the *von Kármán flow* over an infinite, rotating disk (ibid., §5.5). In all cases some appropriate similarity assumption reduces the full PDEs to ODEs. Reliable solution of these is then the task for the computational fluid dynamicist.

Steady state and similarity solutions are interesting because they are frequently observable for some range of the independent variables. One can do quite a bit of fluid mechanics by seeking out such steady states, testing them for linear stability, and then extrapolating from the most unstable mode to how the flow might evolve or disintegrate. The attractiveness of this approach is that the investigation of the linear stability of the steady state or the self-similar solution (obtained by solving the system of boundary value ODEs) can be analytically simplified and reduced to a **two-point eigenvalue problem**. The eigenvalue problem much like the boundary value problem is a system of coupled ODEs, but now they are linear and homogeneous (or unforced). The boundary conditions for the eigenvalue problem are also homogeneous in nature. Therefore, non-trivial natural or eigensolutions to this unforced linear system are possible only for selected eigenvalues of the system. These eigenvalues usually represent the growth (or decay) rate of the instability and its frequency. Algorithms and numerical methodologies needed to solve the eigenvalue problem are somewhat different from those required for the boundary value problem.

Example 1.6.5 Arguably the most celebrated eigenvalue problem in fluid mechanics is given by the following *Orr–Sommerfeld equation*:

$$\frac{1}{\iota\alpha\,\mathrm{Re}}\left(\frac{\mathrm{d}^2}{\mathrm{d}y^2}-\alpha^2\right)^2\phi=(U-c)\left(\frac{\mathrm{d}^2}{\mathrm{d}y^2}-\alpha^2\right)\phi-\frac{\mathrm{d}^2U}{\mathrm{d}y^2}\phi, \tag{1.6.8}$$

for the stability of a one-dimensional unidirectional steady shear flow given by $U(y)$ to two-dimensional disturbances of streamwise wavenumber α. Here Re is the Reynolds number of the flow. The two-dimensional disturbance is represented by a streamfunction, whose structure along the y-direction is given by the eigenfunction $\phi\,(y)$. The disturbance thus takes the form

$$\begin{aligned} &u'=\frac{\partial\psi}{\partial y},\quad v'=-\frac{\partial\psi}{\partial x},\\ &\psi(x,y,z,t)=\phi(y)\exp\left[\iota\,(\alpha x-\alpha ct)\right]+\text{c.c.}, \end{aligned} \tag{1.6.9}$$

where c.c. means complex conjugate. The eigenvalue problem represented by the above equation along with the boundary conditions $\alpha\phi=\mathrm{d}\phi/\mathrm{d}y=0$ at the two boundaries $y=y_1$ and $y=y_2$, admits non-trivial solutions only for certain values of the eigenvalue, c. As can be seen from (1.6.9) the real part of c represents the phase speed of the disturbance, while the imaginary part of αc gives the temporal growth rate of the disturbance.

Apart from boundary value and eigenvalue problems, the third class of ODEs often encountered in fluid mechanics is the **initial value problem**. Now the objective is not to find a steady flow or the eigenstructure of disturbance with confidence, but to compute reliably how the state of a few variables evolves over time. Their evolution is specified by a set of coupled ODEs, the initial value of the variables is also specified.

Several problems in fluid mechanics fall in this category: evolution of the amplitude and phase of a finite number of Galerkin modes arising from a spectral expansion, advection of particles in an Eulerian flow field and motion of interacting point vortices, to mention but a few. For example, the *pathline* of a fluid particle with initial location (x_0, y_0, z_0) at $t = 0$ is obtained by solving the following system of three ODEs

$$\left.\begin{aligned} \frac{\mathrm{d}x}{\mathrm{d}t} &= u(x,y,z,t) \\ \frac{\mathrm{d}y}{\mathrm{d}t} &= v(x,y,z,t) \\ \frac{\mathrm{d}z}{\mathrm{d}t} &= w(x,y,z,t) \end{aligned}\right\} \tag{1.6.10}$$

with initial conditions $x(0) = x_0$, $y(0) = y_0$ and $z(0) = z_0$. The above equations are referred to as the **advection equations**. If the Eulerian velocity field (u, v, w) is steady then the above pathline is also the *streamline* of the flow.

The initial value problem in some sense is not all that different from mappings considered under function evaluation. The initial value problem after discretization in time results in a mapping. In terms of solution procedure, the boundary value problem is also sometimes approached as an initial value problem. For example in the solution of the Blasius equations, (1.6.5), the inhomogeneous boundary condition at $\eta \to \infty$, is replaced with an artificial condition $f''(0) = \beta$, where β is a yet to be determined constant. The three conditions at $\eta = 0$ can be considered as initial values of f and its first and second derivatives, and equation (1.6.5) can be treated as an initial value problem to march forward from $\eta = 0$ to $\eta \to \infty$. The constant β can be varied such that the boundary condition $f'(\eta \to \infty) \to 1$, is satisfied. We shall see more about this in Chapter 5.

1.6.3 Partial Differential Equations

Continuing on in our hierarchy we come next to partial differential equations. There are really a large variety of problems in CFD where flow conditions are such that a steady state solution is anticipated, although the full equations may not collapse all the way down to a few ODEs. In general, the governing steady state equations are coupled three-dimensional PDEs for the three components of velocity and pressure. Depending on the geometry under consideration, the problem might reduce to a set of two-dimensional PDEs. Nevertheless, analogous to their one-dimensional counterpart these multi-dimensional steady state flows result in **boundary value problems**. Figure 1.4 shows an example of steady state motion that is known to considerable precision numerically, but about which much less is known analytically. In this case we are able to compare the computational figure against corresponding experimental results very well and derive satisfaction from the close agreement.

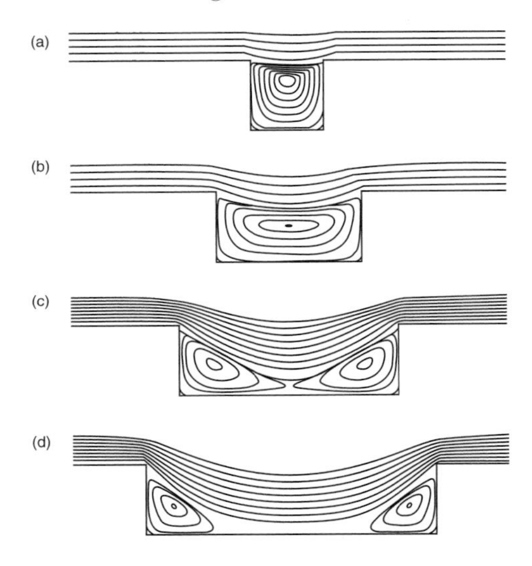

Figure 1.4 The visualizations of flow over a rectangular groove by Higdon (1985), reproduced with permission. The figure shows computed steady-state streamlines.

Figure 1.4 shows the solution of Stokes flow over a groove in a flat plate obtained by Higdon (1985). The computational results support the observed existence of several eddies within the groove. As the groove becomes wider, the single eddy splits in two. As it becomes deeper, a second eddy appears below the first. In this case the determination of the flow configuration amounts to the solution of an integral equation giving a distribution of flow singularities along the solid wall.

The stability of such complex multi-dimensional steady flows can be investigated very much the same way as their one-dimensional counterpart. If the steady flow, whose stability is being investigated, is 2D then the linear stability analysis is termed bi-global. If the base flow is 3D then we have tri-global linear stability analysis (Theofilis, 2011). The resulting equations for the eigenvalues and the structure of the eigenfunctions are coupled PDEs. Although the solution methodology for these multi-dimensional **eigenvalue problems** in principle remains the same, the computational requirement increases many fold over the one-dimensional two-point eigenvalue problem.

Finally we arrive at the problem of computing the time evolution of multi-dimensional flows. In a fully three-dimensional problem we have the time-dependent Navier–Stokes equations. If the problem permits geometric simplifications, the spatial dimensionality of the PDEs can be reduced to two or even to one in some cases. This entire class of problems is referred to as the **initial value problems**.

A time evolution calculation is not that different from the calculation of boundary value problems for the steady state flow just described. Indeed, sometimes the computation of steady flows is done by starting from some initial guess and then letting the flow settle down to the ultimate steady state in a series of steps that can have much in common with time evolution. You may hear a relaxation method to a steady state characterized as *time accurate*, which means that the method used to get to the steady state is a (sometimes crude) time evolution calculation. In nature, of course, one way to get to a steady state is to let the system relax to it. Thus, a useful computational strategy is to follow nature's way and simply monitor this relaxation, sometimes expediting it in various numerically accessible ways that the true relaxation may not be able to use.

1.6.4 Other Types of Computations

While the computation of the evolution of a flow field is the centerpiece of CFD, a large amount of effort goes into aspects of pre- and post-processing, and a few words on these tasks are in order.

Frequently, before any calculation can be done, we need to construct a grid that spans the domain in question. The topic of **grid generation** is a very important one, with its own specialized analysis and literature. We will not really touch upon it in this book, since all the CFD problems we study can be addressed with simple grids. We refer you to the review by Eiseman (1985), the works by Thompson *et al.* (1982, 1985), and to the texts by Fletcher (1991), Edelsbrunner (2001) and Liseikin (2009) for discussions of this topic. Also there are very good commercial and non-commercial mesh generation codes, such as Gridgen, Cubit, now available for automated generation of very complex grids for CFD.

Furthermore, once a solution to a flow problem has been obtained, a lengthy process of analysis and elucidation typically awaits us. The calculation of billions and billions of floating point numbers using a flow code is by no means an end in itself. The real challenge lies in the ability to make sense of these numbers and extract the flow physics or results of interest. Simple statistical processing, such as calculation of mean, rms, power spectra, and probability density functions can greatly condense the data and provide valuable insight. More complex post-processing efforts such as proper orthogonal decomposition, stochastic estimation, and wavelet analysis are nowadays becoming commonplace, as the computed flow becomes more and more complex for simple statistic analysis.

The development of hardware and software for **scientific visualization**, as this subfield of computer graphics has come to be called, now commands its own literature, which on occasion has a lot in common with the experimental flow visualization literature. Although there are more recent works and a growing literature on scientific visualization, Rosenblum (1995) is a good place to start. Here again there are very good commercial and non-commercial visualization tools, such as Paraview and Visit, that allow processing and visualization of very large amounts of data generated in parallel over tens of thousands of computer processors.

It is natural to mention here also other uses of the computer in which the main objectives are entities beyond actual numbers. In particular the realm of **symbol manipulation** or **computer algebra** is an intriguing and not yet extensively explored facet of CFD. Here the idea is to let the computer do some of the tedious algebra instead of just crunching numbers on algebra that we humans have already done. Human intellect is such that it is often easier for us to state precisely the rules for doing a calculation, than it is for us to get the correct answer after a long and tedious application of such rules.

To make a computer do algebra requires programming of a distinctly different flavor than the computations and simulations described so far. What there is in fluid mechanics in this direction mostly consists of the extension of perturbation series (see Van Dyke, 1994). It is possible to have modern computer algebra packages give output in the correct syntax of a language such as FORTRAN, and such interfaces between different modes of computation are likely to become increasingly important in the future.

1.7 Ramification, Turbulence and the Complexity of Fluid Flows

So far in our preliminary discussion it may have appeared that there are really no intrinsic problems in conducting a flow simulation. Surely there are methodological issues to contend with, and there are manual tasks associated with actually writing and running a code on a computer system. But are there any fundamental issues, any items of principle, any obstacles to simulating whatever flow we might wish? The answer to this question is a definite yes, and the most familiar instance of these limitations arises when discussing the simulation of turbulent flows.

Qualitatively speaking what makes a turbulent flow hard to measure, theorize about and ultimately compute is its **ramification**, i.e., the immense range of scales that is excited at large values of Reynolds number, and that must somehow be captured and addressed in a computer simulation. One theme that will be developed as we proceed is that turbulent fluid flow is not alone in having this challenging property. Figure 1.5 shows examples of highly ramified flows containing many scales, that are all coupled. These examples were specifically chosen from areas of fluid mechanics other than turbulence. The problems and issues of dealing with a large variety of coupled scales is common to all these flow situations, although the precise scaling, and hence the problems of storage and runtime, will differ from case to case. There are, to be sure, simpler cases of fluid flow, such as the similarity solutions and the steady states that we spoke about above. For these, miraculously, all the complexities balance out in just a single function or a few algebraic equations. But such cases are in some sense atypical. In most cases we must expect a flow configuration that has many scales of motion. In Section 1.7.1 the scaling of homogeneous, isotropic turbulence is outlined. This provides the best-known case of flow scaling with immediate consequences for CFD. We stress, however, that analogous arguments can be advanced for numerous other flow situations that are not turbulent.

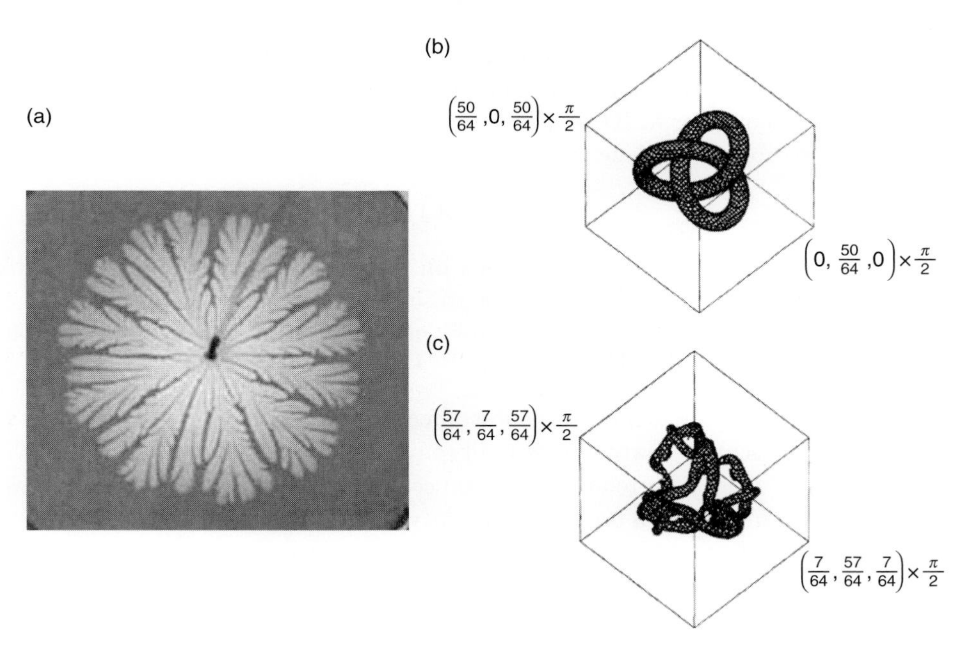

Figure 1.5 Two examples of the notion of ramified flow structures. In (a) an experimental photograph from Ben-Jacob and Garik (1990), reprinted by permission of Macmillan Publishers Ltd., of repeated branching due to viscous fingering in a Hele-Shaw cell in a radial geometry. In (b,c) two images from the computed evolution of a vortex filament started as a knot from the work of Kida and Takaoka (1987), reprinted with permission of AIP Publishing. Note that in both examples a large range of scales are excited by the fluid motion.

In a turbulent flow the velocity at a point is seemingly a random process, and we must address whether that is something we can simulate reliably on a computer. We can generate random numbers on a computer. At least we can generate sequences of quasi-random numbers that pass conventional statistical tests for randomness. The problem with turbulence is that we do not know the probability distribution *a priori*. This is something that is given to us in the rather backhanded manner of complex solutions to certain PDEs. In fact, from experiments we know that turbulence is not always as disorganized as one might like to think. The discovery of coherent structures in turbulent flows has shown that often these flows have a high level of spatial and temporal organization. Numerical simulations must capture this dual, order-disorder nature of turbulence correctly and in the proper proportions.

1.7.1 Kolmogorov Scaling of Homogeneous, Isotropic Turbulence

The conventional picture of homogeneous, isotropic turbulence, pioneered by Richardson and Taylor, involves the notion of an *energy cascade* given mathematical form by Kolmogorov, Obukhov, Onsager and von Weizsäcker. The basic object of the theory is a quantity known as the **energy spectrum**, $E(k)$, which measures the amount of energy per unit mass in the flow per unit interval of wavenumber k. Thus, the total kinetic energy per unit mass of homogeneous, isotropic turbulence is

$$E_{\text{total}} = \int_0^\infty E(k)\, dk\,. \tag{1.7.1}$$

The energy spectrum depends on k itself, but it also depends on the rate at which energy flows from scale to scale, known as the *cascade rate*, ε.

The picture that Kolmogorov and others developed in the early 1940s has the following ingredients:

(i) The flow energy is most effectively dissipated by viscosity at very small scales. Newtonian viscosity involves a term $\nu\nabla^2\mathbf{V}$, where $\mathbf{V}$ is the velocity field. The relative importance of this term in the Navier–Stokes equation depends on the flow scale. At very high Reynolds numbers this term is most important for small scales.

(ii) To generate these small scales the flow sets up a hierarchy of scales. This hierarchy is largely governed by inertial dynamics, i.e., by the Euler equation, for many decades in scale. It is known as the *inertial range*. Hence, energy is exchanged between the larger scales in the hierarchy essentially without dissipation, and is ultimately dissipated at the smallest scales. The rate at which energy is transferred, the *cascade rate*, ε, is a key parameter. It has the units of energy per unit mass per time. Since the same amount of energy flows through each scale in the inertial range, ε is also the rate at which energy is dissipated. For this reason ε is also known as the *dissipation rate*.

(iii) Hence, a functional relation

$$E(k) = \text{function of } (\varepsilon, k) \tag{1.7.2}$$

must hold in the inertial range.

When this relation is subjected to the methods of dimensional analysis,[5] with $E(k)$ having units of energy per unit mass and unit length, and k having the units of reciprocal length, an easy calculation gives the **Kolmogorov spectrum**

$$E(k) = C_k \varepsilon^{2/3} k^{-5/3}, \tag{1.7.3}$$

where C_k is a dimensionless coefficient known as the *Kolmogorov constant.*

Equation (1.7.3) holds until we reach the small scales, where viscous effects become comparable to the inertial terms. At these scales all energy delivered from larger scales is dissipated, and ε can be set equal to the rate of energy dissipation. Let η denote the length scale at which viscous dissipation takes over from the inertial range spectrum (1.7.3). We expect this **Kolmogorov microscale** η to depend on ϵ and on the kinematic viscosity ν (since everything is being done per unit mass). Again use of dimensional analysis gives

$$\eta \approx \left(\frac{\nu^3}{\varepsilon}\right)^{1/4}. \tag{1.7.4}$$

Looking at the large-scale end of the inertial range (1.7.3), we deduce that ϵ must be determined by the velocity on that scale, u, and the overall spatial length scale L of the flow. Viscosity cannot enter here because these scales obey essentially inviscid dynamics. Dimensional analysis now gives the result

$$\varepsilon \approx \frac{u^3}{L}. \tag{1.7.5}$$

Problem 1.7.1 Verify (1.7.3)–(1.7.5) by dimensional analysis.

Finally, combining (1.7.4) and (1.7.5) to eliminate the unknown ϵ, we get the important estimate

$$\frac{L}{\eta} = \mathrm{Re}^{3/4} \qquad \text{where} \qquad \mathrm{Re} = \frac{uL}{\nu}. \tag{1.7.6}$$

This gives the *ratio of largest to smallest length scale* in a turbulent flow as a function of the large-scale Reynolds number, Re.

Let us now address these results of classical turbulence theory from the point of view of CFD. Imagine our flow domain partitioned into grid cells of size η for the purposes of discretization. If you make the cell size larger, you will not have the necessary resolution to compute the dissipation scales. Let the computational domain size be L or more to accommodate the large scales. Then we must have at least L/η cells in each spatial direction in order to capture the entire range of flow scales. Hence in 3D (which is where (1.7.3) is valid) *the number of grid cells required scales as* Re *to the 9/4 power.*

But there is more: we have in mind following the time evolution of the flow. For accuracy we must insist that the time step be of the same order as *the shortest*

[5] References on this subject include Huntley (1967) and Barenblatt (1996).

dynamical time scale in the system. There are at least two suggestions for what this time scale is:

(i) the time scale for vorticity to diffuse a distance η, i.e., η^2/ν; or
(ii) the time to advect something the shortest dynamically significant distance η using u, the largest velocity around, i.e., η/u. The latter is the shorter time, and we designate it τ.

Problem 1.7.2 Show using Kolmogorov scaling that $\eta/u = (\eta^2/\nu)\,\mathrm{Re}^{-1/4}$. Thereby, verify that η/u is much less than η^2/ν at large Reynolds number.

Now, the total time period over which we wish/need to conduct the simulation is set by the largest scales. We really do want to see the largest eddies in the flow turn over at least a few times in order to be assured that we have captured a dynamically significant picture of the flow. Let the time scale on which this happens be T, often called the large eddy turn-over time. We have the estimate $T \approx L/u$, and thus, using (1.7.6),

$$\frac{T}{\tau} \approx \frac{L/u}{\eta/u} \approx \frac{L}{\eta} \approx \mathrm{Re}^{3/4}\ . \tag{1.7.7}$$

This gives the Re-scaling of the *number of time steps* (of size τ) that we must take with our numerical method in order to simulate turbulent flow for a time T.

Finally, we ask: How much *computational work* is required at each of these time steps? How many floating point operations are necessary? Well, at the very least we will have to handle the data at each grid point a few times per time step. We have to compute some differences in grid point values to get approximations for spatial derivatives, run the data through the time-stepping routine, and so on. So a conservative estimate is that *the computational work per time step will scale as the number of grid cells.* Thus, the number of operations for the total calculation will scale as

$$(\#\ \text{grid cells}) \times (\#\ \text{time steps}) = \mathrm{Re}^{9/4} \times \mathrm{Re}^{3/4} = \mathrm{Re}^3\ . \tag{1.7.8}$$

What we really mean by this is that the computational work, W, scales as

$$W = W_{\mathrm{T}} \left(\frac{\mathrm{Re}}{\mathrm{Re}_{\mathrm{T}}}\right)^3 , \tag{1.7.9}$$

where Re_{T} is the value of Reynolds number at the transition to turbulence, and W_{T} is the work required for the calculation at the onset of turbulence.

Several conclusions follow from these estimates. First, the available rapid access memory on the computer must scale by (1.7.6) as $\mathrm{Re}^{9/4}$. Second, given this memory, the computational work measured by the actual time it takes to run the calculation in CPU hours scales according to (1.7.9). The long-term storage requirement, which includes data dumps of flow quantities, such as velocity and temperature, at many different time instances, also scales according to (1.7.9). In particular, we see that in order to even double the value of Reynolds number rapid access memory must increase by slightly more than a factor of 4, and the computational work and data storage required will increase by a factor of 8.

To alleviate the stringent demands on computer performance implied by the scaling in (1.7.6) and (1.7.9) *turbulence modeling* based on *Reynolds-averaged Navier–Stokes equations* is used. Here the philosophy is to only compute the time-averaged flow field. The entire range of fluctuating scales is not directly computed and their effect on the computed time-averaged flow is accounted for through turbulence models. Although this approach is computationally much more affordable than the direct simulations, turbulence modeling, unfortunately, often implies knowledge of the physics of turbulent flows that goes well beyond what we legitimately know at present. *Large eddy simulation* methodology, presents a compromise. In this approach the time evolution of only a small range of large scales, which are likely to be problem dependent, are computed. The rest of the small scales are not directly computed and therefore their influence on the large scale dynamics is accounted for through *subgrid scale models*. Here modeling is delayed to smaller scales, which are believed to be more universal and less problem dependent. Nevertheless, the appropriate definition of large eddies, large eddy equations, and subgrid models are still a matter of active research, let alone the algorithms and numerical methods needed for solving them.

1.8 The Development of Computer Hardware

While many of the techniques to be described in this book are useful for any kind of computer, even hand calculations on occasion – and some early CFD investigations, including the computation of the sessile drop shape by Bashforth and Adams (1883) and the shear layer roll-up calculations of Rosenhead (1931), were, indeed, carried out by hand – the impetus for modern CFD owes a considerable debt to the spectacular evolution of computer hardware that has unfolded in the late twentieth and early twenty-first centuries. For example, in the early 1990s personal microcomputers were capable of handling only a few hundred coupled ODEs in a reasonable time. Mini-computers and workstations were able handle a PDE in one space and one time dimension at reasonable resolution in a reasonable time, and can make some headway with two-dimensional problems if one is willing to wait for the result. For fully three-dimensional flow problems one had to resort to the supercomputers. But hardly a couple of decades later, three-dimensional flow problems are routinely solved on personal computers and workstations. Only the largest of the flow problems require the power of modern massively parallel supercomputers.

Typical graphs of computer performance show speed – both peak speed and effective or sustained speed – increasing dramatically with time (since about

1960), and the size of memory increasing roughly proportional to speed. During the two decades preceding the introduction of the CDC7600 supercomputer in 1969, the first Class IV machine, computer performance increased by about a factor of two every three years (Ledbetter, 1990). The CDC7600 computer had a sustained speed of about 4 MFLOPS[6] and 4×10^5 64-bit words of memory. Computational aerodynamicists judged it capable of calculating the flow about an airfoil using Reynolds-averaged Navier–Stokes equations (i.e., the Navier–Stokes equations augmented by an acceptable turbulence model) with a turn-around time that allows use in an engineering design process.

The CRAY-1, the first Class VI machine, introduced in 1976 utilized a new hardware feature known as vector pipelining, along with strong scalar capabilities, to achieve sustained speeds of about 15 MFLOPS and peak speeds on special problems that could be as much as ten times faster than the CDC7600 (Ledbetter, 1990). The development following the introduction of vector supercomputers has brought us the CRAY X-MP, Y-MP, CRAY-2 and CRAY-C90 with sustained speeds of several hundred MFLOPS, and peak speeds in excess of 1 GFLOP. Computations of high-speed flow around complicated geometries, such as a complete aircraft, can now be attempted, although the full treatment of boundary layer processes is still elusive.

The next era of major development of hardware capabilities involves the notion of parallel processing. The main idea is to have many off-the-shelf processors working in unison and simultaneously on pieces of a problem, and to harvest the results of all processors at fixed times. Hence, the less rapid work of many general purpose processors wins over the speedy work of a few very specialized sophisticated ones. One speaks of *data parallelism*, in contrast to the *functional parallelism* that characterizes the fastest vector supercomputers. The main challenge, of course, is to assure that the many processors cooperate and finish on time. Waiting for one of them slows the whole machine to a crawl. Thinking Machines, Inc., introduced a series of multi-processor supercomputers under the label *Connection Machine*. These initially espoused the so-called SIMD architecture (single-instruction, multiple-data), where all the processors do the same thing to different pieces of data at any one time. Other vendors have assembled supercomputers by connecting together commodity off-the-shelf processors using high-speed interconnects. These parallel machines can operate in the MIMD (multiple-instructions, multiple-data) mode in which independent processors can be programmed to do individual tasks, and each processor has its own memory on which it can operate. Coordination of tasks to complete a computation is then handled by message-passing between processors. As of 2015 parallel machines approach speeds of $O(30)$ petaflops. The major issue in harnessing their full potential still concerns the most efficient way of programming these massively parallel machines. In particular, the issue of allowing the computational scientist to work in a convenient, high-level language while taking advantage of

[6] Floating point operations per second.

the architecture of the underlying machine requires considerable work in basic computer science.

The future of supercomputers promises to be very exciting with unprecedented speeds and memory sizes. The new concept of computer architecture and parallelization challenges our thinking about computational problems at both the conceptual, algorithmic level, and at the level of implementation. The desktop workstation of today acquires the performance level of yesteryear's supercomputer at a fraction of the price. This opens up tremendous opportunity for tackling significant problems of fluid mechanics even without access to the mightiest computers.

1.9 Some Remarks on Software

Two important remarks should be made concerning CFD software. The first is that while the performance of hardware has shown a dramatic and widely appreciated improvement, the gains in efficiency and speed due to algorithmic understanding have been no less impressive, although probably less well known and publicized. It has been estimated that since 1945 algorithmic advances have decreased the operation count for common PDEs on an N^3 grid by about a factor of $N^4/60$. Advances due to algorithms are, of course, harder to quantify than the rather cut-and-dry measurements of clock speeds and petaflops. One sometimes hears the opinion that the biggest bottleneck in CFD is not in the algorithms or the coding of the Navier–Stokes equation, but rather in the inadequate physical understanding of turbulent flow processes. All this points to a concurrent appreciation for numerical methodology and flow physics as being the essential ingredients for mastering CFD.

A second important remark is that there exists today a number of high-quality, commercial and open-source codes for performing CFD investigations. To a certain extent, one can say that the problem of finding an approximate numerical solution to the flow of a Navier–Stokes fluid in an arbitrary domain at intermediate Reynolds numbers has been solved. In many cases impressive agreement with experimental measurements can be demonstrated. In this situation the type of questions that a knowledge of CFD will allow us to ask and (maybe) answer are: How accurate is the solution? Can it be improved? Can it be obtained more efficiently? What is the basic flow physics being revealed by the calculation?

There are still areas of fluid mechanics where good computational procedures are lacking. There are problems that are so big that great compromises must be made to obtain any numerical solution at all, for example in the problem of simulating global climate change. How does one construct codes for such situations so that the predictions are reliable? There are flow problems involving complex constitutive equations, multiple materials, multiple phases (gas, solids and liquids), e.g., in geophysics, chemical engineering and materials science, where adequate CFD codes do not yet exist. There are problems of a conceptual, mathematical nature, such as whether the 3D Euler equation with smooth initial data produces singularities after a finite time, that will presumably first be decided through extensive computations, but require unprecedented care and precise control over numerical errors.

Literature Notes

The literature on numerical analysis is vast. There are a number of good general texts introducing various methods with practical examples such as Carnahan *et al.* (1969). A similar text closely tied in with the IMSL software library is Rice (1983). For a quick overview of certain topics the brief book by Wendroff (1969) is quite useful. Mention should also be made of the "classics", Hamming (1973) and Hildebrand (1974). The two-volume set by Young and Gregory (1988) is a useful and inexpensive reference. The books by Moin (2010a,b) treat several of the topics to be covered in this book, but by not focusing attention only on CFD issues they target a broader engineering audience. The book by Press *et al.* (1986) has an attractive, user-oriented perspective, and gives several ready-to-use codes. It became very popular, and will be referred to frequently in what follows. However, the amount, and the depth, of numerical analysis included varies considerably from topic to topic, and one may be able to do better than relying on the code given in the book. The subtitle of this work – *The Art of Scientific Programming* – is a play on the title of the magnum opus of Knuth (1981), which is well worth perusing. This book is a must for computer scientists. The collection of articles edited by Golub (1984) is similarly inspirational.

General surveys of CFD appear in a number of books, for example, Roache (1976), Holt (1977), Peyret and Taylor (1983), and Fletcher (1991). Books that focus on particular methods will be cited later. There are also several useful review papers treating CFD broadly: these are typically quite different in outlook and approach to the subject. We list here a selection by year of publication: Emmons (1970), Roberts and Christiansen (1972), Orszag and Israeli (1974), Patterson (1978), MacCormack and Lomax (1979), Zabusky (1987), Boris (1989), Glowinski and Pironneau (1992), Moin and Krishnan (1998), Wang *et al.* (2006), Spalart (2009), Pirozzoli (2011), and Fox (2012). An authoritative review of computer algebra applications in fluid mechanics is given by Van Dyke (1994). More recent books on CFD include those by Hirsch (2007), Chung (2010), Pozrikidis (2011), and Karniadakis and Sherwin (2013).

2 Mappings

The PDEs of fluid mechanics deal with the continuum of real numbers. When we discretize these equations in space and time for computational purposes, we project them onto a set of algebraic equations that we hope will capture the behavior of the PDEs in spite of inevitable truncation errors. When we program the discretized equations on a (finite) computer, we incur another projection due to the inevitable effects of round-off errors. To make matters worse, it is relatively easy to construct examples where these different errors accumulate as the calculation proceeds, leading ultimately to results that are completely useless.

In this chapter we attempt to put the best face on things by (i) viewing everything one does in CFD as a discrete mapping. For example, the fluid flow can be considered as a mapping from a velocity field at the current time, $\mathbf{V}(\mathbf{x}, t)$, to a velocity field at a later time $\mathbf{V}(\mathbf{x}, t+\Delta t)$; (ii) discussing some basic, common properties of such mappings; and (iii) turning the whole process to our advantage (at least in part) by arguing that the discretized description to which we are led is very interesting in its own right, and can, in fact, give useful insight!

Mappings form the simplest numerical representation of problems in fluid dynamics – they are simply function evaluation. Numerical methodologies for more complex ODEs and PDEs arising from fluid dynamics will be considered in subsequent chapters. As we shall see in those chapters, even the ODEs and PDEs in most cases after discretization reduce to mappings, albeit more complex in nature than those considered in this chapter.

2.1 Numerical Methods as Mappings

The prototypical problem that we shall consider has the format

$$\xi_{n+1} = F(\xi_n; \lambda)\,, \tag{2.1.1}$$

where ξ_n is interpreted as the state of the system at step n. The state can be anything from a single number to a vector function at every point of a grid (e.g., a velocity field). The step may be just some stage in an iterative process of no particular physical significance, or it may be some instant in time during the computation of the actual time evolution of the system. The right-hand side of

(2.1.1), F, gives the rule for updating ξ, and λ denotes a set of parameters, both physical (such as Reynolds number) and numerical (such as a time step or a grid spacing) that arise from the specification of the problem and our decisions on which numerical method to use.

The format (2.1.1) used with this liberal interpretation does, indeed, embody most numerical methods used in CFD. It follows that we shall be able to say very little about this most general case. To make any headway with (2.1.1) we shall need to consider the very restricted case when ξ is just a single number (or a vector with a few components), and F is a simple function (or a vector of functions with the same small number of components as ξ). If you wish, you can think of this as an attempt to do CFD on a very small computer! You might feel that this is a rather silly thing to do. However, the results that arise from the few-degrees-of-freedom description are fascinating in their own right and turn out to have considerably more to do with fluid mechanics than one might have thought. Furthermore, low-dimensional mappings are fertile ground for writing simple computer programs with non-trivial output. This is useful for getting us started with some computer exercises.

Let us then consider some examples of numerical methods that can be viewed as discrete mappings in the sense of (2.1.1).

Example 2.1.1 In a simple problem we may have a single equation to solve, say

$$\phi(\xi) = 0\,. \tag{2.1.2}$$

One way to do this is by **iteration**: consider rewriting the above equation as, $\xi = \phi(\xi) + \xi$ from which the following simple algorithm for iteration follows:

$$\xi_{n+1} = \phi(\xi_n) + \xi_n\,. \tag{2.1.3}$$

Then, if $\xi_n \to \xi_*$ as we iterate, we have a solution, $\xi = \xi_*$, of our initial equation (2.1.2). In this case the F of (2.1.1) is just $\phi(\xi) + \xi$.

A more sophisticated way of solving the equation (2.1.2) is to use **Newton's method**. This method instructs us to try the iteration

$$\xi_{n+1} = \xi_n - \frac{\phi(\xi_n)}{\phi'(\xi_n)}\,, \tag{2.1.4}$$

where the prime indicates differentiation with respect to the argument. So in this case

$$F(\xi) = \xi - \frac{\phi(\xi)}{\phi'(\xi)}\,. \tag{2.1.5}$$

Instead of a single parameter in F there is a functional dependence on ϕ. The geometric interpretation of Newton's method is to approximate the root by extending the tangent line of ϕ to the x-axis is well known and it is the basis for trying (2.1.4).

Example 2.1.2 When we attempt to integrate a differential equation numerically, we are again faced with generating a sequence of values. Suppose we want to solve

$$\frac{\mathrm{d}\xi}{\mathrm{d}t} = g(\xi;\mu)\,, \tag{2.1.6}$$

where μ represents a set of parameters. In the simplest case, often called **Euler's method** (although it is hard to believe that Euler would not have thought of something better!) we discretize by setting $\xi_n = \xi(n\Delta t)$, where $\xi(t)$ is the solution we are seeking, and then solve successively

$$\xi_{n+1} = \xi_n + \Delta t\, g(\xi_n;\mu)\,. \tag{2.1.7}$$

So the mapping function in this case is

$$F(\xi) = \xi + \Delta t\, g(\xi;\mu)\,. \tag{2.1.8}$$

The parameters λ in (2.1.1) consist of the parameters μ from (2.1.6) augmented by the time step Δt. There are more elaborate integration schemes of higher order in Δt that lead to more complex mappings than (2.1.7). We shall discuss them later in Chapter 3.

2.2 Fixed Points: Stability, Instability and Superstability

Although the mappings we encounter in practice can vary considerably in complexity, and the symbols ξ, F, etc., may take on different meanings in different contexts, there are some general concepts and principles that apply across the board. It is useful to discuss these here using the simplest realization of (2.1.1), *viz.* the case where ξ is a real variable and F is a real function.

The first concept is **fixed point**. A fixed point of (2.1.1) is a solution which remains the same after every iteration. If the fixed point solution at the nth iteration is $\xi_n = \xi_*$, then by definition of fixed point $\xi_{n+1} = \xi_*$, for all n. Thus the fixed point solution satisfies the relation (suppressing any parameters λ for the moment)

$$\xi_* = F(\xi_*)\,. \tag{2.2.1}$$

The next thing to consider is the **stability** of the fixed point. To do this we perturb ξ_* at the start of the iteration slightly by setting $\xi_0 = \xi_* + \delta_0$. We assume that $|\delta_n|$ is "small" for all n and substitute $\xi_n = \xi_* + \delta_n$ in (2.1.1); we then expand F in a Taylor series about ξ_*. Retaining only the linear term in that series, we get

$$\begin{aligned}\xi_{n+1} &= \xi_* + \delta_{n+1} \\ &= F(\xi_* + \delta_n) = F(\xi_*) + \delta_n\, F'(\xi_*) + O(\delta_n)^2\,,\end{aligned}$$

where $O(\delta_n)^2$ represents all other terms of the Taylor series that are of order δ_n^2 and smaller. Thus for small δ_n

$$\delta_{n+1} = F'(\xi_*)\,\delta_n\,. \tag{2.2.2}$$

For brevity let $F'(\xi_*) = \Lambda$. Then (2.2.2) is easily solved by iteration:

$$\delta_n = \Lambda^n\,\delta_0\,, \tag{2.2.3}$$

which says that if $|\Lambda| > 1$, the δ_n will increase exponentially with n. In this case the perturbation grows away from the fixed point, and we say that the fixed point ξ_* is **unstable**. If, on the other hand, $|\Lambda| < 1$, any small perturbation will die out on iteration, and the perturbed solution will decay back to the fixed point ξ_*. In this case we say that the fixed point is **stable**. For the case $|\Lambda| = 1$, which occurs right at the juncture of stable and unstable, we say the fixed point is **neutrally stable** or **marginally stable**. This characterization is only from a linear stability point of view. When $|\Lambda| = 1$ the fixed point can be nonlinearly stable or unstable. To determine this nonlinear behavior one must include additional terms of the Taylor series in (2.2.2) and re-examine the growth or decay of δ_n.

The case $\Lambda = 0$ requires additional considerations. If we continue the Taylor expansion of F, there will in general be some lowest order p at which $F^{(p)}(\xi_*) \neq 0$, where the superscript (p) indicates the pth derivative. In such a situation, in place of (2.2.2) we will have to leading order

$$\delta_{n+1} = \frac{1}{p!} F^{(p)}(\xi_*)\,\delta_n^p\,. \tag{2.2.4}$$

Let us now set

$$\Lambda = \frac{1}{p!} F^{(p)}(\xi_*)\,, \tag{2.2.5}$$

which for $p = 1$ reduces to our previous usage of the symbol Λ, and let us take logarithms on both sides of (2.2.4). We then have

$$\Xi_{n+1} = \log\Lambda + p\Xi_n\,, \tag{2.2.6}$$

where $\Xi_n = \log\delta_n$.

The recursion formula (2.2.6) is *linear* in Ξ_n and is of the *first-order* in the sense that it involves only a unit step in n. Its coefficients are independent of n, and so it may be said to have *constant coefficients*. It is *inhomogeneous* because of the appearance of the term $\log\Lambda$. The theory for solving such recursion formulae is analogous to the theory of integrating linear ODEs with constant coefficients. The general solution of (2.2.6) can be constructed by finding a *particular solution* to the equation and then adding a *homogeneous solution* of *the homogeneous equation*, $\Xi_{n+1} = p\Xi_n$, to it. Both tasks are quite easy: the homogeneous solution is clearly $\Xi_n = Ap^n$, where A is an arbitrary constant. The particular solution to (2.2.6) arises by setting all $\Xi = C$; then C must satisfy $C(1-p) = \log \Lambda$, or $C = \log\{\Lambda^{1/(1-p)}\}$. Thus, the general solution to (2.2.6) is $\Xi_n = \log\{\Lambda^{1/(1-p)}\} + Ap^n$. We must set $A = \Xi_0 - \log\{\Lambda^{1/(1-p)}\}$ in order to match the value for $n = 0$. From these considerations it follows that

$$\log \delta_n = (1 - p^n)\log\left\{\Lambda^{1/(1-p)}\right\} + p^n \log \delta_0, \tag{2.2.7}$$

or that the solution to (2.2.4) is

$$\delta_n = \Lambda^{(1-p^n)/(1-p)}\,\delta_0^{p^n}. \tag{2.2.8}$$

Problem 2.2.1 In (2.2.3), $p = 1$, it is clear that δ_n will flip sign for alternate n if $\Lambda < 0$. Discuss the sign of the iterates δ_n when $p > 1$ and $\Lambda < 0$.

Note that for positive δ_n and $\Lambda>0$, (2.2.8) may be written

$$\frac{\delta_n}{\delta_0} = \left(\frac{\delta_0}{\Lambda^{1/(1-p)}}\right)^{p^n - 1}. \tag{2.2.9}$$

First of all, we see that for δ_n to converge to zero we must have $\left|\delta_0/\Lambda^{1/(1-p)}\right| < 1$. If this is so, the *rate of convergence* depends on δ_0, the magnitude of the initial perturbation. Indeed, if we choose δ_0 smaller and smaller approaching the fixed point, the convergence becomes more and more rapid. This convergence is more rapid than the convergence of $\Lambda^n\delta_0$ in (2.2.3). We call the stability arising out of (2.2.4), with $p \geq 2$, **superstability**.

Example 2.2.2 For $\Lambda = 0.5$ and $\delta_0 = 1$, (2.2.3) gives $\delta_n = 2^{-n}$, whereas with the same value of Λ in (2.2.4) and $p = 2$ the convergence of (2.2.9) is much faster: $\delta_n = 2^{-(2^n-1)}$.

Example 2.2.3 The simple iteration given in Example 2.1.1 leads to

$$F'(\xi) = 1 + \phi'(\xi). \tag{2.2.10}$$

At the fixed point we have $\Lambda = 1 + \phi'(\xi_*)$ and thus in order to converge to the fixed point the following condition must be satisfied:

$$-2 < \phi'(\xi_*) < 0. \tag{2.2.11}$$

Thus, the simple iteration (2.1.3) will converge only when the slope at the fixed point $\phi'(\xi_*)$ is negative and bounded in value.

Example 2.2.4 Newton's method for finding the roots of $\phi(\xi) = 0$ leads to one of the better known examples of a superstable fixed point. For (2.1.5) we have

$$F'(\xi) = 1 - \frac{\phi'(\xi)}{\phi'(\xi)} + \frac{\phi(\xi)\,\phi''(\xi)}{(\phi'(\xi))^2} = \frac{\phi(\xi)\,\phi''(\xi)}{(\phi'(\xi))^2}. \tag{2.2.12}$$

For a fixed point, $F(\xi_*) = \xi_*$ or $\phi(\xi_*) = 0$. Thus, $F'(\xi_*) = 0$, and we must have $p > 1$. In fact, $p = 2$ since the second derivative,

$$F''(\xi_*) = \frac{\phi''(\xi_*)}{\phi'(\xi_*)}, \tag{2.2.13}$$

in general does not vanish.

2.3 Stability of Exact Solution, Implicit and Explicit Approximations

Using these ideas we can also see why an **implicit** method of integrating an ODE may be preferable to an **explicit** method. Let us go back to the earlier numerical approximation to $\mathrm{d}\xi/dt = g(\xi;\,\mu)$ as $\xi_{n+1} = \xi_n + \Delta t\, g(\xi_n;\,\mu)$ and subject it to this type of stability analysis. A "fixed point" is again a solution ξ_* such that $\xi_n = \xi_*$ for all n. This means $g(\xi_*;\,\mu) = 0$; i.e., in the case of the ODE, a fixed point is a steady-state solution of the equation. Note that the numerical discretization correctly identifies the steady state of the ODE. For both the original ODE and its discretization the steady state is a solution of $g(\xi_*;\,\mu) = 0$.

Turning next to the question of stability, there are really two distinct issues. One is that the steady state ξ_* viewed as a solution of the ODE may be stable or unstable to small perturbations. The second is that the numerical method may yield stability properties of ξ_* that are different from those of ODE. We need to investigate the nature of this potential shortcoming of the numerical discretization.

Let us first determine the analytical stability properties of a steady solution to the ODE. We perturb about ξ_* by setting $\xi(t) = \xi_* + \delta(t)$ and again ignore everything but the linear term on the right-hand side:

$$\frac{\mathrm{d}\xi}{\mathrm{d}t} = \frac{\mathrm{d}\delta}{\mathrm{d}t} = g(\xi_*) + g'(\xi_*)\,\delta \tag{2.3.1}$$

(suppressing the parameter(s) μ for the present). Thus, if $g'(\xi_*) > 0$, δ grows exponentially and ξ_* is *unstable*. If $g'(\xi_*) < 0$, δ decays exponentially and ξ_* is *stable*. Finally, the case $g'(\xi_*) = 0$ leads to *neutral stability* or *marginal stability*. Here again neutral or marginal stability is only in the context of linear analysis. When $g'(\xi_*) = 0$, the perturbation δ can nonlinearly grow or decay depending on the sign of $g''(\xi_*)$ and the initial perturbation δ_0. If $g''(\xi_*)$ is also zero, look for the first non-zero derivative at ξ_*.

Now let us consider the numerical approximation. According to (2.1.8) and

(2.2.3) at issue is the numerical value of $\Lambda = F'(\xi_*) = 1 + \Delta t\, g'(\xi_*)$. The stability or instability of the solution ξ_* from the perspective of the numerical method depends on whether $|1 + \Delta t\, g'(\xi_*)|$ is less than 1 or greater than 1. Thus, Euler's method (2.1.7) shows ξ_* to be unstable if $g'(\xi_*) > 0$, in complete accord with the analytical results from the ODE itself. According to linear analysis neutral stability corresponds to Λ=1 or $g'(\xi_*) = 0$. The numerical method can be nonlinearly stable or unstable depending on the value of $g''(\xi_*)$. Again we want the behavior of the numerical solution to be the same as that of the exact solution. Here we will restrict our investigation to linear analysis.

Now, we run into numerical instability in the case $g'(\xi_*) < 0$ if we take large time steps. As long as $0 > g'(\xi_*)\Delta t > -1$ we recover the exact stable behavior of the ODE. The numerical solution becomes unstable if we use a large time step, $-2 > g'(\xi_*)\Delta t$. Over the range $-1 > g'(\xi_*)\Delta t > -2$ even when we have numerical stability, the approach of the numerical solution to ξ_* will be oscillatory, whereas the analytical solution decays monotonically. Oscillatory decay is often called **overstability**.

In summary, the analytical stability properties of a steady-state solution are only reproduced by the numerical method (2.1.7) for sufficiently small time steps. As the time step is increased, the numerical solution first displays overstability, then instability. Table 2.1 compares the stability properties of the numerical scheme with those of the actual ODE.

Table 2.1 Comparison of the stability properties of the forward Euler (explicit) and backward Euler (implicit) schemes with those of the actual ODE $\mathrm{d}\xi/\mathrm{d}t = g(\xi;\, \mu)$.

	Actual Behavior	*Forward Euler (explicit)*	*Backward Euler (implicit)*
$g'(\xi_*)\,\Delta t > 2$	unstable	unstable	overstability
$2 > g'(\xi_*)\,\Delta t > 1$	unstable	unstable	oscillatory instability
$1 > g'(\xi_*)\,\Delta t > 0$	unstable	unstable	unstable
$g'(\xi_*)\,\Delta t = 0$	neutral	neutral	neutral
$0 > g'(\xi_*)\,\Delta t > -1$	stable	stable	stable
$-1 > g'(\xi_*)\Delta t > -2$	stable	overstability	stable
$-2 > g'(\xi_*)\,\Delta t$	stable	unstable	stable

Even if we are just going to be first-order accurate as in (2.1.7), we still have further options. We could set

$$\xi_{n+1} = \xi_n + \Delta t\, g(\xi_{n+1};\, \mu)\,. \tag{2.3.2}$$

We refer to (2.1.7) as the **forward** Euler method, whereas (2.3.2) is called the

backward Euler method. Clearly, (2.1.7) is an **explicit method**, in contrast to (2.3.2), which is **implicit**, i.e., the solution at step $n+1$ is given implicitly by that equation. You have to solve (2.3.2) for ξ_{n+1} to get the solution.

Solving (2.3.2) may be unpleasant in general, but close to our steady state ξ_* we can still do stability analysis. This time we get

$$\delta_{n+1} = \delta_n + \Delta t\, g'(\xi_*)\,\delta_{n+1}\,, \tag{2.3.3}$$

or

$$\delta_{n+1} = \frac{\delta_n}{1 - \Delta t\, g'(\xi_*)}\,. \tag{2.3.4}$$

Now the stability considerations are different: in the unstable case, $g'(\xi_*) > 0$, we must be careful to select a time step Δt smaller than $1/g'(\xi_*)$ to get monotone growth. However, in the more important stable case, $g'(\xi_*) < 0$, we see that the coefficient of δ_n is less than 1 for any value of Δt. Thus, the implicit method (2.3.2) captures the true stability property of the analytical solution in the entire stable region.

We are left with the conclusions that an implicit method may sometimes be more suitable than an explicit method for integration because the range of time step sizes that capture stability is larger. There is a danger here, of course. While (2.3.2) may reproduce stability for arbitrarily large Δt, the accuracy of the method certainly will degrade with increasing Δt. The stability of a solution obtained by using an implicit method should not be mistaken for accuracy of that same solution.

A few words about (2.2.1)–(2.2.3) in the multi-dimensional case may be in order. If the quantity ξ is an N-dimensional vector, and the mapping function F thus an N-dimensional vector function, we have a system of equations of the form

$$\left.\begin{array}{rcl} \xi_{1,(n+1)} &=& F_1(\xi_{1,(n)}, \xi_{2,(n)}, \ldots, \xi_{N,(n)})\,,\\ \xi_{2,(n+1)} &=& F_2(\xi_{1,(n)}, \xi_{2,(n)}, \ldots, \xi_{N,(n)})\,,\\ &\vdots& \\ \xi_{N,(n+1)} &=& F_N(\xi_{1,(n)}, \xi_{2,(n)}, \ldots, \xi_{N,(n)})\,. \end{array}\right\} \tag{2.3.5}$$

The fixed point equations are

$$\xi_{i*} = F_i(\xi_{1*}, \xi_{2*}, \ldots, \xi_{N*}) \qquad \text{for} \quad i = 1, 2, \ldots, N\,. \tag{2.3.6}$$

Linearizing around this solution by setting $\xi_{i,(n)} = \xi_{i,*} + \delta_{i,(n)}$, we now obtain in place of (2.2.2)

$$\delta_{i,(n+1)} = \sum_{k=1}^{N} \frac{\partial F_i}{\partial \xi_{k,*}}\,\delta_{k,(n)}\,. \tag{2.3.7}$$

The Jacobian of the mapping (2.3.5), $\partial F_i/\partial \xi_{k,*}$, evaluated at the fixed point, appears instead of the single parameter $\Lambda = F'(\xi_*)$. The Jacobian has a set of eigenvalues, and if it can be diagonalized, we get N equations of the form (2.2.3) with different Λs, which are just these eigenvalues. The fate of a perturbation under iteration now depends on the moduli of the eigenvalues. In particular, if any of them have modulus greater than unity, the fixed point is unstable. If all eigenvalues have moduli less than unity, the fixed point is stable. There are further classifications possible depending on the number of eigenvalues with modulus greater than, equal to, or less than unity.

2.4 More on Mappings

For the simple case $\xi_{n+1} = F(\xi_n; \lambda)$, where the ξs are real and F is a real function, there is a simple *geometrical construction* for finding the iterates. Figure 2.1(a) illustrates the idea. We have ξ_n along the abscissa, ξ_{n+1} along the ordinate. Two graphs are drawn. One is $\xi_{n+1} = F(\xi_n; \lambda)$, the other $\xi_{n+1} = \xi_n$. Start on the abscissa with ξ_n. Go up vertically to the F-graph and then horizontally to the ordinate to read the value of the next iterate ξ_{n+1}. Now you notice that by going horizontally to the line $\xi_{n+1} = \xi_n$ and then dropping vertically down to the ξ_n-axis you are ready to iterate again, for the point you have found on the ξ_n-axis is just ξ_{n+1}! So you can keep going in this fashion as sketched in Figure 2.1(a): go to the graph, then to the diagonal, to the graph, diagonal, graph, diagonal, etc.

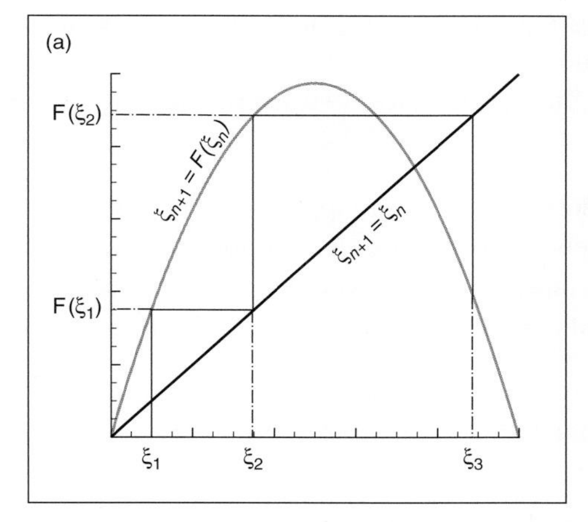

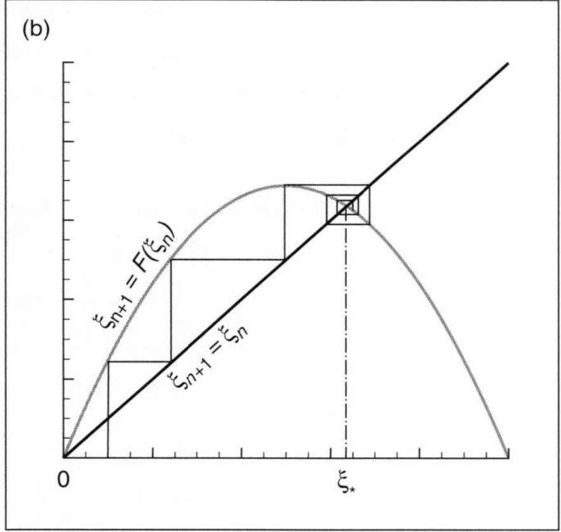

Figure 2.1 (a) Geometrical construction of iterates for a one-bump map. In (b) two fixed points are shown. The one at the origin is unstable since the slope is greater than 45°. The other fixed point is stable since the slope is not as steep as −45°.

If there is a fixed point ξ_*, it must be at the intersection of the graph for F and the line $\xi_{n+1} = \xi_n$. Two such points are shown in Figure 2.1(b). We have let $F(0) = 0$ so $\xi_* = 0$ is, in fact, a fixed point of the mapping, but there is apparently one more. We know from our earlier analysis that stability of a fixed point depends on the value of $F'(\xi_*)$, i.e., on the value of the slope of F at the fixed point. This can also be seen geometrically. If the slope is steeper than 45° or less steep than −45°, then the fixed point is unstable. Thus, in Figure 2.1(b) the fixed point at the origin is unstable (the graph is steeper than 45°), whereas the fixed point at finite ξ is stable. When you iterate in the manner described in Figure 2.1(a), you will wind into the upper fixed point in Figure 2.1(b). That winding in will not happen for the fixed point at the origin when the slope there exceeds 45°. If you steepen the curve $F(\xi)$, the fixed point at finite ξ may also become unstable, but more on that in a moment.

Maps generated by functions F of the general one-bump form shown in Figure 2.1 have a very interesting theory that we shall explore in the following. Let us, however, remark that not all maps of interest have this simple form. For example, the map arising in Newton's method for $\phi(\xi) = \xi^2 - 1$, is $F(\xi) = \frac{1}{2}\left(\xi + \frac{1}{\xi}\right)$. This F-function clearly will not look like the one-bump map shown in Figure 2.1.

2.4.1 Subharmonic Bifurcations

We proceed with elements of the theory of one-bump mappings. In many cases the most interesting feature of the mapping $\xi_{n+1} = F(\xi_n; \lambda)$ will be how the solution depends on some parameter λ. You may think of the mapping as the determining equation for some (presumably quite simple) dynamical system, the analog of the Navier–Stokes equation for a fluid flow, if you wish, and then λ would be analogous to a dimensionless parameter such as the Reynolds number.[1] From the point of view of fluid mechanics, two aspects of the mapping are of particular interest:

(i) the dependence of the solutions to the mapping on the parameter λ displays features that are independent of the exact functional form of $F(\xi)$;
(ii) this type of dependence has been observed in laboratory experiments on fluid systems, e.g., in thermal convection studies, where the role of λ is played by the Rayleigh number.

So we vary λ and ask what happens. One of the prototype Fs that has been studied in great detail is

$$F(\xi) = \lambda\xi(1 - \xi)\,. \tag{2.4.1}$$

This function maps the interval $0 \le \xi \le 1$ into itself as long as $0 \le \lambda \le 4$. Increasing λ steepens the curve and will increase the slope at a fixed point. Ultimately the fixed point will become unstable. This is clear geometrically, but we can be more quantitative given the specific analytic form (2.4.1).

[1] In general, we certainly do not know how to set up such a correspondence in detail. It is mainly a conceptual analogy. We might mention that Lorenz (1963) actually did find a one-bump mapping in his study of the equations that today bear his name, albeit with a cusp at the top of the mapping function.

Consider first the fixed point at $\xi = 0$. Since $F'(\xi) = \lambda(1-2\xi)$, $F'(0) = \lambda$, and so this fixed point becomes unstable when λ is increased beyond 1. The other fixed point, found by solving

$$\xi_* = \lambda\xi_* (1 - \xi_*), \tag{2.4.2}$$

is $\xi_* = 1 - 1/\lambda$. To investigate its stability we calculate

$$F'(1 - 1/\lambda) = \lambda \{1 - 2(1 - 1/\lambda)\} = 2 - \lambda\,. \tag{2.4.3}$$

Hence, the fixed point $1-1/\lambda$ becomes unstable for $\lambda < 1$ and $\lambda > 3$. The question is: what happens next? A stimulating and historically important review paper by May (1976) discusses this.

May's motivation in studying this problem was ecology rather than fluid mechanics. In his case the ξ describe successive values of a population of some species (normalized in some fashion). The population in year $n+1$ is assumed to depend on the population in year n. The population grows exponentially as long as it is small, so that for small values of ξ we want $\xi_{n+1} \approx \lambda\xi_n$. But as the population increases, saturation effects set in. These are modeled by the nonlinear term in (2.4.1).

A simple calculation suggests what happens as λ increases just beyond 3. Inputting any initial value for ξ (except 0 or $1 - 1/\lambda$) leads to a sequence of two values that alternate indefinitely. Such a set of points $\{\xi, F(\xi)\}$, where ξ satisfies $F(F(\xi))$, is called a 2-*cycle*. In general, we define a k-*cycle* as a set of k values $\{\xi_0, \xi_1, \dots, \xi_{k-1}\}$ such that

$$\xi_{j+1} = F(\xi_j) \qquad \text{for} \quad j = 0, 1, \dots, k-1 \quad \text{and} \quad \xi_0 = F(\xi_{k-1})\,. \tag{2.4.4}$$

Periodic solutions of this type play an important role in understanding the iteration given by (2.4.1). In this terminology a fixed point may be thought of as a 1-cycle.

The 2-cycles of F are fixed points of F^2, i.e., solutions to $F^2(\xi) = F(F(\xi)) = \xi$. Clearly, a fixed point of F yields a trivial example of a 2-cycle, but there are others. In the particular case under discussion we should solve the equation $F^2(\xi) = \lambda(\lambda\xi(1-\xi))(1-\lambda\xi(1-\xi)) = \xi$. The quantity $F^2(\xi)$ is a polynomial in ξ of degree 4 and we already know two roots, which are the fixed points $\xi = 0$ and $\xi = 1 - 1/\lambda$. We leave the tedious algebra necessary to obtain the other two roots to the reader. These give the 2-cycle solution.

Problem 2.4.1 Show that the iteration $\xi_{n+1} = \lambda\xi_n (1 - \xi_n)$ has the 2-cycle

$$\left\{ \frac{\sqrt{\lambda+1}(\sqrt{\lambda+1} - \sqrt{\lambda-3})}{2\,\lambda}, \frac{\sqrt{\lambda+1}(\sqrt{\lambda+1} + \sqrt{\lambda-3})}{2\,\lambda} \right\}. \tag{2.4.5}$$

The expression (2.4.5) embodies the main mathematical mechanism that arises as λ is increased. We see that the 2-cycle springs into existence precisely as the fixed point (1-cycle) loses stability. Indeed, both ξ-values of the 2-cycle *bifurcate* from the value of the fixed point at $\lambda = 3$. May (1976) gives a general geometrical argument that shows how for any one-bump mapping a stable 2-cycle springs from a fixed point as it loses stability.

The 2-cycle now becomes the dominant feature of the dynamics of the iteration for a range of λ-values until it too eventually goes unstable.

Problem 2.4.2 Study the instability of the 2-cycle (2.4.5) for the mapping function (2.4.1). Find *analytically* the value of λ for which instability occurs by interpreting the 2-cycle as a fixed point of $F^2(\xi)$.

It is useful for this problem, and for the general theory, to note the following result.

Problem 2.4.3 Each point of a k-cycle $\{\xi_0, \xi_1, \ldots, \xi_{k-1}\}$ is a *fixed point* of F^k, the kth iterate of F. To test for stability we need to compute the derivative of F^k at each of $\xi_0, \xi_1, \ldots, \xi_{k-1}$. Show that all these derivatives have the common value

$$\prod_{i=0}^{k-1} F'(\xi_i)\,. \tag{2.4.6}$$

(The special case $k = 2$ is helpful in the preceding problem.)

A general picture now begins to emerge: we had a fixed point that eventually became unstable and gave way to a 2-cycle at a certain value of λ, let us call it $\lambda = \lambda_1$. The 2-cycle, in turn, will yield to a 4-cycle, each of the ξ-values of the 2-cycle bifurcating to two components of the 4-cycle. This happens at some larger value of λ, which we call λ_2. For the particular $F(\xi)$ given in (2.4.1) we have just computed the value of λ_2. This pattern continues as we increase λ. (You should try this out numerically.) The 4-cycle loses stability to an 8-cycle at $\lambda = \lambda_3$, which loses stability to a 16-cycle at $\lambda = \lambda_4$, and so on. Because of the way cycles of successively doubled periods arise, the phenomenon is referred to as *period doubling* or *subharmonic bifurcation*. The arguments given in May's article show why this behavior is generic for one-bump mappings.

As the parameter λ is increased further and further we reach a range in which the iterates of the map seem to wander randomly over large regions of the interval $0 < \xi < 1$. It appears that the sequence of parameter values $\lambda = \lambda_n$ at which a 2^{n-1}-cycle loses stability to a 2^n-cycle has a *point of accumulation*. For the mapping (2.4.1) this point is at $\lambda_c = 3.5700\ldots$. Beyond this point matters get rather complicated. To quote May (1976):

Beyond this point of accumulation ... there are an infinite number of fixed points with different periodicities, and an infinite number of different periodic cycles. There are also an uncountable number of initial points which give totally aperiodic ... trajectories; no matter how long the ... series generated ... is run out, the pattern never repeats ... Such a situation, where an infinite number of different orbits can occur, has been christened 'chaotic' ...

May continues:

As the parameter increases beyond the critical value, at first all these cycles have even periods ... Although these cycles may in fact be very complicated (having a nondegenerate period of, say, 5,726 points before repeating), they will seem to a casual observer to be rather like a somewhat 'noisy' cycle of period 2. As the parameter value continues to increase, there comes a stage [at $\lambda = 3.6786\ldots$ for (2.4.1)] at which the first odd period cycle appears. At first these odd cycles have very long periods, but as the parameter value continues to increase, cycles with smaller and smaller odd periods are picked up, until at last the three-cycle appears [at $\lambda = 3.8284\ldots$ for (2.4.1)]. Beyond this point, there are cycles with every integer period, as well as an uncountable number of asymptotically aperiodic trajectories: Li and Yorke (1975) entitle their original proof of this result 'Period three implies chaos'.

There is a remarkable theorem due to Sarkovskii on the order in which the various cycles appear (see Devaney, 1984, §1.10, for an exposition). Assume we list the natural numbers in the following way. First we list all the odd numbers except 1, i.e., 3, 5, 7, 9, 11, ... Then we list 2 times these, i.e., 2×3, 2×5, 2×7, 2×9, 2×11, ... Then $2^2 = 4$ times the odds, i.e., $2^2 \times 3$, $2^2 \times 5$, $2^2 \times 7$, $2^2 \times 9$, $2^2 \times 11$, ... Continue with 2^3 times the odds, then 2^4 times the odd numbers, etc. At the very end of the sequence come the powers of 2 in descending order: ... 2^4, 2^3, 2^2, 2^1, 2^0. Define an "order relation," the *Sarkovskii ordering*, by saying that $n \ll m$ (read: n goes before m) if n precedes m in the sequence:

$$3 \ll 5 \ll 7 \ll \cdots \ll 2 \times 3 \ll 2 \times 5 \ll \cdots \ll 2^2 \times 3 \ll 2^2 \times 5 \ll \cdots \ll 2^3 \ll 2^2 \ll 2 \ll .$$

Sarkovskii's theorem states that if the mapping has a p-cycle, with p a prime number, and $p \ll q$ in the above ordering, then the mapping also has a q-cycle. Indeed, this holds for *any continuous mapping* of the real line! It is a very powerful statement.

Hence, existence of a 3-cycle implies existence of all other periods. Furthermore, the existence of a cycle with a length that is not a power of 2 implies the existence of infinitely many cycles. Conversely, if the mapping has only a finite number of cycles, then the lengths of all these cycles must be powers of 2.

It is interesting to compare the behavior in this model problem to the *scenario* for the onset of turbulence in a fluid proposed by L.D. Landau and E. Hopf in the 1940s. Their idea was to extrapolate from the simple linear instability that one can calculate for certain base flows. This instability typically introduces a new frequency and an arbitrary phase to the flow field. Successive instabilities similarly were assumed to add a frequency and phase, such that little by little a very complicated signal is built up. The values of Reynolds number at which these instabilities start are supposed to become more and more closely spaced so that for all practical purposes a seemingly random signal – which is the signature of turbulent flow – arises at a finite, reasonably sharp value of Reynolds number. The period-doubling process does, indeed, have an infinity of such instabilities, and the values of the control parameter at which they occur do have an accumulation point. However, the new frequencies are exact subharmonics of the old. Nevertheless, just beyond the accumulation point for all the subharmonics we are thrown immediately into a chaotic state.

Further developments have added two important additional items. First, as we shall discuss briefly in the next subsection, the cascade of subharmonic bifurcations has some very interesting scaling properties not envisioned or anticipated. Second, there are general theoretical models aimed more closely at fluid flow, and careful experiments to go with them, that point to a generic picture of transition to turbulence in which only a finite number of bifurcations occur before entering a chaotic, weakly turbulent state. The study of model systems, such as (2.4.1), has thus had a major impact on our understanding of the problems of transition to stochastic behavior in deterministic systems – issues related to the turbulence problem. In this way transition to chaos in mappings has taken on a legitimacy and relevance of its own quite separate from any attempts to view such mappings as caricatures of numerical methods.

Computer Exercise 2.4.4 Write a program that produces a plot of iterates ξ_n versus λ for the iteration of (2.4.1). The number of different values λ that you can accommodate will depend on the resolution of your graphics device. The range of λ is nominally $0 < \lambda < 4$, although all the interesting behavior will be found in the interval $3 < \lambda < 4$. For each λ choose an initial value of ξ (randomly in $0 < \xi < 1$) and let the program iterate for some reasonable number of iterations (100 or more) before plotting. Then plot the next several hundred iterates. A sample plot is shown in Figure 2.2. You will see how the subharmonic bifurcations lead to a *cascade* of ever finer two-pronged structures (called *pitchforks*). Also notice the banding in the chaotic region, where there is an inverse cascade of bifurcations. For other examples see Crutchfield *et al.* (1986). The mapping studied there is the *circle map* $\xi_{n+1} = \omega + \xi_n + (k/2\pi)\sin(2\pi\xi_n)$. It isn't often that an entire parametric study of a nonlinear problem can be captured in a single plot!

2.4.2 Universality in One-Bump Maps

One of the major developments in the theory of one-bump mappings is the discovery by Feigenbaum (1978) that not only is the period-doubling *generic* in a topological sense, as May (1976) explains, but it has *universal metric properties* as well. Indeed, if we consider the bifurcation values λ_n introduced in the previous subsection, Feigenbaum showed that the ratios

$$\frac{\lambda_n - \lambda_{n-1}}{\lambda_{n+1} - \lambda_n} = \delta_n \tag{2.4.7}$$

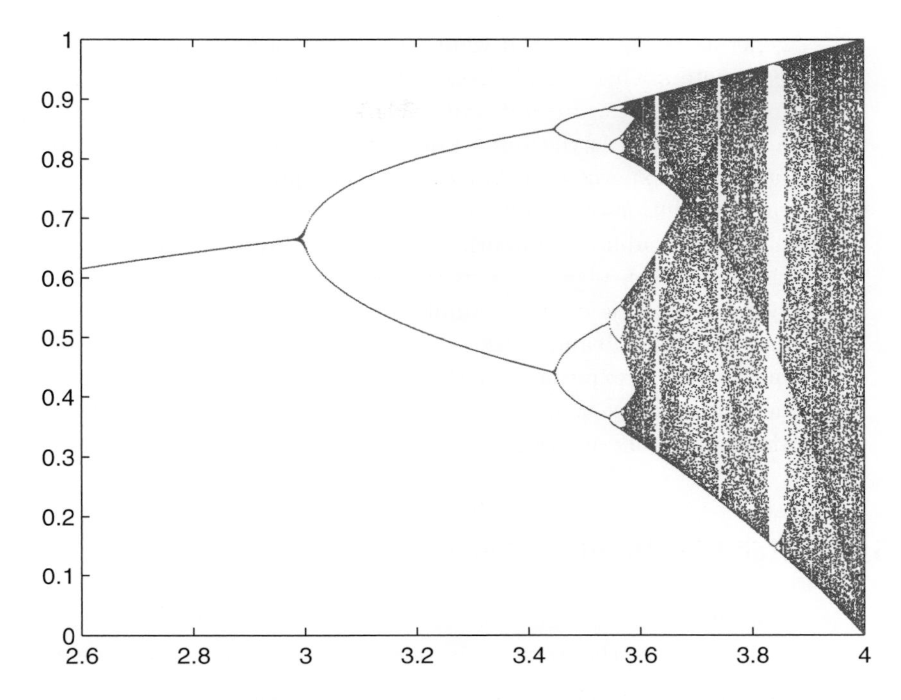

Figure 2.2 The bifurcation diagram for the logistic map (2.4.1). The parameter λ is along the x-axis (recall that not much happens for $\lambda < 3$). Along the ordinate one plots the iterates after letting initial transients settle out, i.e., you iterate for a few hundred iterations and then start plotting iterate values along the ordinate at the chosen value of λ. In this way you pick up the dominant cycle. Thus, one observes a few steps in the repeated subharmonic bifurcations, and a very interesting pattern of stripes and shadows in the chaotic regime. Note the onset of period 3 (the broad clean band around $\lambda = 3.82$).

will converge to $\delta = 4.6692016091029\ldots$, i.e., that the λ_n approach a geometric progression, and, much more remarkably, that this will happen *with the same value of* δ for a wide class of functions F in (2.4.1). (Essentially, F needs to have just one maximum and be approximated by a parabola there.) The value δ is thus *universal* in the sense that it is independent of the actual dynamical rule generating the cascade of subharmonic bifurcations.

Problem 2.4.5 Show that if (2.4.7) is true (note: no n-dependence on the right-hand side) then $\lambda_n = a + bq^n$, where a, b, q are independent of n, and find q in terms of δ.

There is another universal number that describes the scale of the points of a 2^{n+1}-cycle at inception relative to the same scale for points of a 2^n-cycle. As $n \to \infty$ this ratio tends to $1/\alpha$ where $\alpha = 2.502907875\ldots$. The *universal* nature of this number means that it is independent of the function F used to generate it (again over a wide range of functions). The actual values of control parameters where the bifurcations take place, and the actual values of the ξs in the 2^n-cycles, *are not* universal, but the *geometrical scaling* described by α and δ is. This is the important insight provided by Feigenbaum, initially via numerical experiments, but quickly incorporated in an analytical theory. The rapid evolution of understanding is traced in Feigenbaum (1980).

Period-doubling bifurcations and the universal behavior to which they give rise have been observed experimentally in a variety of systems: fluids, optical systems, electric circuits, simple mechanical systems, cardiac cells, etc. See the article by Gollub and Benson (1980) for a discussion of the role of such transitions in convection experiments. We cannot trace all these interesting developments here. Instead let us pursue the idea of complex behavior in iterated mappings in the context of numerical analysis and ultimately CFD.

2.5 Random Number Generation

Consider (2.4.1) again, this time with $\lambda = 4$. The following construction allows us to linearize the iteration in this case (Ulam and von Neumann, 1947). Let

$$\xi_n = \sin^2(\pi\zeta_n)\,. \tag{2.5.1}$$

Substituting in (2.4.1),

$$\sin^2(\pi\zeta_{n+1}) = 4\sin^2(\pi\zeta_n)\,(1-\sin^2(\pi\zeta_n)) = 4\sin^2(\pi\zeta_n)\cos^2(\pi\zeta_n) = \sin^2(2\pi\zeta_n). \tag{2.5.2}$$

Thus,

$$\zeta_{n+1} = 2\,\zeta_n\,, \tag{2.5.3}$$

coupled with (2.5.1) to reconstruct the ξs, solves the iteration!

We can do a bit more. Since in (2.5.1) we can always arrange ζ_n to be in the range $0 \le \zeta_n \le 1$, we replace (2.5.3) by

$$\zeta_{n+1} = (2\zeta_n \bmod 1)\,. \tag{2.5.4}$$

In this way we map the interval $0 \le \zeta_n \le 1$ onto itself. But this is just the equation (1.5.3) studied in Example 1.5.3! Figure 2.3 shows this function and its first two iterates. It is clear from this example how chaotic and thus extremely sensitive to the initial condition the iteration is for $\lambda = 4$. After many iterations we have lost all significant figures of the initial condition, and all we can say is that the iterated value is somewhere between 0 and 1. It is clear, since the function is piecewise linear, that the probability of finding ζ in any small interval $d\zeta$ after many iterations will be uniform, i.e., $P(\zeta)d\zeta$, the probability distribution for ξ, is just

$$P(\zeta)\,d\zeta = d\zeta\,. \tag{2.5.5}$$

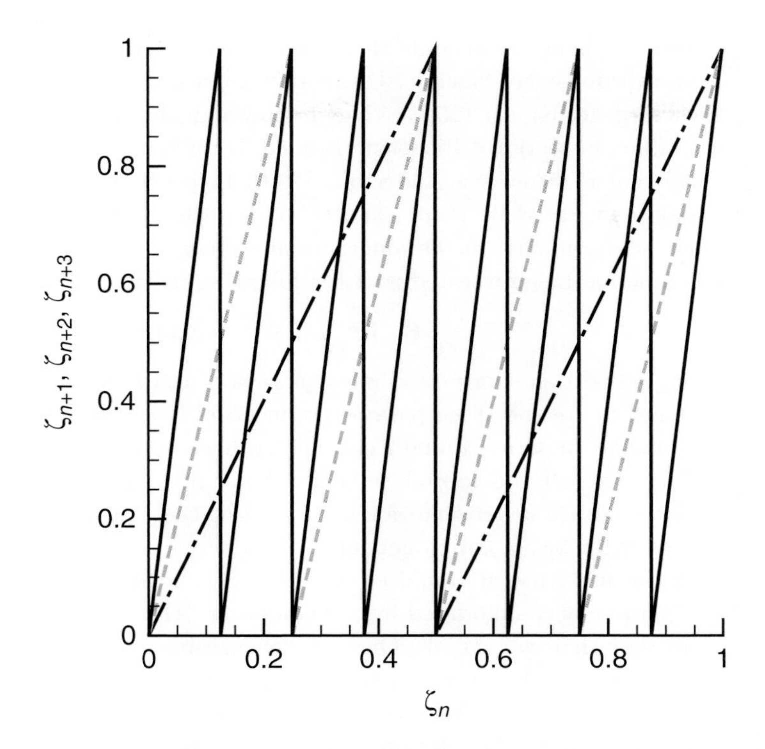

Figure 2.3 First three iterates of the mapping (2.5.4). Dash–dot line: ζ_{n+1}, dash line: ζ_{n+2}; solid line: ζ_{n+3}.

Since ζ and ξ are in a one-to-one correspondence,

$$\mathrm{P}(\xi)\,\mathrm{d}\xi = \mathrm{d}\zeta = \frac{\mathrm{d}\zeta}{\mathrm{d}\sin^2(\pi\zeta)}\,\mathrm{d}\xi = \frac{\mathrm{d}\xi}{2\pi\sqrt{\xi(1-\xi)}}\,. \tag{2.5.6}$$

This is called a β-distribution. Thus, for $\lambda = 4$, iteration of the map (2.4.1) has given us a simple algorithm for generating random numbers with one of the classical probability distributions.

This is conceptually a very interesting result (and, if you need numbers subject to the β-distribution, may even be useful). It illustrates the intriguing aspect that one can generate random numbers on a deterministic machine (the computer) using a deterministic process (or algorithm). It amplifies statements made earlier about inadvertently generating random numbers while attempting to solve a differential equation or other problem with intrinsically “noisy” behavior. In this connection we might note that the Lorenz equations in Figure 1.1 actually have a map very much like (2.5.4) embedded, if one monitors successive maxima of the amplitude variable C. Details may be found in Lorenz’s (1963) paper cited in Chapter 1. Finally, it highlights the pragmatic task of producing reliable algorithms and software for the generation of random numbers.

Because of the finite nature of a computer there are problems in practice of generating long sequences of random numbers, even when we know an algorithm that will do so in principle. Ultimately, on a finite precision machine numbers must repeat. Knuth (1981) gives a thorough discussion of these issues. (The introductory section 3.1 is recommended. For a more pragmatic review of random number generators see Anderson, 1990.) In practice random numbers are not usually generated by nonlinear iterations such as (2.5.1)–(2.5.3) but rather by what is known as a linear congruential sequence. Iteration is still involved, but the function being iterated is of the following nature:

$$\xi_{n+1} \equiv a\,\xi_n + c \pmod{m}\,. \tag{2.5.7}$$

The goal is to generate random sequences of integers, so the iterates ξ_n are all integers as are the three parameters in (2.5.7): a, c and m. One has to choose the "magic" numbers a and c carefully, otherwise very short periods result from (2.5.7). The other compromise is the choice of m. For small m the calculations of the mod are faster than for large m, but the length of the period in (2.5.7) can at most be m. So to get long periods, one wants a large m. It turns out that the most useful values of m are $w \pm 1$, where w is the range of integers that can be accommodated by the computer. Also choosing m to be the largest prime less than w is viable. One can now establish criteria for a and c such that the iterates of (2.5.7) have period m, and implementations of random number generators will use a and c values that satisfy these criteria.

It may seem unlikely for an iteration such as (2.5.7) to arise in CFD, but it is worth recalling that *periodic boundary conditions* can lead to such a mod, and it is thought-provoking that the iteration of such a simple expression can lead to sequences that will pass statistical tests for randomness.

Random number generation can be made the basis of a method for evaluating definite integrals and for integrating certain types of partial differential equations. In this context it is usually referred to as the **Monte Carlo method**. For CFD the most obvious application is the **diffusion equation**. Random number generation comes up also in the so-called *random choice method* for hyperbolic systems (Chorin, 1976), and in creating initial conditions for flow simulations in the turbulent regime.

2.6 Newton's Method in the Complex Plane

The complicated behavior that we studied in Sections 2.4 and 2.5 for mappings of the interval $[0, 1]$ has counterparts in the study of mappings of two- and three-dimensional regions. In the following sections we shall pursue aspects of this that are more obviously relevant to CFD. Here, by way of introduction, and because of the intrinsic beauty of the subject, we study Newton's method when attempted in the complex plane.

Newton's method was introduced in Section 2.1. As it stands it is clearly well defined also for an analytic function ϕ of a *complex variable* z. We are then considering the iteration

$$z_{n+1} = z_n - \frac{\phi(z_n)}{\phi'(z_n)}, \tag{2.6.1}$$

with some initial value for z_0. If ϕ is a polynomial, the right-hand side of (2.6.1) is a rational function. The results of iterating a rational function were considered many years ago by the French mathematicians Julia (1918) and Fatou (1919). Numerical experiments with high-resolution graphics have revealed the extremely intricate global behavior of such maps and led to a resurgence of interest.

If ϕ is a polynomial of degree 1, viz. $\phi(z) = a\,z + b$ with $a \neq 0$, there is just one solution to $\phi(z) = 0$, and it is $z = -b/a$. Newton's method converges in one iteration: $z_1 = -b/a$ for any initial point in the entire complex plane.

If ϕ is a polynomial of degree 2, viz. $\phi(z) = a\,z^2 + b\,z + c$ with $a \neq 0$, we can perform a number of transformations to write (2.6.1) in a simpler form. Straightforward manipulation of (2.6.1) yields

$$z_{n+1} = \frac{z_n^2 - c/a}{2\,(z_n + b/2a)}. \tag{2.6.2}$$

Introducing $z_n + b/2a$ as a new variable, and calling it y_n, gives

$$y_{n+1} = \frac{y_n^2 + (D/2a)^2}{2y_n}, \tag{2.6.3}$$

where D is the *discriminant*:

$$D = \sqrt{b^2 - 4ac}. \tag{2.6.4}$$

If $D = 0$, we see immediately that y_n converges to 0 regardless of where in the complex plane we start the iteration. If not, the problem can be linearized in much the same way as in Section 2.3. We set

$$y_n = \frac{D}{2a} \coth \zeta_n. \tag{2.6.5}$$

Then we see that the recursion (2.6.3) on the y_n can be satisfied by letting $\zeta_{n+1} = 2\zeta_n$. As we iterate ζ_n tends to infinity. Depending on whether the real part of ζ_0 is positive or negative, y_n will tend to either of the two roots of the quadratic: $+D/2a$ or $-D/2a$. Thus, the plane divides into two half-planes, with each half converging to one root of the quadratic.

Things are not always so simple! Already the case of a *cubic* polynomial, such as $z^3 - 1$, leads to a very complicated division of the complex z-plane. There are three separate regions from which the solution is attracted to each of the three roots of unity under Newton iteration, see Figure 2.4. Otherwise the solution drifts to infinity. In general, we define the **attractive basin** $W(p)$ of a fixed point p as the set of initial values z that upon iteration will converge on p. (A minor technical point: the complex plane is closed with a point at infinity for this construction.) The **Julia set** is now defined as the boundary of $W(p)$. In general this set is extremely complicated.

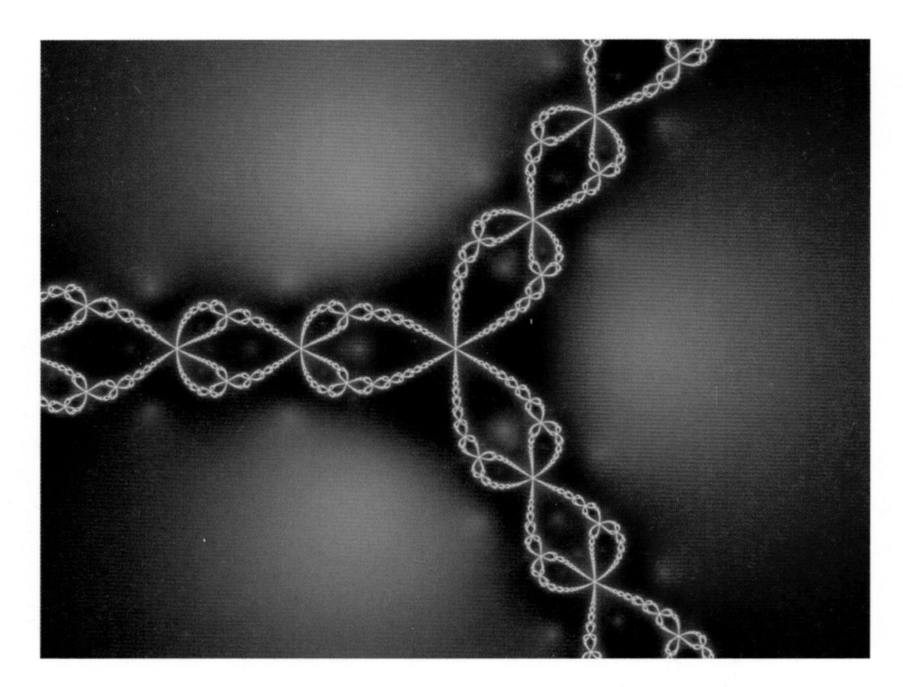

Figure 2.4 Julia set for the solution of the cubic $z^3 - 1$ by Newton's method. This picture was taken from the Wikipedia webpage on the Julia set; similar illustrations of the Julia set are numerous on the internet.

2.6.1 The Mandelbrot set

Another problem of interest is whether spurious solutions to Newton's method can exist, i.e., can the iteration be trapped by something other than a fixed point, such as an attractive, periodic cycle? Fatou proved the following:

Theorem 2.6.1 *If the iteration $z_{n+1} = F(z_n)$, where F is a rational function, has an attractive periodic cycle, then starting the iteration from at least one of the solutions of $F'(z) = 0$ will converge to the periodic cycle.*

The cubic polynomial, $\phi(z) = z^3 + (a - 1)\, z - a$, has three roots and for all values of a, one of the roots is $z = 1$. Thus, to check if, for some value of a, the above polynomial has attractive periodic cycles that can trap Newton iteration aimed at finding the roots, we must check the iterates of points where $F(z) = z - \phi(z)/\phi'(z)$ has vanishing derivative. From (2.2.12) we see that F' is proportional to $\phi\phi''$. The roots of ϕ obviously satisfy $F' = 0$, but Newton iteration starting from these roots will not converge to anything but themselves. We seek points c that satisfy $\phi''(c) = 0$ but have $\phi(c) \neq 0$. The only such point is $c = 0$. Thus, if we iterate starting from $z_0 = 0$, then if there are periodic cycles they can be identified. Otherwise, the iteration will converge to one of the three roots of the cubic polynomial. The points for which we get such convergence constitute a complicated set in the complex a-plane as shown in Figure 2.5. The magnified region on the right is the **Mandelbrot set**, which is traditionally obtained from the quadratic complex mapping $z_{n+1} = z_n^2 + C$. Although the Newton iteration of the cubic polynomial is nothing like the complex quadratic mapping, we recover the Mandelbrot set, illustrating the universal nature of the set.

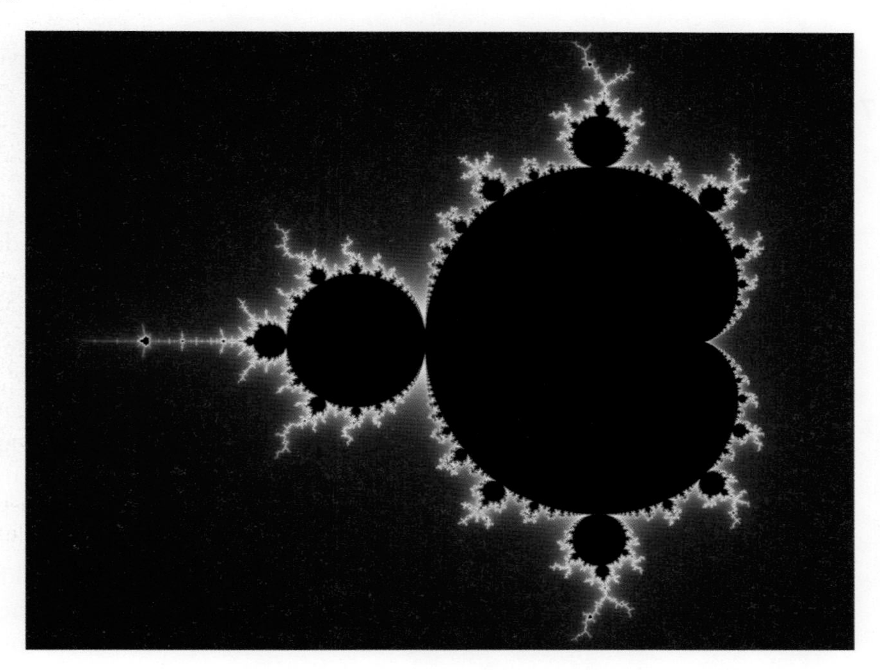

Figure 2.5 Mandelbrot set for the solution of the cubic $z^3 + (a-1)z = a$ by Newton's method. This picture was taken from the Wikipedia webpage on the Mandelbrot set; similar illustrations of the Mandelbrot set are numerous on the internet.

Computer Exercise 2.6.2 Write a program to display the attractive basins of the roots of a cubic polynomial under Newton iteration. Let a square grid of points corresponding to the resolution of your computer display define the initial conditions. For each point proceed to iterate. After each iteration, or after a number of them, check the distances of the iterate from the known roots. Use some criterion to decide whether it is converging on a particular root. If it is converging on one of the roots, turn the pixel corresponding to the initial condition on. When the points in your display have all been tested a black-on-white picture will delineate the attractive basin for the root you chose. Repeat with another root, or use different colors to distinguish basins for different roots. The complex boundaries that you see are (parts of) the Julia set. Similarly, investigate the Mandelbrot set for the cubic $f_a(z) = z^3 + (a-1)\,z - a$. Search in the square $[-2,2] \times [-2,2]$. The set is highly ramified. Much work has been done on this problem, and you can find a multitude of stunning pictures in the literature (and even in some arts and crafts stores!). Suggested references are Mandelbrot (1980, 1983), Curry *et al.* (1983), Peitgen *et al.* (1984), Vrscay (1986), Peitgen and Richter (1986) and Peitgen and Saupe (1988).

2.7 Mappings and Fluid Flows

So far the motivation for our discussion of mappings has been that they provide a generic form for computations on finite, discrete machines. Let us now establish better connections between the concept of a mapping, low-order systems of ODEs, and more conventional problems of fluid mechanics. There are several notions that need to be introduced.

First, let us note the general relation that exists between **flows** and **mappings**. By a flow we mean a continuous time, continuous space evolutionary process and it usually arises from solving a set of differential equations. One must distinguish such *flow of solution* from fluid flow, as one can speak of a flow of solution in the context of ODEs and PDEs outside fluid dynamics. For example, the Lorenz equations, introduced in Figure 1.1, can be thought of as describing a flow of solution in (A, B, C)-space. In this example this flow of solution is supposed to provide a caricature of an actual advecting fluid flow as well. Similar projections and truncations have been investigated for the Euler equations and the Navier–Stokes equations. For example, Franceschini and Tebaldi (1979) studied the system

$$\left.\begin{aligned} \dot{x}_1 &= -2x_1 + 4x_2x_3 + 4x_4x_5\,, \\ \dot{x}_2 &= -9x_2 + 3x_1x_3\,, \\ \dot{x}_3 &= -5x_3 - 7x_1x_2 + r\,, \\ \dot{x}_4 &= -5x_4 - x_1x_5\,, \\ \dot{x}_5 &= -x_5 - 3x_1x_4\,, \end{aligned}\right\} \tag{2.7.1}$$

where r is the surviving remnant of Reynolds number, and the dots indicate time derivatives. The x_i are the amplitudes for certain Fourier modes of the velocity field and the couplings are the same as would arise in the Navier–Stokes equations.

Orszag and McLaughlin (1980) investigated similar systems of the general form

$$\dot{x}_i = a\,x_{i+1}\,x_{i+2} + b\,x_{i-1}\,x_{i-2} + c\,x_{i+1}\,x_{i-1}, \tag{2.7.2}$$

where the amplitudes x_i are counted modulo N, i.e., $x_{i+N} = x_i$. If the constants a, b, c are chosen such that $a + b + c = 0$, then the system conserves energy, i.e.,

$$E = \tfrac{1}{2}\sum_{i=1}^{N} x_i^2 \tag{2.7.3}$$

is invariant in time. Orszag and McLaughlin studied this system for $N = 4$ and $N = 5$.

Problem 2.7.1 Verify that the system (2.7.2) does indeed satisfy energy conservation if $a + b + c = 0$.

It is intuitively clear (and we shall see this in more detail later) that if the velocity field in the incompressible Navier–Stokes equation is expanded in Fourier modes, amplitude equations with *quadratic couplings* arise. These come ultimately from the $(\mathbf{V} \cdot \nabla)\mathbf{V}$ term in the acceleration and from the pressure term.

Systems of ODEs such as (2.7.1)–(2.7.2) define flows in their respective phase spaces $(x_1, x_2, \ldots)$. To create a mapping from such a flow we need the concept of a **surface of section**, often called a **Poincaré section**. By this we mean, loosely speaking, the result of cutting through the lines of the solution flow by a plane. The idea is illustrated in Figure 2.6. We have the flow confined to some portion of phase space. For example, we might know that the state vector $\mathbf{x} = (x_1, x_2, \ldots)$ is constrained by energy conservation as in (2.7.3) where this constraint defines the surface of a sphere in N-dimensional space. To get a lower-dimensional representation of the flow we slice through phase space with a test surface. Through any point P of this surface there must be a flow-line, since we could use the state corresponding to that point as the initial condition of our flow of solution. In general, the flow-line through P will return to the surface after some time. Let the first point of return be Q. We define a mapping of the test surface onto itself by $P \to Q$ for each choice of point P. This mapping, which we say is induced by the flow of solution, will in general be a smooth function on the test surface. In this way there corresponds to each flow of solution a mapping of (more or less) any test surface onto itself. (One can imagine test surfaces chosen tangent to flow lines, such that the above construction will not work, but let us assume that we have avoided such cases.) Usually one is interested in studying the nature of successive intersections of a given flow-line with a Poincaré section, i.e., we have some initial condition P_0 and we study the pattern formed by the set of intersections $P_0 \to P_1 \to P_2 \to \cdots \to P_n \to \cdots$ etc. You keep going for as long as you have computational control on accuracy, and for as long as you have computer resources!

2.7.1 Advection Problems

The preceding discussion was concerned with a somewhat abstract flow of solution in a phase space. Although this can be a very useful point of view for discussions of dynamics, there is a more immediate connection between mappings and flow of solutions that arises when the space is our usual physical space and the flow is simply a conventional fluid velocity field. Suppose we want to elucidate the streamline pattern in some flow. Let us assume for simplicity that

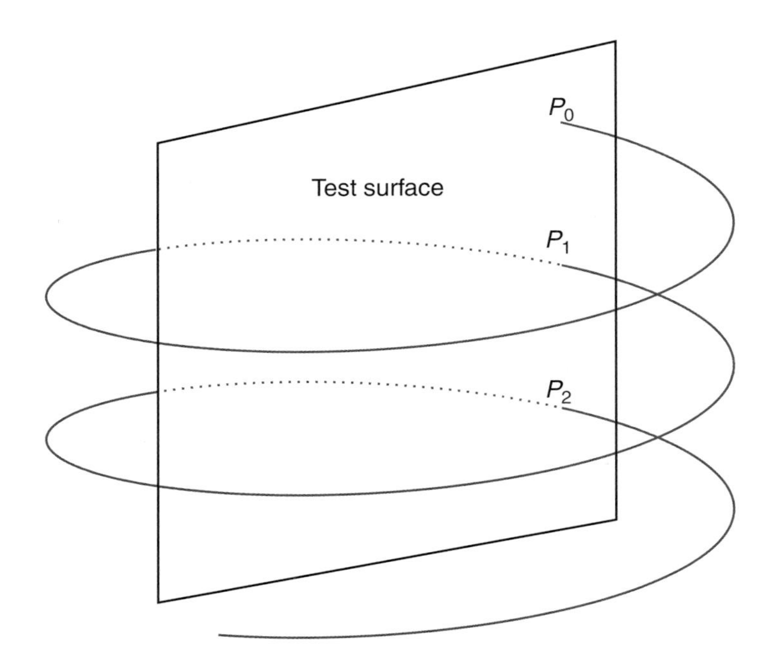

Figure 2.6 Illustration of the idea of a Poincaré section or surface of section. The phase space point executes some orbit during the motion. Each time it pierces a test surface we record the point of intersection.

the flow is steady and periodic in all three spatial directions. An example of long-standing interest is the flow $\mathbf{V} = (u, v, w)$ defined within a cube of side length L with periodic boundary conditions by

$$\left.\begin{aligned} u(x,y,z) &= A\,\sin(2\pi z/L) + C\,\cos(2\pi y/L)\,, \\ v(x,y,z) &= B\,\sin(2\pi x/L) + A\,\cos(2\pi z/L)\,, \\ w(x,y,z) &= C\,\sin(2\pi y/L) + B\,\cos(2\pi x/L)\,, \end{aligned}\right\} \tag{2.7.4}$$

where A, B and C are any set of constants. The field (2.7.4) is referred to as the **ABC-flow** in the literature. It is a solution of the steady Euler equation in three dimensions.

Problem 2.7.2 Show that (2.7.4) is a *Beltrami flow*, i.e., show that $\boldsymbol{\omega} \times \mathbf{V}$ vanishes (although both $\boldsymbol{\omega}$ and $\mathbf{V}$ are finite). Here $\boldsymbol{\omega} = \nabla \times \mathbf{V}$ is the vorticity, as usual. In fact, for these flows $\boldsymbol{\omega} = 2\pi\mathbf{V}/L$, i.e., the vorticity is simply proportional to the velocity. Show that the flows (2.7.4) are *eigenfunctions of the curl operator* with eigenvalue $2\pi/L$ for any values of A, B, C.

Problem 2.7.3 A general formula for a family of Beltrami flows related to the *ABC-flow* has been given by Yakhot *et al.* (1986). In their paper they discuss analogs of steady cellular flows in two dimensions. They are a family of Beltrami flows in three dimensions, defined by finite Fourier sums of the form

$$\mathbf{V}(\mathbf{x}) = \sum_{\mathbf{Q}\in\mathbf{S}} A(\mathbf{Q})\,(\mathbf{n} + \iota\,\mathbf{Q}\times\mathbf{n})\,\exp(\iota 2\pi\mathbf{Q}\cdot\mathbf{x}/L)\,. \tag{2.7.5}$$

Here $\mathbf{S}$ is a finite set of unit vectors $\mathbf{Q}$; for each of these $\mathbf{n}$ is a unit vector perpendicular to $\mathbf{Q}$. The complex amplitudes $A(\mathbf{Q})$ satisfy the reality condition $A(-\mathbf{Q}) = \mathbf{A}^*(\mathbf{Q})$, where the asterisk denotes complex conjugation, and L is again the side length of the periodic cube in which these flows are defined.

Verify that (2.7.5) gives a representation of the flow $\mathbf{V}$ as a sum of Beltrami flows. Can you determine the set $\mathbf{S}$ for which the *ABC-flow* is reproduced? Each term in the sum (2.7.5) is an eigenfunction of the curl operator. What is the eigenvalue?

To study the streamlines in such a flow we can make use of the fact that *in a steady flow particle paths and streamlines coincide.* Thus, we can simply write a program to solve the coupled ODEs

$$\left.\begin{aligned} \dot{x} &= u(x,y,z)\,, \\ \dot{y} &= v(x,y,z)\,, \\ \dot{z} &= w(x,y,z)\,. \end{aligned}\right\} \tag{2.7.6}$$

We shall refer to (2.7.6) as the **advection equations**. In Chapter 3 we explore methods for integrating ordinary differential equations. By and large it is a simple matter that does not present great obstacles, at least not in terms of choosing an algorithm and getting a program to run stably.

So accept for the present that (2.7.6) can be integrated for more or less any choice of u, v and w, in particular for the *ABC-flows* introduced earlier. What do the results look like? Again here is an instance where a surface of section mapping is extremely useful. In the study by Dombre *et al.* (1986) several such sections are shown. In Figure 2.7 we show a slightly different diagram. We consider a single *particle path*, or *streamline*, or – because of the Beltrami property – *a vortex line* in the steady flow (2.7.4) for the case of coefficients $A = B = 1$, $C = 0.65$. Letting the box side length be 2π for convenience, the line was started at $(x,y,z) = (3,\pi,\pi)$. In panels (b), (c) and (d) three views of the path/stream/vortex-line are shown as it would appear if seen through perpendicular faces of the box. In (b) you are looking along the x-axis, in (c) along the y-axis and in (d) along the z-axis. In panel (a) we show a cut through this path/stream/vortex-line in the plane $z = \pi$. The points of intersection appear to scatter throughout some finite region. This is a feature of this particular flow and *not* a manifestation of numerical errors. The origin of this dispersal of points of intersection is the analytical structure of the advection equations for this particular flow. Indeed, the advection equations (2.7.6) are, in general, **nonintegrable** when the advecting flow is the ABC-flow. Numerical experiments on particle paths in the ABC-flows suggesting nonintegrable or chaotic behavior were done by V.I. Arnold and M. Hénon in 1966.

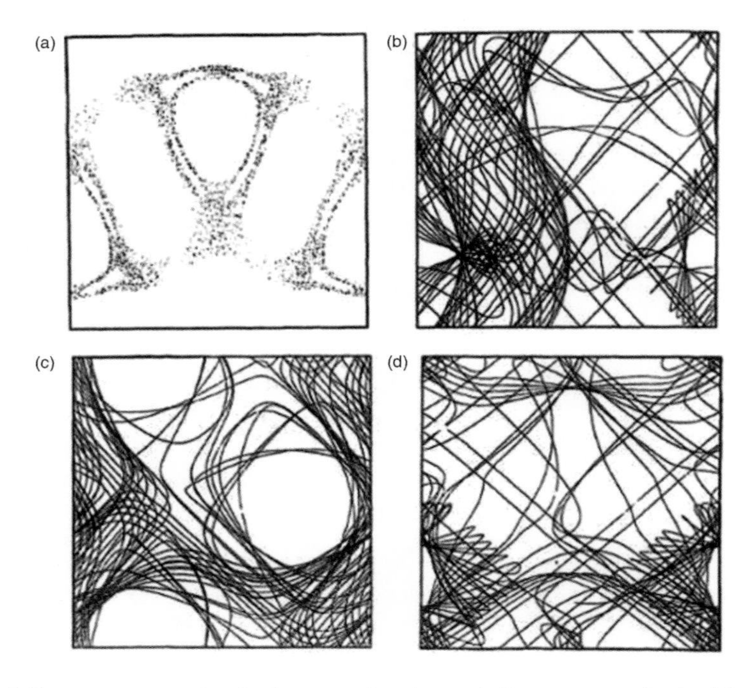

Figure 2.7 Intersection of a single streamline (or path line or vortex line) with test planes perpendicular to the coordinate axes for the ABC-flow with $A = B = 1$, $C = 0.65$. (From Aref *et al.*, 1989.)

The phenomenon that particle paths in regular flows can be chaotic is now known as **chaotic advection** (Aref, 1984, 2002). It has profound consequences for mixing of viscous liquids and for particle dispersion by laminar flows. The subject has received considerable attention (see Ottino, 1989) and provides a number of simple yet interesting topics for computer exercises.

2.7.2 Area-preserving Maps

It is in general impossible to write an explicit formula for the map induced by even a simple flow such as (2.7.1) or (2.7.4). The best approach one can suggest is to integrate the ODEs from intersection to intersection and study the mapping that way. But as we have seen in the context of one-bump maps (Section 2.4) maybe the exact details of the map are not of paramount importance. Maybe, on the one hand, we can learn something about advection and flow kinematics in general by studying maps in their own right without worrying too much about their origin. On the other hand, maybe we can come up with some sufficiently idealized flows where the maps can be written down explicitly. The answers to both these queries are in fact yes, and in this subsection we pursue some of this material.

A very legitimate concern one might have is how to incorporate into a map the constraint of **incompressibility**. We should be able to accommodate this most basic kinematic requirement. In the case of 2D flow one way to assure incompressibility is to introduce a **streamfunction**. This is a function $\psi(x, y)$ from which the velocity components u, v are obtained by differentiation as follows:

$$u = -\frac{\partial \psi}{\partial y}; \quad v = \frac{\partial \psi}{\partial x}. \tag{2.7.7}$$

This works because (2.7.7) assures us that the divergence of the velocity vanishes. The advection equations (2.7.6) now take the form

$$\dot{x} = -\frac{\partial \psi}{\partial y}; \quad \dot{y} = \frac{\partial \psi}{\partial x}. \tag{2.7.8}$$

Formally these are identical to **Hamilton's canonical equations** for a system with a single degree of freedom and with ψ playing the role of the Hamiltonian, and (x, y) being a set of conjugate variables. Phase space is real space for this system.

Close to a stagnation point (i.e., a point of zero velocity), for convenience taken at the origin, we may consider ψ expanded in a Taylor series:

$$\psi(x, y) = \alpha\, x + \beta\, y + \rho\, x^2 + \sigma\, x\, y + \tau\, y^2 + \cdots. \tag{2.7.9}$$

The velocity at the origin is $(-\beta, \alpha)$ according to (2.7.7). We assume it vanishes. Thus we have the linearized equations of motion

$$\dot{x} = -\sigma\, x - 2\tau\, y\,; \quad \dot{y} = 2\rho\, x + \sigma\, y\,, \tag{2.7.10}$$

in the vicinity of a stagnation point. The single component of the vorticity (perpendicular to the xy-flow plane) is

$$\zeta = \frac{\partial v}{\partial x} - \frac{\partial u}{\partial y} = 2(\rho + \tau)\,, \tag{2.7.11}$$

where the last expression is the value of ζ at the origin. The diagonal elements of the rate-of-strain tensor (i.e., the diagonal elements in (2.7.10) viewed as a matrix equation) are $\mp\sigma$. The off-diagonal elements are both $(\rho - \tau)$. The principal rates of strain are thus $\pm\sqrt{\sigma^2 + (\rho - \tau)^2}$. By rotating the coordinate system, and calling the rotated coordinates x and y once again, we may, therefore, reduce (2.7.10) to the form

$$\dot{x} = -\varepsilon\, x - \tfrac{1}{2}\zeta\, y\,; \quad \dot{y} = \tfrac{1}{2}\zeta\, x + \varepsilon\, y\,. \tag{2.7.12}$$

Here ϵ is the principal rate-of-strain and ζ is the vorticity at the stagnation point origin.[2] If $\epsilon > \zeta/2$, the stagnation point will be hyperbolic. For this case particle

[2] We have used the terminology of fluid kinematics in arriving at (2.7.12), but this is, of course, simply a statement of linear algebra in the plane.

paths will not stay close to the stagnation point for a long time. For this reason we will elaborate only on the elliptic case, $\epsilon < \zeta/2$, when particles close to the stagnation point orbit it forever, and a Taylor series expansion may be expected to capture long-time behavior.

The trajectory of a particle is a streamline, given by setting ψ equal to a constant. These curves are ellipses in our approximation:

$$x^2 + (4\varepsilon/\zeta)\, x\, y + y^2 = \text{const}\,. \tag{2.7.13}$$

This comes from (2.7.9) by comparing (2.7.12) to (2.7.10) and writing σ/ρ as $4\epsilon/\zeta$. By rescaling the coordinate axes relative to one another we can make these ellipses into circles.

So much for the part of the velocity field that is linear in the coordinates. The main conclusion is that for an elliptic stagnation point we can rescale the coordinates such that the motion over a finite time is simply a rotation. Now let us look at what happens as we step into the nonlinear problem. We shall follow the analysis by Hénon (1969) and take the nonlinear step to be a quadratic mapping. Thus, instead of the differential form of the trajectory, as in (2.7.10) or (2.7.12), we consider a map, $\mathbf{H}\colon (x, y) \to (x', y')$, connecting the position now with the position some finite time later. We write this map as

$$x' = f(x, y)\,, \quad y' = g(x, y)\,, \tag{2.7.14}$$

and will work with its expansion to second order in the coordinates. The finite-time counterpart of (2.7.8) is that the Jacobian of the transformation (2.7.14) equals unity:

$$\frac{\partial(x', y')}{\partial(x, y)} = 1\,. \tag{2.7.15}$$

Hénon considered mappings (2.7.14) of the general form

$$\left.\begin{array}{rcl} x' & = & a\,x + b\,y + c\,x^2 + d\,x\,y + e\,y^2\,, \\ y' & = & A\,x + B\,y + C\,x^2 + D\,x\,y + E\,y^2\,. \end{array}\right\} \tag{2.7.16}$$

We have already argued that the linear part can, modulo rescaling, be written as a rotation. Thus, we set

$$a = \cos\alpha\,, \quad b = -\sin\alpha\,, \quad A = \sin\alpha\,, \quad B = \cos\alpha\,. \tag{2.7.17}$$

The major subject of interest is the nature of the nonlinear modification of this rotation, which becomes important at larger distances from the stagnation point. We shall see that the condition (2.7.15) severely constrains the form of this nonlinear part. We remark in passing that mappings given by polynomial expressions, such as (2.7.16) are called **entire Cremona transformations** in the mathematical literature. Note that (2.7.16) contain *seven* arbitrary constants at this stage: the angle α and the coefficients c, d, e, C, D, E. We shall see that a dramatic reduction occurs in the number of constants necessary for a complete parametrization of the problem.

Using (2.7.16) in (2.7.15) leads to the condition

$$(a+2\,c\,x+d\,y)(B+D\,x+2\,E\,y)-(b+d\,x+2\,e\,y)(A+2\,C\,x+D\,y)=1\,. \quad (2.7.18)$$

Collecting coefficients of like powers of x and y,

$$\left.\begin{aligned} a\,B-b\,A &= 0\,,\\ a\,D+2\,c\,B-2\,b\,C-d\,A &= 0\,,\\ d\,B+2\,a\,E-2\,e\,A-b\,D &= 0\,,\\ 2\,c\,D-2\,d\,C &= 0\,,\\ 2\,d\,E-2\,e\,D &= 0\,,\\ 4\,c\,E-4\,e\,C &= 0\,. \end{aligned}\right\} \quad (2.7.19)$$

When the coefficients of the linear part, a, b, A and B, are chosen as in (2.7.17), the first of the equalities in (2.7.19) is automatically satisfied.

The last three equalities in (2.7.19) show that $c/C = d/D = e/E$. Denoting the common value of this ratio by $\tan\beta$ we can write

$$\left.\begin{aligned} c=\gamma\,\sin\beta\,,\quad d=\delta\,\sin\beta\,,\quad e=\phi\,\sin\beta\,,\\ C=\gamma\,\cos\beta\,,\quad D=\delta\,\cos\beta\,,\quad E=\phi\,\cos\beta\,. \end{aligned}\right\} \quad (2.7.20)$$

From the second and third equalities of (2.7.19) we then obtain

$$\left.\begin{aligned} 2\,\gamma\sin(\alpha+\beta)+\delta\cos(\alpha+\beta) &= 0\,,\\ \delta\sin(\alpha+\beta)+2\,\phi\cos(\alpha+\beta) &= 0\,. \end{aligned}\right\} \quad (2.7.21)$$

If we now introduce new variables ξ, η via the following rotation:

$$\left.\begin{aligned} x &= \xi\cos(\alpha+\beta)+\eta\sin(\alpha+\beta)\,,\\ y &= -\xi\sin(\alpha+\beta)+\eta\cos(\alpha+\beta)\,, \end{aligned}\right\} \quad (2.7.22)$$

and substitute into (2.7.16) using (2.7.17), (2.7.20) and (2.7.21), we get

$$\left.\begin{aligned} x' &= \xi\cos\beta+\eta\sin\beta+(\gamma+\phi)\,\xi^2\sin\beta\,,\\ y' &= -\xi\sin\beta+\eta\cos\beta+(\gamma+\phi)\,\xi^2\cos\beta\,. \end{aligned}\right\} \quad (2.7.23)$$

Our final transformation consists of the following rescaling of x', y', ξ and η by $-(\gamma+\phi)$:

$$\xi=\frac{-X}{\gamma+\phi},\quad \eta=\frac{-Y}{\gamma+\phi},\quad x'=\frac{-X'}{\gamma+\phi},\quad y'=\frac{-Y'}{\gamma+\phi}\,. \quad (2.7.24)$$

Then we finally have

$$\left.\begin{aligned} X' &= X\cos\beta+Y\sin\beta-X^2\sin\beta\,,\\ Y' &= -X\sin\beta+Y\cos\beta-X^2\cos\beta\,, \end{aligned}\right\} \quad (2.7.25)$$

as the general form for the quadratic, area-preserving mapping (2.7.16). Equations (2.7.25) are known as **Hénon's quadratic map**. Note the reduction in the number of free parameters that has been achieved. In (2.7.25) only the angle β of the rotation appears, certainly the least we would expect.

Problem 2.7.4 Verify the steps leading to (2.7.25).

From the point of view of flow kinematics (2.7.25) has a very simple and interesting interpretation. We see that these formulae can be interpreted as the composition of two elementary maps. The first is $\mathbf{S}$: $(X, Y) \to (X_1, Y_1)$, where

$$X_1 = X\,, \quad Y_1 = Y - X^2\,. \tag{2.7.26}$$

This is a shearing motion parallel to the y-axis. The second is a rotation by the angle β, i.e., $\mathbf{R}$: $(X_1, Y_1) \to (X', Y')$, where

$$\left.\begin{aligned} X' &= X_1 \cos\beta + Y_1 \sin\beta\,, \\ Y' &= -X_1 \sin\beta + Y_1 \cos\beta\,. \end{aligned}\right\} \tag{2.7.27}$$

Hénon's quadratic map, $\mathbf{H}$, then is the composition $\mathbf{H} = \mathbf{R} \circ \mathbf{S}$.

For a fluid mechanician these results have a familiar ring to them. The analysis of the motion relative to a point is a standard piece of flow kinematics (see, e.g., Batchelor, 1967, §2.4). The well-known result is that the motion can be represented as the superposition of a **translation**, a **pure straining motion** and a **rigid-body rotation**. In the case being treated here the translation has been taken out since the origin of coordinates is a stagnation point. The rigid-body rotation component is present, but the pure strain has been replaced by the shear $\mathbf{S}$. The question being asked here is, of course, also a somewhat different one. We are not looking for the instantaneous *local form* of a fluid velocity field. We are after an approximation to the mapping of a point after a finite time (as it moves in the vicinity of a stagnation point).

Numerical investigation of Hénon's quadratic mapping is very interesting and the many details in the original paper (Hénon, 1969) are recommended for study. Figure 2.8 shows a surface of section plot for $\cos\beta = 0.24$ with an expanded view of the intricate structure in the vicinity of one of the unstable period-5 points. The map shows chaotic particle orbits except for trivial values of β.

How should we interpret these results? First, note that there is *no chaos* in advection by a *steady 2D-flow*, since the particles follow a regular family of streamlines. Hence, in such a flow, to second order in the coordinates relative to an elliptic stagnation point, the mapping that sends a point to its new position some time interval later must be represented by one of the regular cases of Hénon's quadratic map. Thus, if chaos appears in the tracking of individual particles in that case it must be ascribed to inaccuracies in the numerical method, such as truncation errors in the time-stepping algorithm.

For time-periodic 2D-flows we can and in general do have chaos. Then the results obtained tell us that in the vicinity of an elliptic stagnation point the mapping (2.7.14) has one of a one-parameter family of chaotic structures (again, to second order in the coordinates). The analysis has captured all the possibilities in terms of a single parameter!

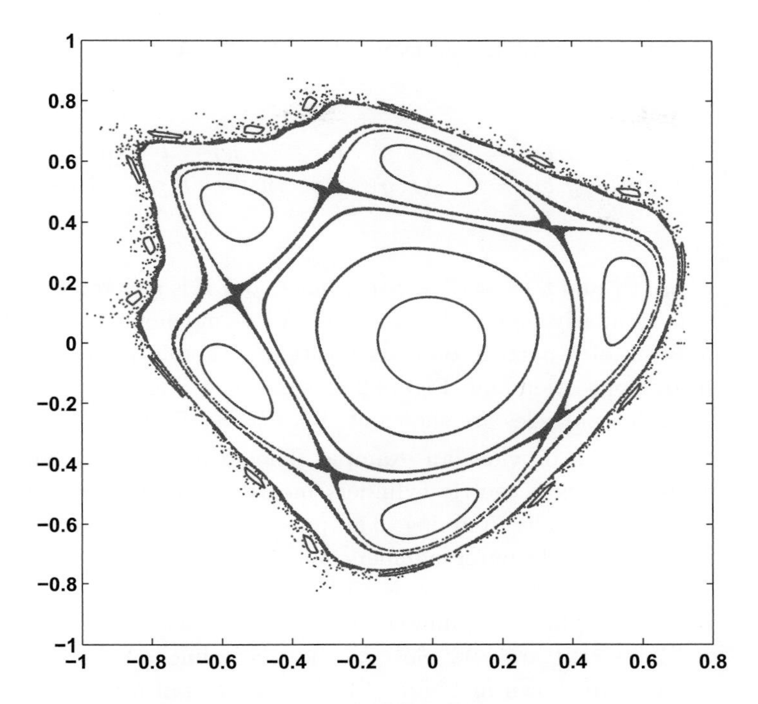

Figure 2.8 Poincaré section of the Hénon mapping with $\cos\beta = 0.24$.

One final point on Hénon's quadratic mapping is worth making: since $\mathbf{H} = \mathbf{R} \circ \mathbf{S}$, the inverse of $\mathbf{H}$ is $\mathbf{R}^{-1} \circ \mathbf{S}^{-1}$. Each of these inverses can be easily written out. The mapping $\mathbf{R}^{-1}$ is just a rotation by $-\beta$, see (2.7.27), and $\mathbf{S}^{-1}$ is the mapping: $(X_1, Y_1) \to (X, Y)$, where

$$X = X_1\,, \qquad Y = Y_1 + X_1^2\,. \tag{2.7.28}$$

Thus, the inverse of $\mathbf{H}$ is given by the formulae $\mathbf{H}^{-1}$: $(X', Y') \to (X, Y)$, where

$$\left.\begin{aligned} X &= X'\cos\beta - Y'\sin\beta\,, \\ Y &= X'\sin\beta + Y'\cos\beta + (X'\cos\beta - Y'\sin\beta)^2\,, \end{aligned}\right\} \tag{2.7.29}$$

which, remarkably, is again a quadratic mapping.

Computer Exercise 2.7.5 Write a program to study surfaces of section of Hénon's quadratic map. For each value of the parameter $\cos\beta$ iterate some 5 to 10 initial points for about 2000 iterations each. Plot all the iterates together to get a global picture of the mapping plane. To investigate iterates of one individual particle you can either color code the points or generate separate plots along with the composite, overall view.

In general, equations (2.7.8) lead to a system of two coupled ODEs that must be solved by time integration. The special case:

$$\left.\begin{array}{rclcc}\psi(x,y,t) & = & \psi_1(x,y) & \text{for} & nT \le t < (n+\frac{1}{2})\,T\,, \\ \psi(x,y,t) & = & \psi_2(x,y) & \text{for} & (n+\frac{1}{2})\,T \le t < (n+1)\,T\,,\end{array}\right\} \qquad (2.7.30)$$

where ψ_1 and ψ_2 describe *steady flows*, and T is a characteristic pulsation period is particularly amenable to study. We would now choose (2.7.14) to map the position of a particle onto its position a time T later. The counterpart of the surface of section plot, Figure 2.6, is now a stroboscopic plot where the positions of chosen particles are shown at times $t = 0,\, T,\, 2T,\, 3T, \ldots$.

Figure 2.9 provides an example. The steady flows ψ_1 and ψ_2 are the Stokes flows between eccentric cylinders mentioned in Example 1.6.2 and Figure 1.3 with one cylinder spinning. The pulsation comes about by alternately turning the inner and the outer cylinders. When the surface of section plot is constructed, one observes that rapid pulsation (short T) approximates the steady flow case with both cylinders spinning. Hence, the stroboscopic points fall along a system of curves that are very close to the streamlines of the steady flow, samples of which were shown in Figure 1.3. For slower pulsation (longer T) the sequence of stroboscopic images blurs, again a manifestation of chaotic particle motion and a hint as to how one can stir fluids efficiently even in this very viscous flow regime. In fact, slow pulsation is essential if one wants to retain the Stokes flow approximation used for obtaining the steady flows. This particular flow has become something of a paradigm for chaotic advection, and has been explored by several authors (Aref and Balachandar, 1986; Chaiken *et al.*, 1986, 1987; Swanson and Ottino, 1990).

Examples such as these lead to simple computational problems with a dynamical systems flavor but with obvious relevance to fluid mechanics. In the simplest cases one can actually explicitly write down the mapping (2.7.14). In general, one must face a numerical integration of the ODEs (2.7.8). We shall discuss how to perform such calculations in the next chapter.

Computer Exercise 2.7.6 Write the mapping (2.7.14) explicitly for the case of two blinking vortices. The flow is assumed to be generated by two point vortices that are fixed at $x = \pm 1$. The velocity field of each is circumferential, and decays inversely proportional to the distance from the vortex. The flow consists of switching these vortices "on" and "off" alternately. Explore the nature of advection in this flow as the switching time is varied. Try stirring blobs and lines made up of many particles. Additional details and examples may be found in Aref (1984) and in the book, Ottino (1989).

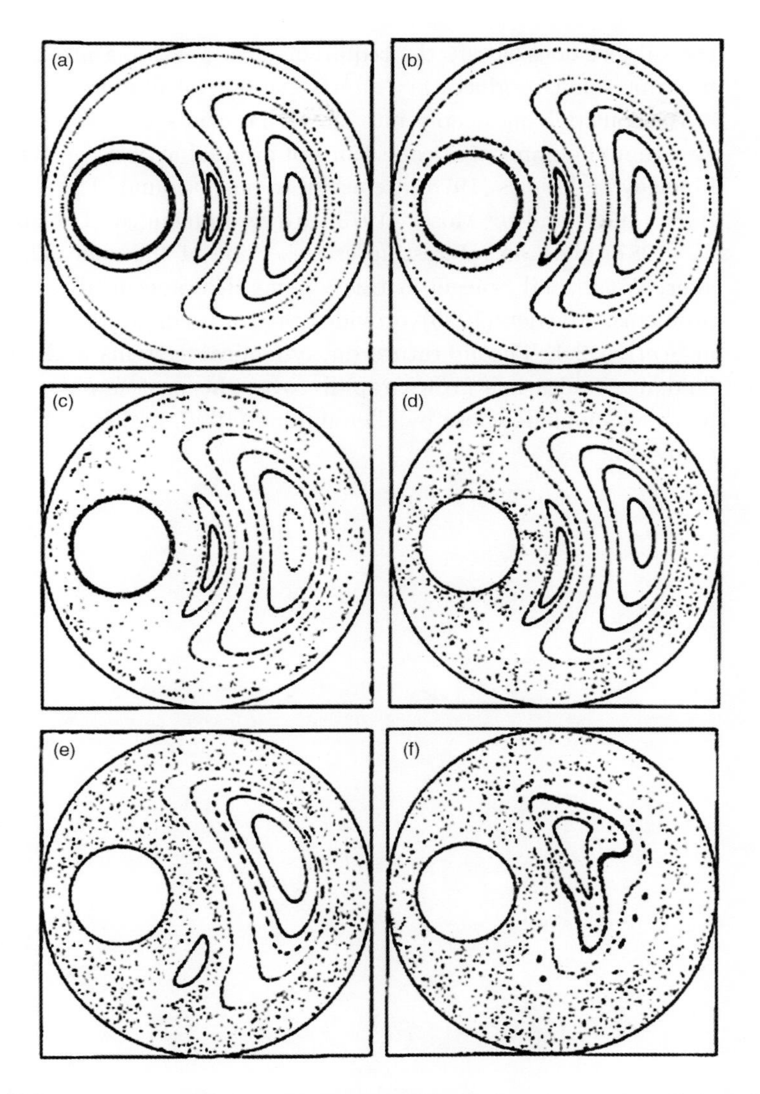

Figure 2.9 Poincaré section of pulsed Stokes flow between eccentric cylinders. The angular speed ratio of outer to inner angular velocity was set to 6, radius ratio was 0.3, and eccentricity was set to 0.75. The period of switching normalized by the angular rotation period of the outer cylinder was varied: (a) 0 (steady flow); (b) 0.1; (c) 0.5; (d) 1.0; (e) 3.0; (f) 9.0. (Taken from Aref and Balachandar, 1986.)

Literature Notes

This chapter is far from traditional in a book on CFD, but it is clear that the subject matter is frequently relevant and becoming more so with the passage of time, and that the numerical exploration of algorithmically simple systems, such as mappings and low-order ODEs is often quite instructive. Furthermore, the subject does satisfy the requirement of providing problems with elementary programs that produce very interesting output in abundance.

Without delving deeply into the dynamical systems literature we would suggest that the annotated collections of original articles by Bai-Lin (1984) and by MacKay and Meiss (1978) are useful to have around. The classics by Arnold and Avez (1968) and by Moser (1973) are recommended. The special issue Berry *et al.* (1987) gives an interesting cross-section of the field. The articles by Libchaber, Weiss, and Spiegel in this volume deal specifically with fluid mechanical problems. Maeder (1995) provides programs in *Mathematica* for the material in Section 2.4. The literature on dynamical systems and their applications to fluid mechanics has grown rapidly over the past few decades. Other relevant textbooks include those by Glendinning (1994) and Hirsch *et al.* (2004), among many others.

3 Ordinary Differential Equations: Initial Value Problem

We now proceed to consider problems in CFD that involve finite systems of **ordinary differential equations** (ODEs). In this chapter we will focus attention on initial value problems and consider algorithms and numerical methods appropriate for this class of ODEs. If one adopts the view that CFD is basically concerned with numerical solutions to PDEs, then a system of a finite number of ODEs is not properly a part of the subject. However, we have a somewhat broader view of CFD in mind, and there are a great number of interesting and important problems in fluid mechanics, where the mathematical formulation and modeling leads finally to the numerical solution of finite systems of ODEs. Even if one insists on solving the full set of PDEs, upon spatial discretization into finite number of grid points or modes, the resulting large, but finite, system of ODEs is an initial value problem. In this respect, our discussion on initial value problems will naturally follow our prior discussions on mappings and take one more step towards solving PDEs.

We will consider the merits and drawbacks of various *time-differencing schemes*, which reduce the system of initial value ODEs to mappings. With this reduction, we can use our knowledge of mappings in exploring the properties of a discretized numerical solution to a system of ODEs. We begin with a quick review of some mathematical theory and numerical analysis of ODEs. We then follow with several fluid mechanical examples of initial value problems leading to a finite system of ODEs and consider their numerical solution. Although we will discuss several numerical schemes and their implementation, the integration of ODEs is a well-studied subject, and you can usually find reliable codes as part of mathematical software libraries, such as MATLAB, IMSL or NAG. Our discussion will provide a somewhat spotty coverage and is not a substitute for a rigorous and comprehensive treatment.

3.1 Some Conventional Wisdom

The **order** of an ODE is the order of the highest derivative that appears in the equation. Thus, the Blasius equation (1.6.5)

$$f''' + \tfrac{1}{2} f\, f'' = 0 \tag{3.1.1}$$

is a third-order ODE. It is a general rule that a single ODE of order k is equivalent to a system of k first-order ODEs. For example, (3.1.1) can be reduced to a system of three first-order ODEs by introducing the auxiliary functions

$$g = f' \quad \text{and} \quad h = f'' = g' \, . \tag{3.1.2}$$

Then (3.1.1) is equivalent to the system

$$\left.\begin{array}{rcl} f' & = & g \, , \\ g' & = & h \, , \\ h' & = & -\frac{1}{2} f \, h \, . \end{array}\right\} \tag{3.1.3}$$

In general, there are several ways of carrying out this reduction. Other considerations might enter in deciding which one to choose for computational purposes. Although not always necessary, the reduction of higher-order ODEs to a system of lower-order ODEs is useful in the description of various numerical time integration schemes to be presented shortly.

In a dynamics problem we typically encounter ODEs of second order arising through Newton's second law, or more generally, from Lagrange's equations. The latter have the form:

$$\frac{\mathrm{d}}{\mathrm{dt}}\left(\frac{\partial L}{\partial \dot{q}_i}\right) = \frac{\partial L}{\partial q_i} \, ; \quad i = 1, \, 2, \, \ldots, \, N \, , \tag{3.1.4}$$

where N is the number of degrees of freedom in the system, and L is a function of the generalized coordinates $q_1, \, \ldots, \, q_N$, the generalized velocities $\dot{q}_1, \, \ldots, \, \dot{q}_N$ and, possibly, the time t. Equations (3.1.4) are second order in time, and they are by construction *linear* in the accelerations $\ddot{q}_1, \, \ldots, \, \ddot{q}_N$. When we want to subject the system (3.1.4) to numerical integration, we must first express $\ddot{q}_1, \, \ldots, \, \ddot{q}_N$ in terms of $q_1, \, \ldots, \, q_N, \, \dot{q}_1, \, \ldots, \, \dot{q}_N$ and (maybe) t. Let the result of this algebra be

$$\ddot{q}_i = f_i(q_1, \, \ldots, \, q_N, \, \dot{q}_1, \, \ldots, \, \dot{q}_N, \, t) \, ; \quad i = 1, \, \ldots, \, N \, . \tag{3.1.5}$$

If L is quadratic in the velocities, as is often the case, we can give specific formulae for the functions f_i appearing in these equations. For numerical purposes we consider the system of $2N$ first-order equations

$$\dot{q}_i = v_i \, ; \quad \dot{v}_i = f_i(q_1, \, \ldots, \, q_N, \, v_1 \ldots, \, v_N, \, t) \, ; \quad i = 1, \ldots, N \, . \tag{3.1.6}$$

In summary, it suffices to address the numerical integration of *systems of first-order ODEs*. Any higher-order ODE, or system of higher-order ODEs, can be decomposed into a system of first-order ODEs as in the examples just given.

We noted above that the reduction to first-order ODEs is in general not unique. Indeed, for a system described by a Lagrangian, instead of proceeding as in (3.1.5)–(3.1.6), we might seek the reduction to Hamiltonian form. That is, we would define generalized momenta

$$p_i = \frac{\partial L}{\partial \dot{q}_i} \, ; \quad i = 1, \, \ldots, \, N \tag{3.1.7}$$

and a Hamiltonian

$$H = \sum_{i=1}^{N} p_i q_i - L . \tag{3.1.8}$$

Following a Legendre transformation from q and the $\dot{q}$ to q and p, we obtain Hamilton's canonical equations:

$$\dot{q}_i = \frac{\partial H}{\partial p_i}; \quad \dot{p}_i = -\frac{\partial H}{\partial q_i}, \quad i = 1, \ldots, N , \tag{3.1.9}$$

where the Hamiltonian, H, depends on the qs and the ps (and possibly time). Equations (3.1.9) are, again, a system of $2N$ first-order ODEs, in general however quite different from the system (3.1.6).

In an initial value problem the necessary data for a unique solution is specified at a single value of the independent variable. The terminology stems from the common case where that variable is "time." That is we have in mind the specification of the "initial" value of all dependent variables at the "initial" value of the independent variable. The solution of the dependent variables then evolves as the independent variable marches forward. For example, in Newtonian mechanics, equations (3.1.4)–(3.1.6), we would specify the "positions and velocities at $t = 0$." In the notation of (3.1.6) we need to know $q_1, \ldots, q_N$ and $v_1, \ldots, v_N$ at $t = 0$. For (3.1.9) we would specify the initial values of $q_1, \ldots, q_N$ and $p_1, \ldots, p_N$.

In a boundary value problem, on the other hand, the conditions that determine the solution are not given at just one value of the independent variable. The most common case is where conditions are specified at two points. The Blasius problem, equations (3.1.1)–(3.1.3), along with the boundary conditions

$$f(\eta = 0) = 0; \quad g(\eta = 0) = 0, \quad \text{and} \quad g(\eta \to \infty) = 1 \tag{3.1.10}$$

is, at least formally, of this type. The desired function f is pinned down by requiring f and f' to vanish at the origin (both conditions at one point), and for f' to tend to unity at infinity (another condition at a second "point," here the "point at infinity"). Such two-point boundary value problems are intrinsically more difficult than initial value calculations, although it turns out that one can treat them in very much the same way as initial value problems. The discussion of two-point boundary and eigenvalue problems will be postponed to later chapters.

3.2 Explicit Euler and Implicit Euler Schemes

The simplest method for solving a first-order differential equation is the explicit Euler method, discussed before in Example 2.1.2. Here the derivative is simply used to extrapolate the solution over an incremental step. Let us write our generic problem as

$$\frac{\mathrm{d}\xi}{\mathrm{d}t} = g(\xi, t) , \tag{3.2.1}$$

where t is the independent variable, g is a known function and $\xi(t)$ is the solution sought. We have suppressed any parameters that occur, and we are now allowing g to depend explicitly on time. Let Δt be the step size with which we wish to effect numerical integration. Then the explicit Euler method is based on the following formula:

$$\xi(t+\Delta t) = \xi(t) + g(\xi(t),t)\,\Delta t + O\,(\Delta t)^2 \ . \tag{3.2.2}$$

For a system of N equations both ξ and g in (3.2.1)–(3.2.2) are vectors with N components. In the above $O(\Delta t)^2$ stands for terms of order Δt^2 or smaller.

Example 3.2.1 Consider the very simple problem of integrating the motion of a particle in a uniformly rotating fluid. We have the equations of motion

$$\frac{\mathrm{d}x}{\mathrm{d}t} = -\Omega\, y\,; \qquad \frac{\mathrm{d}y}{\mathrm{d}t} = \Omega\, x\,. \tag{3.2.3}$$

The analytical solution, of course, is uniform circular motion with angular frequency Ω. What does the explicit Euler method give as the solution for this case? An easy way of conducting the analysis is to set $z = x + \iota\, y$. Equations (3.2.3) then become

$$\frac{\mathrm{d}z}{\mathrm{d}t} = \iota\, \Omega\, z\,. \tag{3.2.4}$$

Now, with $z_k = x_k + \iota\, y_k$ the explicit Euler method gives the mapping

$$z_{k+1} = z_k + \iota\, \Omega\, \Delta t\, z_k\,, \tag{3.2.5}$$

which is easily solved to yield

$$z_k = (1 + \iota\, \Omega\, \Delta t)^k\, z_0\,. \tag{3.2.6}$$

There are deviations with both the phase and the amplitude of this result relative to the analytical solution of the continuous-time problem, (3.2.4). To see this write

$$1 + \iota\Omega\Delta t = \left[1 + (\Omega\Delta t)^2\right]^{1/2} \exp(\iota\alpha)\,; \quad \text{where} \quad \alpha = \tan^{-1}(\Omega\Delta t)\,. \tag{3.2.7}$$

Thus, the angle $k\alpha$ through which the numerical solution turns the initial condition z_0 over k time-steps of size Δt is not quite the same as the analytical result $k\,\Omega\,\Delta t$. The difference is the **phase error** in the numerical solution. For small $\Omega\,\Delta t$ we have

$$k\,\tan^{-1}(\Omega\,\Delta t) - k\,\Omega\,\Delta t = -\frac{k\,(\Omega\,\Delta t)^3}{3} + \frac{k\,(\Omega\,\Delta t)^5}{5} - \cdots\,. \tag{3.2.8}$$

Therefore the phase error goes as $(\Delta t)^3$. More serious, however, is the **amplitude error**. According to (3.2.7), $|z_k|$, which should remain constant according to the analytical solution, will grow as $\left[1 + (\Omega\,\Delta t)^2\right]^{k/2}$. For small $\Omega\,\Delta t$ we have

$$\left[1 + (\Omega\,\Delta t)^2\right]^{k/2} - 1 = \frac{k}{2}(\Omega\,\Delta t)^2 + \frac{k}{2}\left(\frac{k}{2} - 1\right)\frac{(\Omega\,\Delta t)^4}{2} + \cdots\,. \tag{3.2.9}$$

The amplitude error scales as $(\Delta t)^2$ and it compounds exponentially leading to an outward spiraling path of the particle rather than a circle. A cloud of points all moving according to (3.2.3) will be seen to expand, akin to what one might expect if there was some diffusion present, rather than to rotate as a rigid body.

Instead, if an implicit Euler scheme is used to solve (3.2.4), the resulting iteration

$$z_{k+1} = z_k + \iota\,\Omega\,\Delta t\, z_{k+1} \tag{3.2.10}$$

can be easily solved to give

$$z_k = (1 - \iota\,\Omega\,\Delta t)^{-k}\, z_0 = \left[1 + (\Omega\,\Delta t)^2\right]^{-k/2} \exp(\iota\, k\, \alpha)\, z_0 \ , \tag{3.2.11}$$

where α is defined as in (3.2.7). Thus, the phase errors in both the explicit and implicit Euler schemes are the same. Although the amplitude errors are of the same order, it can be seen from (3.2.11) that the amplitude $|z_k|$ decreases for each time-step of the implicit Euler scheme. In this sense, while the implicit Euler scheme provides a stable numerical approximation, the explicit Euler scheme provides an unstable numerical approximation.

3.2.1 Order of Accuracy

The notion of a formal **order of accuracy** of a time integration scheme is evaluated by considering the numerical approximation of the generic first-order ODE: $\mathrm{d}\xi/dt = g(\xi, t)$. The analytical solution at $t+\Delta t$, given $\xi(t)$, can be written

$$\xi(t + \Delta t) = \xi(t) + \int_0^{\Delta t} g(\xi(t + \tau),\, t + \tau)\, \mathrm{d}\tau \ . \tag{3.2.12}$$

Let us expand the integrand in powers of τ assuming that Δt (and thus τ) is small. Then we get

$$\begin{aligned} g(\xi(t + \tau),\, t + \tau) &= g(\xi(t),\, t) + \tau\,\frac{\partial g}{\partial \xi}\frac{\mathrm{d}\xi}{\mathrm{d}t} + \tau\,\frac{\partial g}{\partial t} + O(\tau)^2 \\ &= g(\xi(t),\, t) + \tau\,\frac{\partial g}{\partial \xi} g(\xi(t),\, t) + \tau\,\frac{\partial g}{\partial t} + O(\tau)^2 \ . \end{aligned} \tag{3.2.13}$$

Integrating this term by term as directed by (3.2.12) gives

$$\xi(t + \Delta t) = \xi(t) + g(\xi(t),\, t)\,\Delta t + \tfrac{1}{2}(\Delta t)^2 \left[\frac{\partial g}{\partial \xi}\, g + \frac{\partial g}{\partial t}\right] + O(\Delta t)^3 \ . \tag{3.2.14}$$

A more transparent way of arriving at (3.2.14) may be to consider the Taylor series expansion of the exact solution:

$$\xi(t + \Delta t) = \xi(t) + \Delta t\, \dot{\xi}(t) + \tfrac{1}{2}(\Delta t)^2\, \ddot{\xi}(t) + \cdots \ . \tag{3.2.15}$$

But, $\dot{\xi}(t) = g(\xi, t)$, and thus

$$\ddot{\xi}(t) = \frac{\mathrm{d}g}{\mathrm{d}t} = \frac{\partial g}{\partial t} + \frac{\partial g}{\partial \xi}\dot{\xi} = \frac{\partial g}{\partial t} + \frac{\partial g}{\partial \xi} g \,. \tag{3.2.16}$$

Substituting into (3.2.15) gives the same result as (3.2.14).

In the case of the explicit Euler scheme, the above is approximated as

$$\xi(t + \Delta t) \approx \xi(t) + g(\xi(t),\, t)\, \Delta t \,. \tag{3.2.17}$$

The numerical and analytical solutions agree up to terms of order (Δt); therefore we say that the explicit Euler scheme is *first-order accurate*. Similarly, the implicit Euler scheme approximates $\xi(t + \Delta t)$ as

$$\left.\begin{aligned} \xi(t + \Delta t) &\approx \xi(t) + g(\xi(t + \Delta t),\, t + \Delta t)\, \Delta t \\ &= \xi(t) + g(\xi(t), t)\, \Delta t + (\Delta t)^2 \left[\frac{\partial g}{\partial \xi} g + \frac{\partial g}{\partial t} \right] + O(\Delta t)^3 . \end{aligned}\right\} \tag{3.2.18}$$

Agreement with the exact integration is again only up to terms of order (Δt) and therefore the implicit Euler scheme is also *first-order accurate.*

A comparison of (3.2.17) and (3.2.18) with the exact solution (3.2.14) also gives a measure of the **local truncation error** of the schemes. Local truncation error is defined as the leading-order term in the difference between $\mathrm{d}\xi/dt$ (i.e., $g(\xi, t)$) and its numerical approximation $[\xi(t + \Delta t) - \xi(t)]\,/\Delta t$. For both the explicit and implicit Euler schemes the local truncation is $\frac{1}{2}\Delta t\, dg/dt$ or $\frac{1}{2}\Delta t\, [g\, \partial g/\partial\xi + \partial g/\partial t]$. The order (Δt) behavior of the local truncation error is consistent with their first-order accuracy. The local truncation error gives the asymptotic behavior of the error in the numerical approximation as we approach the limit $\Delta t \to 0$. As the name suggests, the local error provides a measure of error in the numerical solution over a single time-step. When the ODE is integrated over several time-steps, starting from an initial condition $\xi(t = 0)$, local truncation error can accumulate to give **global error**. For instance, in Example 3.2.1 both the amplitude and phase errors accumulate each time-step and increase monotonically with time. The global error is the quantity of ultimate interest when integrating an ODE over long time. Unfortunately, the global error depends not only on the numerical scheme, but also on the ODE being integrated. Thus, local truncation error is what is usually used in gauging the accuracy of a numerical scheme.

3.2.2 Stability

Stability of a time integration scheme is usually investigated in the context of the following simple linearized ODE

$$\frac{\mathrm{d}\xi}{\mathrm{d}t} = c\,\xi \,, \tag{3.2.19}$$

where $c = c_R + \iota\, c_I$ is some complex constant and in general ξ is also complex. For non-trivial c the only fixed point of the above equation is $\xi_* = 0$ and therefore

ξ itself can be considered a perturbation away from the fixed point. The exact solution to the above linear equation is $\xi(t+\Delta t) = \xi(t)\exp(c\Delta t)$. Thus the real part of the constant, c_R, being positive or negative, represents the exponentially growing or decaying nature of the exact solution to (3.2.19). The stable and unstable regions of the exact solution are shown in Figure 3.1(a). As seen in Example 3.2.1, a non-zero imaginary part corresponds to rotation of ξ in the complex plane with angular frequency given by c_I.

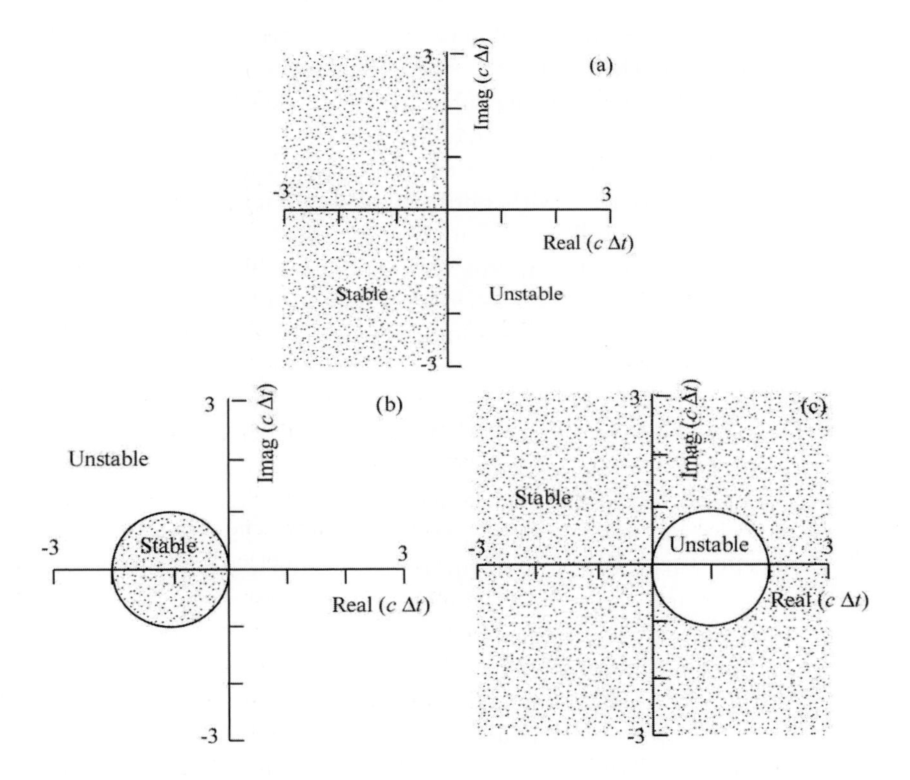

Figure 3.1 Stability diagrams for the different schemes. (a) The actual stability region for the equation (3.2.19); (b) region of stability and instability for the explicit Euler scheme; (c) region of stability and instability for the implicit Euler scheme.

Later in the context of partial differential equations we shall see that the advection (wave) and diffusion equations of interest reduce to (3.2.19) upon spatial discretization, with the advection operator resulting in a pure imaginary c and the diffusion operator resulting in a real negative c.

From the point of stability, the exact solution to (3.2.19) remains stable (in the sense that the solution decays to 0) over the entire left half of the complex c-plane, and unstable (in the sense that the solution grows beyond all bounds) over the right half of the complex c-plane. This behavior of the exact solution can be compared with that of the numerical solution. For example, using an explicit

Euler scheme we get

$$\xi_{k+1} = (1 + c\,\Delta t)\,\xi_k\,. \tag{3.2.20}$$

Provided $|1 + c\,\Delta t| \leq 1$, the term ξ_k will remain bounded by the initial condition, ξ_0, and the explicit Euler scheme can be considered stable. This **region of absolute stability** for the explicit Euler scheme is plotted on the complex $c\,\Delta t$ plane in Figure 3.1(b). The region of absolute stability, $|1 + c\,\Delta t| < 1$, corresponds to the interior of a unit circle whose origin is on the negative real axis at $c\,\Delta t = -1$. The unstable nature of the exact solution is captured by the explicit Euler scheme over the entire right half of the complex plane. However, the stable nature of the exact solution over the left half plane is guaranteed only for sufficiently small Δt such that $c\,\Delta t$ will be within the region of absolute stability. A curious point to note is that the explicit Euler scheme is always unstable to wave-like problems with pure imaginary c, no matter how small a Δt is chosen. This was well evident in Example 3.2.1, where particles spiral out for all $\Delta t > 0$, instead of going round on a circular path. Thus, for a wave equation it is important to consider other time-integration schemes whose region of absolute stability includes at least a finite segment of the imaginary axis.

The implicit Euler scheme is stable provided $\left|(1 - c\,\Delta t)^{-1}\right| < 1$. The region of absolute stability now corresponds to the region outside of a unit circle whose origin is on the positive real axis at $c\,\Delta t = 1$: see Figure 3.1(c). In particular, the stability region includes the entire left half of the complex plane. Such schemes are called **A-stable** and are unconditionally stable for diffusion problems (with real part of c less than zero) irrespective of Δt. The implicit schemes are therefore ideally suited for rapidly converging towards a steady solution with large Δt, if time-accuracy as to how the solution converges toward the steady state is not of interest.

The question of stability for a general (possibly nonlinear) ODE of the form $d\xi/dt = g(\xi, t)$ is not simple. Here the border between stability and instability is very much problem dependent. Note that even for (3.2.19) the stability boundary is defined in terms of $c\,\Delta t$ and therefore, strictly speaking, is problem dependent – linearly dependent on the coefficient c. Of course the scenario is likely to be more complex in the case of a more general ODE. The question of stability, however, can be addressed in the neighborhood of any point $\xi_k(t_k)$ by linearizing about that point as follows: $\delta = \xi - \xi_k$ and $\tau = t - t_k$. The resulting linearized ODE is

$$\frac{d\delta}{dt} \approx a + b\tau + c\delta\,, \tag{3.2.21}$$

where $a = g(\xi_k, t_k)$, $b = \frac{\partial g}{\partial t}(\xi_k, t_k)$ and $c = \frac{\partial g}{\partial \xi}(\xi_k, t_k)$. The exact solution to the above linearized equation can be written as

$$\delta = \frac{b + ac + \delta_0 c^2}{c^2} \exp[c\tau] - \frac{b}{c^2}(1 + c\tau) - \frac{a}{c}\,, \tag{3.2.22}$$

where δ_0 is the initial perturbation away from the point of linearization. The solution has a linear part and an exponential part. The local exponential growth or decay of the solution is governed by the last term of (3.2.21). It is a simple matter to see that the rapid growth or decay of numerical approximations, such as those resulting from explicit and implicit Euler schemes, is also governed by the last term of (3.2.21). Thus the stability diagram of a time integration scheme, though obtained based on the simple linear equation (3.2.19), provides a good approximation to the local stability behavior even in the case of nonlinear equations. Of course the algebraic term in (3.2.22) and the higher-order terms neglected in (3.2.21) will influence long-term stability. Thus, the stability diagrams must be used with some caution in the case of nonlinear ODEs.

3.3 Runge–Kutta Methods

It does not take long to realize that Euler's methods (both explicit and implicit) are not the last word on taking an integration step in a (system of) first-order ODE(s). Let us now discuss the well-known higher-order methods for solving ODEs known as Runge–Kutta methods. Since we are concerned with computational fluid dynamics it may be appropriate to mention that M. Wilhelm Kutta was an early twentieth-century aerodynamicist, who wrote on this subject in 1901. You have undoubtedly come across his name in connection with the Kutta–Zhukovskii condition in airfoil theory. C.D.T. Runge, whose paper on the methods that today bear his name was published in 1895, is sometimes called the "father of numerics." He was a German mathematician, a contemporary and close friend of L. Prandtl.

Our problem is to integrate $\mathrm{d}\xi/dt = g(\xi, t)$. The *second-order* Runge–Kutta method arises by taking two steps across an interval of length Δt instead of just one. First we obtain an intermediate solution value at $\Delta t/2$ using the explicit Euler scheme. Thus, we first compute

$$\tilde{\xi} = \xi_k + \tfrac{1}{2}\Delta t\, g(\xi_k, t_k)\,. \tag{3.3.1}$$

Using this intermediate value we then compute

$$\xi_{k+1} = \xi_k + \Delta t\, g(\tilde{\xi}, t_{k+1/2}) = \xi_k + \Delta t\, g(\xi_k + \tfrac{1}{2}\Delta t\, g(\xi_k, t_k), t_k + \tfrac{1}{2}\Delta t)\,. \tag{3.3.2}$$

The numerical solution ξ_{k+1} and the analytical solution $\xi(t_k + \Delta t)$ agree through terms of order $(\Delta t)^2$ (hence the attribute "second-order" in the name). The increased order of accuracy of the above method over the Euler methods can be explained as follows. The explicit Euler scheme uses the slope at the starting point, $g(\xi_k, t_k)$, to advance the solution from t_k to t_{k+1} (see Figure 3.2), while the implicit Euler scheme uses the slope at the end point, $g(\xi_{k+1}, t_{k+1})$. By the mean value theorem the slope that will yield the exact answer is somewhere in between. In the second-order Runge–Kutta method, $g(\tilde{\xi}, t_{k+1/2})$ can be considered as an approximation to the slope at the midway point in time.

Consider a numerical approximation to $\xi(t + \Delta t)$ of the general form

$$\xi_{k+1} = \xi_k + w_0\,\Delta t\, g(\xi_k, t_k) + w_1\,\Delta t\, g(\xi_k + \beta\,\Delta t\, g(\xi_k, t_k), t_k + \alpha\,\Delta t)\,, \tag{3.3.3}$$

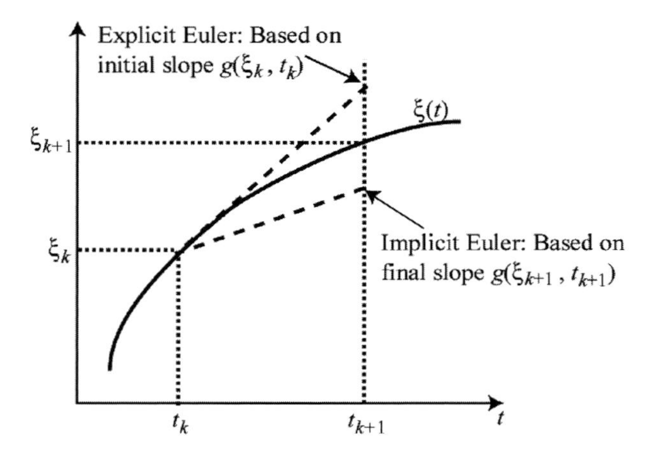

Figure 3.2 Graphical interpretation of the explicit and implicit Euler schemes. Runge–Kutta higher-order schemes can then be interpreted as higher-order approximations to the correct slope that would yield the exact value of ξ at the end of the time-step.

where w_0, w_1, α and β are weights, i.e., certain numbers to be determined: for (3.3.2), $w_0 = 0$, $w_1 = 1$, $\alpha = \beta = 1/2$. The above numerical solution needs to be compared with the exact solution (3.2.14). Let us inquire into the constraints imposed on w_0, w_1, α and β by requiring (3.2.14) and (3.3.3) to agree through terms of order $(\Delta t)^2$. Expanding (3.3.3) in powers of Δt gives

$$\xi_{k+1} = \xi_k + (w_0 + w_1)\,\Delta t\, g(\xi_k, t_k) + w_1\,(\Delta t)^2 \left[\alpha \frac{\partial g}{\partial t} + \beta \frac{\partial g}{\partial \xi} g\right]_{\xi_k, t_k} + O(\Delta t)^3 . \tag{3.3.4}$$

So we must insist on

$$\begin{aligned} w_0 + w_1 &= 1, \\ w_1\,\alpha = w_1\,\beta &= \tfrac{1}{2}, \end{aligned} \tag{3.3.5}$$

or

$$w_1 \neq 0; \quad w_0 = 1 - w_1; \quad \alpha = \beta = (2w_1)^{-1} . \tag{3.3.6}$$

The second-order Runge–Kutta method (3.3.1)–(3.3.2) arises, as we have already seen, for $w_1 = 1$. Another favorite, known as **Heun's method**, arises for $w_1 = \frac{1}{2}$:

$$\xi_{k+1} = \xi_k + \tfrac{1}{2}\Delta t\, \{g(\xi_k, t_k) + g(\xi_k + \Delta t\, g(\xi_k, t_k),\, t_k + \Delta t)\} . \tag{3.3.7}$$

Example 3.3.1 Return for a moment to the simple example of particle motion in a uniformly rotating flow. With the notation of Example 3.2.1 we see that the method (3.3.7) leads to

$$z_{k+1} = z_k + \iota\,\Omega\,\Delta t\,\left(z_k + \tfrac{1}{2}\,\iota\,\Omega\,\Delta t\, z_k\right) = \left(1 - \frac{1}{2}(\Omega\,\Delta t)^2 + \iota\,\Omega\,\Delta t\right) z_k . \tag{3.3.8}$$

Again the modulus of the prefactor is not quite unity, but the deviation from 1 is now of the *fourth*-order in Δt even though the method is only second-order accurate:

$$\left|1 - \tfrac{1}{2}(\Omega\,\Delta t)^2 + \iota\,\Omega\,\Delta t\right| = \left[1 + \tfrac{1}{4}(\Omega\,\Delta t)^4\right]^{\frac{1}{2}} = 1 + \frac{(\Omega\,\Delta t)^4}{8} - \frac{(\Omega\,\Delta t)^8}{128} + \cdots . \tag{3.3.9}$$

Amplitude errors accumulate much less rapidly than for Euler's method.

Problem 3.3.2 Obtain the local truncation error for the second-order Runge–Kutta and Heun's methods as follows:

$$\textbf{Runge–Kutta II:}\ (\Delta t)^2\left[\frac{1}{6}\left(\frac{d^2g}{dt^2}\right)_{\xi_k,t_k} - \frac{1}{8}\left(g^2\frac{\partial^2 g}{\partial\xi^2} + \frac{\partial^2 g}{\partial t^2} + 2g\frac{\partial^2 g}{\partial\xi\partial t}\right)_{\xi_k,t_k}\right],$$

$$\textbf{Heun:}\ (\Delta t)^2\left[\frac{1}{6}\left(\frac{d^2g}{dt^2}\right)_{\xi_k,t_k} - \frac{1}{4}\left(g^2\frac{\partial^2 g}{\partial\xi^2} + \frac{\partial^2 g}{\partial t^2} + 2g\frac{\partial^2 g}{\partial\xi\partial t}\right)_{\xi_k,t_k}\right].$$

Problem 3.3.3 Compare the Taylor expansion through terms of order $(\Delta t)^3$ for $\xi(t+\Delta t)$ to the expansion of the same order for a Runge–Kutta method of the form

$$\left.\begin{aligned}
\xi_{k+1} &= \xi_k + \Delta t\,(w_0 K_0 + w_1 K_1 + w_2 K_2)\ ,\\
K_0 &= g(\xi_k, t_k)\ ,\\
K_1 &= g(\xi_k + \beta\Delta t\,K_0, t_k + \alpha\,\Delta t)\ ,\\
K_2 &= g(\xi_k + \gamma\Delta t\,K_0 + \delta\Delta t\,K_1, t_k + (\gamma+\delta)\Delta t)\ ,
\end{aligned}\right\} \tag{3.3.10}$$

with parameters w_0, w_1, w_2, α, β, γ and δ. Here K_0 is the slope of ξ versus t at the starting point, t_k, and for appropriate values of the constants (α, β, γ, δ), we can interpret K_1 and K_2 as approximations to the slopes at points in between t_k and t_{k+1}. By tailoring the weights, w_0, w_1 and w_2, the mix of the three different slopes, $w_0K_0 + w_1K_1 + w_2K_2$ can be used to obtain an $O(\Delta t)^2$ accurate approximation to the exact slope.

Show that agreement through terms of order $(\Delta t)^3$ requires

$$w_0 + w_1 + w_2 = 1, \tag{3.3.11a}$$

$$\alpha\, w_1 + (\gamma+\delta) w_2 = \frac{1}{2}, \tag{3.3.11b}$$

$$\frac{1}{2}\left[w_1\,\alpha^2 + w_2(\gamma+\delta)^2\right]\left(g_{tt} + 2\,g\,g_{t\xi} + g^2 g_{\xi\xi}\right) + \alpha\,\delta\,w_2(g_t\,g_\xi + g\,g_\xi^2)$$
$$= \frac{1}{6}\left(g_{tt} + 2\,g\,g_{t\xi} + g^2 g_{\xi\xi} + g_t\,g_\xi + g\,g_\xi^2\right), \tag{3.3.11c}$$

where we have used the abbreviations g_ξ for $\partial g/\partial\xi$, g_t for $\partial g/\partial t$, etc. Hence, if (3.3.11a)–(3.3.11c) is to hold independently of the form of g, we must have

$$w_1\alpha^2 + w_2(\gamma+\delta)^2 = \frac{1}{3}\,, \qquad \alpha\delta w_2 = \frac{1}{6}\,, \tag{3.3.12}$$

in addition to (3.3.11a), (3.3.11b).

Consider the special case $\alpha = 1/2$, $(\gamma+\delta) = 1$. Show that the method

$$\left.\begin{aligned} \xi_{k+1} &= \xi_k + \frac{\Delta t}{6}\,[K_0 + 4\,K_1 + K_2]\;, \\ K_0 &= g(\xi_k, t_k)\;, \\ K_1 &= g(\xi_k + \frac{\Delta t}{2}\,K_0, t_k + \tfrac{1}{2}\,\Delta t)\;, \\ K_2 &= g(\xi_k - \Delta t\,K_0 + 2\Delta t\,K_1, t_k + \Delta t)\;, \end{aligned}\right\} \tag{3.3.13}$$

known as the **third-order Runge–Kutta method**, results.

The conventional way of writing out the general Runge–Kutta method is to take a time-step from $t_k = k\Delta t$ to $t_{k+1} = (k+1)\,\Delta t$ and advance the solution from $\xi_k = \xi(t_k)$ to $\xi_{k+1} = \xi(t_{k+1})$. Following (3.3.10) we get

$$\xi_{k+1} = \xi_k + \Delta t\,(w_0K_0 + w_1K_1 + \cdots + w_nK_n)\;, \tag{3.3.14}$$

where

$$\left.\begin{aligned} &K_0 = g(\xi_k, t_k)\;, \\ &K_1 = g(\xi_k + \beta_{1,0}\Delta t\,K_0, t_k + \alpha_1\,\Delta t)\;, \\ &K_2 = g(\xi_k + \beta_{2,0}\Delta t\,K_0 + \beta_{2,1}\Delta t\,K_1, t_k + \alpha_2\Delta t)\;, \\ &\quad\vdots \\ &K_n = g(\xi_k + \beta_{n,0}\Delta t\,K_0 + \beta_{n,1}\Delta t\,K_1 + \cdots + \beta_{n,n-1}\Delta t\,K_{n-1},\, t_k + \alpha_n\Delta t)\;. \end{aligned}\right\} \tag{3.3.15}$$

The numerical solution (3.3.14)–(3.3.15) obtained in this way is supposed to agree with the analytical solution through terms of order $(\Delta t)^{n+1}$. This is achieved by adjusting the constants $\alpha_1, \ldots, \alpha_n, \beta_{1,0}, \beta_{2,0}, \beta_{2,1}, \ldots, \beta_{n,0}, \beta_{n,1}, \ldots, \beta_{n,n-1}, w_0, \ldots, w_n$. The cases $n = 1$ and 2, which correspond to second- and third-order methods, were considered in detail above.

With all these adjustable parameters around one might wonder whether a method that is accurate through terms of order $(\Delta t)^M$ is always guaranteed for $n = M-1$, or even better can order $(\Delta t)^M$ accuracy be obtained for some value of n less than $M-1$. It would be an advantage to minimize the number of stages, n, required to obtain a certain order of accuracy, since each line of (3.3.15) corresponds to a new computation of the right-hand side in the differential equation, and it would be nice to keep that number as small as possible. In most problems, evaluation of g is the expensive part of the calculation. It turns out that for $n \le 3$ the constants in (3.3.14) and (3.3.15) can be adjusted to obtain an order of accuracy of $n+1$; for $n = 4$, 5 or 6 the best possible order of accuracy that can be achieved is only n, and for $n \geq 7$, the best possible order of accuracy further falls to $n-1$. Thus, $M = n+1 = 4$ is a natural breakpoint of the hierarchy.

The following **fourth-order Runge–Kutta** method

$$\begin{aligned}
\xi_{k+1} &= \xi_k + \frac{\Delta t}{6}\left[K_0 + 2K_1 + 2K_2 + K_3\right] , \\
K_0 &= g(\xi_k, t_k) , \\
K_1 &= g(\xi_k + \frac{\Delta t}{2} K_0, t_k + \frac{1}{2}\Delta t) , \\
K_2 &= g(\xi_k + \frac{\Delta t}{2} K_1, t_k + \frac{1}{2}\Delta t) , \\
K_3 &= g(\xi_k + K_2, t_k + \Delta t) ,
\end{aligned}$$

is widely used. It has the property that if g does not depend on ξ (an admittedly artificial case), it reduces to **Simpson's rule** for integrating the function $g(t)$ over the interval from t_k to t_{k+1} since in this case $K_1 = K_2$. (The same is true for the third-order Runge–Kutta method, (3.3.13).) Furthermore, it is a simple matter to see that a fourth-order Runge–Kutta method of step size Δt is better than a second-order Runge–Kutta method of step size $\Delta t/2$ or an Euler scheme of step size $\Delta t/4$. Although the work load in terms of the number of g evaluations is the same in all three cases, the accuracy of the fourth-order Runge–Kutta method will far outweigh the other two.

3.3.1 Adaptive Time-Stepping

It is useful to note the following property of the local truncation error of a Runge–Kutta method of order M. To leading order the local truncation error for a step of size Δt from t_k to $t_k + \Delta t$ is of the form $C_k(\Delta t)^{M+1}$. Now let us consider the following two numerical solutions both starting with ξ_k at t_k:

(1) we take one step of length Δt producing $\xi_{k+1}^{(\Delta t)}$;
(2) we take two steps, each of length $\Delta t/2$ producing $\xi_{k+2}^{(\Delta t/2)}$.

We want to compare these two results to $\xi(t_k + \Delta t)$, the true value of the solution after Δt. If we assume that C_k is independent of the step, i.e., independent of the subscript k, we get to leading order in Δt:

$$\left.\begin{aligned}
\xi(t_k + \Delta t) - \xi_{k+1}^{(\Delta t)} &= C\,(\Delta t)^{M+1} + O(\Delta t)^{M+2}, \\
\xi(t_k + \Delta t) - \xi_{k+2}^{(\Delta t/2)} &= 2\,C\left(\frac{\Delta t}{2}\right)^{M+1} + O(\Delta t)^{M+2}.
\end{aligned}\right\} \tag{3.3.16}$$

Hence,

$$C = \frac{\xi_{k+2}^{(\Delta t/2)} - \xi_{k+1}^{(\Delta t)}}{(1-2^{-M})\,(\Delta t)^{M+1}} + O(\Delta t) \tag{3.3.17}$$

and

$$\xi(t_k + \Delta t) = \xi_{k+1}^{(\Delta t)} + \frac{1}{(1-2^{-M})}\left[\xi_{k+2}^{(\Delta t/2)} - \xi_{k+1}^{(\Delta t)}\right] + O(\Delta t)^{M+2} . \tag{3.3.18}$$

The above formulae can be used in the following way: the value of C evaluated from (3.3.17) can be used in (3.3.16) to adjust Δt such that the estimate for the truncation error is below a certain desired tolerance. We can then use (3.3.18) to formally gain one extra order of accuracy. This approach goes by the name **Richardson extrapolation**. Richardson extrapolation is a very powerful tool, since it can be used in the context of any time integration scheme (without even knowing its inner workings) to obtain an extra order of accuracy. All one needs to know is the order of accuracy of the underlying scheme and take one step of size Δt, followed by two steps of size $\Delta t/2$. Note that (3.3.18) is a method of order $M+1$ and one can apply Richardson extrapolation to this. In other words, Richardson extrapolation can thus be applied recursively to improve the order of accuracy of an underlying time integration scheme to any desired level.

A somewhat rough-and-ready procedure, that is often used in practice, is to perform the numerical integration of the ODE with a fixed step size but then repeat the calculation with a smaller step size, say $\Delta t/2$. If the results agree with those of the coarser step size, some level of convergence has been ascertained. Although not very sophisticated, this kind of check defines a minimum standard. Unfortunately, even published research in reputable journals may be found that does not meet this modest requirement.

More sophisticated methods for implementing such adaptive step-size techniques are available and are discussed in books on numerical analysis, e.g. Press *et al.* (1986), Chapter 15, or Gear (1971). One popular adaptive-step-size method is the **Runge–Kutta–Fehlberg** method. Here, instead of comparing the truncation errors of one step of size Δt with two smaller steps of size $\Delta t/2$, truncation errors of fourth- and fifth-order Runge–Kutta methods are compared to evaluate the step size, Δt, required to maintain a certain level of local truncation error. Good implementations of such error control techniques can be found in many mathematical software packages.

Let us now consider the stability of the Runge–Kutta methods. The stability analysis follows the same lines as the Euler methods. The numerical solution of (3.2.19) using the second-order Runge–Kutta method can be written as

$$\xi_{k+1} = \left(1 + c\,\Delta t + \tfrac{1}{2}(c\,\Delta t)^2\right)\xi_k\,, \tag{3.3.19}$$

and for the fourth-order Runge–Kutta method (after some algebra) it can be shown that

$$\xi_{k+1} = \left(1 + c\,\Delta t + \tfrac{1}{2}(c\,\Delta t)^2 + \tfrac{1}{6}(c\,\Delta t)^3 + \tfrac{1}{24}(c\,\Delta t)^4\right)\xi_k\,. \tag{3.3.20}$$

The region of absolute stability in the complex plane is given by $|\xi_{k+1}/\xi_k| < 1$. Figure 3.3 shows the region of absolute stability for the second-, third- and fourth-order Runge–Kutta methods.

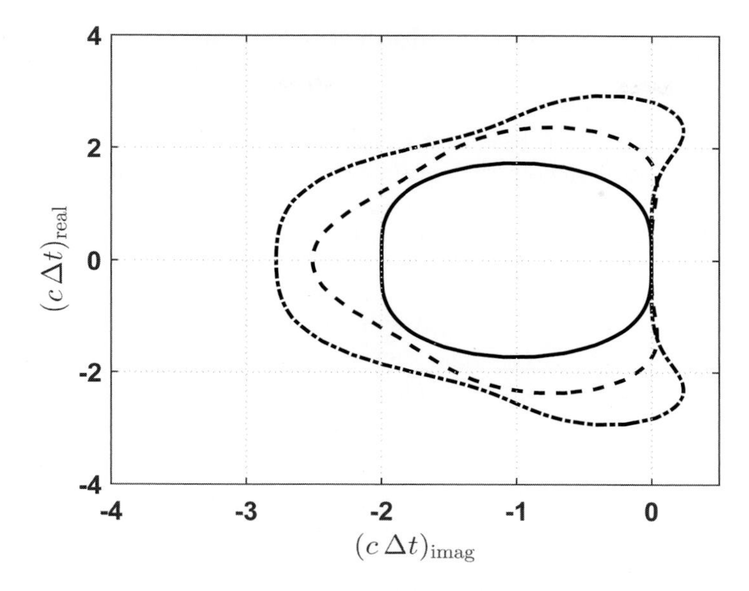

Figure 3.3 Stability diagrams for the RK family of schemes. Solid line: RK2 scheme; Dashed line: RK3 scheme; Dash–dot line: RK4 scheme. Note that the stability region increases with increasing order of the RK scheme. Also note that RK2 stability region does not include any portion of the imaginary axis and therefore will be unstable for a pure advection problem.

Problem 3.3.4 For the second- and fourth-order Runge–Kutta methods obtain (3.3.19) and (3.3.20), respectively.

Computer Exercise 3.3.5 Use a Runge–Kutta method to integrate the motion of an advected particle in the ABC-flow to produce plots such as the one shown in Figure 2.7. Once you are convinced of the soundness of your program, experiment with the values of A, B and C, and/or try one of the other Beltrami flows treated in Problem 2.7.3. A problem of principle arises in determining points of intersection of a pathline with a particular plane. This can be solved by interpolation. A more sophisticated yet easily implementable solution is given by Hénon (1982). In this particular case an independent method of accuracy verification is possible. It is easily seen that if A, B and C are replaced by −A, −B and −C, flow along the same path but in the opposite direction is produced. Thus, after integrating for N steps, the computation can be stopped and N "backward" steps can be taken, and the results compared to the initial point. While a true pathline is reversible, any numerical deviation from the true pathline may not be reversible. In this way one can determine whether a true pathline of the system is being followed or not.

3.4 Adams–Bashforth–Moulton Methods

The methods to be discussed in this section are also known as **multistep methods**. They are based on rather different ideas from the Runge–Kutta methods. We first note that given any set of $n+1$ distinct data points (t_{k-j}, g_{k-j}), $j = 0, \ldots, n$, there exists exactly one polynomial, P_n, of degree n such that $P_n(t_{k-j}) = g_{k-j}$. This is intuitive for $n = 1$ or 2 but for larger n needs a formal proof, which we give now.

Proof (a) *Uniqueness*: Assume there were two such polynomials P_n and Q_n. The difference, $P_n - Q_n$, then would be a polynomial of degree no more than n with $n+1$ roots, i.e., identically 0.

(b) *Existence*: This can be established by using an ingenious construction attributed to Lagrange. For $m = 0, \ldots, n$ define the polynomials

$$D_m(t) = \prod_{j=0}^{n}{}' \frac{t - t_{k-j}}{t_{k-m} - t_{k-j}}, \tag{3.4.1}$$

where the prime on $\prod$ signifies that $j \neq m$. These are $n+1$ distinct polynomials, each of degree n. The polynomials D_m have the important property

$$D_m(t_{k-j}) = \delta_{mj}, \tag{3.4.2}$$

where δ_{mj} is the usual *Kronecker delta*, equal to 1 if $m = j$ and 0 otherwise. That is, the polynomial, D_m, is collocated through all t_{k-j}, $j = 0, 1, \ldots, n$ and D_m is 1 at t_{k-m} and 0 at all other t_{k-j}. Thus, the polynomial

$$P_n(t) = \sum_{m=0}^{n} g_{k-m} D_m(t) \tag{3.4.3}$$

satisfies the conditions. It is known as the **Lagrange interpolating polynomial** (see Figure 3.4 for a schematic of the Lagrange interpolating polynomial for the case of $n = 2$). □

We make use of this result in the following way: suppose we have integrated $d\xi/dt = g(\xi, t)$ numerically for a while, and we have solution values ξ_m at a number of instants t_m. We have advanced to some time t_k and we have kept the solution values at the earlier times t_{k-m} for $m = 0, 1, \ldots, n$. Using these solution values ξ_{k-m}, $m = 0, 1, \ldots, n$, we now calculate the corresponding values of g_{k-m}, $m = 0, 1, \ldots, n$, and from them we construct the Lagrange interpolating polynomial of degree n that takes on the value g_{k-m} at $t = t_{k-m}$, $m = 0, 1, \ldots, n$.

Now we evaluate the solution at the next time t_{k+1} by using in the exact integral (3.2.12) the interpolating polynomial in place of the actual g. Formally this predicted value is then given by

$$\xi_{k+1} = \xi_k + \int_{t_k}^{t_{k+1}} \sum_{m=0}^{n} g_{k-m} D_m(\tau)\, d\tau = \xi_k + \sum_{m=0}^{n} g_{k-m} \int_{t_k}^{t_{k+1}} D_m(\tau)\, d\tau . \tag{3.4.4}$$

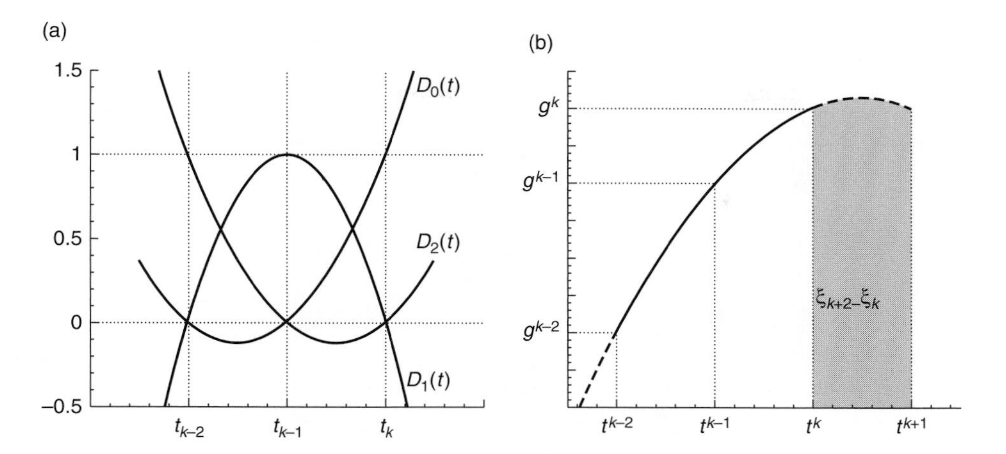

Figure 3.4 (a) The collocating polynomials, $D_0(t)$, $D_1(t)$ and $D_2(t)$, for the case of $n = 2$. (b) The Lagrange interpolating polynomial interpolated through $g_k = g(t_k)$, $g_{k-1} = g(t_{k-1})$ and $g_{k-2} = g(t_{k-2})$ has been extrapolated from t_k to t_{k+1}. The area under the curve, $\xi_{k+1} - \xi_k = \int_{t_k}^{t_{k+1}} g(\tau)\,\mathrm{d}\tau$, can then be evaluated using the extrapolation.

The point is that all the integrals over the polynomials $D_m(\tau)$ can be worked out ahead of time in terms of the breakpoints t_{k-m} for $m = 0, 1, \ldots, n$. This gives the **Adams–Bashforth** family of methods. For $n = 0$, the interpolation polynomial collocates through only one point and therefore it is a constant: $D_0(\tau) = 1$. Thus for $n = 0$, the **one-step Adams–Bashforth method** reduces to the explicit Euler scheme

$$\xi_{k+1} = \xi_k + \Delta t\, g_k \,. \tag{3.4.5}$$

For $n = 1$, the polynomials D_0 and D_1 are collocated at t_k and t_{k-1}. Both of them are straight lines and according to (3.4.2) take the following values:

$$D_0(t_k) = 1, \quad D_0(t_{k-1}) = 0, \quad D_1(t_k) = 0, \quad D_1(t_{k-1}) = 1.$$

Integrating using (3.4.4) yields the following **two-step Adams–Bashforth method**

$$\xi_{k+1} = \xi_k + \Delta t \left[\tfrac{3}{2} g_k - \tfrac{1}{2} g_{k-1}\right] . \tag{3.4.6}$$

Taylor expanding g_{k-1} about t_k on the right-hand side, and comparing with the result of exact integration (3.2.14), it is a simple matter to show that the local truncation error for the two-step Adams–Bashforth method is $\frac{5}{12}(\Delta t)^2 \frac{\mathrm{d}^3\xi}{\mathrm{d}t^3}$. This process can be continued to obtain higher-order Adams–Bashforth methods. Since, in general, information is required not only at t_k but also at previous time-steps in order to obtain ξ_{k+1}, these are collectively known as **multistep methods**. By contrast, the Runge–Kutta methods falls in the category of **one-step methods**. Table 3.1 provides a list of multistep Adams–Bashforth (AB) methods and their local truncation error.

Table 3.1 Multistep Adams–Bashforth schemes and their local truncation error.

n	*Name*	*Scheme*	*Local truncation error*
0	One-step AB scheme	$\xi_{k+1} = \xi_k + \Delta t\, g_k$	$\frac{1}{2}(\Delta t)\frac{\mathrm{d}^2\xi}{\mathrm{d}t^2}$
1	Two-step AB scheme	$\xi_{k+1} = \xi_k + \Delta t\left[\frac{3}{2}g_k - \frac{1}{2}g_{k-1}\right]$	$\frac{5}{12}(\Delta t)^2\frac{\mathrm{d}^3\xi}{\mathrm{d}t^3}$
2	Three-step AB scheme	$\xi_{k+1} = \xi_k + \frac{\Delta t}{12}\left[23g_k - 16g_{k-1} + 5g_{k-2}\right]$	$\frac{3}{8}(\Delta t)^3\frac{\mathrm{d}^4\xi}{\mathrm{d}t^4}$
3	Four-step AB scheme	$\xi_{k+1} = \xi_k + \frac{\Delta t}{24}\left[55g_k - 59g_{k-1} + 37g_{k-2} - 9g_{k-3}\right]$	$\frac{251}{720}(\Delta t)^4\frac{\mathrm{d}^5\xi}{\mathrm{d}t^5}$

Problem 3.4.1 Obtain the truncation error for the first few multistep Adams–Bashforth methods.

Another class of multistep method, known as the **Adams–Moulton method**, is implicit in nature. This method uses the $n+2$ times $t_{k-n}, \ldots, t_k$ and t_{k+1}, and constructs an $(n+1)$th-order Lagrange interpolating polynomial, P_{n+1}. We substitute this into a formula of the type (3.4.4) to numerically evaluate ξ_{k+1} as:

$$\xi_{k+1} = \xi_k + \int_{t_k}^{t_{k+1}} P_{n+1}(\tau)\,\mathrm{d}\tau\,. \tag{3.4.7}$$

For $n = 0$, the Lagrange interpolating polynomial, P_1, is simply a straight line passing through g_k at t_k and g_{k+1} at t_{k+1}. Upon integration we get the following one-step Adams–Moulton method:

$$\xi_{k+1} = \xi_k + \tfrac{1}{2}\Delta t\,\left(g_k + g_{k+1}\right)\,. \tag{3.4.8}$$

This popular method is also known as the **Crank–Nicolson method**. It is semi-implicit, since it depends on g at both the current and the next time instant. For $n = 1$, the Lagrange interpolating polynomial is a parabola passing through g_{k-1} at t_{k-1}, g_k at t_k and g_{k+1} at t_{k+1}. The corresponding two-step and higher-order Adams–Moulton methods are presented in Table 3.2 along with their local truncation errors.

Problem 3.4.2 Obtain the truncation errors for the Crank–Nicolson scheme and the two-step Adams–Moulton method.

From the local truncation errors presented in Tables 3.1 and 3.2 it is clear

Table 3.2 Multistep Adams–Moulton schemes and their local truncation error.

n	*Name*	*Scheme*	*Local Truncation Error*
0	Crank–Nicolson scheme	$\xi_{k+1} = \xi_k + \frac{\Delta t}{2}\,[g_{k+1} + g_k]$	$-\frac{1}{12}(\Delta t)^2 \frac{\mathrm{d}^3\xi}{\mathrm{d}t^3}$
1	Two-step AM scheme	$\xi_{k+1} = \xi_k + \frac{\Delta t}{12}\,[5g_{k+1} + 8g_k - g_{k-1}]$	$-\frac{1}{24}(\Delta t)^3 \frac{\mathrm{d}^4\xi}{\mathrm{d}t^4}$
2	Three-step AM scheme	$\xi_{k+1} = \xi_k + \frac{\Delta t}{24}\,[9g_{k+1} + 19g_k - 5g_{k-1} + g_{k-2}]$	$-\frac{19}{720}(\Delta t)^4 \frac{\mathrm{d}^5\xi}{\mathrm{d}t^5}$
3	Four-step AM scheme	$\xi_{k+1} = \xi_k + \frac{\Delta t}{720}\,[251g_{k+1} + 646g_k - 264g_{k-1} + 106g_{k-2} - 19g_{k-3}]$	$-\frac{3}{160}(\Delta t)^5 \frac{\mathrm{d}^6\xi}{\mathrm{d}t^6}$

that for the same order of accuracy the implicit Adams–Moulton methods are superior to their explicit Adams–Bashforth counterparts. For example, the local truncation error of the two-step Adams–Moulton method is about nine times smaller than that of three-step Adams–Bashforth method. Both these methods formally have third-order accuracy and require the value of g at three time instances. In spite of its improved accuracy, the direct application of the Adams–Moulton methods is limited to only very simple problems, owing to its implicit nature.

3.4.1 Predictor–Corrector Methods

In a more general problem, the Adams–Moulton family of methods finds application in the context of **predictor–corrector methods**. Here, ξ_{k+1} is first *predicted* with an explicit method, such as one of the Adams–Bashforth family of methods given in Table 3.1. Once we have the predicted ξ_{k+1}, the value of g_{k+1} can be evaluated based on this predicted value. We then *correct* the value of ξ_{k+1} according to one of the Adams–Moulton family of methods given in Table 3.2. Typically, if an n-step Adams–Bashforth method is used as the predictor, an n-step Adams–Moulton method is used as the corrector. Then, a step in the predictor–corrector method consists of:

(i) the predictor step, symbolically designated **P**; followed by
(ii) an evaluation step, **E** (the new value of g has to be added to the list); then
(iii) the corrector step, **C**; and
(iv) another evaluation step, **E**.

Because of this pattern the name **PECE formulas** can be found in the literature.

You might think of simply doing several "**P** steps" in succession. That is not a good idea, since the prediction is the most unstable part of the entire procedure. You could also think of doing the **EC** part over and over. That is sometimes done, leading to methods that we could write as $\mathbf{P}(\mathbf{EC})^n\mathbf{E}$, where n is the number of times **EC** is repeated. However, the advantages of using $n > 1$ are not necessarily great enough to merit the additional work. For more discussion see the texts on numerical analysis cited previously. A detailed exposition of predictor–corrector methods with discussion of the performance of a well-known package is given by Shampine and Gordon (1975).

As a historical aside let us mention that the predictor component of predictor–corrector methods is associated with the names of F. Bashforth and J.C. Adams, who did their work in the context of a fluid mechanical problem. Their 1883 report was entitled *An Attempt to Test the Theories of Capillary Action by Comparing the Theoretical and Measured Forms of Drops of Fluid, with an Explanation of the Method of Integration Employed in Constructing the Tables which give the Theoretical Forms of such Drops.* Thus, in both the classical methods for integrating ODEs, the Runge–Kutta and Adams–Bashforth methods, fluid mechanics and fluid mechanicians have been intimately involved!

Problem 3.4.3 Determine the region of absolute stability of the Adams–Bashforth, Adams–Moulton and predictor–corrector family of multistep methods. *Hint*: Consider the numerical solution of (3.2.19). Assume that $\xi_{k+1}/\xi_k = \xi_k/\xi_{k-1} = \cdots = r$ and obtain an algebraic equation for r. Stability of the numerical method is guaranteed if $|r| < 1$ for all the roots of the algebraic equation. The stability diagrams for the Adams–Bashforth and Adams–Moulton family of methods are shown in Figures 3.5 and 3.6 respectively.

Figure 3.6 and a comparison of Figures 3.1(b) and 3.1(c) leads to a general conclusion that the region of absolute stability for an implicit scheme extends over a larger portion of the left half plane than that of the corresponding explicit scheme. This allows a larger time step to be chosen with the implicit scheme – the main reason for the popularity of implicit schemes. However three factors must be taken into consideration. Implicit schemes require the solution of an implicit equation as in (3.4.8) (also see Table 3.2), which may not be computationally easy except for very simple $g(\xi, t)$. For more complex problems one must settle for explicit schemes in order to avoid solving implicit equations. Secondly, even if an implicit scheme can be implemented, the choice of Δt may be restricted from accuracy considerations rather than from stability considerations. Finally, there are a number of problems in fluid mechanics, where the solution of interest (or the exact solution) is unstable and growing. In which case, one may face the strange possibility of numerically stabilizing the solution with a large Δt by using an implicit scheme.

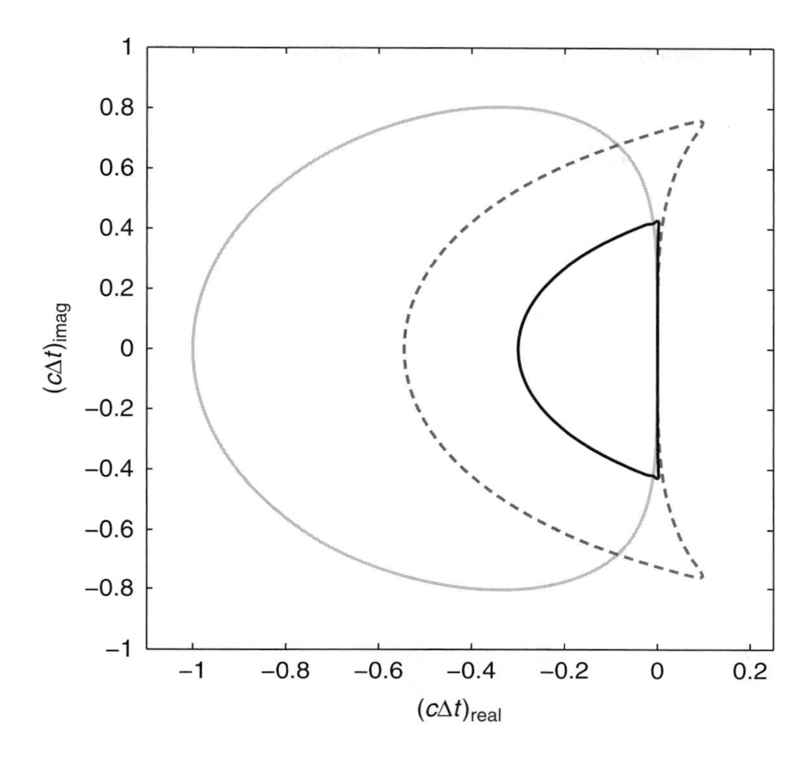

Figure 3.5 Stability diagrams for the Adams–Bashforth (AB) family of schemes. Gray line: AB 2-step scheme; dashed line: AB 3-step scheme; black line: AB 4-step scheme. Note that the AB 1-step scheme is the same as the explicit Euler scheme and the stability region decreases with increasing order of the scheme. Also note that the AB 2-step stability region does not include any portion of the imaginary axis and therefore will be unstable for a pure advection problem.

We conclude with some remarks on the comparison between Runge–Kutta (RK) methods and the Adams–Bashforth–Moulton predictor–corrector (PC) methods just described. RK methods are self-starting in the sense that you give the same type of data that you would give to the analytical initial value problem, and RK method begins integrating. PC methods are not self-starting. You need to generate some points in a solution before the PC method can work. Thus, in a numerical integration package the PC method will typically have an RK option to use on start-up.

For order M an RK method *requires at least M evaluations of the derivatives.* The PC method in the form PECE *requires only two, regardless of order.* The standard rule of thumb is to use PC methods when the evaluation of g is expensive. RK methods are fine when g is cheap to compute. The small number of g evaluations in a PC method comes, of course, at a price in *memory.* You have to keep as many previous data values as needed to achieve the desired order. While this may not be an issue while integrating a single ODE or a system of a few ODEs, memory can become an issue when PC methods are extended to the very large-scale solution of PDEs involving billions of grid points.

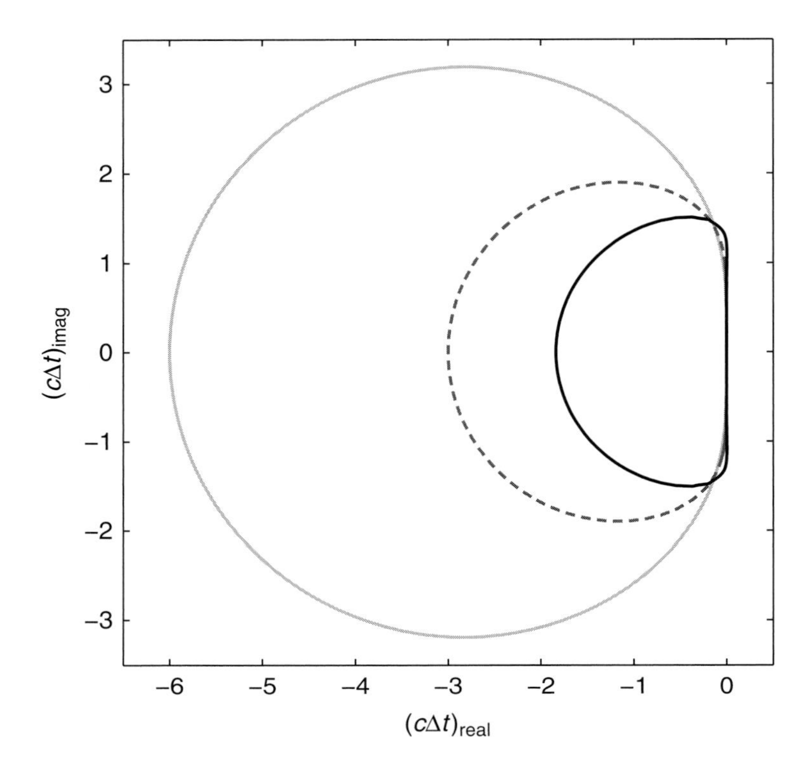

Figure 3.6 Stability diagrams for the Adams–Moulton (AM) family of schemes. Gray line: AM 2-step scheme; dashed line: AM 3-step scheme; black line: AM 4-step scheme. Note that the AM 1-step scheme is commonly known as the Crank–Nicolson scheme and it is stable over the entire left half of the complex plane and the stability region decreases with increasing order of the AM scheme. Also note that the AM 2-step and 3-step schemes' stability regions do not include any portion of the imaginary axis and therefore are unstable for a pure advection problem.

3.5 Other Methods and Considerations

All the methods considered so far for the numerical integration of $\mathrm{d}\xi/dt = g(\xi, t)$ involve relating the value of ξ at t_{k+1} to the value of ξ at t_k at each time step. In all these methods $\mathrm{d}\xi/dt$ is approximated by $\{(\xi_{k+1} - \xi_k)/\Delta t\}$ and the right-hand side is approximated either by repeated evaluation of g as in RK methods or by interpolation through multiple time levels of g as in PC methods. Other methods can be obtained by changing the approximation to the time derivative. The simplest and the most popular is the **leap-frog scheme**, which can be written as:

$$\frac{\xi_{k+1} - \xi_{k-1}}{2\Delta t} = g(\xi_k, t_k)\,. \tag{3.5.1}$$

In terms of computational effort, this scheme is just about as efficient as the explicit Euler scheme, but due to the central differencing of the leap-frog scheme it is second-order accurate. While this may be a very desirable property in terms of computational efficiency, the region of stability of the leap-frog scheme covers no area: it extends only between $+1$ and -1 along the imaginary axis in the complex $c\Delta t$-plane!

Along this line we obtain a class of implicit integration schemes called **Backward Differentiation** (BD) methods, which in general can be written as

$$\frac{\alpha_{p,0}\,\xi_{k+1} + \alpha_{p,1}\,\xi_k + \alpha_{p,2}\,\xi_{k-1} + \cdots + \alpha_{p,p}\,\xi_{k-p+1}}{\Delta t} = g(\xi_{k+1}, t_{k+1})\,. \quad (3.5.2)$$

Here p is the order of accuracy of the scheme and the left-hand side provides an accurate representation of the time derivative. For $p = 1$ one obtains the implicit Euler scheme and for $p = 2$ one obtains a second-order backward differentiation method with $\alpha_{2,0} = \frac{3}{2}$, $\alpha_{2,1} = -2$ and $\alpha_{2,2} = \frac{1}{2}$. Owing to their implicit nature these methods find limited application, but their real advantage is that like the implicit Euler scheme, the region of absolute stability for the above BD methods for $p \leq 6$ is outside of a closed region, which is primarily on the right half of the complex $c\Delta t$-plane. For $p \geq 7$ the BD methods are unconditionally unstable.

For a system of N first-order equations, ξ and g in (3.2.1) are replaced by their N-component vector equivalents $\boldsymbol{\xi}$ and $\mathbf{g}$. Furthermore $\mathbf{g}$ has $N+1$ arguments, the N components of $\boldsymbol{\xi}$ and time. In the implementation of Runge–Kutta methods, in (3.3.14) and (3.3.15) the quantities $K_0, \ldots, K_n$ are now N-component vectors. The constants $\alpha_1, \ldots, \alpha_n, \beta_{1,0}, \beta_{2,0}, \beta_{2,1}, \ldots, \beta_{n,0}, \beta_{n,1}, \ldots, \beta_{n,n-1}$, $w_0, \ldots, w_n$ multiply these entire vectors. Similarly, extension of other integration schemes to a system of ODEs is straightforward. The order of accuracy remains the same, but the stability property deserves some attention. While the stability diagrams for the different integration schemes remain the same, the appropriate value of c in (3.2.19) for a system is not straightforward. Stability of the system is controlled by

$$\frac{\mathrm{d}\boldsymbol{\xi}}{\mathrm{d}t} = \mathbf{c}\boldsymbol{\xi}\,, \quad (3.5.3)$$

where the matrix $\mathbf{c}$ is the Jacobian $\partial g_i/\partial \xi_j$. Under the transformation $\boldsymbol{\psi} = \mathbf{M}^{-1}\boldsymbol{\xi}$, the above coupled equations for the N components of $\boldsymbol{\xi}$ reduce to N decoupled equations of the form

$$\frac{\mathrm{d}\psi_i}{\mathrm{d}t} = \lambda_i\,\psi_i \qquad \text{for} \qquad i = 1, \ldots, N\,, \quad (3.5.4)$$

where the λs are the N eigenvalues of $\mathbf{c}$; the corresponding matrix formed by the N eigenvectors is $\mathbf{M}$ and its inverse is $\mathbf{M}^{-1}$. Stability of the system is thus guaranteed if $\lambda_i\,\Delta t$ is within the region of absolute stability for all $i = 1, 2, \ldots, N$.

In Example 3.2.1

$$\mathbf{c} = \begin{bmatrix} 0 & -\Omega \\ \Omega & 0 \end{bmatrix}, \quad (3.5.5)$$

and the two eigenvalues are $\pm\iota\,\Omega$. The two eigenvalues multiplied by Δt are outside the region of absolute stability of the explicit Euler scheme, but are inside the stability region of the implicit Euler scheme. This explains the unstable outward spiraling path of particles for explicit Euler scheme and the stable inward spiral for the implicit Euler scheme. Furthermore, second-order RK and AB methods are unsuitable for this problem since their region of absolute stability does not include any segment of the imaginary axis. On the other hand, third- and fourth-order RK and AB methods guarantee stability with a suitable choice of Δt such that $\pm \iota\, \Omega\, \Delta t$ lies within their respective regions of stability.

Problem 3.5.1 Show that the leap-frog scheme for (3.2.4) is not only stable but has zero amplitude error. Show that the leading-order term in the phase error is $\frac{1}{6}(\Omega\, \Delta t)^3$ consistent with the second-order accuracy of the leap-frog scheme.

There are many further exciting numerical analysis aspects of integrating ODEs. The methods we have described are the workhorses of day-to-day computations. We shall not pursue the topic further, but turn our attention to problems of fluid mechanics that can be reduced to ODEs, and that require numerical solution. As we shall see there is considerable variety, and a rich array of physical situations.

3.6 Bashforth's Problem: Sessile Drop on a Flat Plate

For our first example consider the problem that led to the introduction of the Adams–Bashforth part of the predictor–corrector methods just discussed: the determination of the shape of a fluid drop resting on a horizontal flat plate that it does not wet. This is a non-trivial problem in fluid statics that leads to a second-order ODE for the drop shape.

Let the plane of the plate be parallel to the xy-plane of a Cartesian xyz system. Let the axis of z be along the vertical, directed downwards, and let the origin of coordinates be chosen coincident with the highest point of the drop (see Figure 3.7). If R_1 and R_2 denote the two principal radii of curvature of the drop, which vary from point to point along the drop surface, then the Young–Laplace law (see Landau and Lifshitz (1987), §61) states that the pressure just inside the drop is larger than the atmospheric pressure outside by an amount

$$\Delta p = \sigma \left(\frac{1}{R_1} + \frac{1}{R_2} \right) , \tag{3.6.1}$$

where σ is the surface tension coefficient. We shall be concerned with an axisymmetric drop, so $R_1 = R_2 = R$.

On the other hand, from hydrostatics the difference in pressure, Δp, depends on the z-coordinate as

$$\Delta p(z) = \Delta p(0) + \Delta \rho\, g\, z \, , \tag{3.6.2}$$

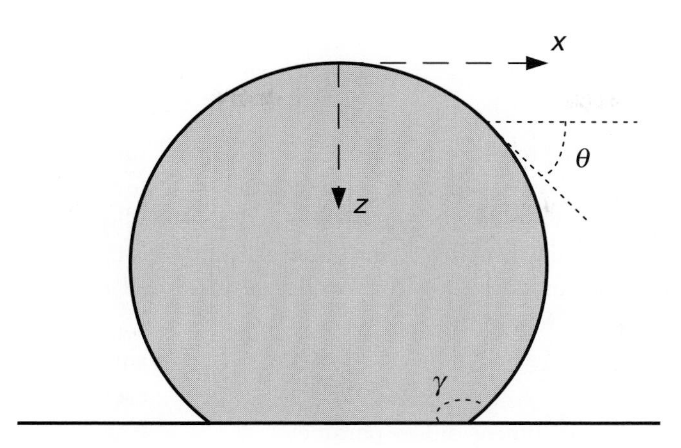

Figure 3.7 Schematic of a fluid droplet resting on a flat plate. The contact angle is denoted by γ.

where $\Delta\rho = \rho_{\text{liq}} - \rho_{\text{gas}}$ is the difference in densities between the liquid (inside the drop) and the gas (surrounding it), and g is the acceleration of gravity. The pressure jump $\Delta p(0)$ at the apex of the drop equals $2\sigma/b$, where b is the common value of the two principal radii of curvature at the apex of the axisymmetrical drop surface. Note that in (3.6.2) $z = 0$ corresponds to the apex of the drop.

Introducing the angle θ between the tangent to the drop surface and the x-axis (see Figure 3.7) we have by simple geometry

$$\frac{\mathrm{d}x}{\mathrm{d}s} = \cos\theta\,; \qquad \frac{\mathrm{d}z}{\mathrm{d}s} = \sin\theta\,; \qquad \frac{\mathrm{d}\theta}{\mathrm{d}s} = \frac{1}{R}\,; \qquad x = R\sin\theta\,, \tag{3.6.3}$$

where s is the arclength along the drop profile. (For additional background on the differential geometry of a surface of revolution see, e.g., Struik, 1961.)

Combining (3.6.3), (3.6.2) and (3.6.1) we now have the following system of three first-order ODEs for determining the drop shape:

$$\frac{\mathrm{d}x}{\mathrm{d}s} = \cos\theta\,, \tag{3.6.4a}$$

$$\frac{\mathrm{d}z}{\mathrm{d}s} = \sin\theta\,, \tag{3.6.4b}$$

$$\frac{\mathrm{d}\theta}{\mathrm{d}s} = \frac{2}{b} + \frac{g\,\Delta\rho}{\sigma}z - \frac{\sin\theta}{x}\,. \tag{3.6.4c}$$

This is probably the most transparent form of the determining equations. However, it is possible to condense these three equations to one second-order equation (not third order as one might have thought). This can be done in a variety of ways. For example, division of (3.6.4c) by (3.6.4a) gives the result

$$\frac{\mathrm{d}\sin\theta}{\mathrm{d}x} = \frac{2}{b} + \frac{g\,\Delta\rho}{\sigma}z - \frac{\sin\theta}{x}\,, \tag{3.6.5}$$

or

$$\frac{1}{x}\frac{\mathrm{d}(x\sin\theta)}{\mathrm{d}x} = \frac{2}{b} + \frac{g\,\Delta\rho}{\sigma}z\,. \tag{3.6.6}$$

Taking one more derivative with respect to x, and using that $\mathrm{d}z/\mathrm{d}x = \tan\theta$ from (3.6.3), gives a nonlinear, second-order ODE for $\sin\theta$.

We may non-dimensionalize equations (3.6.4a)–(3.6.4c) by choosing b as our unit of length. This changes nothing in (3.6.4a), (3.6.4b) (assuming that we use the same symbols for the dimensional and non-dimensional variables) but replaces (3.6.4c) by

$$\frac{\mathrm{d}\theta}{\mathrm{d}x} = 2 + B\,z - \frac{\sin\theta}{x}\,, \tag{3.6.7}$$

where the dimensionless parameter

$$B = \frac{g\,\Delta\rho\,b^2}{\sigma} \tag{3.6.8}$$

is known as the *Bond number.*

The system (3.6.4a)–(3.6.4c) is to be integrated starting from $s = 0$ where x, z and θ all vanish. If we take the limit $s \to 0+$ in (3.6.3), we see that $\mathrm{d}\theta/\mathrm{d}s$ and $(\sin\theta)/x$ both tend to $1/b$. Hence, the limiting form of (3.6.7) as $s \to 0+$ is to be read as $1 = 2 + 0 - 1$. We continue the integration until the angle θ reaches the contact angle of the drop surface with the flat plate (γ in the notation of Figure 3.7). This determines the entire solution in terms of the parameters B and γ.

The original report (Bashforth and Adams, 1883) may be consulted for historical interest. There is, however, a more modern and much more comprehensive collection of formulas, tables and even computer programs for this and a series of related fluid–fluid interface problems. The reference work Hartland and Hartley (1976) discusses sessile and pendant drops in a large variety of cases.

There is a rich mathematical literature on the type of problem we have considered. We can compute the drop volume and height as functions of the size of the wetted area. It turns out that there is a simple correspondence between sessile drops and capillary rise in a tube. The problem of pendant drops is particularly rich and the equations for this case admit solutions, first found by Kelvin more than a century ago, with multiple bulges. We cannot pursue all this material here but refer the interested reader to the elegant monograph by Finn (1986) in which many further references may be found.

Computer Exercise 3.6.1 Write a program to integrate equations (3.6.4a)–(3.6.4c) and produce a solution accurate to four digits using the value $\gamma = \pi/2$ for the contact angle and $B = 0.01$, 0.1 and 1.0. Plot these three drop profiles. Explore aspects of the numerical solution, such as drop height, drop volume, and the wetted area on the plate in terms of B and γ.

Problem 3.6.2 The case of a 2D sessile drop (a long cylinder of fluid, of uniform cross-section, lying on a solid surface) is simpler since $R_2 = \infty$. (The expression for x in (3.6.3) is really the expression for the *radial* distance of the drop surface from the axis, and this relation is not meaningful in the 2D case.) Write out the differential equations for the 2D sessile drop, and show that a second-order ODE, similar to the equation for the pendulum, arises for θ. Find an analytical expression for the height of the 2D drop in terms of B and γ.

3.7 Flow Due to a Collapsing Bubble

This is a charming chapter of fluid mechanics with a surprisingly extended literature and a host of applications. In particular, the solution to be described here for a spherically symmetric bubble is the idealized problem to which one turns initially in theories of cavitation (Knapp *et al.*, 1979) and nucleation. The problem was formulated in 1859 by Besant in the following way:

> An infinite mass of homogeneous, incompressible fluid, acted upon by no forces, is at rest, and a spherical portion of the fluid is suddenly annihilated; it is required to find the instantaneous alteration of pressure at any point of the mass, and the time in which the cavity will be filled up, the pressure at an infinite distance being supposed to remain constant.

In this form the problem was solved by Lord Rayleigh in 1917, a solution that may be found in many textbooks on fluid mechanics (e.g., Lamb, 1932, §91a; Batchelor, 1967, p. 479; Landau and Lifshitz, 1987, §10, Problem 7) and that we shall write out in Section 3.7.1 below. Rayleigh found the time T for the cavity to fill up, the *collapse time*, to be given by

$$T \approx 0.915\, R_0 \sqrt{\frac{\rho}{p_\infty}}. \tag{3.7.1}$$

Here R_0 is the initial radius of the cavity, ρ is the fluid density, and p_∞ is the constant pressure at infinity. The numerical prefactor, given approximately in (3.7.1), can be expressed in terms of the Γ-function (see Section 3.7.2).

We shall refer to the substance inside the bubble as the *gas*. The surrounding fluid medium will be referred to as the *liquid*. The spherical interface between gas and liquid will be called the *bubble wall*. The bubble does not move, and always has its center at the origin of coordinates. There is no flow inside the bubble, and at any instant the pressure is the same everywhere within it. However, this spatially uniform pressure may change in time.

3.7.1 Constant Pressure Inside the Bubble

We assume the liquid is incompressible. From this and the spherical symmetry of the problem we obtain a purely radial flow with a radial component of velocity, u, that by mass conservation must satisfy

$$u(r,t)\,r^2 = U(t)\,[R(t)]^2\,. \tag{3.7.2}$$

The right-hand side is simply the value of the left-hand side at the bubble wall. Note that the right-hand side of (3.7.2) may also be written as $\dot{R}\,R^2$. At $r = R$ we have $u = U$ as must be true by continuity of the velocity.

For now we take the governing dynamical law to be Euler's equation for inviscid fluid:

$$\frac{\partial \mathbf{V}}{\partial t} + (\mathbf{V}\cdot\nabla)\mathbf{V} = -\frac{\nabla p}{\rho}\,. \tag{3.7.3}$$

Again invoking the spherical symmetry of the problem only the radial component of this equation is of significance. Thus we get

$$\frac{\partial u}{\partial t} + u\frac{\partial u}{\partial r} = -\frac{1}{\rho}\frac{\partial p}{\partial r}\,. \tag{3.7.4}$$

On the left-hand side of this equation we substitute the form (3.7.2) for u. This gives

$$\frac{\ddot{R}\,R^2}{r^2} + 2\frac{R\,\dot{R}^2}{r^2} - 2\frac{\dot{R}^2 R^4}{r^5} = -\frac{1}{\rho}\frac{\partial p}{\partial r}\,. \tag{3.7.5}$$

The terms on the left correspond to those in (3.7.4) after substituting (3.7.2) and carrying out the designated derivatives. Also, we have consistently written $\dot{R}$ for U and $\ddot{R}$ for $\dot{U}$. Note that according to (3.7.5) the left-hand side of (3.7.4) evaluated at $r = R$ is simply $\ddot{R}$, as one would expect.

We now integrate (3.7.5) at a fixed time from $r = R$, the instantaneous radius of the bubble, to $r = \infty$. This is easy to do on the left-hand side since only simple powers of r appear. On the right-hand side we use the boundary conditions on the pressure, that $p = p_0$ at $r = R$ (constant pressure inside the bubble; the pressure is continuous from liquid to gas since we are ignoring forces due to surface tension or fluid viscosity) and $p = p_\infty$ at $r = \infty$ (constant pressure at infinity). The integration then gives

$$(\ddot{R}\,R^2 + 2R\,\dot{R}^2)\left[\frac{-1}{r}\right]_{r=R}^{r=\infty} + \frac{1}{2}\dot{R}^2R^4\left[\frac{1}{r^4}\right]_{r=R}^{r=\infty} = -\frac{p_\infty - p_0}{\rho}\,, \tag{3.7.6}$$

or, completing the evaluation of the left-hand side,

$$R\,\ddot{R} + \frac{3}{2}\dot{R}^2 = -\frac{p_\infty - p_0}{\rho}\,. \tag{3.7.7}$$

This is *Rayleigh's equation* in its simplest version. It is to be integrated subject to the initial conditions $R = R_0$, $\dot{R} = 0$ at $t = 0$.

Problem 3.7.1 Assume $p_\infty > p_0$, i.e., that the bubble collapses, and introduce non-dimensional variables by scaling velocities by $\sqrt{(p_\infty - p_0)/\rho}$ and lengths by the initial bubble size R_0. What is the resulting scaling for non-dimensional time? Show that the equation

$$\beta\,\ddot{\beta} + \frac{3}{2}\dot{\beta}^2 + 1 = 0 \tag{3.7.8}$$

is obtained for $\beta = R/R_0$. Here dots indicate differentiation with respect to non-dimensional time τ. What are the initial conditions in terms of the non-dimensional variables?

3.7.2 Solution by Analogy to an Oscillator Problem

In order to solve (3.7.8) it is useful to note the substitution

$$x = \beta^{5/2}\,. \tag{3.7.9}$$

An easy calculation then shows that

$$\left.\begin{aligned} \dot{x} &= \frac{5}{2}\,\beta^{3/2}\,\dot{\beta}\,, \\ \ddot{x} &= \frac{5}{2}\,\beta^{1/2}\left(\beta\,\ddot{\beta} + \frac{3}{2}\dot{\beta}^2\right) = \frac{5}{2}x^{1/5}\left(\beta\,\ddot{\beta} + \frac{3}{2}\dot{\beta}^2\right)\,. \end{aligned}\right\} \tag{3.7.10}$$

Thus, if this substitution is made in (3.7.8), the problem becomes one of solving

$$\ddot{x} = -\frac{5}{2}\,x^{1/5} \tag{3.7.11}$$

with the initial conditions $x = 1$, $\dot{x} = 0$ at $\tau = 0$.

This is a particularly convenient form to consider since it looks like Newton's second law of motion for a particle moving along the x-axis. The left-hand side is the acceleration (think of the particle mass as being unity) and the right-hand side is the force. In (3.7.11) the force is always directed towards the origin, $x = 0$, and the bubble must always collapse. (Since $x \geq 0$ according to the definition of β and (3.7.9), we must abandon the analogy as soon as x reaches zero.)

Equation (3.7.11) may be integrated analytically by using the conservation of "energy" for the "oscillator" that it represents. Multiplying by $\dot{x}$ and integrating once we see that (3.7.11) has the integral of motion

$$\dot{x}^2 + \frac{25}{6}\,x^{6/5} = \frac{25}{6}\,, \tag{3.7.12}$$

where the value of the integral given on the right-hand side follows from substituting the initial values into the left-hand side. Hence,

$$\dot{x} = \pm\frac{5}{\sqrt{6}}\,\sqrt{1 - x^{6/5}}\,. \tag{3.7.13}$$

Only the minus sign is acceptable here, since x starts at $x = 1$ and must decrease. Integration of (3.7.13) gives

$$\int_x^1 \frac{\mathrm{d}\xi}{\sqrt{1 - \xi^{6/5}}} = \frac{5}{\sqrt{6}}\,\tau\,. \tag{3.7.14}$$

This equation gives the solution $x(\tau)$ albeit in a not very transparent form. It may become a bit more familiar if we set $\eta = \xi^{6/5}$. Carrying out the substitution we find

$$\int_{x^{6/5}}^1 \eta^{-1/6}\,(1-\eta)^{-1/2}\,\mathrm{d}\eta = \sqrt{6}\,\tau\,. \tag{3.7.15}$$

If we set $x = 0$ on the left-hand side of (3.7.15), we get the value of τ corresponding to collapse. Let us call it τ_*. This is the value of τ that corresponds to $t = T$ in (3.7.1). We recognize (see Abramowitz and Stegun, 1965) the integral as the beta function $\mathrm{B}(\frac{5}{6}, \frac{1}{2})$, i.e., the exact value of τ_* is:

$$\tau_* = \sqrt{\frac{3\pi}{2}}\,\frac{\Gamma\left(\frac{5}{6}\right)}{\Gamma\left(\frac{1}{3}\right)}\,, \tag{3.7.16}$$

where Γ is the generalized factorial or Γ-function.

Problem 3.7.2 For general x we can rewrite the left-hand side of (3.7.15) as an integral from 0 to 1 minus an integral from 0 to $x^{6/5}$. The first is the beta function just considered. The second is an incomplete beta function (see Abramowitz and Stegun, 1965). Show that the general solution of (3.7.14) or (3.7.15) may thus also be written

$$\mathrm{B}_{x^{6/5}}\left(\tfrac{5}{6}, \tfrac{1}{2}\right) = \sqrt{6}\,(\tau - \tau_*) \tag{3.7.17}$$

where τ_* is given by (3.7.16).

Although the functions involved may be somewhat unfamiliar, the solution as expressed here is clearly specified, and we consider $x(\tau)$ to be given by (3.7.17) where a computation or table look-up may be required to obtain actual numerical values. The function $x(\tau)$ yields $R(t)$ by use of (3.7.10) and return to dimensional variables.

We have pursued this analysis at length in order to have a solid basis from which to proceed with numerical explorations. It is possible to add several further physical effects to the description while retaining the interpretation of an "oscillator analogy." The "restoring force" on the right-hand side of the equation will, of course, become progressively more complicated, and obtaining analytical solutions will no longer be possible. It is in this kind of situation that accurate numerical computations become a useful research tool. Additional analytical features of the Rayleigh problem are pursued in Problem 3.7.4 below.

Computer Exercise 3.7.3 Write a computer program to integrate (3.7.11) using a Runge–Kutta method of at least second order. Also write a program that computes the analytical solution (3.7.17). Code for evaluating the incomplete beta function may, for example, be found in Press *et al.* (1986). Choose numerical parameters, such as integration step size, such that several digit agreement is obtained between the (computer-assisted) analytical and the direct numerical solutions.

Problem 3.7.4 Show that:

$$\left.\begin{aligned} \dot{\beta}^2 &= \frac{2}{3}\left(\beta^{-3}-1\right)\,, \\ \ddot{\beta} &= -\beta^{-4}\,, \end{aligned}\right\} \tag{3.7.18}$$

or in terms of dimensional variables:

$$\left.\begin{aligned} \frac{3}{2}\dot{R}^2 &= \frac{p_\infty - p_0}{\rho}(z-1)\,, \\ R\ddot{R} &= -\frac{p_\infty - p_0}{\rho}\,z\,, \end{aligned}\right\} \tag{3.7.19}$$

where

$$z = \left(\frac{R_0}{R}\right)^3 . \tag{3.7.20}$$

With these results return to (3.7.5) and seek the pressure field $p(r,t)$. Integrate (3.7.5) from some arbitrary radius $r > R$ to $r = \infty$ to obtain

$$(\ddot{R}R^2 + 2R\dot{R}^2)\frac{1}{r} - \frac{1}{2}\dot{R}^2R^4\frac{1}{r^4} = \frac{p(r,t) - p_\infty}{\rho}\,, \tag{3.7.21}$$

or introducing (3.7.19) and simplifying

$$\frac{p(r,t) - p_\infty}{p_\infty - p_0} = \frac{R}{3r}(z-4) - \frac{R^4}{3r^4}(z-1)\,. \tag{3.7.22}$$

This formula shows the (r,t)-dependence of the pressure field very clearly. At $t = 0$, $R = R_0$, $z = 1$, and the right-hand side of (3.7.22) is simply $-R_0/r$. The pressure rises monotonically from p_0 inside the bubble to p_∞ at a great distance from it. Show that the pressure has a maximum at a moving location $r = r_m(t)$ given by

$$\frac{r_m(t)}{R(t)} = \left(\frac{4(z-1)}{z-4}\right)^{1/3}, \tag{3.7.23}$$

where the time-dependence of z follows from its definition (3.7.20). This location is within the liquid for $z > 4$, i.e., the pressure remains monotonic until $R(t)$ has decreased to $R_0/4^{1/3}$. Beyond this time $r_m(t)$ moves in from $r = \infty$. At late stages of the collapse, when z is very large, $r_m(t) \approx 4^{1/3}\,R(t)$.

If, in analogy with (3.7.20), we introduce

$$z_m = \left(\frac{R_0}{r_m(t)}\right)^3 , \tag{3.7.24}$$

show that (3.7.23) may be restated as

$$z_m = \frac{z(z-4)}{4(z-1)} . \tag{3.7.25}$$

3.7.3 Additional Physical Effects

Let us next explore the addition of other physical effects to the model (3.7.7). The resulting ODE will quickly become so complicated that analytical theory is difficult and numerical exploration is helpful, not to say essential.

Surface Tension at the Bubble Surface

To include the effect of surface tension at the liquid–gas interface of the cavity wall we must set the pressure in the liquid as $r \to R+$ equal to $p_0 - 2\sigma/R$, according to the Young–Laplace law. Here σ is the surface tension coefficient. Repeating the derivation (3.7.6) with this change we now obtain Rayleigh's equation in the form

$$R\ddot{R} + \frac{3}{2}\dot{R}^2 + \frac{2\sigma}{\rho R} = -\frac{p_\infty - p_0}{\rho} . \tag{3.7.26}$$

Non-dimensionalization proceeds as before (see Problem 3.7.1):

$$\beta\ddot{\beta} + \frac{3}{2}\dot{\beta}^2 + 2D\beta^{-1} + 1 = 0 , \tag{3.7.27}$$

where D, the new non-dimensional surface tension parameter, is given by

$$D = \frac{\sigma}{R_0\,(p_\infty - p_0)} . \tag{3.7.28}$$

Equation (3.7.27) is the generalization of (3.7.8).

The equation of motion in the oscillator analogy, corresponding to (3.7.11) is

$$\ddot{x} = -\frac{5}{2}x^{1/5} - 5\,D\,x^{-1/5} . \tag{3.7.29}$$

The "energy equation" within this analogy is now

$$\dot{x}^2 + \frac{25}{6}\,x^{6/5} + \frac{25}{2}\,D\,x^{4/5} = \frac{25}{6}\,(1+3D) , \tag{3.7.30}$$

corresponding to (3.7.12).

From these results it is clear that surface tension simply enhances the tendency of the cavity to collapse.

Gas Pressure in the Bubble

In reality we have a bubble rather than a cavity, and the notion that the pressure within this bubble is a constant regardless of the volume requires refinement. A natural – though not necessarily terribly accurate – *ansatz* is to replace p_0 by the "polytropic" relation $p_0(R_0/R)^{3\gamma}$, where $\gamma = 1$ for isothermal compression/expansion, and $\gamma = 5/2$ for adiabatic transformations.

The generalized form of the Rayleigh equation (3.7.7) is then

$$R\ddot{R} + \frac{3}{2}\dot{R}^2 = -\frac{p_\infty - p_0(R_0/R)^{3\gamma}}{\rho}. \tag{3.7.31}$$

Introducing the non-dimensional pressure ratio,

$$\kappa = \frac{p_\infty}{p_0}, \tag{3.7.32}$$

we have

$$\beta\ddot{\beta} + \frac{3}{2}\dot{\beta}^2 + \frac{\kappa - \beta^{-3\gamma}}{\kappa - 1} = 0. \tag{3.7.33}$$

This corresponds to an oscillator of the form

$$\ddot{x} = -\frac{5}{2}\frac{\kappa - x^{-(6\gamma/5)}}{\kappa - 1}x^{1/5}, \tag{3.7.34}$$

which is qualitatively different from what we have seen previously in that it has an equilibrium point, $x = \kappa^{-5/6\gamma}$, other than $x = 0$. In fact, as $x \to 0$ we now have a restoring force that grows arbitrarily (as we would expect), and the bubble will rebound. In this model R oscillates.

Effect of Liquid Viscosity

Equation (3.7.2) is based only on incompressibility of the liquid and must therefore hold regardless of whether the liquid is treated as an inviscid or a viscous fluid. The Rayleigh equation and its extensions considered hitherto have tacitly assumed that the liquid is inviscid. The viscous stresses within the liquid, represented through the term $\nu\nabla^2\mathbf{V}$ of the Navier–Stokes equation, vanish for the flow field (3.7.2).

Problem 3.7.5 Verify that $\nu\nabla^2\mathbf{V}$ vanishes for a velocity field of the form (3.7.2).

Thus, the effect of liquid viscosity will enter our solution only in the treatment of the liquid–gas boundary condition, i.e., in the force balance at the instantaneous position of the bubble wall. In the presence of viscous forces this condition is no longer simply one of equality of pressure but now involves equating the normal stress in the liquid to the pressure in the gas.

The balance of stress just outside the bubble with pressure just inside takes the form

$$p - 2\mu \frac{\partial u}{\partial r} = p_0 \,. \tag{3.7.35}$$

The term involving the liquid viscosity, μ, arises from the diagonal component of the stress tensor. We substitute (3.7.2) for u in this balance, perform the differentiation, and set $r = R$. This gives

$$p = p_0 - 4\mu \frac{\dot{R}}{R} \tag{3.7.36}$$

as the bubble wall is approached from the liquid side. The integration from $r = R+$ to $r = \infty$ is now performed as in (3.7.6) yielding

$$R\,\ddot{R} + \frac{3}{2}\dot{R}^2 + \frac{4\mu\dot{R}}{\rho\,R} = -\frac{p_\infty - p_0}{\rho} \,. \tag{3.7.37}$$

Non-dimensionalizing as before we obtain the equation

$$\beta\,\ddot{\beta} + \frac{3}{2}\dot{\beta}^2 + 4\,C\,\dot{\beta}\,\beta^{-1} + 1 = 0 \,, \tag{3.7.38}$$

where the new non-dimensional parameter C is given by

$$C = \frac{\mu}{R_0\sqrt{\rho\,(p_\infty - p_0)}} \,. \tag{3.7.39}$$

The oscillator equation corresponding to (3.7.39) is

$$\ddot{x} = -\frac{5}{2}x^{1/5} - 4\,C\,\dot{x}\,x^{-4/5} \,. \tag{3.7.40}$$

The $\dot{x}$-dependent term on the right-hand side of this equation has the nature of a dissipative force, in this case a velocity and position dependent friction. This qualitatively new feature obviates the use of an energy equation to solve the oscillator problem.

We have pursued this particular problem in so much detail since it provides an example of a fluid mechanical situation of considerable importance that can be reduced to an ODE with a solution space of considerable richness. You are invited to explore this space in the following problems and exercises.

There are additional physical effects that can be introduced into the equation. For example, when the oscillations of the bubble become rapid compared to the speed of sound in the liquid, c, correction terms of order $\dot{R}/c$ must be considered. This occurs during underwater explosions. For the older literature the review by Plesset and Prosperetti (1977) is very useful.

The general format of the collapsing cavity or bubble problem is that of a damped oscillator. If the ambient pressure, p_∞, is allowed to become time-dependent, the models considered above can display chaotic behavior. This has been studied by several authors (Smereka *et al.*, 1987; Kamath and Prosperetti, 1989). The details of the chaos are, unfortunately, quite sensitive to the assumptions made regarding the complex thermo-mechanical processes in the bubble interior.

Problem 3.7.6 Consider the model of a collapsing cavity with viscous effects included, but no surface tension. Show that for sufficiently large values of C the bubble will not collapse in a finite time (Shu, 1952).

Problem 3.7.7 Consider the model of a collapsing bubble (cavity filled with gas) without surface tension or viscous forces. Derive a general formula for the period of pulsation of the bubble. Consider in particular small oscillations about the equilibrium point in (3.7.34), and find this period in a linearized approximation. Compare to the known result (see Lighthill, 1978) for the natural frequency

$$\omega = \frac{c_g}{R_0}\sqrt{\frac{3\rho_g^{(0)}}{\rho}}. \tag{3.7.41}$$

In this formula c_g is the speed of sound in the gas, $\rho_g^{(0)}$ is the density of the gas when $R = R_0$, and ρ is the liquid density.

Problem 3.7.8 Generalize the method for finding the pressure field presented earlier to the cases considered here where additional physical effects are present.

Computer Exercise 3.7.9 Augment the code of Computer Exercise 3.7.3 to include effects of surface tension, liquid viscosity and gas equation of state. Explore the space of parameters. You can, for example, verify the period of oscillation calculated in Problem 3.7.7 and the criterion for no collapse in Problem 3.7.6. It is instructive to compute the pressure field at various stages of the collapse using the formalism developed in Problem 3.7.4. It is interesting to follow the motion in a phase plane by plotting $\dot{\beta}$ versus β.

3.8 Motion of a Solid in Ideal Fluid

It is a remarkable and sometimes overlooked fact that the equations of motion for a solid body moving in an inviscid, incompressible fluid "otherwise at rest"[1] reduce to a system of *ordinary* differential equations. This simplification comes about because the fluid motion satisfying Laplace's equation is determined by the

[1] By which one means that the only fluid motion is the irrotational motion set up through the motion of the body.

linear and angular velocities of the solid body producing that motion. A particularly elegant formulation of the equations of motion was found by G.R. Kirchhoff in 1869, and given in Chapter 19 of his lectures on mechanics (Kirchhoff, 1876). The subject occupies Chapter VI in Lamb (1932). This basic problem of fluid mechanics was also treated in considerable detail by Thomson (the later Lord Kelvin) and Tait in their well-known if somewhat dated book *Natural Philosophy.* The equations to be described here are often cited as the *Kelvin–Kirchhoff equations* in the literature.

In terms of the vectors of velocity, $\mathbf{U}$, and angular velocity, $\boldsymbol{\Omega}$, of the body referred instantaneously to a frame of coordinates moving with the body the equations of motion are:

$$\left.\begin{aligned} \frac{\mathrm{d}}{\mathrm{d}t}\left(\frac{\partial T_{\mathrm{tot}}}{\partial \mathbf{U}}\right) + \boldsymbol{\Omega}\times\frac{\partial T_{\mathrm{tot}}}{\partial \mathbf{U}} &= 0\,, \\ \frac{\mathrm{d}}{\mathrm{d}t}\left(\frac{\partial T_{\mathrm{tot}}}{\partial \boldsymbol{\Omega}}\right) + \boldsymbol{\Omega}\times\frac{\partial T_{\mathrm{tot}}}{\partial \boldsymbol{\Omega}} + \mathbf{U}\times\frac{\partial T_{\mathrm{tot}}}{\partial \mathbf{U}} &= 0\,. \end{aligned}\right\} \tag{3.8.1}$$

Here T_{tot} is the kinetic energy of the solid body and the surrounding fluid:

$$T_{\mathrm{tot}} = \frac{1}{2}\mathbf{U}\cdot(\mathbf{T}+m\,\mathbf{1})\;\mathbf{U} + \frac{1}{2}\boldsymbol{\Omega}\cdot(\mathbf{J}+\mathbf{I})\;\boldsymbol{\Omega} + \mathbf{U}\cdot\mathbf{S}\,\boldsymbol{\Omega}\,, \tag{3.8.2}$$

where m is the mass and $\mathbf{I}$ is the inertia tensor of the solid body (see Whittaker, 1937; Landau and Lifshitz, 1987), $\mathbf{1}$ is the unit 3×3 identity tensor, and $\mathbf{T}$, $\mathbf{S}$ and $\mathbf{J}$ make up the 6×6 added mass tensor, $T_{\alpha\beta}$, as follows:

$$T_{\alpha\beta} = \left\{\begin{array}{cc} \mathbf{T} & \mathbf{S} \\ \mathbf{S}^{\mathrm{T}} & \mathbf{J} \end{array}\right\}, \tag{3.8.3}$$

where $\mathbf{S}^{\mathrm{T}}$ is the transpose of $\mathbf{S}$ (see Lamb, 1932). The elements of $\mathbf{T}$, $\mathbf{S}$ and $\mathbf{J}$ depend only on the body shape and the density of the fluid. For certain simple body shapes, such as ellipsoids, these constants are given analytically in the literature. In writing (3.8.2) the usual assumption of rigid body dynamics, that the origin of the moving coordinate system coincides with the center of mass of the body, has been made.

We may write the equations of motion more explicitly as follows: from (3.8.2) we introduce a generalized momentum (the linear impulse; Lamb, 1932)[2]

$$\mathbf{P} = \frac{\partial T_{\mathrm{tot}}}{\partial \mathbf{U}} = (\mathbf{T}+m\,\mathbf{1})\,\mathbf{U} + \mathbf{S}\,\boldsymbol{\Omega}\,, \tag{3.8.4}$$

and a generalized angular momentum (the angular impulse; Lamb 1932)

$$\mathbf{L} = \frac{\partial T_{\mathrm{tot}}}{\partial \boldsymbol{\Omega}} = (\mathbf{J}+\mathbf{I})\,\boldsymbol{\Omega} + \mathbf{S}^{\mathrm{T}}\,\mathbf{U}\,. \tag{3.8.5}$$

Equations (3.8.1) then take the form

$$\dot{\mathbf{P}} + \boldsymbol{\Omega}\times\mathbf{P} = \mathbf{0}\,, \tag{3.8.6}$$

[2] This is not quite the "generalized momentum" in the sense of Lagrangian mechanics. It is, however, closely related.

or

$$(\mathbf{T} + m\,\mathbf{1})\,\dot{\mathbf{U}} + \mathbf{S}\,\dot{\boldsymbol{\Omega}} + \boldsymbol{\Omega} \times \{(\mathbf{T} + m\,\mathbf{1})\;\mathbf{U} + \mathbf{S}\boldsymbol{\Omega}\} = \mathbf{0}\,, \tag{3.8.7}$$

and

$$\dot{\mathbf{L}} + \boldsymbol{\Omega} \times \mathbf{L} + \mathbf{U} \times \mathbf{P} = \mathbf{0}\,, \tag{3.8.8}$$

or

$$\begin{aligned} &(\mathbf{J} + \mathbf{I})\dot{\boldsymbol{\Omega}} + \mathbf{S}^{\mathrm{T}}\,\dot{\mathbf{U}} + \\ &\quad \boldsymbol{\Omega} \times \left\{(\mathbf{J} + \mathbf{I})\;\boldsymbol{\Omega} + \mathbf{S}^{\mathrm{T}}\,\mathbf{U}\right\} + \mathbf{U} \times \{(\mathbf{T} + m\mathbf{1})\,\mathbf{U} + \mathbf{S}\,\boldsymbol{\Omega}\} = \mathbf{0}\,. \end{aligned} \tag{3.8.9}$$

There is no particular reason to recount the theory further here, since it may be found in standard references. Several integrable cases of the above equations have been found in work by Kirchhoff, Clebsch, Steklov and others. In general, one expects the Kirchhoff–Kelvin equations to display chaos (Kozlov and Onischenko, 1982), as is, indeed, what numerical experiments indicate (Aref and Jones, 1993). To obtain the full motion of the body, the Kirchhoff–Kelvin equations must be supplemented by integrations that determine the position of the center of mass and the orientation of the body.

Problem 3.8.1 Show that the Kirchhoff–Kelvin equations have the integrals $\mathbf{P}$, $\mathbf{P}\cdot\mathbf{L}$, and T_{tot}. (*Hint*: It may be useful to establish the identity $\mathbf{U}\cdot\mathbf{P}+\boldsymbol{\Omega}\cdot\mathbf{L} = 2\,T_{\mathrm{tot}}$.)

Computer Exercise 3.8.2 Write a program to integrate the Kirchhoff–Kelvin equations. Consider the case of an ellipsoid with semi-axis in the ratio $a : b : c$. The non-zero elements of the added mass tensor can be expressed as (Lamb, 1932)

$$T_{1,1} = -\frac{\alpha_0}{2-\alpha_0}\frac{4}{3}\pi\rho_f abc\,, \tag{3.8.10}$$

$$T_{2,2} = -\frac{\beta_0}{2-\beta_0}\frac{4}{3}\pi\rho_f abc\,, \tag{3.8.11}$$

$$T_{3,3} = -\frac{\gamma_0}{2-\gamma_0}\frac{4}{3}\pi\rho_f abc\,, \tag{3.8.12}$$

$$T_{4,4} = -\frac{1}{5}\frac{(b^2-c^2)^2(\gamma_0-\beta_0)}{2(b^2-c^2)+(b^2+c^2)(\beta_0-\gamma_0)}\frac{4}{3}\pi\rho_f abc\,, \tag{3.8.13}$$

$$T_{5,5} = -\frac{1}{5}\frac{(c^2-a^2)^2(\alpha_0-\gamma_0)}{2(c^2-a^2)+(c^2+a^2)(\gamma_0-\alpha_0)}\frac{4}{3}\pi\rho_f abc\,, \tag{3.8.14}$$

$$T_{6,6} = -\frac{1}{5}\frac{(a^2-b^2)^2(\beta_0-\alpha_0)}{2(a^2-b^2)+(a^2+b^2)(\alpha_0-\beta_0)}\frac{4}{3}\pi\rho_f abc\,. \tag{3.8.15}$$

All other terms of the 6×6 added mass tensor are zero and the constants α_0,

β_0 and γ_0 are given by

$$\alpha_0 = abc \int_0^\infty \frac{d\lambda}{\Delta(a^2+\lambda)}\,,\ \beta_0 = abc \int_0^\infty \frac{d\lambda}{\Delta(b^2+\lambda)}\,,\ \gamma_0 = abc \int_0^\infty \frac{d\lambda}{\Delta(c^2+\lambda)}\,, \tag{3.8.16}$$

where $\Delta = \left[(a^2+\lambda)(b^2+\lambda)(c^2+\lambda)\right]^{1/2}$. Check the program by comparison with analytical solutions, such as the existence of three axes of permanent translation and Kirchhoff's solution (see Lamb, ibid., for this material), and by conservation of the integrals discussed in Problem 3.8.1.

3.9 The Point-Vortex Equations

3.9.1 Background and Formulation

In 1858 H. von Helmholtz published a seminal paper on the evolution of vorticity in fluid flows. Starting from the momentum equation for inviscid fluid, usually referred to as the *Euler equation*, he derived the **vorticity equation**

$$\frac{\partial \boldsymbol{\omega}}{\partial t} + \mathbf{V}\cdot\nabla\boldsymbol{\omega} = \boldsymbol{\omega}\cdot\nabla\mathbf{V}\,, \tag{3.9.1}$$

where, as usual, the vorticity $\boldsymbol{\omega} = \nabla \times \mathbf{V}$. This equation gives an intuitive interpretation of fluid motion in terms of the evolution of Lagrangian elements carrying vorticity. The left-hand side of (3.9.1) is the **material derivative** of ω. The operator on the right-hand side describes the tilting and stretching of vortex lines in three dimensions.

The right-hand side of (3.9.1) vanishes for planar flows, i.e., for

$$\mathbf{V} = (u(x,y,t), v(x,y,t), 0).$$

Hence, in 2D inviscid hydrodynamics (3.9.1) becomes the statement that *the vorticity of each fluid particle is an integral of the motion.* This leads to a very interesting possibility for modeling two-dimensional flow, which was also considered by Helmholtz. Let us introduce a "generalized solution" of the equations of two-dimensional inviscid, incompressible flow where the vorticity $\boldsymbol{\omega} = (0, 0, \zeta(x,y,t))$ has the form

$$\zeta(x,y,t) = \sum_{\alpha=1}^{N} \Gamma_\alpha\, \delta(x - x_\alpha(t))\, \delta(y - y_\alpha(t))\,. \tag{3.9.2}$$

In this type of solution the vorticity is assumed to be concentrated in a finite number of points $\alpha = 1, \ldots, N$. Everywhere else the vorticity vanishes. These N points are singularities of the flow. They are referred to as **point vortices**. The parameters Γ_α give the *circulations* or *strengths* of the point vortices. According to **Kelvin's circulation theorem** they are invariant in time for an inviscid, barotropic flow.

Helmholtz showed that (3.9.1) leads to the general result that vortex lines are material lines. Thus, the point vortices in (3.9.2) must move in such a way that their velocity at any instant equals the local fluid velocity. Their motion follows the advection equations (2.7.6), but, with the important additional remark that the velocity on the right-hand side of (2.7.6) is not arbitrarily prescribed. It is coupled to the instantaneous value of $\zeta(x, y, t)$ by the kinematic constraint that $\boldsymbol{\omega} = \nabla \times \mathbf{V}$, or in two dimensions

$$\zeta = \frac{\partial v}{\partial x} - \frac{\partial u}{\partial y}. \tag{3.9.3}$$

So instead of obtaining a system such as (2.7.6), where the right-hand sides need to be specified by other considerations, we obtain for N point vortices a system of $2N$ coupled ODEs. These are most conveniently written by thinking of the two-dimensional flow plane as the complex plane and concatenating the position coordinates (x_α, y_α) into complex positions $z_\alpha = x_\alpha + \imath\, y_\alpha$. The equations of motion for the z_α then are:

$$\frac{\mathrm{d}\, z_\alpha^*}{\mathrm{d}\, t} = \frac{1}{2\pi\imath} \sum_{\beta=1}^{N}{}' \frac{\Gamma_\beta}{z_\alpha - z_\beta}; \tag{3.9.4}$$

the prime on the summation symbol indicates omission of the singular term $\beta = \alpha$, and the asterisk on the left-hand side denotes complex conjugation.

Equations (3.9.4) provide a system of ODEs that can be subjected to the methods outlined earlier. Given an initial distribution of vorticity, the solution of (3.9.4) will trace its evolution in time as governed by the Euler equation for two-dimensional flow. The streamfunction $\psi(x, y, t)$ at any instant is obtained by combining equations (2.7.7) with (3.9.3) to get

$$\nabla^2 \psi = \zeta \tag{3.9.5}$$

i.e., ψ can be obtained by solving a *Poisson equation*. For the simple unbounded flow case treated so far ψ is just a sum of logarithms.

The above is a very sketchy exposition of aspects of vortex dynamics. For added detail consult one or more of the following references: Lamb (1932), Chapter 7; Batchelor (1967), Chapter 7; Sommerfeld (1964), Chapter 4.

3.9.2 Regular and Chaotic Few-Vortex Motion

Birkhoff and Fisher (1959) raised an important point in their provocative paper by pursuing the observations:

(1) that point-vortex dynamics described by equations (3.9.4) can be cast in the form of **Hamilton's canonical equations**; and thus,
(2) that statistical mechanics should govern the long-time behavior of a point-vortex system.

The first observation had been made already nearly a century before by Kirchhoff (1876). The second had been pursued, a decade earlier and in a somewhat different direction, in a classic paper, Onsager (1949), of great power and vision.

It turns out that the point-vortex equations (3.9.4) can be cast in the form of Hamilton's canonical equations. Thus, if we set

$$H = \frac{1}{4\pi} \sum_{\alpha,\beta=1}^{N}{}' \; \Gamma_\alpha \Gamma_\beta \log |z_\alpha - z_\beta| \,, \tag{3.9.6}$$

where the prime on the summation symbol again means $\alpha \neq \beta$, equations (3.9.4) can be written as

$$\Gamma_\alpha \frac{\mathrm{d}x_\alpha}{\mathrm{d}t} = -\frac{\partial H}{\partial y_\alpha} \qquad \Gamma_\alpha \frac{\mathrm{d}y_\alpha}{\mathrm{d}t} = -\frac{\partial H}{\partial x_\alpha} \,. \tag{3.9.7}$$

Defining $q_\alpha = x_\alpha$, $p_\alpha = \Gamma_\alpha y_\alpha$, these equations become Hamilton's canonical equations in their usual form.

From statistical mechanics one expects systems governed by the dynamics (3.9.7) to relax to some kind of thermal equilibrium. As argued by Birkhoff and Fisher (ibid.), it is unlikely that the statistical equilibrium for a collection of vortices, even if they are started on a regular curve or other pattern, is a neat pattern. One would expect rather that the vortices should form a large cluster with no well-defined structure. Indeed, according to an analysis by Lundgren and Pointin (1977) the vortices should in general equilibrate to *a circularly symmetric cloud*, calculable from the principles of equilibrium statistical mechanics. Of course, there is an important question of time scales involved here.

This suggestion of *ergodic motion* of point vortices, argued on the basis of Hamiltonian mechanics and statistical equilibrium, has as its precursor the onset of chaos as the number of vortices is increased. It can be shown that equations (3.9.4) are integrable, i.e., can be reduced to quadratures, for $N = 1, 2$ and 3 and any values of the circulations. This observation goes back at least a century to work by W. Gröbli (who studied with Kirchhoff) and H. Poincaré (the intriguing history of the early work on the three-vortex problem is described by Aref *et al.* (1992)). For $N \geq 4$, however, numerical experiments indicate that the motion is chaotic (Aref, 1983). One standard way to test for chaotic behavior is to compute a *Poincaré section* as discussed in Section 2.7. Figure 3.8 shows the Poincaré section and a sample trajectory for three and four identical vortices. The change from regular, in this case quasi-periodic, to chaotic (aperiodic) motion is readily apparent.

Chaotic vortex motion in the sense of sensitive and highly complex dependence on initial conditions can be seen also in the interaction of vortices that are not all of the same sign, notably in the interaction of vortex pairs. Figure 3.9 shows sample trajectories of two colliding vortex pairs. In one pair the circulations are $\pm\Gamma$; in the other they are $\pm 0.9\Gamma$. The changes in initial conditions necessary to go from one interaction sequence to the next are very slight. This manifests itself if one plots the "scattering time" versus the "impact parameter" describing

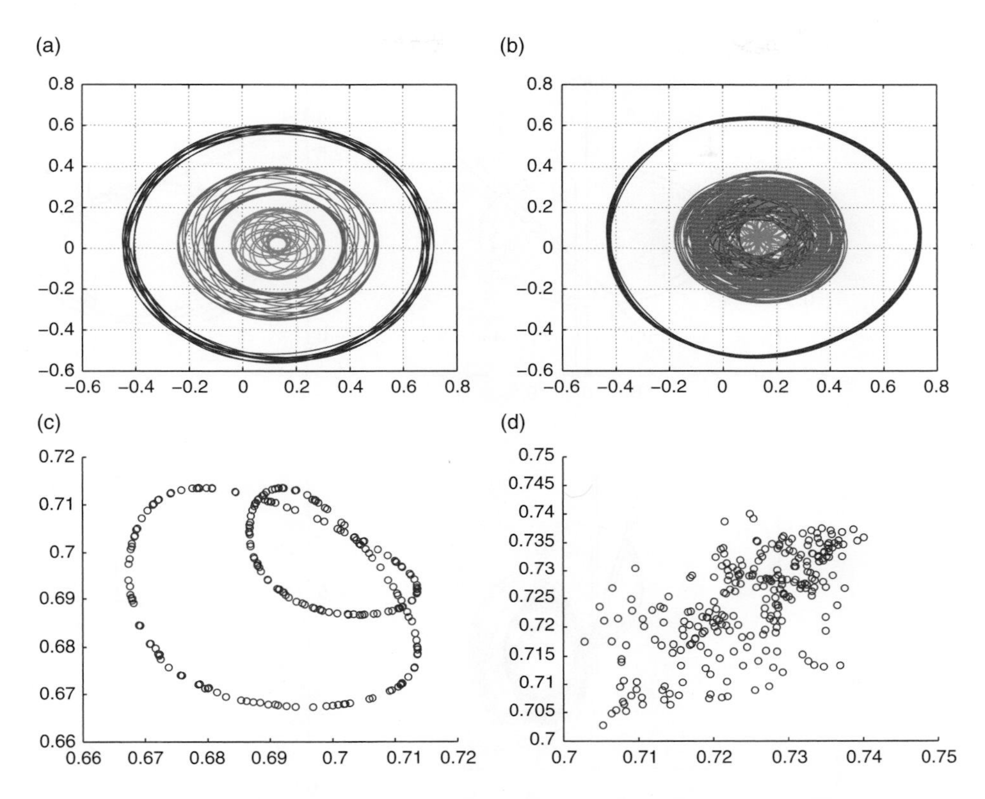

Figure 3.8 Examples of regular and chaotic point-vortex motion. (a) Trajectories of three-point vortices whose initial positions are $(-0.42, 0.0)$, $(-0.1, 0.1)$, and $(0.3, 0.0)$, whose strengths were 10π, 20π, and 60π, respectively. The trajectories were computed with a short, 30-line, MATLAB code that used the RK3 scheme with $\Delta t = 0.0005$. (b) Trajectories of four-point vortices whose initial positions are $(-0.42, 0.0)$, $(-0.1, 0.1)$, $(0.3, 0.0)$, and $(0.2, 0.2)$, whose strengths were 10π, 20π, 60π, and 20π, respectively. (c) A Poincaré section of the second vortex in the three-vortex problem was taken along the positive x-axis. That is, every time the second vortex crossed the positive x-axis its position was noted down. A plot of the ith crossing versus the $(i+1)$th crossing shows that the path is complex but regular. (d) The Poincaré section of the second vortex in the four-vortex problem. The chaotic nature of the vortex trajectories is clear.

the initial state. The function giving the scattering time is believed, on the basis of numerical experiments, to be extremely complicated with infinitely many arbitrarily tall spikes corresponding to tiny intervals of impact parameter. A function of this type is often referred to as a devil's staircase. For more details the reader is referred to the original paper (Eckhardt and Aref, 1988).

As well as the chaotic states of motion there are also many fascinating regular states that can be explored numerically (quite apart from the steadily rotating states considered in Chapter 1). As already mentioned all three-vortex states are regular. For example, one can explore numerically the collision of a $\pm$vortex pair with a single vortex. There are three-vortex states in which the three vortices

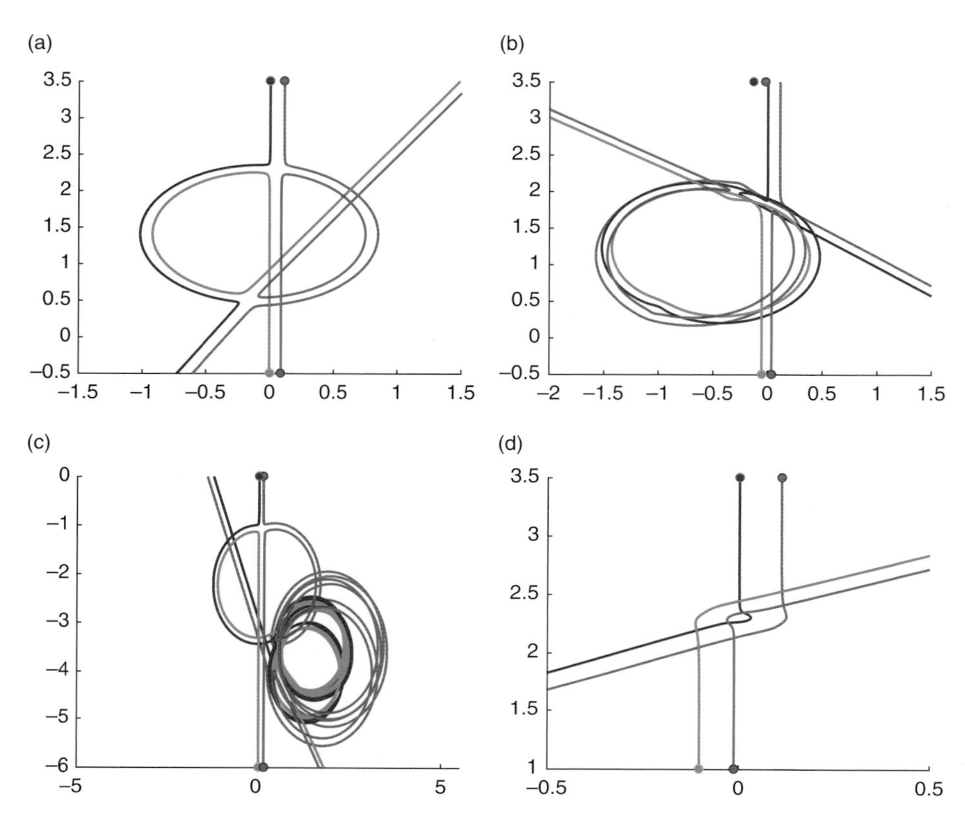

Figure 3.9 Trajectories of two colliding vortex pairs. In all cases the first pair that starts at the top was initially located at (0.1122,15.0) and (0.0016,15.0). The strengths of the first pair are 0.9 and −0.9. The starting y location of the lower pair was always −15.0, but their x location was varied. The strengths of the second pair are 1.0 and −1.0. The x location of the lower pair are (a) −0.0021 and 0.0882, (b) −0.0521 and 0.0382, (c) −0.0021 and 0.14 and (d) −0.1021 and −0.0118. Initially, the top pair moves down and bottom pair moves up vertically and their entry points are marked. In all cases after an elaborate dance and exchange of partners, the original pairs leave the computational domain. For a detailed discussion of colliding vortex pairs see Eckhardt and Aref (1988), from which this figure is taken. They discuss extreme sensitivity to the initial location of the colliding vortices.

collapse to a point in a finite time. There are regular states of four-vortex motion also when the vortices are arranged subject to certain constraints. Two pairs with a common axis provide an interesting example. Multiple-vortex systems with a center of symmetry provide other interesting solutions to the point-vortex equations. These solutions have been discussed in Aref (1982). An example with four vortex pairs is shown in Figure 3.10. You can see the pairs merge into two approximate circular orbits. The initial condition of the four vortices in this figure does not have a perfect center of symmetry and as a result the vortex pairs will eventually leave their near-circular orbit. But systems with a perfect center of symmetry will lead to stable final circular orbits.

It is possible to introduce simple boundaries by adding *image contributions* to (3.9.4). More complicated boundaries can be handled by conformal mappings, as explained by Lin (1943). Boundaries reduce the number of vortices required for chaotic motion. An interactive program that allows exploration of the dynamics of few-vortex motion is not too difficult to write for a laptop. You will be invited to write your own below!

Computer Exercise 3.9.1 Write a program to evolve an arbitrary number of point vortices of arbitrary circulations forward in time given an initial configuration. Use an RK or PC time integrator. Monitor the value of the Hamiltonian as a check on the accuracy of time-stepping. Other quantities of interest are the components of fluid impulse Q, P and the angular impulse I given by

$$Q = \sum_{\alpha=1}^{N} \Gamma_\alpha \, x_\alpha \,, \qquad P = \sum_{\alpha=1}^{N} \Gamma_\alpha \, y_\alpha \,, \qquad I = \sum_{\alpha=1}^{N} \Gamma_\alpha \, (x_\alpha^2 + y_\alpha^2) \,. \tag{3.9.8}$$

These are constants of the motion for interacting point vortices on the unbounded plane.

Try one or more of the following:

(1) Three-vortex motion, e.g., "vortex collapse" – see Aref (1979) on how to set these up.
(2) Chaos in the problem of four identical vortices.
(3) Vortex motion inside a circular boundary (add one image per vortex).
(4) Vortex sheet "roll-up" simulation for a line of vortices simulating a vortex sheet (see Section 3.10 for further discussion).
(5) Collision of two vortex pairs. The most interesting case is when the two initial pairs are of slightly different absolute strengths.
(6) Two or more vortex pairs with a common axis, or equivalently vortices close to a plane wall.

3.10 Vortex Sheet Roll-up

The decomposition of the vorticity field into a finite sum of contributions from individual vortices can be and has been used extensively to simulate two-dimensional, inviscid, incompressible flow. This kind of computation began, in fact, with some early hand calculations by Rosenhead in 1931 in which he attempted to follow the evolution of a dozen identical vortices simulating the roll-up of a vortex sheet. Batchelor (1967) presents a 1936 calculation by Westwater of the same kind. The expectation from experiment and from various similarity solutions

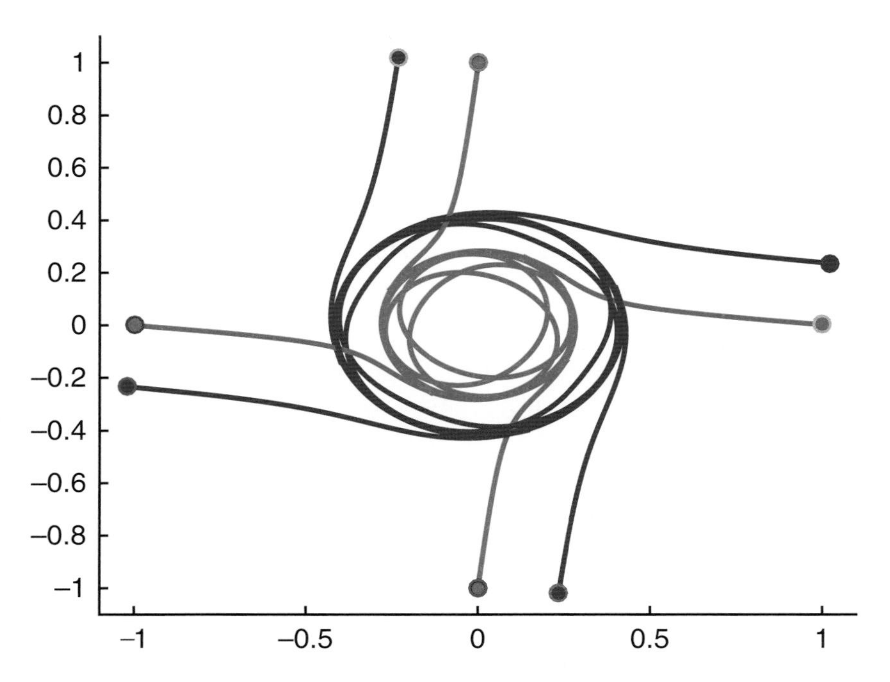

Figure 3.10 Motion of four symmetrically placed vortex pairs is shown here. You can see the pairs merge into two approximate circular orbits. The initial condition of the four vortices in this figure does not have a perfect center of symmetry and as a result the vortex pairs will eventually leave their near-circular orbit. But systems with a perfect center of symmetry will lead to stable final circular orbits. Refer to Aref (1982) for additional details.

for vortex sheet roll-up was that the vortices should roll up in a spiral pattern and some reasonable approximation to this state is expected in the numerical calculations.

Naively one expects to do better and better as the number of vortex elements used in the calculations is increased. Higher spatial resolution should lead to a more accurate calculation. Experience, however, indicated otherwise. Increasing the number of vortices used to discretize the sheet led to instabilities at ever earlier stages of the calculation. It was observed, for example, that the polygon line connecting the vortices would self-intersect as the roll-up proceeded. Indeed, in a provocative 1959 paper Birkhoff and Fisher were led to question on formal grounds whether vortex sheet roll-up really would occur at all. (Experimentally there is little doubt that the phenomenon does occur.) They made several observations. One was that in the continuum problem of the instability of a vortex sheet to infinitesimal perturbations, the classical Kelvin–Helmholtz instability problem, one finds a dispersion relation where the amplification rate of a wave of wavelength λ grows proportionally to λ^{-1}. Thus, shorter wavelengths grow faster, and arbitrarily short wavelengths grow arbitrarily fast. Clearly, then, unless the nonlinear interactions in the equations can provide a substantial "saturation" mechanism, one must expect some analytic "catastrophe" to occur after a finite time. Birkhoff and Fisher suggested in effect that the formal, mathematical vortex sheet problem would have a singularity after a finite time.

In the ensuing years several ingenious devices were introduced to force the vortex sheet, defined as the polygon line that results from connecting the point-vortex elements sequentially, to remain smooth as it rolled up. To understand some of the issues let us briefly state the mathematical problem of vortex sheet roll-up. It arises by taking a continuum limit of (3.9.4) assuming infinitely many vortex elements infinitely closely spaced on a plane curve. Thus, the individual circulations, which now become infinitesimal, are replaced by a distribution of vorticity along the sheet, viz. $\mathrm{d}\Gamma = \gamma(s,t)\,\mathrm{d}s$. The sum in (3.9.4) is replaced by an integral. The omission of a self-contribution $\beta \neq \alpha$ in (3.9.4) translates into a principal value integral. Finally we have

$$\frac{\mathrm{d}z^*}{\mathrm{d}t} = \frac{1}{2\pi\iota} \int \frac{\gamma(s,t)}{z - \zeta(s,t)} \mathrm{d}s\,, \tag{3.10.1}$$

where the vortex sheet at time t is assumed parametrized by arclength s as $\zeta(s,t)$, and z is a field point along the vortex sheet.

Equation (3.9.4) may be thought of as a discretization of (3.10.1) by using Lagrangian vortex elements. There are clearly several available options in making the discrete case approximate the continuum. We can declare at the outset that the initial configuration of the sheet is to be split up into N vortex elements, and we can stick with those elements for all time. That is not such a good idea, because the sheet stretches non-uniformly as it evolves, and sections of it that stretch a lot will become less well represented than sections that stretch less. Or we can decide to increase the number of vortex elements as the sheet stretches. That is a good idea, but we must then face a plethora of options on how to insert the additional elements. Also, the computational work will increase as the calculation proceeds. Finally, we can stick with a fixed number of vortex elements, but drop the constraint that they be the same from start to finish. In this mode of operation we monitor the distribution of the points along the sheet, and whenever certain sections become too stretched out, we pause to redistribute points uniformly along the sheet again, interpolating to get accurate values for the new circulations. Thus, (3.10.1), discretized in the manner of (3.9.4), is really just the equation that tells one how to take a single step. All the decisions on point selection and redistribution are extraneous considerations that can affect the results in various ways.

One avenue of development is exemplified by the work of van de Vooren (1980) and Fink and Soh (1978). In these works approximations of higher order than (3.9.4) are derived for the instantaneous approximation of (3.10.1). Fink and Soh's method uses such a higher-order approximation as the basis for a repositioning algorithm so the same Lagrangian points are not followed for all time.

However, if the vortex sheet problem really is ill-posed in the sense that the sheet will lose analyticity after a finite time, as we now believe it to be, the use of approximations that assume the existence of additional derivatives is probably intrinsically incorrect. While these methodological advances were taking place new insight on the development of the singularity was also being gained.

The case on which most progress has been made is the roll-up of the infinite, periodic sheet, as studied by Rosenhead. The case of roll-up of a semi-infinite sheet, as in the study by Westwater is, of course, closely related. Moore (1979) developed an asymptotic analysis for the vortex sheet roll-up problem by considering a sinusoidal perturbation and retaining a certain family of modes. He found an explicit, albeit approximate, formula for the "critical time" at which a singularity would form on the sheet. At this time the sheet curvature becomes infinite at every other node of the sine wave. A complementary analysis was performed by Meiron *et al.* (1964). Instead of perturbing the shape of the vortex sheet these authors considered perturbations of the vortex sheet strength γ for a flat sheet. Working from (3.10.1) they were able to develop an algorithm for computing the coefficients of the Taylor series in time governing the evolution of the sheet. This in itself involves substantial numerical work. The resulting series was then subjected to analysis by Padé approximant methods, developed primarily in the context of phase transition theory, which allowed an estimate of the nature of the singularity and its time of occurrence. A description of the details of all this would not fit well into our development at this stage. The interested reader is directed to the references given. In summary, two rather independent methods of analysis produced convincing evidence that something traumatic does happen to the analyticity properties of a vortex sheet evolving under (3.10.1) after a finite time.

Later work (Krasny, 1986a,b) gives several interesting examples of how otherwise reliable numerical procedures can be ruined by attempting a problem with pathological analytical behavior. Figure 3.11 shows calculations of very early stages of vortex sheet roll-up. The sheet has been represented by a string of 50 equidistant point vortices and an initial, small-amplitude sine wave has been imparted to this configuration. The point-vortex equations are used to follow the evolution, not quite in the form (3.9.4) because periodic boundary conditions have been imposed in x. (This leads to the interaction $(z_a - z_b)^{-1}$ being replaced by $\cot\{\pi(z_a - z_b)\}$ in a strip of unit width.) The time-stepping is done by a fourth-order Runge–Kutta method. In the units used for Figure 3.11 the time-step was chosen as $\Delta t = 0.01$ for $t \leq 0.25$ and $\Delta t = 0.001$ for $t > 0.25$. The four stages shown in each panel of Figure 3.11 are at the same times and differ only in the machine precision used for the calculation! In (a) "single precision" (7 decimal digits) was used; in (b) "double precision" (16 decimal digits). Figure 3.12 is an analogous sequence. The number of vortices has now been increased to 100. Panels (a) and (b) of Figure 3.12 are directly comparable to panels (a) and (b) of Figure 3.11. Clearly the increase in the number of vortices has made matters worse. Increasing machine precision again helps to control the scrambling of vortex positions. In Figure 3.12 the precisions used were (a) 7 digits; (b) 16 digits; (c) 29 digits.

The irregular vortex distributions seen in Figures 3.11 and 3.12 are related to

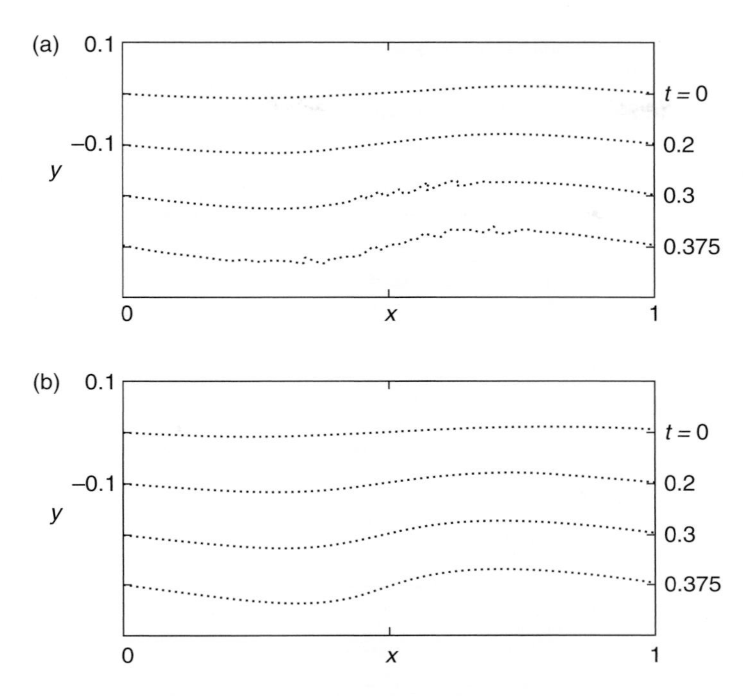

Figure 3.11 Numerical calculations of very early stages of vortex sheet roll-up. The sheet has been represented by a string of 50 equidistant point vortices and an initial, small-amplitude sine wave has been imparted to this configuration. The point-vortex equations are used to follow the evolution. (a) Results from single precision (7 decimal digits) calculation; (b) results from double precision (16 decimal digits) calculation. The effect of higher round-off error in the single precision quickly shows up as the small-scale numerical instability. From Krasny (1986a).

short wavelength components in a Fourier mode representation of the sheet configuration. There are three sources for the production of such short wavelengths: (i) nonlinear interactions as described by equation (3.10.1); (ii) truncation errors describing the deviation between (3.10.1) and equations of the form (3.9.4); and (iii) machine round-off errors. Nonlinear interactions should not lead to irregular distributions of vortices. Even though the vortex sheet evolution has a singularity after a finite time, it is regular up to that time. In this problem truncation errors act opposite to what one is used to. Increasing the number of vortices, N, leads to a better instantaneous representation, and also to an improved resolution of short wavelengths. Since shorter wavelengths grow faster, spurious amplitude contributions will evolve more rapidly for large N. Thus, we have the paradoxical situation that increasing N decreases the truncation error, and places us closer to the true continuum problem, but that the computation becomes more sensitive to inaccuracies at short wavelengths. This is why parts (a) and (b) of Figure 3.12 look "worse" than their counterparts, with half as many vortices, in Figure 3.11, and, ultimately, why Rosenhead's few-vortex representation looked

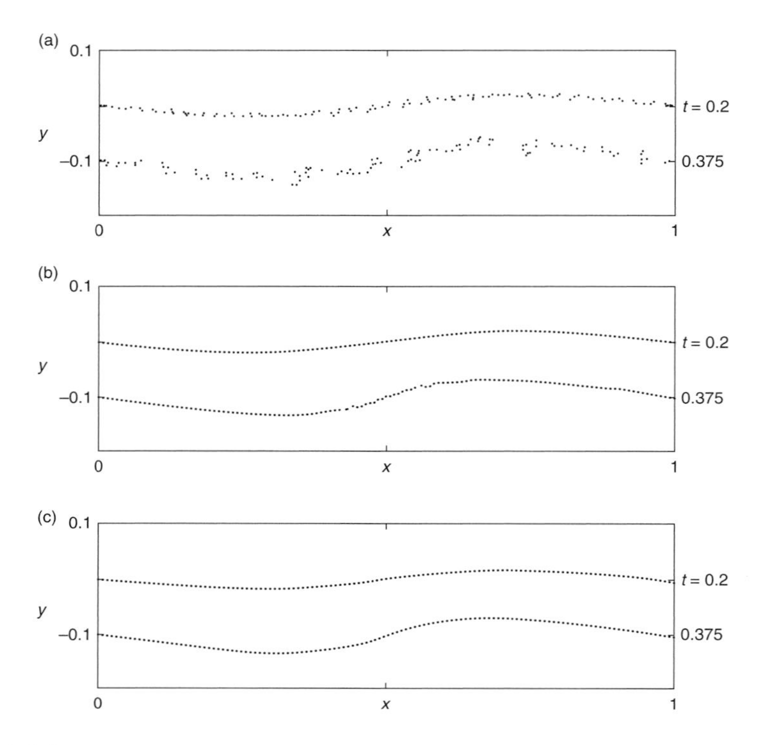

Figure 3.12 The same results as in the previous figure, but with 100 equidistant point vortices. (a) Results from single-precision (7 decimal digits) calculation; (b) results from double-precision (16 decimal digits) calculation; (c) results from 29 decimal digits calculation. From Krasny (1986a).

so reasonable. Round-off error is a background "noise source" constantly trying to trigger spurious waves on the sheet. Increasing the machine accuracy controls this effect. Thus, Figures 3.11 and 3.12 improve as we progress from (a) to (b) (to (c)). As N increases ever better control on round-off is required.

How might one avoid the singularity associated with the idealized model of (3.10.1) and push the calculation to the physically interesting regime of substantial roll-up? The suggestion from what we have seen so far is to damp out the short wavelengths in some effective way. Experience indicates that a variety of methods will do. A common feature of what has been done in this direction is to "soften" the interaction in (3.9.4) or (3.10.1) at close range. Consider augmenting the denominator of (3.9.4) or (3.10.1) by a term δ^2. Then the growth rate for waves of wavelength λ will no longer be proportional to λ^{-1} for all λ but will "roll off" when λ becomes of order δ. From this vantage point one might think of the continuum limit as attempting to produce infinite roll-up in a finite time. The size of δ used in calculations is quite large, and strictly speaking the discretization is no longer an exact representation of the Euler equation. Such *ad hoc* modifications of the governing equations are not uncommon in CFD, and this one is more precise and transparent than many you will encounter.

The idea of discretizing the Euler equation by a finite assembly of vorticity carrying elements is the basis for a family of numerical methods commonly referred to as *vortex methods*. We have pursued one particular example here because of its interesting analytical properties, and because of the many numerical studies of it to be found in the literature.

4 Spatial Discretization

In the previous chapter we considered initial value ODEs, where the interest was in the computation of time evolution of one or more variables given their starting value at some initial time. There is no inherent upper time limit in integrating these initial value ODEs. Therefore numerical methods for their solution must be capable of accurate and stable long time integration. By contrast, in the case of two-point boundary value and eigenvalue problems for ODEs arising in fluid mechanics, the independent space variable has two well-defined end points with boundary conditions specified at both ends. The spatial domain between the two boundary points can be infinite, as in the case of Blasius boundary layer: see (1.6.5), where the spatial domain extends from the wall ($\eta = 0$) out to infinity ($\eta \to \infty$). For such a problem it is possible to treat the space variable much like time in an initial value problem, and proceed with integration from one boundary to the other and then subsequently verify the boundary conditions at the other end. We shall consider numerical methods of this sort in the next chapter.

An alternative approach is to **discretize** the entire domain between the two boundaries into a finite number of grid points and to approximate the dependent variables by their grid point values. This leads to a system of equations that can be solved simultaneously. Much like the time integration error considered in the previous chapter, here one encounters a **discretization error**. The discretization error arises from several sources: *interpolation errors* arise from approximating the function between grid points; *differentiation errors* arise in the approximation of first-, second- and higher-order derivatives; and *integration errors* arise from the numerical integration of a function based on its discretized values at the grid points. These errors are indeed interrelated and depend on the discretization scheme. This chapter will consider various discretization schemes. In particular, discrete approximations to the first- and second-derivative operators will be obtained. Errors arising from the different discretization schemes will be considered. The concept of discrete approximation to the first and second derivatives as matrix operators will be introduced. Finally, we will consider spatial discretization as a means to numerically integrate functions. The actual solution methodologies for two-point boundary value and eigenvalue problems for ODEs, using the tools developed in this chapter, are treated in Chapter 5.

4.1 Forward, Backward and Central Difference

Let us first consider the simple case of a finite spatial domain, where the x-axis from 0 to L_x ($0 \leq x \leq L_x$) is discretized into $N+1$ equi-spaced grid points: $x_j = j\,\Delta x$ for $j = 0, 1, \ldots, N$, with grid spacing $\Delta x = L_x/N$. Given the value of a function f at the grid points: $f_j = f(x_j)$, the corresponding value of its first derivative at the grid points can be approximated in a number of different ways. The three most straightforward approximations to f' are

$$\textbf{Forward difference: } f_j' = f'(x_j) \approx \frac{f_{j+1} - f_j}{\Delta x}\,, \tag{4.1.1a}$$

$$\textbf{Backward difference: } f_j' = f'(x_j) \approx \frac{f_j - f_{j-1}}{\Delta x}\,, \tag{4.1.1b}$$

$$\textbf{Central difference: } f_j' = f'(x_j) \approx \frac{f_{j+1} - f_{j-1}}{2\,\Delta x}\,. \tag{4.1.1c}$$

Forward and backward differencing schemes in space are analogous to the implicit and explicit Euler time integration schemes, respectively, while central differencing is analogous to the leap-frog scheme. With Taylor series expansions for f_{j+1} and f_{j-1} about x_j it is a simple matter to show that

$$\frac{f_{j+1} - f_j}{\Delta x} = f_j' + \frac{1}{2}\Delta x\, f_j'' + O(\Delta x)^2\,, \tag{4.1.2a}$$

$$\frac{f_j - f_{j-1}}{\Delta x} = f_j' - \frac{1}{2}\Delta x\, f_j'' + O(\Delta x)^2\,, \tag{4.1.2b}$$

$$\frac{f_{j+1} - f_{j-1}}{2\Delta x} = f_j' + \frac{1}{6}(\Delta x)^2\, f_j''' + O(\Delta x)^3\,. \tag{4.1.2c}$$

The second terms on the right-hand sides are the leading-order error in the respective approximation of the first derivative. Thus while the forward and backward difference schemes are first-order accurate, the central difference scheme is second-order accurate. This is consistent with the accuracy of the implicit Euler, explicit Euler and leap-frog schemes.

The forward and backward difference schemes involve a **stencil** of only two grid points – the grid point at which the derivative is evaluated and one other point to its right or left. The central difference scheme, on the other hand, involves a stencil of three points and this may be crudely considered as the reason for its higher accuracy. Even higher-order schemes can be constructed by expanding the stencil. Construction of a finite difference scheme on a five-point stencil consisting of points $j-2, j-1, j, j+1$ and $j+2$ begins with the following approximation to the first derivative:

$$f_j' \approx a f_{j-2} + b f_{j-1} + c f_j + d f_{j+1} + e f_{j+2}\,, \tag{4.1.3}$$

where coefficients a through e are yet to be determined. Expanding the right-hand side in Taylor series about x_j we obtain

$$\begin{aligned} f_j' = {} & (a+b+c+d+e)f_j + \Delta x\,(-2a-b+d+2e)f_j' \\ & + \frac{(\Delta x)^2}{2!}(4a+b+d+4e)f_j'' + \frac{(\Delta x)^3}{3!}(-8a-b+d+8e)f_j''' \\ & + \frac{(\Delta x)^4}{4!}(16a+b+d+16e)f_j'''' + O(\Delta x)^5 \,. \end{aligned} \tag{4.1.4}$$

A family of second-order methods with two parameters (say a and b) can be obtained by requiring that the left- and right-hand sides agree to terms of order $(\Delta x)^2$; that is,

$$\left.\begin{aligned} a+b+c+d+e &= 0\,, \\ -2a-b+d+2e &= 1/\Delta x\,, \\ 4a+b+d+4e &= 0\,. \end{aligned}\right\} \tag{4.1.5}$$

In terms of the two parameters, a and b, one obtains $c = -(\frac{3}{2\Delta x} + 6a + 3b)$, $d = \frac{2}{\Delta x} + 8a + 3b$ and $e = -(\frac{1}{2\Delta x} + 3a + b)$. For third-order accuracy it is additionally required that

$$-8a - b + d + 8e = 0\,, \tag{4.1.6}$$

and as a result a family of third-order methods with one parameter (say a) is obtained where now $b = -(\frac{1}{3\Delta x} + 4a)$. With a five-point stencil fourth-order accuracy can be obtained by requiring that

$$16a + b + d + 16e = 0\,, \tag{4.1.7}$$

in addition to (4.1.5) and (4.1.6). Then $a = -e = \frac{1}{12\Delta x}$, $d = -b = \frac{2}{3\Delta x}$ and $c = 0$. The resulting fourth-order accurate five-point central difference representation of the first derivative is

$$f_j' \approx \frac{f_{j-2} - 8f_{j-1} + 8f_{j+1} - f_{j+2}}{12\Delta x}\,. \tag{4.1.8}$$

Problem 4.1.1 Show that the leading-order error for the above five-point central difference scheme is $\frac{(\Delta x)^5}{5!} f^{\mathrm{v}}(-32a - b + d + 32e) = -\frac{1}{30}(\Delta x)^4 f^{\mathrm{v}}$.

The above symmetric stencil becomes inappropriate close to boundaries, but higher-order finite difference schemes on asymmetric and one-sided stencils can be developed in a similar manner. We leave the algebraic details to the reader and simply present below two other fourth-order accurate finite difference schemes for the first derivative on forward-biased asymmetric stencils:

$$f_j' \approx \frac{-3f_{j-1} - 10f_j + 18f_{j+1} - 6f_{j+2} + f_{j+3}}{12\Delta x}\,, \tag{4.1.9a}$$

$$f_j' \approx \frac{-25f_j + 48f_{j+1} - 36f_{j+2} + 16f_{j+3} - 3f_{j+4}}{12\Delta x}\,. \tag{4.1.9b}$$

Problem 4.1.2 Show that the corresponding leading-order errors of the above forward-biased schemes are $\frac{1}{20}(\Delta x)^4 f^{\mathrm{v}}$ and $-\frac{1}{5}(\Delta x)^4 f^{\mathrm{v}}$, respectively.

In general, in a finite difference approximation of the first derivative with an M-point stencil the highest order of accuracy attainable is $M-1$. For example, all the three five-point stencil schemes given in (4.1.8) and (4.1.9a)–(4.1.9b) are fourth-order accurate. But, the magnitude of the pre-factor multiplying the leading-order error term increases with increasing asymmetry of the stencil.

4.1.1 Fourier or von Neumann Error Analysis

One way to interpret the above finite difference schemes is that an appropriate Lagrange interpolation polynomial of order $M-1$ is fit through the stencil (refer back to Section 3.4) and used in the evaluation of the derivative. For example, with the three-point stencil used in the central difference scheme, the following quadratic polynomial interpolates the values f_{j-1}, f_j and f_{j+1}:

$$f_{j-1}\left(\frac{x-x_j}{x_{j-1}-x_j}\right)\left(\frac{x-x_{j+1}}{x_{j-1}-x_{j+1}}\right)+f_j\left(\frac{x-x_{j-1}}{x_j-x_{j-1}}\right)\left(\frac{x-x_{j+1}}{x_j-x_{j+1}}\right)$$
$$+f_{j+1}\left(\frac{x-x_{j-1}}{x_{j+1}-x_{j-1}}\right)\left(\frac{x-x_j}{x_{j+1}-x_j}\right). \tag{4.1.10}$$

The first derivative of the above interpolant, evaluated at x_j, reproduces the central difference formula in (4.1.1a)–(4.1.1c). This suggests that if the function f being differentiated is a polynomial of order $M-1$ or less, then the numerical results are in fact exact. For example, if f is a quadratic then a three-point finite difference scheme can provide the exact derivative independent of Δx. For polynomials of order greater than $M-1$, the error in numerical differentiation will usually be non-zero and increase with the order of the polynomial. As a rule the higher the oscillatory nature of f, the larger will be the error. A good characterization of this dependence of error on the oscillatory nature of f can be obtained from a **von Neumann error analysis**. This analysis considers the accuracy of the numerical scheme in differentiating sinusoidal functions of increasing wavenumber and therefore is also known as a **Fourier error analysis**.

Consider a sequence of test functions

$$f_k(x)=\exp\left[\iota\, k\,\frac{2\pi\, x}{L_x}\right] \qquad \text{for} \quad k=0,\,\ldots,\,\frac{N}{2}\,. \tag{4.1.11}$$

Here the wavenumber, k, gives the number of waves spanning the domain $0\le x\le L_x$. While $k=0$ corresponds to a constant (non-oscillatory) function, $k=N/2$ corresponds to a highly oscillatory function with one wave for every two points. By the **Nyquist–Shannon theorem** k cannot be greater than $N/2$. That is, you need at least two points to resolve every wave. The above complex exponential form of the test functions is just a convenient compact notation. From Euler's formula, $\exp(\iota\theta) = \cos(\theta) + \iota\sin(\theta)$, we see that the test functions, $f_k(x)$, are nothing but combinations of sine and cosine functions. The complex notation makes the following error analysis simple and elegant.

Introducing a scaled wavenumber, $K = k(2\pi/N)$, and using the definition $L_x = N\Delta x$, the function and its exact first derivative may be written as

$$f_k(x) = \exp\left[\iota K \frac{x}{\Delta x}\right] \qquad \text{and} \qquad f_k'(x) = \frac{\iota K}{\Delta x}\exp\left[\iota K \frac{x}{\Delta x}\right] . \tag{4.1.12}$$

We see that K can be interpreted as the number of waves for every 2π points, and it ranges from 0 to π as k ranges from 0 to $N/2$. In a numerical calculation we are only concerned about discrete values at the grid points. From the definition $x_j = j\Delta x$, in terms of the scaled wavenumber these grid point values are

$$f_{k,j} = f_k(x_j) = \exp\left[\iota K j\right] \qquad \text{and} \qquad f_{k,j}' = f_k'(x_j) = \frac{\iota K}{\Delta x}\exp\left[\iota K j\right] . \tag{4.1.13}$$

Now we move to numerical approximations to the first derivative. From the grid point values of the function (i.e., from $f_{k,j}$) using any one of the schemes given in (4.1.1a)–(4.1.1c), (4.1.8), and (4.1.9a)–(4.1.9b), we can obtain numerically evaluated first derivatives at the grid points. We now claim that the numerical approximation can be expressed in the following form:

$$f_{k,j}' \approx \frac{\widetilde{K}}{\Delta x}\exp\left[\iota K j\right] , \tag{4.1.14}$$

where $\widetilde{K}$ is called the **modified wavenumber**. This claim can be easily confirmed by substituting $f_{k,j}$ from (4.1.13) into any of the expressions (4.1.1a)–(4.1.1c), (4.1.8) or (4.1.9a)–(4.1.9b) (examples will be provided below). The modified wavenumber is a function of K and the functional form of $\widetilde{K}(K)$ depends on the numerical scheme. $\widetilde{K}$ is complex with a real part, $\widetilde{K}_{\mathrm{R}}$, and an imaginary part, $\widetilde{K}_{\mathrm{I}}$. Comparing the second equality in (4.1.13) and (4.1.14) the exact first derivative is obtained if and only if $\widetilde{K}_{\mathrm{I}} = K$ and $\widetilde{K}_{\mathrm{R}} = 0$; any deviation from this behavior is indicative of the error in the numerical approximation.

The error can be separately identified as **amplitude** and **phase error** by considering the numerical solution of the wave equation

$$\frac{\partial f}{\partial t} + c\frac{\partial f}{\partial x} = 0 . \tag{4.1.15}$$

The exact solution to the wave equation with the test function (4.1.12) as initial condition is

$$f_{k,\mathrm{ex}}(x,t) = \exp\left[\iota\frac{K}{\Delta x}(x - ct)\right] , \tag{4.1.16}$$

and the grid point values of the exact solution are

$$f_{k,j,\mathrm{ex}}(t) = \exp\left[\iota\frac{K}{\Delta x}(x_j - ct)\right] . \tag{4.1.17}$$

We have used the subscript "ex" in the above to denote the "exact" nature of the solution, which will be compared against the numerical approximation (which for simplicity will not be distinguished with any subscript). According to the exact solution, depending on the sign of c, the wave travels to the right (in the direction of increasing x) or to the left (in the direction of decreasing x) without any change in its amplitude. The phase speed of the wave remains the same, c, irrespective of the wavenumber. Now, if one approximates the spatial derivative, $\partial f/\partial x$, in (4.1.15) by (4.1.14) one obtains at each grid point the ODE:

$$\frac{df_{k,j}}{dt} = -c\frac{\widetilde{K}}{\Delta x} f_{k,j}\,. \tag{4.1.18}$$

Now integrating the above equations exactly (i.e., without any time integration errors) with the initial condition $f_{k,j}(0) = \exp(\iota K x_j/\Delta x)$ we obtain

$$f_{k,j}(t) = \exp\left[\iota\frac{K}{\Delta x}x_j - c\frac{\widetilde{K}}{\Delta x}t\right]. \tag{4.1.19}$$

With $\widetilde{K} = \widetilde{K}_{\rm R} + \iota\,\widetilde{K}_{\rm I}$ this may be written as

$$f_{k,j}(t) = A(t)\exp\left[\iota\frac{K}{\Delta x}\left(x_j - \tilde{c}t\right)\right], \tag{4.1.20}$$

where the amplitude and phase speed of the numerical approximation are $A(t) = \exp\left[-c\widetilde{K}_{\rm R}\,t/\Delta t\right]$ and $\tilde{c} = c\widetilde{K}_{\rm I}/K$, respectively. In general, these will differ from the values of $A = 1$ and $\tilde{c} = c$ of the exact solution (4.1.17). It is now clear that $\widetilde{K}_{\rm R} \neq 0$ contributes to an error in amplitude and leads to growing or decaying solutions depending on the sign of $c\,\widetilde{K}_{\rm R}$. On the other hand, $\widetilde{K}_{\rm I} \neq K$ contributes to phase error, with waves traveling faster if $\widetilde{K}_{\rm I} > K$ and slower if $\widetilde{K}_{\rm I} < K$.

As an example, the amplitude and phase errors for the forward difference scheme can be determined by substituting the left equality of (4.1.13) into the right-hand side of (4.1.1a)

$$\begin{aligned}\frac{\exp\left[\iota K(j+1)\right] - \exp\left[\iota K j\right]}{\Delta x} &= \frac{1}{\Delta x}\left\{\exp\left[\iota K\right] - 1\right\}\exp\left[\iota K j\right] \\ &= \frac{1}{\Delta x}\left\{\iota\sin K - (1-\cos K)\right\}\exp\left[\iota K j\right]. \end{aligned} \tag{4.1.21}$$

Comparing the right-hand side of (4.1.21) with (4.1.14) we read off the real and imaginary parts of the modified wavenumber $\widetilde{K}_{\rm I} = \sin K$ and $\widetilde{K}_{\rm R} = \cos K - 1$. For small K, the phase error $\widetilde{K}_{\rm I} - K$ is $\mathrm{O}\left(K^3\right)$. The amplitude error $\widetilde{K}_{\rm R}$ is $\mathrm{O}\left(K^2\right)$ and so more serious. These are consistent with the amplitude and phase errors of the Euler time-integration schemes considered in Section 3.2. An analysis similar to that leading to (4.1.21) for the second-order accurate central difference scheme yields $\widetilde{K}_{\rm I} = \sin K$ and $\widetilde{K}_{\rm R} = 0$. Thus, as in the leap-frog scheme (see Problem 3.5.1) there is no amplitude error and the phase error is $\mathrm{O}\left(K^3\right)$. It is a general property of central difference schemes that they are free of amplitude error. By contrast, asymmetric stencils involve both amplitude and phase errors.

Problem 4.1.3 Show that for the five-point central difference scheme given in (4.1.8),

$$\widetilde{K}_{\mathrm{I}} = \tfrac{4}{3}\sin K - \tfrac{1}{6}\sin(2K) \qquad \text{and} \qquad \widetilde{K}_{\mathrm{R}} = 0\,. \tag{4.1.22}$$

In Figures 4.1(a) and 4.1(b), $\widetilde{K}_{\mathrm{R}}$ and $\widetilde{K}_{\mathrm{I}}$ for the different finite difference approximations to the first derivative are plotted as functions of K. Both amplitude and phase errors are small for small values of K, but the errors rapidly increase for large K. In particular, the phase error can become 100% as K approaches π. To contain the error it is important to maintain the number of grid points per wavelength much larger than 2 for all waves of importance. The phase error, $\widetilde{K}_{\mathrm{I}} - K$ is a function of K. Therefore the phase velocity of the numerical solution depends on the wavenumber. However, according to the exact solution (4.1.17) all waves travel at the same speed c. Thus, in effect, the numerical simulation tends to disperse a group of waves over time. As a result phase error is also known as **dispersion error**.

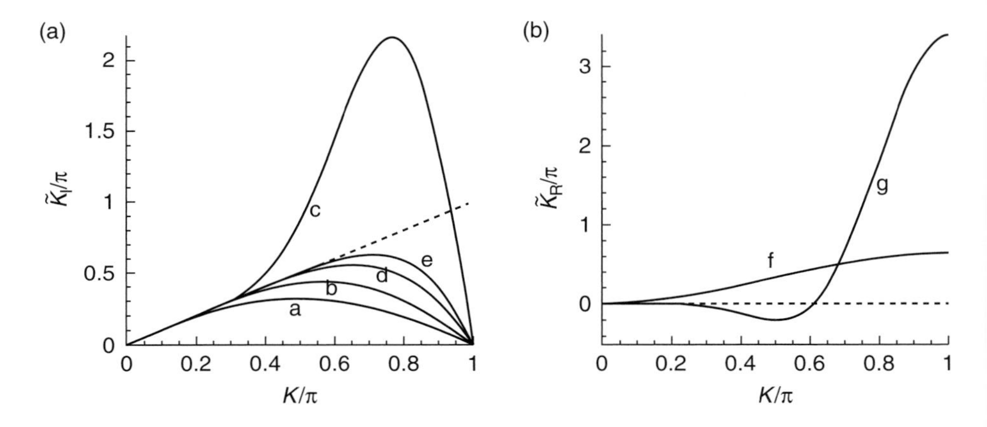

Figure 4.1 A plot of the real and imaginary parts of the modified wavenumber for different finite and compact difference schemes. The dashed lines give the desired exact answer. Curves marked (a) and (f) are for the forward difference scheme given in (4.1.1a). The curve marked (b) is for the three-point central difference scheme given in (4.1.1c). Curves marked (c) and (g) are for the five-point forward one-sided scheme given in (4.1.9b). The curve marked (d) is for the fourth-order tridiagonal compact scheme (Padé scheme) and the curve marked (e) is for the sixth-order 3–5 compact scheme. For the central and compact difference schemes, $\widetilde{K}_{\mathrm{R}} \equiv 0$. For backward schemes the real part $\widetilde{K}_{\mathrm{I}}$ remains the same, but the sign of the imaginary part $\widetilde{K}_{\mathrm{R}}$ reverses.

4.1.2 Approximation for the Second Derivative

Construction of finite difference approximations for the second derivative proceeds along the same lines as outlined for the first derivative. For a five-point symmetric stencil instead of (4.1.5) we now have

$$\left. \begin{aligned} a+b+c+d+e &= 0\,, \\ -2a-b+d+2e &= 0\,, \\ 4a+b+d+4e &= \frac{2}{(\Delta x)^2}\,, \end{aligned} \right\} \tag{4.1.23}$$

while relations (4.1.6) and (4.1.7) still hold. A three-point central difference scheme can be obtained as a subset by setting $a = e = 0$ and forgoing relations (4.1.6) and (4.1.7). Then $2b = 2d = -c = 2/(\Delta x)^2$ and the resulting three-point central difference approximation to the second derivative is

$$f_j'' \approx \frac{f_{j-1} - 2f_j + f_{j+1}}{(\Delta x)^2}\,. \tag{4.1.24}$$

Interestingly, owing to the symmetry of the central difference, (4.1.6) is also automatically satisfied. Five-point central difference provides the following approximation:

$$f_j'' \approx \frac{-f_{j-2} + 16f_{j-1} - 30f_j + 16f_{j+1} - f_{j+2}}{12\,(\Delta x)^2}\,. \tag{4.1.25}$$

Finite difference approximations for the second derivative can be constructed on asymmetric stencils similar to (4.1.9a)–(4.1.9b). Two such forward-biased approximations are

$$f_j'' \approx \frac{11f_{j-1} - 20f_j + 6f_{j+1} + 4f_{j+2} - f_{j+3}}{12\,(\Delta x)^2}\,, \tag{4.1.26a}$$

$$f_j'' \approx \frac{35f_j - 104f_{j+1} + 114f_{j+2} - 56f_{j+3} + 11f_{j+4}}{12\,(\Delta x)^2}\,. \tag{4.1.26b}$$

Problem 4.1.4 Show that the leading-order errors of schemes (4.1.24), (4.1.25) and (4.1.26a)–(4.1.26b) are respectively, $\frac{1}{12}(\Delta x)^2 f^{\text{iv}}$, $-\frac{1}{90}(\Delta x)^4 f^{\text{vi}}$, $-\frac{1}{12}(\Delta x)^3 f^{\text{v}}$ and $\frac{10}{12}(\Delta x)^3 f^{\text{v}}$.

In the case of the second derivative, central difference schemes are clear winners as their order of accuracy is one higher than the corresponding asymmetric differences. In general, for the second derivative an M-point symmetric stencil can lead to $(M-1)$th-order accuracy, while the highest order of accuracy with an M-point asymmetric stencil is only $M-2$.

Problem 4.1.5 Obtain the backward-biased five-point finite difference schemes corresponding to (4.1.9a)–(4.1.9b) and (4.1.26a)–(4.1.26b) for the first and second derivatives. Explore the symmetry patterns of the resulting coefficients.

The amplitude and phase errors of the second-derivative approximation can be investigated in much the same way as before using the sequence of test functions (4.1.11). The exact derivatives of the test functions at the grid points are

$$f''_{k,j,\mathrm{ex}} = -\frac{1}{(\Delta x)^2} K^2 \exp\left[\iota K j\right] , \tag{4.1.27}$$

while the numerical solution can be written in a similar form

$$f''_{k,j} \approx -\frac{1}{(\Delta x)^2} \tilde{\tilde{K}} \exp\left[\iota K j\right] . \tag{4.1.28}$$

The amplitude and phase errors can be separated by considering numerical solution of the diffusion equation

$$\frac{\partial f}{\partial t} = \alpha \frac{\partial^2 f}{\partial x^2} . \tag{4.1.29}$$

The exact solution to (4.1.29) with the test functions (4.1.12) as initial condition can be easily obtained and can be written at the grid points as

$$f_{k,j,\mathrm{ex}}(t) = \exp\left[-\frac{\alpha K^2}{(\Delta x)^2} t\right] \exp\left[\iota \frac{K}{\Delta x} x_j\right] . \tag{4.1.30}$$

According to the exact solution the amplitude of the waves decays exponentially in time, but they are now standing waves and do not travel to the right or left. The rate of dissipation depends quadratically on K, with rapidly oscillating waves of large K decaying faster. Instead, if one approximates the spatial derivative, $\partial^2 f/\partial x^2$, in (4.1.29) by (4.1.28) and performs the time integration exactly (without any time integration error) the resulting approximate solution can be expressed as

$$f_{k,j}(t) \approx A(t) \exp\left[\iota \frac{K}{\Delta x} \left(x_j - \tilde{\tilde{c}}\, t\right)\right] , \tag{4.1.31}$$

where the amplitude and phase speed of the numerical solution are

$$A(t) = \exp\left[-\frac{\alpha \tilde{\tilde{K}}_{\mathrm{R}}}{(\Delta x)^2 t}\right] \quad \text{and} \quad \tilde{\tilde{c}} = \frac{\tilde{\tilde{K}}_{\mathrm{I}}}{K\, \Delta x} . \tag{4.1.32}$$

In general, these will differ from the corresponding values of the exact solution (4.1.30). A value $\tilde{\tilde{K}}_{\mathrm{I}} \neq 0$ contributes to phase error and the waves spuriously travel to the right or left depending on the sign of $\tilde{\tilde{K}}_{\mathrm{I}}$. On the other hand, $\tilde{\tilde{K}}_{\mathrm{R}} \neq K^2$ now contributes to the amplitude error, with $\tilde{\tilde{K}}_{\mathrm{R}} > K^2$ indicating faster decay than the exact solution.

The amplitude and phase errors for the central difference scheme (4.1.24) can be extracted by substituting the left equality in (4.1.13) into the right-hand side of (4.1.24)

$$\begin{aligned}&\frac{\exp\left[\iota K(j-1)\right]-2\exp\left[\iota Kj\right]+\exp\left[\iota K(j+1)\right]}{(\Delta x)^2}\\&=\frac{1}{(\Delta x)^2}\left\{\exp\left[-\iota K\right]-2+\exp\left[\iota K\right]\right\}\exp\left[\iota Kj\right]\\&=\frac{1}{(\Delta x)^2}\left\{2(\cos K-1)\right\}\,\exp\left[\iota Kj\right]\qquad.\end{aligned}\tag{4.1.33}$$

By comparing the right-hand side of (4.1.33) with (4.1.28), the real and imaginary parts can be identified as $\widetilde{\widetilde{K}}_{\mathrm{R}}=2(1-\cos K)$ and $\widetilde{\widetilde{K}}_{\mathrm{I}}=0$, Central difference approximations to the second derivative are free of phase error, while forward- and backward-biased schemes on asymmetric stencils involve both amplitude and phase errors.

Problem 4.1.6 Obtain the amplitude and phase errors of the forward-biased five-point stencil schemes given in (4.1.26a)–(4.1.26b).

In Figures 4.2(a) and 4.2(b), $\widetilde{\widetilde{K}}_{\mathrm{R}}$ and $\widetilde{\widetilde{K}}_{\mathrm{I}}$ for the different finite difference approximations to the second derivative are plotted for K in the range 0 to π. Both amplitude and phase errors are small for small values of the scaled wavenumber, but as with the first derivative, the errors rapidly increase for large K.

4.2 Matrix Derivative Operators

The process of differentiation can be thought of as a linear mapping in infinite-dimensional space, where the linear operators $\mathrm{d}/\mathrm{d}x$, $\mathrm{d}^2/\mathrm{d}x^2$, etc., map the function $f(x)$ on to $f'(x)$, $f''(x)$, etc. Upon discretization, the set of grid point values of the function, f_j for $j=0,\ldots,N$, is finite dimensional and differentiation is now a finite-dimensional linear mapping that returns the grid point values of the derivative, f'_j, f''_j, etc., for $j=0,\ldots,N$. If the function and its derivative are represented as vectors of length $N+1$, then the discrete differentiation operator corresponding to the pth derivative is a matrix, $\mathbf{D}_{\mathrm{p}}$, of size $(N+1)\times(N+1)$ with elements $[D_{\mathrm{p}}]_{jm}$:[1]

$$f_j^{\mathrm{p}}=[D_{\mathrm{p}}]_{jm}\,f_m\,.\tag{4.2.1}$$

Simply put, the process of numerical differentiation is nothing but multiplication of the data vector by an appropriate derivative operator matrix. It produces

[1] Henceforth we will use the notation that a discrete variable in bold font, such as $\mathbf{D}_1$, indicates a matrix. A vector of discrete values will be denoted by an overhead arrow, such as $\vec{f}$. The elements of the matrix or the vector are denoted in italic font with appropriate subscripts, for example as $[D_1]_{jm}$ or f_m.

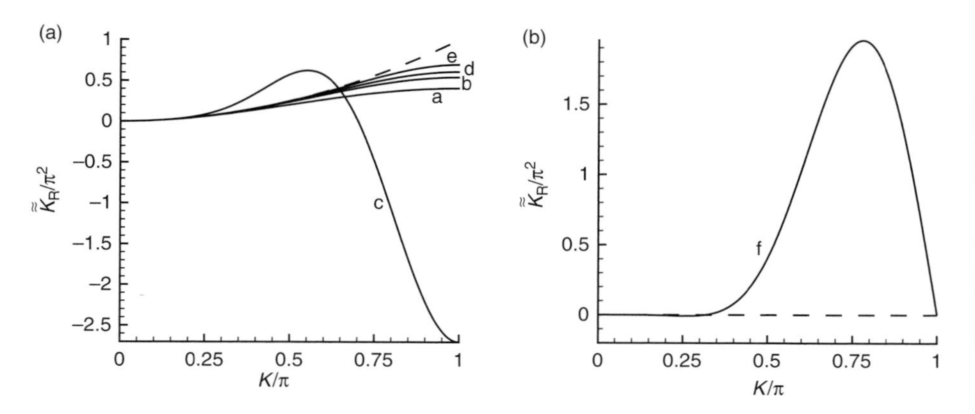

Figure 4.2 A plot of the real and imaginary parts of $\widetilde{\widetilde{K}}_{\rm R}$ (left) and $\widetilde{\widetilde{K}}_{\rm I}$ for different finite and compact difference approximations to the second-derivative operator (right). In each case the dashed lines give the exact answer. The curve marked (a) is for the three-point central difference scheme given in (4.1.24). The curve marked (b) is for the five-point central difference scheme given in (4.1.25). Curves marked (c) and (f) are for the five-point forward one-sided scheme given in (4.1.26b). The curve marked (d) is for the fourth-order tridiagonal compact scheme and the curve marked (e) is for the sixth-order 3–5 compact scheme. For the central and compact difference schemes $\widetilde{\widetilde{K}}_{\rm I} \equiv 0$. For the backward schemes the real part $\widetilde{\widetilde{K}}_{\rm R}$ remains the same, but the sign of the imaginary part $\widetilde{\widetilde{K}}_{\rm I}$ reverses.

an output vector of grid point values of the desired derivative. Different finite-difference approximations to the pth derivative yield different $\mathbf{D}_{\rm p}$, and thereby provide different discrete finite difference approximations to $\mathrm{d}^{\rm p}/\mathrm{d}x^{\rm p}$.

As an example let us construct the matrix operator corresponding to the three-point central difference approximation to the first derivative given in (4.1.1c). If the function f is periodic then the construction of the matrix operator becomes easy, since the three-point central difference can be uniformly applied at all points. Periodicity implies $f_N = f_0$, $f_{N+1} = f_1$ and so on. Thus it is sufficient to consider only points x_0 through x_{N-1}. In this case the discrete derivative operator can be rewritten as an $N \times N$ matrix for the computation of f_0' through f_{N-1}':

$$\begin{bmatrix} f_0' \\ f_1' \\ \vdots \\ f_{N-1}' \end{bmatrix} = \underbrace{\frac{1}{2\Delta x}\begin{bmatrix} 0 & 1 & 0 & \cdots & -1 \\ -1 & 0 & 1 & 0 & \cdots \\ 0 & \ddots & \ddots & \ddots & \ddots \\ \cdots & 0 & -1 & 0 & 1 \\ 1 & \cdots & 0 & -1 & 0 \end{bmatrix}}_{\mathbf{D}_1:\ \text{Matrix of size } N\times N} \begin{bmatrix} f_0 \\ f_1 \\ \vdots \\ f_{N-1} \end{bmatrix}. \tag{4.2.2}$$

The above matrix is banded and has zeros along the diagonal. It is known as an asymmetric *circulant* matrix. A circulant matrix is a special matrix where each row is a cyclic shift of the row above it.

If periodicity does not hold, then two changes are required. First, all $N+1$ points must be considered and as a result the derivative operator is a matrix of size $(N+1)\times(N+1)$. Second, one-sided approximations to the first derivative must be employed at the end points, x_0 and x_N. Correspondingly the first and last rows of the matrix differ from the rest. Implementation of the forward- and backward-biased three-point stencil at x_0 and x_N, respectively, will result in

$$\begin{bmatrix} f_0' \\ f_1' \\ \vdots \\ \\ f_N' \end{bmatrix} = \underbrace{\frac{1}{2\Delta x}\begin{bmatrix} -3 & 4 & -1 & 0 & \cdots \\ -1 & 0 & 1 & 0 & \cdots \\ 0 & \ddots & \ddots & \ddots & \ddots \\ \cdots & 0 & -1 & 0 & 1 \\ \cdots & 0 & 1 & -4 & 3 \end{bmatrix}}_{\mathbf{D}_1:\ \text{Matrix of size } (N+1)\times(N+1)} \begin{bmatrix} f_0 \\ f_1 \\ \vdots \\ \\ f_N \end{bmatrix}. \tag{4.2.3}$$

The above provides a second-order accurate discrete approximation to the first derivative at all points, including the boundary points. Except for the top and bottom rows, owing to the three-point central difference, the matrix is tridiagonal in the interior. With the top and bottom rows included, (4.2.3) becomes a pentadiagonal matrix. Higher-order finite difference approximations similarly yield banded matrices with the width of the band dependent on the finite difference stencil.

The second-order accurate, three-point central difference scheme for the second derivative of a periodic function results in the following $N\times N$ symmetric circulant matrix:

$$\begin{bmatrix} f_0'' \\ f_1'' \\ \vdots \\ \\ f_{N-1}'' \end{bmatrix} = \underbrace{\frac{1}{(\Delta x)^2}\begin{bmatrix} -2 & 1 & 0 & \cdots & 1 \\ 1 & -2 & 1 & 0 & \cdots \\ 0 & \vdots & \vdots & \vdots & \vdots \\ \cdots & 0 & 1 & -2 & 1 \\ 1 & \cdots & 0 & 1 & -2 \end{bmatrix}}_{\mathbf{D_2}:\ \text{Matrix of size } \mathbf{N}\times\mathbf{N}} \begin{bmatrix} f_0 \\ f_1 \\ \vdots \\ \\ f_{N-1} \end{bmatrix}. \tag{4.2.4}$$

If the function is not periodic then one-sided schemes are needed at the end points. A four-point stencil is required at the boundary points in order to maintain second-order accuracy. The resulting second-order accurate second derivative matrix is heptadiagonal

$$\begin{bmatrix} f_0'' \\ f_1'' \\ \vdots \\ \\ f_N'' \end{bmatrix} = \underbrace{\frac{1}{(\Delta x)^2} \begin{bmatrix} 2 & -5 & 4 & -1 & 0\cdots \\ 1 & -2 & 1 & 0 & \cdots \\ 0 & \ddots & \ddots & \ddots & \ddots \\ \cdots & 0 & 1 & -2 & 1 \\ \cdots 0 & -1 & 4 & -5 & 2 \end{bmatrix}}_{\mathbf{D}_2:\ \text{Matrix of size } (N+1)\times(N+1)} \begin{bmatrix} f_0 \\ f_1 \\ \vdots \\ \\ f_N \end{bmatrix} . \tag{4.2.5}$$

4.2.1 Eigenvalues of the Matrix Derivative Operator

The eigenvalues of the derivative matrix are very useful. First of all, the eigenvalues of the discrete derivative operator can be compared with those of the exact (continuous) derivative operator to gauge the accuracy of the finite difference approximation. Later in the context of PDEs the eigenvalues of the spatial operator will be considered along with the stability diagram of the time-integration scheme to evaluate the stability of the numerical solution to the PDE.

The eigenmodes of the first-derivative operator, $\mathrm{d}/\mathrm{d}x$, are given by $\mathrm{d}\xi/\mathrm{d}x = \lambda\xi$. With periodic boundary condition over the finite domain $0 \le x \le L_x$ the eigensolutions are $\exp[2\pi\iota kx/L_x]$ for $k = 0, \pm 1, \ldots, \pm\infty$ (this can be easily verified by substitution). The corresponding discrete spectrum of pure imaginary eigenvalues is

$$\iota\frac{2\pi k}{L_x} = \iota\frac{1}{\Delta x}\frac{2\pi k}{N} \quad \text{for} \quad k = 0, \pm 1, \ldots, \pm\infty . \tag{4.2.6}$$

For eigenvalues of the discretized approximation, we need to consider the following theorem for circulant matrices.

Theorem 4.2.1 *The eigenvalues of an $N \times N$ circulant matrix of the form*

$$\mathbf{C} = \begin{bmatrix} c_0 & c_1 & c_2 & \cdots & c_{n-1} \\ c_{n-1} & c_0 & c_1 & \cdots & c_{n-2} \\ \vdots & \vdots & \vdots & \vdots & \vdots \\ c_1 & c_2 & c_3 & \cdots & c_0 \end{bmatrix} \tag{4.2.7}$$

are given by

$$\lambda^{(k)} = \sum_{m=0}^{N-1} c_m \exp\left[-\iota\frac{2\pi mk}{N}\right] \text{ for } k = 0, 1, \ldots, N-1.$$

Proof The eigenvalues, λ, and eigenvectors, $\mathbf{v}$, are solutions of $\mathbf{Cv} = \lambda\mathbf{v}$. Written out in components they are

$$\sum_{m=0}^{N-1-p} c_m v_{m+p} + \sum_{m=N-p}^{N-1} c_m v_{m-(N-p)} = \lambda v_p \quad \text{for} \quad p = 0, 1, \ldots, N-1 . \tag{4.2.8}$$

Try a solution of the form $v_m = \rho^m$. Substituting into the above equation and canceling ρ^m yields

$$\sum_{m=0}^{N-1-p} c_m \rho^m + \rho^{-N} \sum_{m=N-p}^{N-1} c_m \rho^m = \lambda \,. \tag{4.2.9}$$

Choose $\rho^N = 1$, i.e., let ρ be a complex Nth root of unity. There are N such roots: $\rho^{(k)} = \exp\left[-2\pi\iota k/N\right]$ for $k = 0, 1, \ldots, N-1$. Substituting these roots in (4.2.9) we obtain the eigenvalues. The corresponding kth eigenvector (normalized to have length 1) is then

$$\mathbf{v}^{(k)} = \frac{1}{\sqrt{N}} \left[1, \mathrm{e}^{-2\pi\iota k/N}, \ldots, \mathrm{e}^{-2\pi\iota k(N-1)/N}\right] . \tag{4.2.10}$$

□

Applying Theorem 4.2.1 to the three-point central difference first-derivative operator $\mathbf{D}_1$ in (4.2.2) we obtain the following eigenvalues:

$$\iota \frac{1}{\Delta x} \sin\left(\frac{2\pi\, k}{N}\right) \quad \text{for} \quad k = -\frac{N}{2} + 1, \ldots, -1, 0, 1, \ldots, \frac{N}{2} \,. \tag{4.2.11}$$

Because of the periodic nature of the discretization, the eigenvalues of $\mathbf{D}_1$ are precisely the same as the modified wavenumbers of the von Neumann error analysis. Only for small values of k are the true eigenvalues well approximated. The error increases progressively as $k \to \pm N/2$. For the three-point central scheme the approximate eigenvalues remain pure imaginary and therefore there is no amplitude error. For one-sided discretization of the first derivative, the corresponding eigenvalues will be complex – indicating both amplitude and phase errors.

Eigenvalues of the second-derivative operator over a finite domain $0 \le x \le L_x$ with periodic boundary condition can be examined in a similar manner. The eigensolutions of the continuous operator $\mathrm{d}^2/\mathrm{d}x^2$ are again $\exp\left[2\pi\iota k x/L_x\right]$ for $k = 0, \pm 1, \ldots, \pm\infty$. The corresponding discrete spectrum of eigenvalues is

$$-\left(\frac{2\pi\, k}{L_x}\right)^2 = -\left(\frac{1}{\Delta x}\right)^2 \left(\frac{2\pi\, k}{N}\right)^2 \quad \text{for} \quad k = 0, \pm 1, \ldots \,. \tag{4.2.12}$$

Theorem 4.2.1 can be used to obtain the eigenvalues of the finite difference approximation. For the three-point central difference, the eigenvalues of $\mathbf{D}_2$, given in (4.2.4), are

$$\frac{2}{(\Delta x)^2}\left\{\cos\left(\frac{2\pi\, k}{N}\right) - 1\right\} \quad \text{for} \quad k = -\frac{N}{2} + 1, \ldots, -1, 0, 1, \ldots, \frac{N}{2} \,. \tag{4.2.13}$$

These eigenvalues are again consistent with the modified wavenumbers of the von Neumann error analysis. For small values of k the difference between the exact eigenvalues and their numerical approximation is small, but the difference increases as k increases and approaches $\pm N/2$. The eigenvalues of the three-point central difference are real and negative, just like the exact eigenvalues. Thus the error in this finite difference scheme is entirely amplitude error. As seen in the von Neumann analysis, for an asymmetric stencil there will also be phase error. Correspondingly the matrix operator will have complex eigenvalues.

For periodic problems the eigenvalues add nothing more to what we already knew from the von Neumann error analysis. The attractiveness of considering the eigenvalues arises for non-periodic problems. When solving ODEs and PDEs, the derivative operators must be used in conjunction with boundary conditions. In periodic problems, the boundary condition is simply the requirement that the function be periodic and therefore the periodic operators (4.2.2) and (4.2.4) already have the boundary condition incorporated. A non-periodic problem requires the application of a boundary condition either at x_0 or at x_N or both. The boundary condition must be included in the derivative matrix before the eigenvalues are computed. As we shall see later in Chapters 7 and 8, the eigenvalues of the matrix operator with the boundary condition incorporated operators play a role in the analysis of stability of numerical approximations to PDEs.

Dirichlet Boundary Conditions

First let us consider Dirichlet boundary conditions at both ends: $f(0) = f_0 = a$ and $f(L_x) = f_N = b$. The eigensolutions of the continuous second-derivative operator, $\mathrm{d}^2/\mathrm{d}x^2$, with homogeneous Dirichlet boundary conditions ($a = b = 0$) are $\sin[\pi\iota k x/L_x]$ for $k = 1, 2, \ldots$. The corresponding discrete spectrum of eigenvalues is

$$-\left(\frac{\pi k}{L_x}\right)^2 = -\left(\frac{1}{\Delta x}\right)^2\left(\frac{\pi k}{N}\right)^2 \quad \text{for} \quad k = 1, 2, \ldots . \tag{4.2.14}$$

The boundary conditions have two effects on the derivative matrix. First, when solving an ODE or a PDE, typically the derivative need not be computed at the boundary points, since only the boundary values are enforced there. Second, in the computation of the derivative at the interior points, the known value of the function at the boundary points can be used. For the three-point central difference scheme the interior portion of (4.2.5) can then be written as

$$\begin{bmatrix} f_1'' \\ f_2'' \\ \vdots \\ \\ f_{N-1}'' \end{bmatrix} = \underbrace{\frac{1}{(\Delta x)^2}\begin{bmatrix} -2 & 1 & 0 & \cdots & 0 \\ 1 & -2 & 1 & 0 & \cdots \\ 0 & \vdots & \vdots & \vdots & \vdots \\ \cdots & 0 & 1 & -2 & 1 \\ 0 & \cdots & 0 & 1 & -2 \end{bmatrix}}_{\mathbf{D}_{2\mathrm{I}}:\ \text{Matrix of size } (N-1)\times(N-1)} \begin{bmatrix} f_1 \\ f_2 \\ \vdots \\ \\ f_{N-1} \end{bmatrix} + \underbrace{\frac{1}{(\Delta x)^2}\begin{bmatrix} a \\ 0 \\ \vdots \\ \\ b \end{bmatrix}}_{\text{Boundary correction}} . \tag{4.2.15}$$

Here $\mathbf{D}_{2\mathrm{I}}$ is the **interior second-derivative matrix**, which for the three-point central difference scheme is a tridiagonal matrix with constant entries along the diagonals. Since the boundary points are avoided, the three-point central difference scheme is uniformly used at all the interior points. Note that the non-zero boundary values appear only as an additive vector. For the eigenvalues of the derivative matrix $\mathbf{D}_{2\mathrm{I}}$ we have the following theorem.

Theorem 4.2.2 *The eigenvalues of an $(N-1)\times(N-1)$ tridiagonal matrix, with constant triplet $\{\alpha, \beta, \gamma\}$ along the sub-diagonal, diagonal and super-diagonal, are $\lambda_k = \beta + 2\sqrt{\alpha\gamma}\cos[\pi k/N]$ for $k = 1, 2, \ldots, N-1$.*

Proof The eigenvalues of a matrix $\mathbf{A}$ are the roots of $\mathrm{Det}[\mathbf{A} - \lambda\mathbf{I}] = \mathbf{0}$, where $\mathbf{I}$ is the identity matrix. Define Det_{N-1} to be the determinant of the $(N-1)\times(N-1)$ tridiagonal matrix with constant triplet $\{\alpha, \beta - \lambda, \gamma\}$ along the diagonals. We can then obtain the recurrence relation: $\mathrm{Det}_{N-1} = (\beta - \lambda)\,\mathrm{Det}_{N-2} - \alpha\gamma\;\mathrm{Det}_{N-3}$. The recurrence relation can be iterated resulting in the following equation:

$$\mathrm{Det}_{N-1} = (\alpha\gamma)^{N-1/2}(\sin\theta)^{-1}\sin(N\theta), \quad \text{where} \quad \cos(\theta) = (\beta - \lambda)/2\sqrt{\alpha\gamma}. \tag{4.2.16}$$

Then $\mathrm{Det}_{N-1} = 0$ yields $\sin(N\theta) = 0$, i.e., $N\theta = k\pi$, from which the eigenvalues follow. Note that the solution $\lambda_k = \beta - 2\sqrt{\alpha\gamma}\cos[\pi k/N]$ is the same as in the theorem. □

The above theorem applies to $\mathbf{D}_{2\mathrm{I}}$ with $\alpha = \gamma = 1/\Delta x^2$ and $\beta = -2/\Delta x^2$, and thus the eigenvalues are

$$\frac{2}{(\Delta x)^2}\left\{\cos\left(\frac{\pi k}{N}\right) - 1\right\} \quad \text{for} \quad k = 1, \ldots, N-1\,. \tag{4.2.17}$$

With a Taylor series expansion it can be readily seen that for small values of k the leading error between the exact eigenvalues (4.2.14) and their numerical approximation (4.2.17) goes as $\pi^4 k^4/(12\Delta x^2 N^4)$. Thus for lower-order eigenmodes (i.e., for $k \ll N$) the derivatives are well approximated by the three-point central difference scheme. The accuracy of the numerical approximation degrades as k increases and approaches N. Furthermore, as with the periodic problem, the central difference introduces only amplitude error and involves no phase error. This property of central differencing is not affected by the Dirichlet boundary conditions.

Neumann Boundary Conditions

Now let us discuss the Neumann boundary conditions at both ends: i.e., $f_0' = a$ and $f_N' = b$. The eigensolutions of the continuous second-derivative operator, $\mathrm{d}^2/\mathrm{d}x^2$, with the homogeneous Neumann boundary conditions ($a = b = 0$) are $\cos[\iota\pi k x/L_x]$ for $k = 0, 1, \ldots$. The corresponding discrete spectrum of eigenvalues is

$$-\left(\frac{\pi k}{L_x}\right)^2 = -\left(\frac{1}{\Delta x}\right)^2\left(\frac{\pi k}{N}\right)^2 \quad \text{for} \quad k = 0, 1, \ldots\,. \tag{4.2.18}$$

The eigenvalues with the Neumann boundary conditions are the same as those with the Dirichlet boundary conditions, except for the important difference that with the Neumann boundary condition there is a **null mode** corresponding to a zero eigenvalue. The eigenfunction of the null mode is a constant vector.

Three-point one-sided approximations to the Neumann boundary conditions are

$$\frac{-3f_0 + 4f_1 - f_2}{2\Delta x} = a \quad \text{and} \quad \frac{3f_N - 4f_{N-1} + f_{N-2}}{2\Delta x} = b\,.$$

These two conditions can now be used to write f_0 and f_N in terms of the interior points as

$$f_0 = \frac{4}{3}f_1 - \frac{1}{3}f_2 - \frac{2}{3}a\Delta x \quad \text{and} \quad f_N = \frac{4}{3}f_{N-1} - \frac{1}{3}f_{N-2} + \frac{2}{3}b\Delta x\,.$$

Going back to the three-point central difference second-derivative operator given in (4.2.5), the first and the last rows that correspond to the second derivative at the boundary points are not needed. The first and last columns of the second-derivative operator given in (4.2.5) operate on f_0 and f_N. Using the above equations we can substitute these boundary values f_0 and f_N in terms of the interior values, and the interior portion of (4.2.5) can then be written as

$$\begin{bmatrix} f_1'' \\ f_2'' \\ \vdots \\ \\ f_{N-1}'' \end{bmatrix} = \underbrace{\frac{1}{(\Delta x)^2}\begin{bmatrix} -2/3 & 2/3 & 0 & \cdots & 0 \\ 1 & -2 & 1 & 0 & \cdots \\ 0 & \vdots & \vdots & \vdots & \vdots \\ \cdots & 0 & 1 & -2 & 1 \\ 0 & \cdots & 0 & 2/3 & -2/3 \end{bmatrix}}_{\mathbf{D}_{2\mathrm{I}}:\ \text{Matrix of size } (N-1)\times(N-1)} \begin{bmatrix} f_1 \\ f_2 \\ \vdots \\ \\ f_{N-1} \end{bmatrix} + \underbrace{\frac{2}{3\Delta x}\begin{bmatrix} -a \\ 0 \\ \vdots \\ \\ b \end{bmatrix}}_{\text{Boundary correction}}\,. \tag{4.2.19}$$

Here $\mathbf{D}_{2\mathrm{I}}$ is the interior second-derivative matrix that incorporates the Neumann boundary conditions. It is still a tridiagonal matrix with constant entries along the diagonals, except in the first and the last row. Note that second-order accuracy is maintained at all the points, including the boundaries.

The spectrum of eigenvalues of the continuous second-derivative operator with Dirichlet or Neumann boundary conditions is shown in Figure 4.3. The spectrum is plotted as a function of k/N. For all values of k and N the discrete spectrum of eigenvalues (4.2.14) or (4.2.18) falls on the continuous curve shown in the figure. Also shown are the eigenvalues of the three-point central difference approximation to the second-derivative operator. For the case of Dirichlet boundary conditions the eigenvalues are explicitly given by (4.2.17).

For Neumann boundary conditions the eigenvalues of $\mathbf{D}_{2\mathrm{I}}$ given in (4.2.19) are evaluated numerically. Unlike the earlier cases the discrete spectrum of eigenvalues is not given by a simple expression valid for all k and N. For large N the effect of boundary condition decreases and the spectrum approaches that of Dirichlet boundary conditions. However, the discrete spectrum for Neumann boundary conditions includes the zero eigenvalue (null mode), which is absent in the case of Dirichlet boundary conditions. In this respect the discrete operator accurately captures the behavior of the continuous operator. The implication of

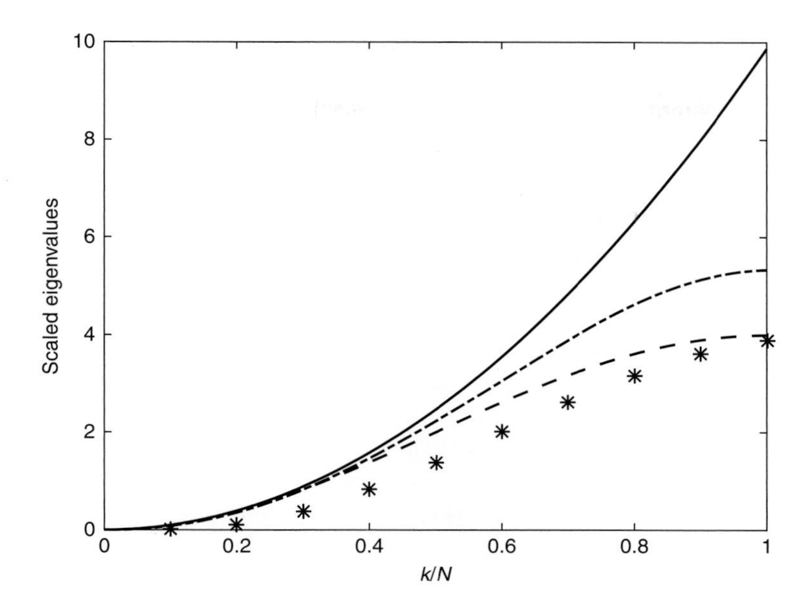

Figure 4.3 A plot of the eigenvalues of the second-derivative operator with Dirichlet and Neumann boundary conditions. The eigenvalues, λ_k, are real and negative for all cases considered. Here we plot $-\lambda_k(L_x/N)^2$. Solid line: the scaled eigenvalues of the continuous second-derivative operator with Dirichlet–Dirichlet or Neumann–Neumann boundary conditions at the two ends. Dashed line: scaled eigenvalues of the three-point central difference approximation to the second derivative with Dirichlet–Dirichlet boundary conditions. Symbol: scaled eigenvalues of the three-point central difference approximation to the second derivative with Neumann–Neumann boundary condition for $N = 11$. As $N \to \infty$ the result for Neumann–Neumann boundary conditions approaches that for Dirichlet–Dirichlet boundary conditions (dashed line). Dash–dot line: scaled eigenvalues of the five-point central difference approximation to the second derivative with Dirichlet–Dirichlet boundary conditions. Here a five-point one-sided scheme is used as we approach the end points.

this zero eigenvalue in the context of solving ODEs and PDEs with Neumann boundary conditions will be discussed in Chapters 5, 7 and 8. For comparison, the eigenvalues of $\mathbf{D}_{2\mathrm{I}}$ (given in (4.2.19)) are shown for $N = 11$ in Figure 4.3. Note that the eigenvalues of the finite difference approximation span only the range $0 \leq k/N \leq 1$. The eigenvalues of the continuous second-derivative operator, however, extend beyond $k = N$. These higher eigenmodes are not representable with a finite N-point discretization.

Computer Exercise 4.2.3 Consider the five-point central difference approximation to the second derivative (4.1.25) at the interior points, with corresponding five-point one-sided discretization (4.1.26a)–(4.1.26b) at x_1 and x_{N-1}. Construct the interior (boundary incorporated) second-derivative operator for Dirichlet boundary conditions at both ends. Compute the eigenvalues. For large N your result must be in agreement with that shown in Figure 4.3.

Computer Exercise 4.2.4 Von Neumann error analysis is strictly valid only for periodic problems. The presence of boundaries can have a significant influence on the actual error. This was well evident in the distribution of eigenvalues of the first- and second-derivative operators. Furthermore, the distribution of error need not be homogeneous along x; the level of error may vary significantly with distance from the boundaries. In this computer exercise we will numerically investigate the error analysis of finite difference schemes with the inclusion of boundary conditions. Without loss of generality consider a domain of $0 \leq x \leq 1$ (i.e., $L_x = 1$) and define a sequence of *test functions* in this domain through the following recurrence relation:

$$T_0(x) = 1\,;\ \ T_1(x) = (2x-1)\,;\ \ T_k(x) = (4x-2)T_{k-1}(x) - T_{k-2}(x) \ \text{ for } \ k \geq 2. \tag{4.2.20}$$

These are nothing but the Chebyshev polynomials, here simply scaled to fit within the domain. From (4.2.20) it is easy to see that T_k is a kth-order polynomial and, from the property of Chebyshev polynomials, $|T_k(x)| \leq 1$ over the domain $0 \leq x \leq 1$ for all k. The above sequence of polynomials will replace the sequence of periodic functions (4.1.12) in our numerical error analysis. Consider two different numerical schemes:

(1) Uniform grid of N points (x_0 to x_N) with a three-point central difference in the interior and three-point one-sided schemes at the end points for the first derivative (matrix (4.2.3)).
(2) Uniform grid of N points with a five-point central difference in the interior (4.1.8) and five-point one-sided schemes (4.1.9b) at the end point, and (4.1.9a) at the next-to-end point for the first derivative.

Use the above two schemes to evaluate the derivative of the test functions at the grid points $T'_{k,j}$ from the grid-point values of the test functions. The corresponding exact derivatives are given by the recursion

$$\left.\begin{aligned} T'_0(x) = 0\,;\ \ T'_1(x) &= 2\,; \ \ T'_2(x) = 16x - 8\,; \\ \frac{T'_k(x)}{k} &= 4T_{k-1}(x) + \frac{T'_{k-2}(x)}{k-2} \ \text{ for } \ k > 2\,. \end{aligned}\right\} \tag{4.2.21}$$

An approximate, but simple measure of relative error can now be defined as

$$E_k = \frac{1}{[I_k]^{\frac{1}{2}}} \left[\sum_{j=1}^{N-1} \frac{\left(T'_k(x_j) - T'_{k,j}\right)^2}{\sqrt{x_j(1-x_j)}} \right]^{\frac{1}{2}}, \tag{4.2.22}$$

where $T'_k(x_j)$ is the exact derivative and $T'_{k,j}$ is the numerical approximation.

The normalization factor $I_k = \int_0^1 \frac{(T_k'(x))^2}{\sqrt{x(1-x)}} dx$ is given by the recurrence

$$I_0 = 0\,; I_1 = 4\pi\,; I_2 = 32\pi \quad \text{and} \quad I_k = 8\pi\, k^2 + \frac{k^2}{(k-2)^2} I_{k-2}\,. \tag{4.2.23}$$

Note that the end points have been avoided in the sum. For the two different schemes plot E_k versus k. Try $N = 32$ and k in the range 0 to 32. For a few different values of k, plot the exact derivative along with its numerical approximation as a function of x. Plot the error as well as a function of x. Can you observe any trend in the error as the boundaries are approached? Also plot, say E_{10} versus N, for N in the range 10 to 200 on a log–log plot and comment on the slope of the resulting curves.

Computer Exercise 4.2.5 Repeat the above error analysis for the second derivative.

Problem 4.2.6 In the section to follow on compact differences, and later in Chapter 6 on spectral methods, we will define more accurate second-derivative matrices that will not be banded or sparse. In this problem you will obtain the interior operator for a general non-banded second-derivative matrix. Start by defining the vector of function values at the interior nodes and the vector of second-derivative values at the interior nodes as

$$\vec{f}_I = \begin{bmatrix} f_1 \\ f_2 \\ \vdots \\ \\ f_{N-1} \end{bmatrix} \quad \text{and} \quad \vec{f}_I'' = \begin{bmatrix} f_1'' \\ f_2'' \\ \vdots \\ \\ f_{N-1}'' \end{bmatrix}\,. \tag{4.2.24}$$

Let the Dirichlet boundary conditions at the two ends be $f_0 = a$ and $f_N = b$. These vectors are of length $(N-1)$. Show that the generalization of (4.2.15) for any second-derivative matrix can be written as

$$\vec{f}_{\mathrm{I}}'' = \mathbf{D}_{2\mathrm{I}}\, \vec{f}_{\mathrm{I}} + a\,\vec{C}_{\mathrm{LI}} + b\,\vec{C}_{\mathrm{RI}}\,, \tag{4.2.25}$$

where $\mathbf{D}_{2\mathrm{I}}$ is the interior second-derivative matrix of size $(N-1)\times(N-1)$, $\vec{C}_{\mathrm{LI}}$ and $\vec{C}_{\mathrm{RI}}$ are left and right correction vectors of size $(N-1)$. Show that $\mathbf{D}_{2\mathrm{I}}$, $\vec{C}_{\mathrm{LI}}$ and $\vec{C}_{\mathrm{RI}}$ are independent of the boundary values a and b and can be defined in terms of the elements of the regular second-derivative matrix $\mathbf{D}_2$ as

$$[D_{2\mathrm{I}}]_{lm} = [D_2]_{lm} \quad \text{for} \quad l, m = 1, 2, \ldots, N-1 \tag{4.2.26}$$

and

$$[C_{\mathrm{LI}}]_l = [D_2]_{l0} \quad \text{and} \quad [C_{\mathrm{RI}}]_l = [D_2]_{lN} \quad \text{for} \quad l = 1, 2, \ldots, N-1\,. \tag{4.2.27}$$

Problem 4.2.7 In this problem you will obtain the interior operator for a general second-derivative matrix with Neumann boundary conditions at one end and Dirichlet boundary conditions at the other end. Let the boundary conditions at the two ends be $f'_0 = a$ and $f_N = b$. Show that (4.2.25) still applies. Furthermore, show that $\mathbf{D}_{2\mathrm{I}}$, $\vec{C}_{\mathrm{LI}}$ and $\vec{C}_{\mathrm{RI}}$ are still independent of the boundary values a and b. They can be defined in terms of the elements of the regular first- and second-derivative matrices as

$$[D_{2\mathrm{I}}]_{lm} = [D_2]_{lm} - \frac{[D_2]_{l0}\,[D_1]_{0m}}{[D_1]_{00}} \quad \text{for} \quad l, m = 1, 2, \ldots, N-1 \tag{4.2.28}$$

and

$$[C_{\mathrm{LI}}]_l = \frac{[D_2]_{l0}}{[D_1]_{00}} \quad \text{and} \quad [C_{\mathrm{RI}}]_l = [D_2]_{lN} - \frac{[D_2]_{l0}\,[D_1]_{0N}}{[D_1]_{00}} \quad \text{for } l = 1, 2, \ldots, N-1. \tag{4.2.29}$$

Problem 4.2.8 Repeat Problem 4.2.7 for Neumann boundary conditions at both ends: $f'_0 = a$ and $f'_N = b$. Show that (4.2.25) still applies and obtain the corresponding expressions for $\mathbf{D}_{2\mathrm{I}}$, $\vec{C}_{\mathrm{LI}}$ and $\vec{C}_{\mathrm{RI}}$ in terms of the elements of the regular first- and second-derivative matrices.

4.3 Compact Differences

With finite difference approximations for accuracy higher than fourth order it is necessary to expand the stencil beyond five points. One way to obtain higher-order accuracy is to consider **compact difference schemes** of the form

$$\alpha f'_{j-1} + f'_j + \alpha f'_{j+1} = a\,[f_{j+1} - f_{j-1}] + b\,[f_{j+2} - f_{j-2}]\,. \tag{4.3.1}$$

Based on central difference approximation the coefficients have been chosen to be symmetric on the left-hand side and antisymmetric on the right-hand side. For non-zero α, the derivative at each point is coupled to the derivative at its two neighbors and in this fashion the derivative at any point is globally dependent on the value of the function over the entire domain.

From Taylor series expansion of the left-hand side one obtains

$$(1 + 2\alpha)\, f'_j + \alpha\, f'''_j\, (\Delta x)^2 + \frac{\alpha}{12}\, f^{\mathrm{v}}_j\, (\Delta x)^4 + \cdots \tag{4.3.2}$$

and from a Taylor series expansion of the right-hand side

$$2(a+2b)\, f_j'\, \Delta x + \frac{(a+8b)}{3}\, f_j'''\, (\Delta x)^3 + \frac{(a+32b)}{60}\, f_j^{\mathrm{v}}\, (\Delta x)^5 + \cdots . \tag{4.3.3}$$

Matching the left- and right-hand sides, the following three relations between the coefficients in (4.3.1) can be obtained:

$$\left.\begin{aligned} (1+2\alpha) &= 2(a+2b)\,\Delta x, \\ \alpha &= \frac{(a+8b)}{3}\Delta x, \\ \frac{\alpha}{12} &= \frac{(a+32b)}{60}\Delta x . \end{aligned}\right\} \tag{4.3.4}$$

Matching only the first two of (4.3.4), a one-parameter family of fourth-order methods can be obtained with α as the parameter and $a = (\alpha+2)/(3\Delta x)$, $b = (4\alpha - 1)/(12\Delta x)$. In particular, a fourth-order-accurate, purely tridiagonal system on a three-point stencil results for $\alpha = \frac{1}{4}$:

$$f_{j-1}' + 4\, f_j' + f_{j+1}' = \frac{3}{\Delta x}\, [f_{j+1} - f_{j-1}] . \tag{4.3.5}$$

The above fourth-order tridiagonal scheme goes by the name **Padé scheme**. All three equations in (4.3.4) are satisfied with $\alpha = 1/3$, $a = 7/9\Delta x$ and $b = 1/36\Delta x$ and the resulting sixth-order-accurate compact difference scheme can be written as

$$f_{j-1}' + 3\, f_j' + f_{j+1}' = \frac{7}{3\Delta x}\, [f_{j+1} - f_{j-1}] + \frac{1}{12\Delta x}\, [f_{j+2} - f_{j-2}] . \tag{4.3.6}$$

The local truncation error of (4.3.5) is due to the mismatch in the last term of the Taylor series expansions shown in (4.3.2) and (4.3.3). The error can be estimated as $-f^{\mathrm{v}}\, (\Delta x)^4/120$, which is a factor four smaller than that of the fourth-order accurate five-point central difference scheme (4.1.8). The local truncation of the sixth-order scheme (4.3.6) is $-f^{\mathrm{vii}}\, (\Delta x)^6/1260$, a very small value indeed.

Problem 4.3.1 Apply von Neumann error analysis to the compact difference schemes of the form (4.3.1) and show that

$$\widetilde{K}_{\mathrm{I}} = \frac{2a\,\Delta x \sin(K) + 2b\,\Delta x \sin(2K)}{1 + 2\alpha\cos(K)}, \qquad \widetilde{K}_{\mathrm{R}} = 0 . \tag{4.3.7}$$

Thus (4.3.1) is free of amplitude error and has only phase or dispersion error (see Figure 4.1).

Both (4.3.5) and (4.3.6) can be written as a matrix equation for the grid point values of the function and its first derivative

$$\mathbf{L}\, \vec{f'} = \mathbf{R}\, \vec{f} . \tag{4.3.8}$$

In the above compact schemes $\mathbf{L}$ is a tridiagonal matrix, while $\mathbf{R}$ is a tridiagonal matrix in the fourth-order compact scheme (4.3.5) and pentadiagonal in the sixth-order compact scheme (4.3.6). An effective-derivative matrix for the compact scheme can be defined as $\mathbf{D}_1 = (\mathbf{L}^{-1}\mathbf{R})$, which when multiplied by the vector of grid point values of the function yields the corresponding vector of first derivatives. The matrix $\mathbf{D}_1$ thus defined, although diagonally dominant, is not strictly banded since the inversion operation destroys the bandedness of $\mathbf{L}$. Calculation of $\vec{f}'$ from the banded linear system (4.3.8) is often computationally more efficient than explicitly calculating it from the first-derivative matrix $\mathbf{D}_1$. Computational effort in solving (4.3.8) scales as order N, while multiplication of $\vec{f}$ by the dense (non-banded) matrix $\mathbf{D}_1$ scales as N^2.

Higher-order compact difference schemes can be obtained with broader stencil on both the left- and right-hand sides of (4.3.1). Following the procedure outlined above, asymmetric and one-sided compact difference schemes can be obtained for usage near the boundaries. Further information on compact difference schemes can be found in Lele (1992) and in classic texts on numerical methods (Collatz, 1960; Kopal, 1961) under the title **Hermitian methods**, since compact difference schemes can be viewed as arising from a Hermite interpolation polynomial as opposed to the Lagrange interpolation polynomial for the finite difference schemes. We particularly recommend taking a look at the various finite difference and compact-difference approximations given in the appendix of Collatz (1960).

Compact difference approximations to the second derivative of up to sixth-order accuracy can be expressed in a form similar to (4.3.1):

$$\alpha f''_{j-1} + f''_j + \alpha f''_{j+1} = a\,[f_{j+1} - 2f_j + f_{j-1}] + b\,[f_{j+2} - 2f_j + f_{j-2}]\,. \quad (4.3.9)$$

By expanding both the left- and right-hand sides of (4.3.9) in Taylor series and equating like terms one obtains three algebraic relations between the constants a, b and α. A family of fourth-order schemes can be obtained with α as the parameter. Of particular interest is the following fourth-order accurate tridiagonal scheme which is obtained by setting $b = 0$:

$$f''_{j-1} + 10\,f''_j + f''_{j+1} = \frac{12}{(\Delta x)^2}\,[f_{j+1} - 2f_j + f_{j-1}]\,. \quad (4.3.10)$$

Also of interest is the following sixth-order scheme that results for $\alpha = 2/11$:

$$2f''_{j-1} + 11\,f''_j + 2f''_{j+1} = \frac{12}{(\Delta x)^2}\,[f_{j+1} - 2f_j + f_{j-1}] + \frac{3}{4(\Delta x)^2}\,[f_{j+2} - 2f_j + f_{j-2}]\,. \quad (4.3.11)$$

The local truncation errors for (4.3.10) and (4.3.11) are respectively $f^{\mathrm{vi}}\,(\Delta x)^4/20$ and $-23f^{\mathrm{viii}}\,(\Delta x)^6/5040$; here and henceforth a roman numeral superscript indicates the order of the derivative. For example, f^{vi} indicates the sixth derivative.

Problem 4.3.2 Apply von Neumann error analysis to compact difference schemes of the form (4.3.9) and show that

$$\widetilde{\widetilde{K}}_{\rm R} = \frac{2a\,(\Delta x)^2\,\{1-\cos(K)\} + 2b\,(\Delta x)^2\,\{1-\cos(2K)\}}{1+2\alpha\cos(K)}\,, \qquad \widetilde{\widetilde{K}}_{\rm I} = 0\,. \quad (4.3.12)$$

Thus (4.3.9) is free of phase error and has only amplitude error (see Figure 4.2).

Computer Exercise 4.3.3 Construct the effective-derivative matrix for fourth- and sixth-order compact-difference schemes. Form the interior matrix for Dirichlet boundary conditions at both ends. Compute the eigenvalues for $N = 100$ and compare them with those in Figure 4.3.

Computer Exercise 4.3.4 Repeat the error analysis of Computer Exercises 4.2.4 and 4.2.5 for the fourth-order and sixth-order compact schemes and compare with the results of the finite-difference approximations.

4.4 Non-uniform Discretization

The equal spacing (uniform distribution) of grid points that we have been using is often not the optimal choice. It was convenient to consider a uniform distribution of grid points in order to address discretization errors. Intuitively, however, it makes more sense to bunch grid points in regions where there is rapid variation of the flow quantities and fewer points in regions of relative inactivity. For example, in the case of a boundary layer the most rapid variation in velocity is confined to the region close to the wall. Away from the wall the variation decreases monotonically and by definition all variation ceases at distances from the wall that are greater than the boundary layer thickness. It is then natural to cluster grid points close to the wall and spread them out away from the wall. In fact, if the entire computational domain is discretized with a finite number of grid points, it is imperative that the grid spacing increase as one approaches infinity. More often than not the spatial domain will be truncated at a large but finite distance, which is greater than several boundary layer thicknesses. Even in this case it is computationally more advantageous to vary the grid distribution smoothly over the computational domain.

Internal flows such as flow through pipes and ducts have solid boundaries along two directions (in 3D). A sensible approach in these problems is to cluster grid points along the boundaries with relatively fewer points in the interior. By contrast in the case of free shear layers such as mixing layers, jets and wake flows, the most rapid variation in the velocity field occurs in the interior of the computational domain, while away from the shear layer relatively quiet conditions prevail. In these problems the clustering of grid points should be in the interior of the computational domain and the "free-stream" can be handled using a more sparse distribution of grid points.

Apart from the above fluid mechanical arguments, there are compelling numerical reasons for choosing a non-uniform grid. It can be shown rigorously that for periodic problems, where there is spatial homogeneity, the optimum distribution of grid points is indeed uniform! For non-periodic problems, on the other hand, it is necessary to cluster grid points in order to maintain a uniform level of accuracy over the entire computational domain. Several arguments can be presented to illustrate this point. First of all, discretization error tends to increase close to boundaries as can be seen from the five-point stencil for the first derivative: the central difference is about six times more accurate than the one-sided scheme. This effect is more evident for the second derivative, where the order of accuracy is one less for the one-sided difference. Computer Exercises 4.2.4, 4.2.5 and 4.3.4 are very instructive in this regard. In particular, the progressive decrease in accuracy towards the end points of the interval will be well evident in the plots of normalized error versus x.

On an equi-spaced grid it becomes difficult to construct a stable, very high-order finite difference approximation. The finite difference weights such as a, b, c, etc. in (4.1.3) increase for the one-sided schemes near the boundaries. As a result the derivative matrix becomes increasingly sensitive and prone to large errors especially near the boundaries. Thus, it can be argued that Δx must decrease close to the boundaries in order to compensate for the less accurate one-sided schemes. A quadratic clustering of points with grid spacing that decreases as N^{-2} close to the boundaries, where N is the number of grid points, is often sufficient to obtain uniform accuracy over the entire domain. For a very interesting discussion of the optimal distribution of grid points see the book on pseudo-spectral methods by Fornberg (1998).

The most convenient way to describe a non-uniform distribution of points x_j over the domain $0 \leq x \leq L_x$ is in terms of a uniform distribution of points $\tilde{x}_j = j\, L_x/N$, through a mapping $x_j = h(\tilde{x}_j)$, where h is the mapping function. For example, a Chebyshev–Gauss–Lobatto distribution of grid points is obtained with the mapping

$$x = \frac{L_x}{2}\left[1 - \cos\left(\frac{\pi\,\tilde{x}}{L_x}\right)\right] \tag{4.4.1}$$

and it provides a quadratic clustering of points near $x = 0$ and $x = L_x$. As we will see later in the context of spectral methods, the above mapping provides a widely used distribution of grid points. Several good theoretical reasons for its use in spectral methods are provided by Fornberg (1998). In the context of finite and compact difference schemes, other clusterings of points are also possible.

Finite difference and compact difference schemes on non-uniform grids can be obtained in a straightforward manner using Taylor series expansion. The three-point central difference approximation to the first derivative becomes

$$f'_j = -\frac{\Delta x_+ \, f_{j-1}}{\Delta x_- (\Delta x_+ + \Delta x_-)} + \frac{(\Delta x_+ - \Delta x_-) \, f_j}{\Delta x_+ \, \Delta x_-} + \frac{\Delta x_- \, f_{j+1}}{\Delta x_+ (\Delta x_+ + \Delta x_-)}, \quad (4.4.2)$$

where $\Delta x_- = x_j - x_{j-1}$ and $\Delta x_+ = x_{j+1} - x_j$. Higher-order and asymmetric approximations on a non-uniform grid can be obtained in a similar manner for both the first- and second-derivative operators.

Problem 4.4.1 Show that the leading-order error for (4.4.2) is $\Delta x_+ \Delta x_- f'''/6$.

An alternative approach is to make use of the chain rule

$$\mathrm{d}f/\mathrm{d}x = (\mathrm{d}\tilde{x}/\mathrm{d}x)\,(\mathrm{d}f/\mathrm{d}\tilde{x})$$

and define the derivative approximation on the non-uniform grid $(\mathrm{d}/\mathrm{d}x)$ based on that of the uniform grid $(\mathrm{d}/\mathrm{d}\tilde{x})$ along with the metric of the mapping between the two: $\mathrm{d}\tilde{x}/\mathrm{d}x$ or $(\mathrm{dx}/\mathrm{d}\tilde{x})^{-1}$. In other words, if the discrete derivative matrix on the non-uniform grid is defined by analogy with (4.2.1) as

$$f'(x_j) = [D_{1,\mathrm{NU}}]_{jm} f(x_m)\,, \quad (4.4.3)$$

then it can be expressed in terms of the first-derivative matrix of a uniform grid $\mathbf{D}_1$ through the following matrix multiplication:

$$[D_{1,\mathrm{NU}}]_{jm} = [M]_{jp}\,[D_1]_{pm}\,, \quad (4.4.4)$$

where $\mathbf{M}$ is the diagonal mapping matrix whose elements are given by

$$[M]_{jp} = \begin{cases} \left(\dfrac{\mathrm{d}x}{\mathrm{d}\tilde{x}}\right)^{-1}_{x_j} & \text{for} \quad j = p\,, \\ 0 & \text{for} \quad j \neq p\,. \end{cases} \quad (4.4.5)$$

Thus, in this simple construction, each row of the standard derivative matrix is multiplied by the metric of the local grid transformation to obtain the discrete derivative on a non-uniform grid. Special care is needed if $\mathrm{d}\tilde{x}/\mathrm{d}x$ approaches zero as in the case of (4.4.1) at $x = 0$ and at $x = L_x$.

The derivative matrix thus defined will clearly be different from those defined based on Taylor series expansion. For example, (4.1.2c), even after multiplication by the metric of local grid transformation, is fundamentally different from the approximation given in (4.4.2). The performance of the two approaches can be compared by computing their eigenvalue distribution and also by performing the error analysis of Computer Exercises 4.2.3 and 4.2.4. But we leave this as an exercise for the reader.

4.5 Numerical Interpolation

We now consider the problem of **numerical interpolation**, where given the function values at a set of grid points our task is to find the function value at an off-grid point. In other words, given the value of a function at only a set of grid points as f_j for $j = 0, 1, \ldots, N$ through interpolation we can reconstruct the function $f(x)$, by knowing the value of the function at all the points within the domain. In fact, we have already implicitly considered interpolation. It is only through the construction of the interpolated function $f(x)$ that we were able to obtain the derivatives at the grid points.

The simplest interpolation is linear interpolation between neighboring grid points. Here the function $f(x)$ between the grid points x_j and x_{j+1} is interpolated by a straight line fit between f_j and f_{j+1}. In a general framework we can interpret this as the first-order Lagrange interpolation polynomial that can be represented as

$$f_I(x) = f_j \left(\frac{x - x_{j+1}}{x_j - x_{j+1}} \right) + f_{j+1} \left(\frac{x - x_j}{x_{j+1} - x_j} \right) \quad \text{for} \quad x_j \le x \le x_{j+1} \,. \tag{4.5.1}$$

The above defines the interpolation approximation $f_I(x)$ only over the interval between the jth and the $(j+1)$th grid point.

Problem 4.5.1 Taylor series expand both the left- and right-hand sides of (4.5.1) and x_j and show the leading-order interpolation error is

$$\frac{\xi(1-\xi)}{2!} \Delta x^2 f_j'' \,. \tag{4.5.2}$$

The interpolation error can also be evaluated with von Neumann error analysis. We again consider the sequence of test functions given in (4.1.11). These test functions are sampled at the $(N+1)$ equi-spaced grid points and interpolated in between the grid points using (4.5.1). In terms of the scaled wavenumber $K = k(2\pi/N)$ the exact and the interpolated test functions are the same at the jth and the $(j+1)$th grid points:

$$f_k(x_j) = f_{I,k,j} = \exp[\iota K j] \quad \text{and} \quad f_k(x_{j+1}) = f_{I,k,j+1} = \exp[\iota K (j+1)] \,. \tag{4.5.3}$$

However, the exact and the interpolated functions will differ in between the grid points. If we define $\xi = (x - x_j)/\Delta x$ as the scaled distance from the jth grid point, in terms of ξ the exact test function in between the jth and the $(j+1)$th grid points can be expressed as

$$f_k(x) = \exp[\iota K (j + \xi)] \,, \tag{4.5.4}$$

and the numerical approximation will be accordingly expressed as

$$f_{I,k}(x) = \zeta \exp[\iota K (j + \xi)] \,, \tag{4.5.5}$$

where ζ is a complex coefficient. If $\zeta = 1$ then interpolation is exact. Otherwise, $|1-\zeta|^2$ provides a measure of mean square interpolation error. From (4.5.1) the interpolated polynomial becomes

$$f_{I,k}(x) = (1-\xi)\,\exp[\iota K j] + \xi\,\exp[\iota K(j+1)]\,, \tag{4.5.6}$$

and by equating (4.5.5) and (4.5.6) we obtain

$$\zeta = \frac{(1-\xi) + \xi(\cos K + \iota \sin K)}{\cos(K\xi) + \iota \sin(K\xi)}\,. \tag{4.5.7}$$

The interpolation error is identically zero at the grid points (i.e., at $\xi = 0$ and 1).

The mean square interpolation error can be defined as

$$\frac{1}{L_x}\int_0^{L_x} \left[f_k(x) - f_{I,k}(x)\right]^2\,\mathrm{d}x\,, \tag{4.5.8}$$

where the integration is over the entire domain. For the linear interpolation between the grid points being discussed here the mean square error can be written as

$$\frac{1}{L_x}\sum_{j=0}^{N-1} \Delta x \int_0^1 \left[1-\zeta(\xi)\right]^2\,\mathrm{d}\xi\,. \tag{4.5.9}$$

Substituting (4.5.7) for $\zeta(\xi)$ we obtain the following final expression for the mean square interpolation error:

$$-\frac{4}{K^2} + \left(\frac{4}{K^2} + \frac{1}{3}\right)\cos(K) + \frac{5}{3}\,. \tag{4.5.10}$$

As can be expected, the error is identically zero for $K = 0$ and increases quadratically for small K to reach its maximum value of $(4/3) - (8/\pi^2)$ at the Nyquist limit of $K = \pi$.

Instead of linear interpolation between f_j and f_{j+1}, one can consider quadratic interpolation as given in (4.1.10). This quadratic interpolation was obtained by including f_{j-1} at x_{j-1} as the third function value. For $x_j \le x \le x_{j+1}$, we also have the option of considering f_{j+2} at x_{j+2} as the third function value. Both approximations are one-sided, therefore let us proceed to the next level of approximation, where the function within the gap $x_j \le x \le x_{j+1}$ is interpolated with the function values at two points on either side. The resulting cubic polynomial can be written in terms of ξ as

$$\begin{aligned} f_I(x) \;=\;& f_{j-1}\frac{\xi(\xi-1)(\xi-2)}{(-1)(-2)(-3)} + f_j\frac{(\xi+1)(\xi-1)(\xi-2)}{(1)(-1)(-2)} \\ &+ f_{j+1}\frac{(\xi+1)\xi(\xi-2)}{(2)(1)(-1)} + f_{j+2}\frac{(\xi+1)\xi(\xi-1)}{(3)(2)(1)}\,. \end{aligned} \tag{4.5.11}$$

Problem 4.5.2 For the cubic interpolation given in (4.5.11) obtain an expression similar to (4.5.7). Obtain an expression for the interpolation error and obtain an expression similar to (4.5.10) for the integrated mean square error. Plot both (4.5.10) and the new expression as a function of K.

The interpolation given in (4.5.11) will not work close to the left and right boundaries of the domain. In the region $x_0 \le x \le x_1$, a third-order accurate interpolation will involve f_0, f_1, f_2 and f_3. As we saw in the case of first and second derivatives, we can expect these one-sided interpolations to be less accurate than their centered counterparts. In fact, the higher error in the derivative evaluation is due to the higher interpolation error. Clearly the two are related.

4.5.1 Particle Tracking

One of the biggest uses of interpolation in computational fluid dynamics is in the tracking of particles. Often fluid or tracer particles are tracked within a complex flow to obtain the desired Lagrangian statistics. The advection equations given in (1.6.10) are solved to obtain the pathline of a tracer particle. Note that in (1.6.10) the position of the tracer particle $(x(t), y(t), z(t))$ is time dependent. If the flow in which the tracer is being advected is analytically known, as in the case of ABC-flow, then the tracer velocity is readily known at any instance and (1.6.10) can be easily integrated. However, more often the flow in which the tracer is being advected is solved on a fixed Eulerian grid using one of the CFD methods to be discussed in Chapter 8. In which case, at any given time, the position of the tracer particle in general will not coincide with the Eulerian grid points on which the flow field is being solved. The evaluation of the fluid velocity at the tracer location, i.e., evaluation of $u(x(t), y(t), z(t), t)$, $v(x(t), y(t), z(t), t)$ and $w(x(t), y(t), z(t), t)$ requires the interpolation of the three components of velocity from the grid point to an off-grid point where the tracer is located.

Particle tracking in 2D and 3D flows will require multi-dimensional interpolation. In the case of 2D or 3D structured Cartesian grids, interpolation to an off-grid point is straightforward. It can be performed one direction at a time. For example, the above outlined linear or cubic interpolation can be first carried out along the x-direction, followed by an interpolation along the y-direction and then along the z-direction. In the case of linear interpolation, such an approach will be called **tri-linear** interpolation, where the interpolation uses the value of the function at the eight grid points that surround the off-grid point where the function value is required. The accuracy of different multi-dimensional interpolation methods for particle tracking has been investigated in detail (see Yeung and Pope, 1988; Balachandar and Maxey, 1989). As a general guideline, the level of accuracy of interpolation must be the same as that used in solving the fluid flow.

Interpolation and particle tracking becomes complex in the case of curvilinear grids and unstructured meshes. Also, the interpolation of fluid properties, such as fluid velocity and temperature, from neighboring grid points to the particle location is an important step in modern Eulerian–Lagrangian multiphase flow simulations. Here the particles being tracked are not the fluid or tracer particles. The inertial effect of particles is often important and their velocity differs from that of the local fluid velocity. At the simplest level the particle advection equations become

$$\frac{\mathrm{d}\mathbf{x}}{\mathrm{d}t} = \mathbf{v} \quad \text{and} \quad \frac{\mathrm{d}\mathbf{v}}{\mathrm{d}t} = \frac{\mathbf{u} - \mathbf{v} - \mathbf{w}}{\tau_p}, \tag{4.5.12}$$

where $\mathbf{v}$ is the velocity of the inertial particle and it is different from $\mathbf{u}$, which is the fluid velocity at the particle position. In the above, $\mathbf{w}$ is the velocity at which a particle settles in a still fluid due to gravity, and τ_p, the particle time scale, is given by

$$\tau_p = \frac{(2\rho + 1)d^2}{36\nu}, \tag{4.5.13}$$

where ρ is the particle-to-fluid density ratio, d is particle diameter and ν is the kinematic viscosity of the fluid (Balachandar and Eaton, 2010). The important point of relevance to the present discussion is that in the time integration of (4.5.12) for the particle motion, the evaluation of fluid velocity at the particle position requires interpolation. The **equation of motion** for an inertial particle given in (4.5.12) is appropriate only in the limit when the particle and fluid motions are steady and spatially uniform. Under more complex unsteady and non-uniform conditions it must be replaced by the **Basset–Boussinesq–Oseen** (BBO) (Basset, 1888; Boussinesq, 1885; Oseen, 1927) or **Maxey–Riley–Gatignol** (MRG) equation of motion (Maxey and Riley, 1983; Gatignol, 1983).

4.5.2 Interpolation Matrix

One of the main purposes of this section is to develop interpolation operators that will be needed later in Chapter 8 when we address numerical methods for partial differential equations. In the context of a uniform distribution of grid points, if we define $x_j = j\Delta x$, for $j = 0, \ldots, N$, as the regular points, then we can define the center points as $x_{j+1/2} = (j + 1/2)\Delta x$ for $j = 0, \ldots, N-1$. Thus, the number of center points will be one less than the number of regular points. We define the interpolation matrix as the linear operator that will take the function values from the regular points and interpolate them onto the center points. If we use linear interpolation for this purpose the resulting regular-to-center point

$(C \to R)$ interpolation can be written as

$$\begin{bmatrix} f_{1/2} \\ f_{3/2} \\ \vdots \\ \\ f_{N-1/2} \end{bmatrix} = \underbrace{\begin{bmatrix} 1/2 & 1/2 & 0 & \cdots & 0 \\ 0 & 1/2 & 1/2 & 0 & \cdots \\ 0 & \ddots & \ddots & \ddots & \vdots \\ \cdots & 0 & 1/2 & 1/2 & 0 \\ 0 & \cdots & 0 & 1/2 & 1/2 \end{bmatrix}}_{\mathbf{I}_{R\to C}:\ \text{Matrix of size } (N-1)\times(N)} \begin{bmatrix} f_0 \\ f_1 \\ \vdots \\ \\ f_N \end{bmatrix}. \tag{4.5.14}$$

If instead we use the cubic interpolation defined in (4.5.11) then the resulting regular-to-center interpolation operator can be written out as

$$\mathbf{I}_{R\to C} = \begin{bmatrix} 3/8 & 6/8 & -1/8 & 0 & \cdots & 0 \\ -1/16 & 9/16 & 9/16 & -1/16 & 0 & \cdots \\ 0 & \ddots & \ddots & \ddots & \ddots & \vdots \\ \cdots & 0 & -1/16 & 9/16 & 9/16 & -1/16 \\ 0 & \cdots & 0 & -1/8 & 6/8 & 3/8 \end{bmatrix}. \tag{4.5.15}$$

Note that the standard cubic interpolation that was used in the interior cannot be used at the two end center points. In the above we have used quadratic interpolations for $f_{1/2}$ and $f_{N-1/2}$. Therefore, the accuracy of these end point interpolations is lower than in the interior. It is an easy matter to develop one-sided cubic interpolations for $f_{1/2}$ and $f_{N-1/2}$. We leave this as an exercise for the reader.

We sometimes need to interpolate from the center points to the regular points. As we will see later, in Chapter 8, in the solution of the Navier–Stokes equations, pressure is solved at the center points and must be interpolated to the regular points where velocity information is solved. For the linear interpolation this center-to-regular (C $\to$ R) interpolation operator can be expressed as

$$\begin{bmatrix} f_0 \\ f_1 \\ \vdots \\ \\ f_N \end{bmatrix} = \underbrace{\begin{bmatrix} 3/2 & -1/2 & 0 & \cdots & 0 \\ 1/2 & 1/2 & 0 & 0 & \cdots \\ 0 & \ddots & \ddots & \ddots & \vdots \\ \cdots & 0 & 0 & 1/2 & 1/2 \\ 0 & \cdots & 0 & -1/2 & 3/2 \end{bmatrix}}_{\mathbf{I}_{\mathrm{C}\to\mathrm{R}}:\ \text{Matrix of size } (N)\times(N-1)} \begin{bmatrix} f_{1/2} \\ f_{3/2} \\ \vdots \\ \\ f_{N-1/2} \end{bmatrix}. \tag{4.5.16}$$

Technically, the first row of $\mathbf{I}_{\mathrm{C}\to\mathrm{R}}$ is an extrapolation from $f_{1/2}$ and $f_{3/2}$ to the end point value f_0 and similarly the last row of $\mathbf{I}_{\mathrm{C}\to\mathrm{R}}$ is also an extrapolation to obtain f_N.

Problem 4.5.3 Write out the center-to-regular interpolation matrix for cubic interpolation, with appropriate one-sided cubic extrapolations for f_0 and f_N, and one-sided cubic interpolations for f_1 and f_{N-1}.

Problem 4.5.4 A linear interpolation from regular-to-center points followed by a linear interpolation from center-to-regular points will not return the original function that we started with at the regular points. In other words, the errors will not cancel. In fact the error in the regular-to-center interpolation adds on to the error in center-to-regular interpolation. We can use von Neumann error analysis to investigate this cumulative error. Again consider the sequence of test functions given in (4.1.11). Since the functions to be interpolated are periodic, the end points of the domain can be treated just like the interior points. Obtain the mean square error and plot it as a function of K.

4.6 Numerical Integration

Here we are faced with the task of integrating a function numerically over a specified range. Following the format of our earlier discussion on numerical differentiation we will consider a function $f(x)$ defined over the range $0 \leq x \leq L_x$. As before we will first consider a simple discretization using $N+1$ equi-spaced points: $x_j = j\Delta x$ for $j = 0, 1, 2, \ldots, N$, where the grid spacing $\Delta x = L_x/N$. Integration over the entire interval can now be expressed as a sum of integrals over the N gaps between the $N+1$ points:

$$\begin{aligned} I &= \int_0^{L_x} f(x)\,\mathrm{d}x \\ &= \int_{x_0}^{x_1} f(x)\,\mathrm{d}x + \int_{x_1}^{x_2} f(x)\,dx + \cdots + \int_{x_{N-1}}^{x_N} f(x)\,\mathrm{d}x \\ &= I_0 + I_1 + \cdots + I_{N-1}\,. \end{aligned} \tag{4.6.1}$$

Thus an algorithm to approximate the integral over a single gap will be sufficient and we can repeat the algorithm over all the gaps to evaluate the overall integral, I.

The simplest approximation to I_j is the **midpoint rule**:

$$I_j = \int_{x_j}^{x_{j+1}} f(x)\,\mathrm{d}x \approx \Delta x\, f_{j+1/2} = \Delta x\, f(x_j + \Delta x/2)\,, \tag{4.6.2}$$

where the integral, which has the geometric interpretation as the area under curve – see Figure 4.4(a) – is approximated by a rectangular strip of width Δx and height corresponding to the value of the function at the midpoint, $f_{j+1/2}$ – see Figure 4.4(b). Repeating the midpoint rule for all the intervals gives the following approximation to the overall integral:

$$I \approx \Delta x \left\{ f_{1/2} + f_{1+1/2} + \cdots + f_{j+1/2} + \cdots + f_{N-1/2} \right\}\,. \tag{4.6.3}$$

An alternative approach is where the integral I_j is approximated by the trape-

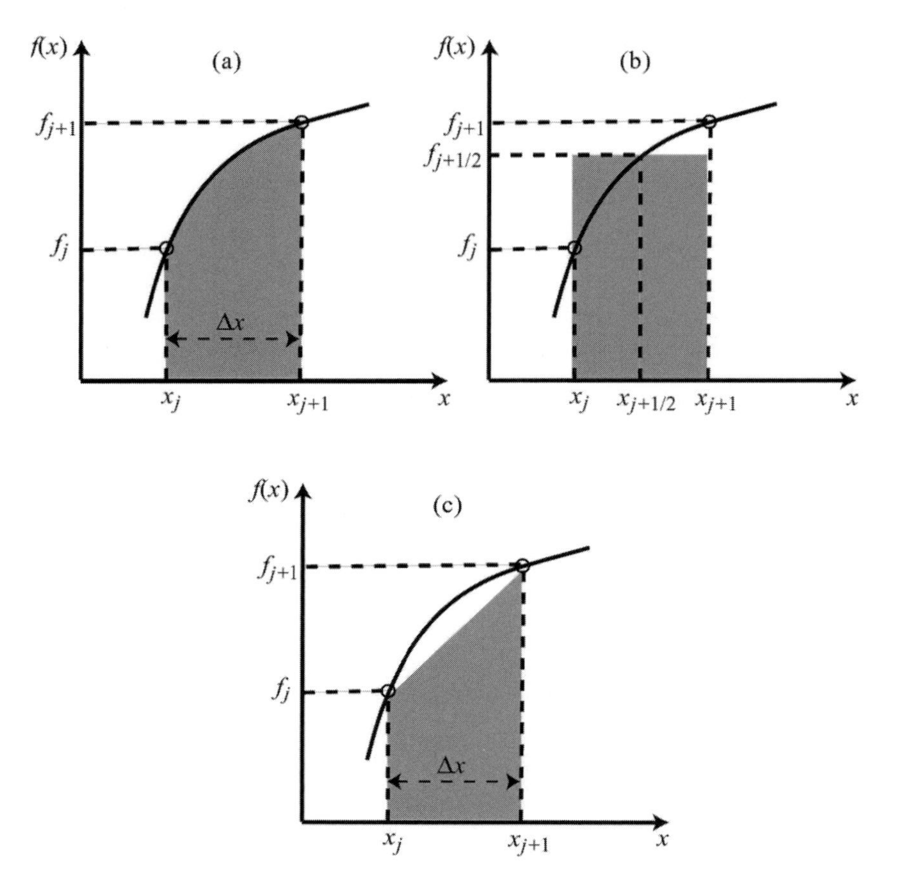

Figure 4.4 Schematic of the exact and numerical approximation to integration of $\int_{x_j}^{x_{j+1}} f(x)\,\mathrm{d}x$. In each case the area of the shaded region represents the integral. (a) Exact; (b) midpoint rule; and (c) trapezoidal rule.

zoidal area shown in Figure 4.4(c):

$$I_j \approx \frac{\Delta x}{2}\left(f_j + f_{j+1}\right) . \tag{4.6.4}$$

The above approximation when repeated for all the intervals gives rise to the **trapezoidal rule**:

$$I \approx \frac{\Delta x}{2}\left\{f_0 + 2f_1 + \cdots + 2f_j + \cdots + 2f_{N-1} + f_N\right\} . \tag{4.6.5}$$

This approximation has a simple geometric interpretation. Each of the interior points is associated with a strip of width Δx, while the two end points at x_0 and x_N are assigned strips of width only $\Delta x/2$, viz. 0 to $\Delta x/2$ and $L_x - \Delta x/2$ to L_x. Together, all these strips cover the entire range $0 \le x \le L_x$. With this interpretation we see that what started out as the trapezoidal approximation has turned out to be very similar to the midpoint rule: compare (4.6.3) to (4.6.5).

4.6.1 Error Analysis

The error for the midpoint and the trapezoidal rules can be obtained from a Taylor series expansion. The Taylor series of $f(x)$ about the point x_j is used to obtain the following exact result for the integral I_j as

$$\begin{aligned} I_j &= \int_0^{\Delta x} f(x_j+\zeta)\,\mathrm{d}\zeta = \int_0^{\Delta x}\left(f_j+\zeta f_j'+\frac{\zeta^2}{2!}f_j''+\frac{\zeta^3}{3!}f_j'''+\cdots\right)\mathrm{d}\zeta \\ &= \Delta x\, f_j+\frac{(\Delta x)^2}{2!}f_j'+\frac{(\Delta x)^3}{3!}f_j''+\frac{(\Delta x)^4}{4!}f_j'''+\cdots. \end{aligned} \tag{4.6.6}$$

A Taylor series expansion for $f_{j+1/2}$ about the point x_j in the midpoint rule yields the following result:

$$\begin{aligned} \Delta x\, f_{j+1/2} &= \Delta x\left(f_j+(\Delta x/2)\,f_j'+\frac{(\Delta x/2)^2}{2!}f_j''+\frac{(\Delta x/2)^3}{3!}f_j'''+\cdots\right) \\ &= I_j-\frac{\Delta x^3}{24}f_j''-\frac{\Delta x^4}{48}f_j'''+\cdots. \end{aligned} \tag{4.6.7}$$

Thus, the leading-order error (exact minus approximate integral) for the midpoint rule is $(\Delta x)^3 f_j''/24$. The error in the trapezoidal approximation can be similarly evaluated. A Taylor series expansion about the point x_j in the trapezoidal rule yields:

$$\begin{aligned} \frac{\Delta x}{2}(f_j+f_{j+1}) &= \frac{\Delta x}{2}\left(f_j+f_j+\Delta x\, f_j'+\frac{(\Delta x)^2}{2!}f_j''+\frac{(\Delta x)^3}{3!}f_j'''+\cdots\right) \\ &= I_j+\frac{(\Delta x)^3}{12}f_j''+\frac{(\Delta x)^4}{24}f_j'''+\cdots. \end{aligned} \tag{4.6.8}$$

Thus, the leading-order error in the trapezoidal rule is $-(\Delta x)^3 f_j''/12$, which is twice that of the midpoint rule. The larger error in the trapezoidal rule is consistent with the geometric interpretation shown in Figure 4.4, although it may be intuitively surprising.

The global error of numerical integration over the entire range of x from 0 to L_x is obtained by summing the local errors over all the intervals. The leading-order error for the trapezoidal rule can be expressed as

$$\text{Leading-order error} = -\frac{(\Delta x)^3}{12}\sum_{j=0}^{N-1} f_j'' = -\frac{L_x(\Delta x)^2}{12}f''(\zeta)\,, \tag{4.6.9}$$

where we make use of the mean value theorem from calculus, which states that for a smooth function $\sum_{j=0}^{N-1} f_j'' = N f''(\zeta)$ for some ζ that lies within the interval $0 \le x \le L_x$. It is interesting to note that the trapezoidal rule is third-order accurate (error goes as $(\Delta x)^3$) for an individual interval, but it is only second-order accurate (error goes as $(\Delta x)^2$) for the entire range. As Δx decreases the number of intervals (N) that are required to cover the range $0 \le x \le L_x$ proportionately increases and this accounts for the reduction in the overall order of accuracy by one.

Problem 4.6.1 The error estimates for the midpoint and trapezoidal rules given in (4.6.7) and (4.6.8) are asymmetric in the sense they are based on derivatives evaluated at the left-hand end of the interval (i.e., based on f_j'', f_j''', etc.). A more transparent description of the error is to base it on the midpoint of the interval. Use a Taylor series expansion of f_j'', f_j''', etc. about the midpoint to rewrite (4.6.7) and (4.6.8) for the midpoint and trapezoidal rules, respectively, as

$$\Delta x\, f_{j+1/2} = I_j - \frac{(\Delta x)^3}{24} f''_{j+1/2} - \frac{(\Delta x)^5}{1920} f^{iv}_{j+1/2} + \cdots \,, \tag{4.6.10a}$$

$$\frac{\Delta x}{2}(f_j + f_{j+1}) = I_j + \frac{(\Delta x)^3}{12} f''_{j+1/2} + \frac{(\Delta x)^5}{480} f^{iv}_{j+1/2} + \cdots \,. \tag{4.6.10b}$$

Note that in this symmetric description of the error the even powers of Δx disappear.

Problem 4.6.2 Show that the following simple end-correction to the trapezoidal rule will increase its overall accuracy from second- to fourth-order:

$$I = \frac{\Delta x}{2}\{f_0 + 2f_1 + \cdots + 2f_j + \cdots + 2f_{N-1} + f_N\} + \frac{\Delta x^2}{12}(f_0' - f_N') + O(\Delta x)^4 \,. \tag{4.6.11}$$

Note that in (4.6.10b) the second term on the right can be rewritten as

$$\frac{\Delta x^3}{12}\left(\frac{f'_{j+1} - f'_j}{\Delta x} - \frac{\Delta x^2}{24} f^{iv}_{j+1/2} + O(\Delta x)^4\right) \,. \tag{4.6.12}$$

It is then clear that the end-correction accounts for the leading-order error in the trapezoidal rule and thus the overall accuracy improves to fourth-order.

Problem 4.6.3 Numerically evaluate the following integral:

$$\int_0^{2\pi} \frac{\mathrm{d}x}{1+x} \,, \tag{4.6.13}$$

which can be easily shown to be equal to $\ln(1 + 2\pi)$. Define relative error as

$$\frac{|\text{exact integral} - \text{numerical integral}|}{|\text{exact integral}|} \,.$$

Use the trapezoidal rule and plot relative error as a function of N on log–log scale for $(N+1) = 6$, 10, 14, etc., until you observe the asymptotic second-order accuracy. Repeat the above with the end-point correction. Verify fourth-order accuracy.

4.6.2 Simpson's Rule

The trapezoidal rule approximates the function as piecewise-continuous straight-line segments connecting the function value at the grid points. The next-order approximation will then be to consider a piecewise-continuous quadratic fit through the grid point values. A quadratic fit will require at least three points and the Lagrangian interpolant through the three points f_{j-1}, f_j and f_{j+1} is given by (see Section 3.4)

$$\frac{f_{j-1}}{2\Delta x^2}(x-x_j)(x-x_{j+1})-\frac{f_j}{\Delta x^2}(x-x_{j-1})(x-x_{j+1})+\frac{f_{j+1}}{2\Delta x^2}(x-x_{j-1})(x-x_j)\,. \tag{4.6.14}$$

Integrating the above interpolant we obtain Simpson's rule

$$\int_{x_{j-1}}^{x_{j+1}} f(x)\,\mathrm{d}x \approx \frac{\Delta x}{3}(f_{j-1}+4f_j+f_{j+1})\,, \tag{4.6.15}$$

with the leading-order error given by $-(\Delta x)^5 f_j^{\mathrm{iv}}/90$. Simpson's rule for the entire domain is obtained by summing over (4.6.15) applied to contiguous strips of width $2\Delta x$:

$$I \approx \frac{\Delta x}{3}(f_0+4f_1+2f_2+4f_3+2f_4+\cdots+2f_{N-2}+4f_{N-1}+f_N)\,. \tag{4.6.16}$$

Note that the above formula assumes the total number of grid points $(N+1)$ to be odd, so that the even number of gaps between these points can be grouped into pairs and a quadratic interpolant can be fit within each pair. The overall accuracy of Simpson's rule is fourth-order.

4.6.3 Richardson Extrapolation

The Richardson extrapolation that we saw earlier under time integration applies equally well in the present context. The idea remains the same: the integral is first numerically approximated with an $(N+1)$-point grid of spacing Δx. The numerical approximation is then repeated with a finer $(2N+1)$-point grid of spacing $\Delta x/2$. As we shall see below, by combining these two approximations an improved estimation of the integral, which increases the order of accuracy by 2, is obtained.

Consider the trapezoidal rule to be the starting point. From (4.6.10a)–(4.6.10b) the trapezoidal approximation can be expressed as

$$I_{\mathrm{T}}^{\Delta x} = I + c_2\Delta x^2 + c_4\Delta x^4 + \cdots\,, \tag{4.6.17}$$

where $I_{\mathrm{T}}^{\Delta x}$ denotes the trapezoidal approximation to the integral with grid spacing Δx. In the above equation c_2 is the coefficient of the leading-order error term, c_4 is the coefficient of the higher-order error term, and so on. The trapezoidal approximation with half the grid spacing is then:

$$I_{\mathrm{T}}^{\Delta x/2} = I + \frac{c_2}{4}(\Delta x)^2 + \frac{c_4}{16}(\Delta x)^4 + \cdots\,. \tag{4.6.18}$$

The above two estimates can be combined to obtain a fourth-order approximation to the integral as follows:

$$\frac{4I_{\mathrm{T}}^{\Delta x/2} - I_{\mathrm{T}}^{\Delta x}}{3} = I - \frac{c_4}{4}(\Delta x)^4 + \cdots . \tag{4.6.19}$$

Observe that when the left-hand side is written out it reduces to Simpson's rule applied on the finer grid of spacing $\Delta x/2$.

It is impressive that we arrived at Simpson's rule from the trapezoidal rule simply by applying Richardson's extrapolation (without any quadratic fit, etc.). Similar to (4.6.19), one can combine trapezoidal rules applied with grid spacings $\Delta x/2$ and $\Delta x/4$ to obtain a fourth-order accurate approximation

$$\frac{4I_{\mathrm{T}}^{\Delta x/4} - I_{\mathrm{T}}^{\Delta x/2}}{3} = I - \frac{c_4}{64}(\Delta x)^4 + \cdots , \tag{4.6.20}$$

which can then be combined with (4.6.19) to obtain a sixth-order accurate approximation:

$$\frac{64I_{\mathrm{T}}^{\Delta x/4} - 20I_{\mathrm{T}}^{\Delta x/2} + I_{\mathrm{T}}^{\Delta x}}{45} = I + O((\Delta x)^6) . \tag{4.6.21}$$

Richardson extrapolation is thus a very powerful tool and it can be applied repeatedly to obtain ever higher-order accuracy. It is, of course, not restricted to the trapezoidal rule and it can be applied on top of any other scheme to progressively increase its accuracy to any desired level. Numerical integration with repeated application of Richardson extrapolation as described above until a desired level of accuracy is achieved goes by the name **Romberg integration**.

Problem 4.6.4 Apply Richardson extrapolation to Problem 4.6.3 and obtain a sixth-order accurate approximation by using the trapezoidal rule results for $(N + 1) = 6, 11, 21$.

4.6.4 Gauss-type Quadrature

The above integration schemes can be viewed as weighted sums of the function values evaluated at the chosen grid points. The trapezoidal (4.6.5) and Simpson's (4.6.16) rules differ only in the weights. Even higher-order approximations can be constructed either by fitting higher-order polynomials as in (4.6.14) or by adopting Richardson extrapolation. While the weights are appropriately optimized for higher-order accuracy, it is natural to question the choice of equi-spaced points. It turns out that if we allow ourselves the freedom to choose the location of the $N + 1$ points, x_j, and the associated weights, w_j, even better approximations to the integral can be obtained. In particular, the integral can be made to be exact for all polynomials of order $(2N + 1)$ or less. This gives rise to what is known as integration by **Gauss quadrature**. Considering that the trapezoidal and Simpson rules are exact only for polynomials of order less than 1 and 3, respectively, Gauss quadrature offers significant improvement.

Legendre Quadrature

The notion of quadrature has its origin in the concept of expanding a given function in orthogonal polynomials. Depending on the choice of orthogonal polynomials different quadratures can be constructed. The quadrature based on Legendre polynomials is particularly appealing for integrating polynomials over a finite domain. Legendre polynomials are defined over the interval $-1 \leq x \leq 1$. The first few are

$$L_0(x) = 1\,; \quad L_1(x) = x\,; \quad L_2(x) = \frac{3}{2}x^2 - \frac{1}{2}\,, \tag{4.6.22}$$

and higher-order ones are given by the recurrence

$$L_{k+1}(x) = \frac{2k+1}{k+1} x\, L_k(x) - \frac{k}{k+1} L_{k-1}(x)\,. \tag{4.6.23}$$

Legendre polynomials satisfy the following orthogonality relation:

$$\int_{-1}^{1} L_k(x) L_l(x)\, \mathrm{dx} = \begin{cases} 0 & \text{if} \quad k \neq l\,, \\ \left(k + \frac{1}{2}\right)^{-1} & \text{for} \quad k = l\,. \end{cases} \tag{4.6.24}$$

Legendre–Gauss-type quadratures are given by the following equality between the desired integral and a weighted sum:

$$\int_{-1}^{1} f(x)\, \mathrm{d}x \approx \sum_{j=0}^{N} f(x_j) w_j\,. \tag{4.6.25}$$

Table 4.1 The first few Legendre–Gauss quadrature points and weights.

$N=1$		$N=2$		$N=3$	
x_j	w_j	x_j	w_j	x_j	w_j
$\pm\frac{1}{3}\sqrt{3}$	1.0	0.0	8/9	$\pm\frac{1}{35}\sqrt{525 - 70\sqrt{30}}$	$\frac{1}{36}(18 + \sqrt{30})$
		$\pm\frac{1}{5}\sqrt{15}$	5/9	$\pm\frac{1}{35}\sqrt{525 + 70\sqrt{30}}$	$\frac{1}{36}\left(18 - \sqrt{30}\right)$

The grid points, x_j, and their associated weights, w_j, are determined by demanding (4.6.25) be exact for polynomial $f(x)$ of as high an order as possible.

The above optimization problem can be easily solved. The resulting **Legendre–Gauss quadrature** is given by (for details see Canuto *et al.*, 2006)

$$\begin{aligned} &x_0,\, x_1, \ldots,\, x_N \quad \text{are the zeros of } L_{N+1}(x) \\ &w_j = \frac{2}{(1 - x_j^2)\left(L'_{N+1}(x_j)\right)^2} \quad \text{for } j = 0,\, 1, \ldots,\, N\,, \end{aligned} \tag{4.6.26}$$

where $L_{N+1}(x)$ is the $(N+1)$th Legendre polynomial whose zeros (or roots) are the quadrature points. Although an explicit expression for these points is not available, it can be easily shown that the points quadratically cluster near the ends (grid spacing goes as N^{-2} close to $x = \pm 1$), while they are spaced farther apart in the middle. In the definition of the weights, $L'_{N+1}(x)$ is the first derivative of the $(N+1)$th Legendre polynomial. Table 4.1 gives the first few quadrature points and weights.

The Legendre–Gauss quadrature, with points and weights defined according to (4.6.26), is exact for polynomials $f(x)$ of order $(2N+1)$ or lower. This result can be easily proven starting with the following observation. If $f(x)$ is a polynomial of order $(2N+1)$ then it can be written as

$$f(x) = L_{N+1}(x) g_N(x) + h_N(x)\,, \tag{4.6.27}$$

where $g_N(x)$ and $h_N(x)$ are polynomials of order at most N. Substituting this on the left-hand side of (4.6.25)

$$\int_{-1}^{1} f(x)\,\mathrm{d}x = \int_{-1}^{1} g_N(x) L_{N+1}(x)\,\mathrm{d}x + \int_{-1}^{1} h_N(x)\,\mathrm{d}x\,, \tag{4.6.28}$$

where the first term on the right is zero since $g_N(x)$ can be expanded in terms of the first N Legendre polynomials, which are orthogonal to the $(N+1)$th Legendre polynomial. The right-hand side of (4.6.25) can be similarly rewritten as

$$\sum_{j=0}^{N} f(x_j) w_j = \sum_{j=0}^{N} g_N(x_j) L_{N+1}(x_j) w_j + \sum_{j=0}^{N} h_N(x_j) w_j\,. \tag{4.6.29}$$

Here the choice of x_j as the zeros of L_{N+1} makes the first term on the right vanish and the weights w_j are computed such that

$$\int_{-1}^{1} h_N(x)\,\mathrm{d}x = \sum_{j=0}^{N} h_N(x_j) w_j\,, \tag{4.6.30}$$

for any Nth-order polynomial $h_N(x)$.

It is important to note that all of the Gauss quadrature points lie inside the domain $-1 \le x \le 1$ (i.e., $-1 < x_0$ and $x_N < 1$). It is sometimes desired (not necessarily for integration, but for other reasons such as satisfying boundary conditions in the context of differential equations) that the end points of the quadrature coincide with the end points of the domain. In that case we fix $x_0 = -1$, $x_N = 1$ and optimize on the location of the other $(N-1)$ points and the $(N+1)$ weights associated with all the points. This gives rise to the **Legendre–Gauss–Lobatto quadrature**:

$$\left.\begin{aligned} x_0 &= -1,\, x_N = 1,\ x_1, \ldots, x_{N-1} \quad \text{are the zeros of } L'_N(x)\,, \\ w_j &= \frac{2}{N(N+1)(L_N(x_j))^2} \quad \text{for } j = 0,\, 1, \ldots, N\,. \end{aligned}\right\} \tag{4.6.31}$$

The fixing of the end points in the Gauss–Lobatto quadrature results in the loss of two degrees of freedom. As a result (4.6.25) in conjunction with (4.6.31) is exact only for polynomials of order $(2N-1)$ or lower (which is still very accurate indeed).

The proof for the above statement goes as follows. If $f(x)$ is a polynomial of order $(2N-1)$ then it can be written as

$$f(x) = (1-x^2)L'_N(x)g_{N-2}(x) + h_N(x)\,, \tag{4.6.32}$$

where $g_{N-2}(x)$ and $h_N(x)$ are polynomials of order $N-2$ and N, respectively. Substituting this on the left-hand side of (4.6.25)

$$\int_{-1}^{1} f(x)\,\mathrm{d}x = \int_{-1}^{1} (1-x^2)g_{N-2}(x)L'_N(x)\,\mathrm{d}x + \int_{-1}^{1} h_N(x)\,\mathrm{d}x\,, \tag{4.6.33}$$

where it can be shown that the first term on the right is zero. The right-hand side of (4.6.25) can be similarly rewritten as

$$\sum_{j=0}^{N} f(x_j)w_j = \sum_{j=0}^{N} (1-x^2)g_N(x_j)L'_N(x_j)w_j + \sum_{j=0}^{N} h_N(x_j)w_j\,. \tag{4.6.34}$$

With the interior points given as the zeros of L'_N the first term on the right is identically zero and the weights w_j are again given by (4.6.31).

Chebyshev Quadrature

Quadrature based on Chebyshev polynomials is appropriate for integrating polynomials with weight $w(x) = \left(1-x^2\right)^{-1/2}$. Chebyshev polynomials are defined over the interval $-1 \le x \le 1$; the first few are

$$T_0(x) = 1\,; \quad T_1(x) = x\,; \quad T_2(x) = 2x^2 - 1\,, \tag{4.6.35}$$

and higher-order ones are given by the recurrence

$$T_{k+1}(x) = 2x\,T_k(x) - T_{k-1}(x)\,. \tag{4.6.36}$$

Chebyshev polynomials satisfy the following orthogonality relation with respect to the weight $w(x) = \left(1-x^2\right)^{-1/2}$:

$$\int_{-1}^{1} T_k(x)T_l(x)w(x)\,\mathrm{d}x = \begin{cases} 0 & \text{if} \quad k \neq l, \\ \pi & \text{if} \quad k = l = 0, \\ \dfrac{\pi}{2} & \text{if} \quad k = l \geq 1. \end{cases} \tag{4.6.37}$$

As before, Chebyshev–Gauss-type quadratures are expressed as

$$\int_{-1}^{1} f(x)w(x)\,\mathrm{d}x \approx \sum_{j=0}^{N} f(x_j)w_j\,, \tag{4.6.38}$$

where again the grid points, x_j, and the associated weights, w_j, are determined by demanding that the above approximation be exact for polynomial $f(x)$ of as high an order as possible.

The **Chebyshev–Gauss quadrature** is given by

$$\left.\begin{aligned} x_j &= \cos\frac{(2j+1)\pi}{2N+2} \quad \text{for} \quad j = 0,\, 1, \ldots,\, N\,, \\ w_j &= \frac{\pi}{N+1} \quad \text{for} \quad j = 0,\, 1, \ldots,\, N\,. \end{aligned}\right\} \tag{4.6.39}$$

Unlike Legendre quadrature, here the grid points are explicitly defined and the quadratic clustering of the grid points near the ends can be readily seen. With the above definition of quadrature points and weights (4.6.38) is exact provided $f(x)$ is a polynomial of order $(2N+1)$ or lower. **Chebyshev–Gauss–Lobatto quadrature** can be similarly defined with the following grid points and weights:

$$\left.\begin{aligned} x_j &= \cos\frac{\pi j}{N} \quad \text{for} \quad j = 0,\, 1, \ldots,\, N, \\ w_j &= \begin{cases} \dfrac{\pi}{2N} & j = 0,\, N, \\ \dfrac{\pi}{N} & j = 1, 2, \ldots, N-1 \end{cases} . \end{aligned}\right\} \tag{4.6.40}$$

The quadrature is exact for polynomials of order $(2N-1)$ or lower. The above statements can be proved along the lines of Legendre quadrature, but with the weight function being $w(x) = \left(1 - x^2\right)^{-1/2}$ instead of unity.

Quadratures based on other orthogonal polynomials are similarly useful for integrating functions involving other weight functions. Furthermore, the domain of integration extends to infinity in some cases. In such cases, while other approaches to integration may face difficulty in coping with the infinite domain, quadratures yield highly accurate results with little effort. Table 4.2 lists some of the orthogonal polynomials suited for quadrature and their associated weight functions and range of integration. Golub and Kautsky (1983) provide a general algorithm where the user can specify a weight function and choose the location of some grid points and the algorithm will then determine the optimal location of the other grid points and all the weights associated with the grid points.

Table 4.2 Some of the orthogonal polynomials suited for quadrature and their associated weight functions and range of integration.

Polynomial	$w(x)$	**Range**
Legendre	1	$(-1, 1)$
Chebyshev	$\left(1 - x^2\right)^{-1/2}$	$(-1, 1)$
Laguerre	e^{-x}	$(0, \infty)$
Hermite	e^{-x^2}	$(-\infty, \infty)$

Example 4.6.5 Compute the integral

$$\int_0^\infty (x^7 + 4x^6 + 4x^5 - x^4 - 3x^3 + 2x^2 + 3x + 5)\, \mathrm{e}^{-x}\, \mathrm{d}x$$

using Laguerre–Gauss quadrature. Table 4.3 lists the node locations and the weights for $N = 1$, 2 and 3. The integral, as evaluated by the quadrature for $N =$ 1, 2 and 3 are 1966.0, 7326.1 and 8370.0, respectively. For the above integrand, quadrature with $N \geq 4$ gives exact result.

Problem 4.6.6 Compute the integral

$$\int_0^3 (x^5 - x^4 - 3x^3 + 2x^2 + 3x + 5)\, \mathrm{d}x$$

whose exact value is 8370, using quadrature. The polynomial integrand is ideally suited for Legendre quadrature. Using $N = 2$, 3, 4 show that the exact integral can be recovered very rapidly. Use Chebyshev–Gauss and Chebyshev–Gauss–Lobatto quadratures for the same problem and show that for the same level of convergence N in the range of several hundreds will be required.

Table 4.3 The first few Laguerre–Gauss quadrature points and weights.

$N = 1$		$N = 2$		$N = 3$	
x_j	w_j	x_j	w_j	x_j	w_j
0.585786	0.853553	0.415775	0.711093	0.322548	0.603154
3.41421	0.146447	2.29428	0.278518	1.74576	0.357419
		6.28995	0.0103893	4.53662	0.0388879
				9.39507	0.000539295

4.6.5 Prandtl–Meyer Flow

In the rest of the chapter we will consider a few examples in fluid mechanics that require integration of complicated functions. The first example is a classic problem in compressible fluid mechanics that goes by the name Prandtl–Meyer flow, after the distinguished fluid dynamicist Ludwig Prandtl. When a supersonic flow parallel to a wall encounters a convex corner, an expansion fan forms across which the velocity of the flow and the flow Mach number increases: see Figure 4.5(a). Here the flow evolves from the incoming state of Mach number M_1

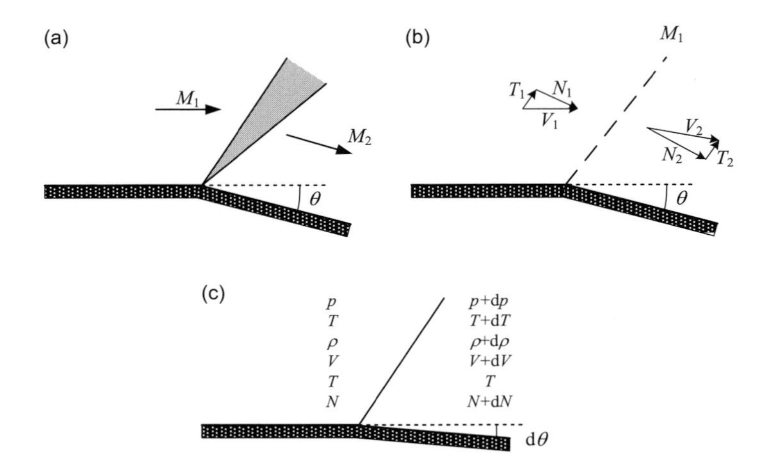

Figure 4.5 Schematic of the Prandtl–Meyer flow where a supersonic flow parallel to a wall forms an expansion fan around a convex corner.

to the final state of Mach number M_2 through an infinite number of infinitesimal Mach waves. In the case of a sharp corner shown in the figure, a centered expansion fan results with all the Mach waves diverging from the sharp corner. In the case of a smooth corner, the waves emerge from the entire curved section, but can be extended backwards to meet at a point.

Note that the flow cannot turn the convex corner with a shock. This is illustrated in Figure 4.5(b), where it can first be noted that the velocities V_1 and V_2 are parallel to the respective walls. Furthermore, the tangential velocity component must remain unaltered across the shock which requires $T_1 = T_2$. From geometrical arguments we conclude that we require the normal component downstream of the shock to be larger than the upstream normal component, i.e., $N_1 < N_2$. This condition is contrary to the well-known Rankine–Hugoniot shock relations. In other words, this will be in violation of the second law of thermodynamics, and as a result we require a centered expansion fan to turn a supersonic flow around a convex corner.

The question can then be posed as follows: given the upstream velocity (V_1), pressure (p_1), density (ρ_1), speed of sound (a_1), with an upstream supersonic Mach number ($M_1 = V_1/a_1 > 1$), and the angle of the convex turn (θ), how can we predict the downstream velocity (V_2), pressure (p_2), density (ρ_2) and speed of sound (a_2)? In order to obtain these relations let us first consider a simple Mach wave resulting when the flow turns an infinitesimal angle ($\mathrm{d}\theta$). Conditions upstream and downstream are illustrated in Figure 4.5(c). Equations of continuity and normal momentum give:

$$\left.\begin{aligned} \rho N &= (\rho + \mathrm{d}\rho)(N + \mathrm{d}N) \\ p + \rho N^2 &= (p + \mathrm{d}p) + (\rho + \mathrm{d}\rho)(N + \mathrm{d}N)^2 \,. \end{aligned}\right\} \tag{4.6.41}$$

From these relations and the geometry of the wave shown in Figure 4.5(c), two facts can be established. First, the upstream normal velocity is the same as the speed of sound; and second, the angle of the Mach wave is related to the upstream Mach number as given by

$$N = \frac{\mathrm{d}p}{\mathrm{d}\rho} = a \qquad \text{and} \qquad \mu = \sin^{-1}\left(\frac{1}{M}\right). \tag{4.6.42}$$

The tangential momentum and energy balances give

$$\left.\begin{aligned} V\cos\mu &= (V+\mathrm{d}V)\cos(\mu-\mathrm{d}\theta)\,,\\ \left(\frac{2\gamma}{\gamma-1}\right)\frac{p}{\rho}+V^2 &= \left(\frac{2\gamma}{\gamma-1}\right)\frac{p+\mathrm{d}p}{\rho+\mathrm{d}\rho}+(V+\mathrm{d}V)^2\,, \end{aligned}\right\} \tag{4.6.43}$$

where γ is the specific heat ratio. The details of the subsequent algebraic steps can be found in many books on elementary gas dynamics. The bottom line is the above equations can be massaged to obtain the following relation connecting the upstream and downstream Mach numbers with the angle θ as

$$\theta = \int_{M_1}^{M_2} \frac{\sqrt{M^2-1}}{1+\frac{\gamma-1}{2}M^2}\frac{\mathrm{d}M}{M}. \tag{4.6.44}$$

In this case, the integration can be performed analytically and the answer can be explicitly written as

$$\theta = \left[\sqrt{\frac{\gamma+1}{\gamma-1}}\tan^{-1}\sqrt{\frac{\gamma-1}{\gamma+1}(M^2-1)} - \tan^{-1}\sqrt{M^2-1}\right]_{M_1}^{M_2}. \tag{4.6.45}$$

Prandtl–Meyer flow presents a very interesting feature. If we set $M_1 = 1$ and $M_2 = \infty$, then we consider the limiting case of maximum supersonic expansion from an initial sonic flow to infinite Mach number. The corresponding maximum angle can be obtained as

$$\theta_{\max} = \frac{\pi}{2}\left(\sqrt{\frac{\gamma+1}{\gamma-1}} - 1\right), \tag{4.6.46}$$

which for air ($\gamma = 1.4$) yields the limiting angle of 130.5°. For convex corners with a greater angle, this will result in flow stopping at 130.5° and a vacuum existing between this angle and the wall! Of course, rest assured that this will not happen in reality since some of the model assumptions we made in obtaining the above results will break down under such extreme conditions.

Problem 4.6.7 Air flow ($\gamma = 1.4$) at a Mach number of 1.5 with a pressure of 100 kPa and a temperature of 25°C turns a corner of 10° away from the flow leading to the formation of a centered expansion fan. Calculate the Mach number of the flow downstream of the expansion fan, by numerically integrating (4.6.44). Then compare the results with that of (4.6.45). The other properties of the flow downstream of the fan can be computed from

$$\frac{T_2}{T_1} = \frac{1+(\gamma-1)M_1^2/2}{1+(\gamma-1)M_2^2/2}, \quad \frac{p_2}{p_1} = \left(\frac{T_2}{T_1}\right)^{\gamma/(\gamma-1)} \quad \text{and} \quad \frac{\rho_2}{\rho_1} = \left(\frac{T_2}{T_1}\right)^{1/(\gamma-1)}. \tag{4.6.47}$$

Problem 4.6.8 Integrate (4.6.44) for $M_1 = 1$, $M_2 = \infty$ and obtain the result (4.6.46) for $\gamma = 1.4$. Use different integration schemes and compare their accuracy.

4.6.6 Inertial Lift on a Small Sphere in a Linear Shear Flow

Here we discuss the solution for the lift force on a small particle immersed in an unbounded linear shear flow, originally expounded in a seminal paper of Saffman (1965). The statement of the problem is quite simple and is shown in Figure 4.6. Consider a sphere of diameter d immersed in a linear shear flow of velocity gradient G. For simplicity, consider the sphere to be stationary and the ambient flow velocity at the particle center to be V. Due to ambient shear the particle will experience, in addition to the drag force, a lift force that is directed perpendicular to the streamlines of the ambient flow. Saffman derived a general relation for the magnitude of this lift force that resulted in a complicated integral that he evaluated numerically. This is the problem that we wish to examine.

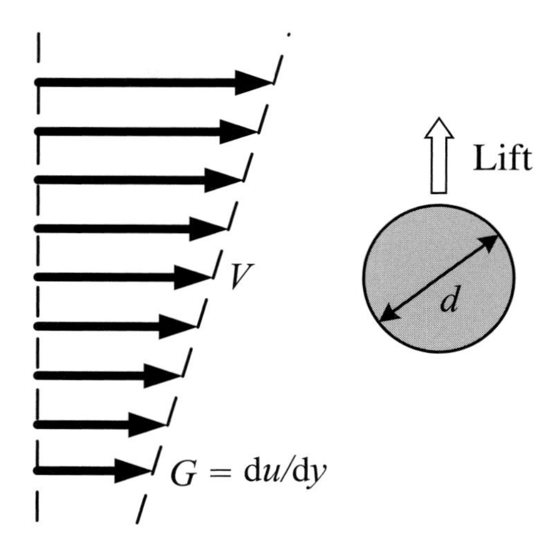

Figure 4.6 Schematic of the Saffman lift force on a spherical particle subjected to a linear shear flow.

Before we plunge into the analysis, let us first consider some preliminaries.

In this problem two different Reynolds numbers can be defined. First is the Reynolds number based on relative velocity: $\mathrm{Re}_V = dV/\nu$ and the second one based on shear: $\mathrm{Re}_G = d^2G/\nu$, where ν is the kinematic viscosity of the fluid. If the flow around the particle is in the Stokes regime, in other words in the limit $\mathrm{Re}_V = \mathrm{Re}_G = 0$, it can be proved that the lift force on the particle will be identically zero. Saffman's analysis is valid for small but non-zero inertia, i.e., for $\mathrm{Re}_V \ll 1$ and $\mathrm{Re}_G \ll 1$, and hence the lift force is called inertial lift.

The theoretical analysis of Saffman is far too technical and involved for a complete discussion here. Saffman used a technique similar to that introduced by Proudman and Pearson (1957). They recognized the fact that no matter how small the Reynolds number, sufficiently far from the sphere the inertial effects will balance the viscous effects. This requires a matched asymptotic technique, where the flow in the *inner* region close to the sphere is obtained as an inner expansion and the flow in the *outer* region is solved as an outer expansion and the two are matched term by term in the middle. Skipping lots of details, we can write the resulting simple expression for the lift force on the sphere in a linear shear flow obtained by Saffman as

$$\text{Lift} = KV\left(\frac{d}{2}\right)^2\left(\frac{G}{\nu}\right)^{1/2}. \tag{4.6.48}$$

The key point of relevance to this chapter is that the constant K is given only in terms of an integral as

$$K = \frac{9\,(1/4)!}{4\,(3/4)!}\int_0^\infty\int_0^\infty \frac{\left(3t^2 + (\eta^2/4) + 1\right)\quad dt\,d\eta}{\eta^{1/2}(t^2+(\eta^2/12)+1)^{3/2}\left[(t^2+(\eta^2/4)+1)^2 - \eta^2 t^2\right]}. \tag{4.6.49}$$

The above integral cannot be evaluated analytically and one must resort to numerical techniques. Saffman did just that and obtained $K = 6.46$.

Problem 4.6.9 Carry out the numerical integration of (4.6.49) using Laguerre quadrature and obtain $K = 6.46$. Two new aspects make this problem a bit more challenging than usual. First, (4.6.49) is a double integral. One of them should be considered inner and the other outer. For each value of the outer variable, the inner integral must be evaluated. Second, the kernel of the integral is singular as $\eta \to 0$, but this singularity is integrable.

It is quite impressive that an exact analytical result (almost analytical, except for the numerical evaluation of the integral) is obtained for a fairly complex fluid mechanical problem. In fact this line of research has been improved in several different ways. For example Saffman's analysis is valid only in the limit $\mathrm{Re}_V \ll \mathrm{Re}_G^{1/2} \ll 1$. This analysis was extended to a more general low Reynolds number limit of Re_V, $\mathrm{Re}_G \ll 1$, without any constraint on their relative values, by McLaughlin (1991). Again the analysis reduces to evaluation of a complicated integral. When the Reynolds numbers Re_V and Re_G are not small, one needs to resort to full numerical simulation of the governing PDEs (Navier–Stokes equations), about which we shall see more in Chapter 8. Finite Reynolds number expressions for lift force on a spherical particle in an ambient shear flow have been obtained by Kurose and Komori (1999) and by Bagchi and Balachandar (2002a).

The analyses of Saffman and of Mclaughlin are for a particle immersed in an unbounded shear flow. A related problem of great practical interest is the lift force on a spherical particle in a wall-bounded shear flow. The problem of inertial lift on a stationary particle in contact with a flat wall in a linear shear flow was analytically considered by Leighton and Acrivos (1985). This analysis was extended to the case of a translating and rotating particle in contact with the wall by Krishnan and Leighton (1995). It was shown that the lift force can be separated into six contributions. Three of the contributions arise from the ambient shear, and the translational and rotational motions of the particle. The other three contributions arise from shear–translation, translation–rotation and shear–rotation binary couplings. Analytical evaluation of these six different lift contributons required numerical integration of six different integrals. Lee and Balachandar (2010) extended the analytical results of Krishnan and Leighton by numerically solving the Navier–Stokes equations. They considered the problem of a spherical particle translating and rotating on a flat wall in a linear shear flow and obtained results in the finite-Reynolds number regime. An important point must be stressed here. Even though solutions of PDEs with a full-fledged Navier–Stokes solver are needed in most problems of practical interest, the analytical results, which reduce to evaluation of an integral or integration of an ODE, are of great value. They serve as a **benchmark** and the full-fledged Navier–Stokes solver, when run at the appropriate conditions, must yield results that can be rigorously compared against the analytical results. In scientific computing this process is called **verification**, whereby you guarantee that the code (Navier–Stokes solver) is indeed solving the correct governing equations with the right boundary conditions and so on. For example, before discussing their finite Reynolds number results, Lee and Balachandar first verified that in the limit of low Reynolds number their results are in agreement with those of Krishnan and Leighton (1995).

Literature Notes

There is an extensive literature on finite difference schemes and their accuracy. There are three categories of reference available. First, there are excellent textbooks on numerical analysis, where finite difference methods are covered as a popular class of numerical methods for ODEs and PDEs. In this category the reader can refer to Hildebrand (1974), Chapra (2002), Moin (2010a,b) and many more. There are also books on CFD that are either fully or partly based on finite difference methods: Pletcher *et al.* (2012), Chung (2010), Hirsch (2007), Ferziger and Peric (2012), to mention just a few. These books typically present an excellent introduction to finite difference methods before launching on to CFD. The third category of books are focused on finite difference methods and they are typically for the solution of partial differential equations. They are not focused only on the numerical solution of fluid flows, and therefore CFD is not their primary subject matter. We refer the reader to texts by Smith (1985), LeVeque (2007), Evans *et al.* (2012), Thomas (2013) – and there are several more – for analysis of higher-order forward, backward and central difference finite difference methods for first-, second- and higher-order derivatives. Further information on compact difference schemes can be found in Lele (1992) and in classic texts on numerical methods by Collatz (1960), and Kopal (1961) under the title Hermitian methods.

5 Boundary Value and Eigenvalue ODEs

We will begin this chapter with a discussion of two-point boundary value ODEs and then move on to two-point eigenvalue problems. The Blasius equation in Example 1.6.4 is one of the most celebrated two-point boundary value problems in fluid dynamics.[1] Two-point boundary value problems form a challenging class of numerical computations with a wide literature. For an exposition of the theoretical ideas on two-point boundary value problems see Press *et al.* (1986), Chapter 16, or Keller (1968). Broadly speaking there are two classes of methods for such problems, **shooting methods** and **relaxation methods**. Shooting methods are related to integration schemes for initial value problems, while relaxation methods are based on discrete approximations to derivative operators. Thus, we are ready to tackle both these methods. In a shooting method you start from one end as in an initial value calculation using the actual boundary conditions specified at that boundary supplemented with a few other assumed (or guessed) ones which replace the actual boundary conditions of the ODE specified at the other end, and integrate (or "shoot") towards the other boundary. In general when you eventually reach the other boundary, the boundary conditions there will not be satisfied (in other words you "miss" the target). Hence, you modify your guessed boundary conditions at the starting boundary ("aim" again) and integrate forward ("shoot" again). In this way you have in effect generated a mapping of initially guessed boundary conditions onto errors in matching the actual boundary conditions to be enforced at the other end point. Iterative procedures can now be invoked to converge to the desired value of the boundary conditions at the other end.

In a relaxation method you write the ODE(s) as a system of finite difference equations, guess a solution, and then iteratively improve it by using a root finding procedure for a system of coupled nonlinear equations. This can obviously be a rather formidable task. Often a good initial guess is needed to converge towards the correct answer. Press *et al.* (1986) give the whimsical recommendation for dealing with two-point boundary value problems to "shoot first and relax later." As we shall see below, in most problems of interest in fluid dynamics the relaxation method is very powerful, exhibiting rapid convergence towards the

[1] We shall see shortly that a particular feature of the Blasius equation disqualifies it as a genuine two-point boundary value problem.

correct answer. Furthermore, unlike the shooting method, the relaxation method is easily extendible to boundary value problems in multi-dimensions.

5.1 Linear Boundary Value Problems

We start with simpler linear problems and then progress to nonlinear ODEs. The shooting method will be considered first followed by the relaxation method. Both these methods greatly simplify for a linear problem. We plan to illustrate these methods by way of a simple fluid dynamical example.

Consider the classic problem of flow between two infinite parallel plates $2h$ apart, with the bottom plate stationary and the top plate moving at speed U relative to the bottom plate (see Figure 5.1). A pressure gradient also exists along the x-direction. This is the classic plane Poiseuille–Couette problem. The velocity between the plates in the absence of a pressure gradient is linear. In the presence of a pressure gradient, but with the top plate stationary, the velocity profile is parabolic. When both the top plate moves and the pressure gradient is simultaneously applied, the velocity profile is a superposition of the linear and parabolic parts (for further discussion see White, 1974, §3.2).

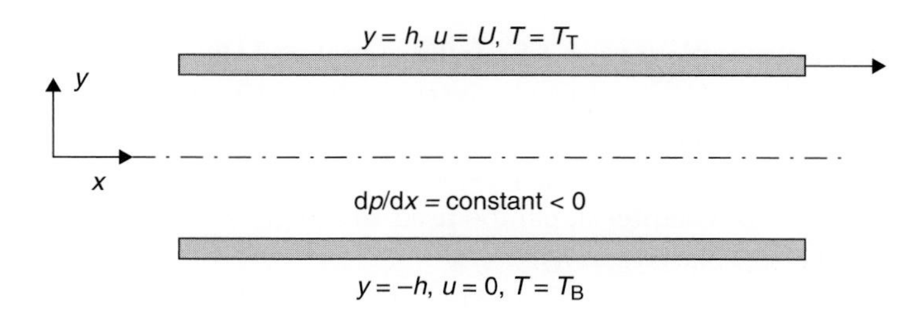

Figure 5.1 Schematic of the Poiseuille–Couette problem of flow between two parallel plates. Here the two plates are separated by $2h$. The bottom wall is stationary at temperature T_B, while the top wall is moving at a constant speed of U and is at temperature T_T. A constant streamwise pressure gradient is applied along the x-direction.

Here we add a small twist to this problem by allowing the flow to be non-isothermal. If the bottom and top plates are maintained at different temperatures, say T_B and T_T, respectively, then under thermally fully developed condition a linear temperature profile exists between the plates. Variation in thermal conductivity and viscous heating are neglected. To make the problem somewhat more interesting, let us consider viscosity to be strongly dependent on temperature. Therefore in the evaluation of the velocity field, viscosity, μ, must be considered a function of the wall-normal direction, y. Under fully developed conditions, here we will crudely approximate the streamwise component of velocity to follow[2]

$$0 = -\frac{\mathrm{d}p}{\mathrm{d}x} + \mu(y)\frac{\mathrm{d}^2 u}{\mathrm{d}y^2}, \tag{5.1.1}$$

which simply states that the applied streamwise pressure gradient ($\mathrm{d}p/\mathrm{d}x = G$) balances the viscous stress. The appropriate boundary conditions are

$$u(y = -h) = 0 \qquad \text{and} \qquad u(y = h) = U. \tag{5.1.2}$$

For constant viscosity the above problem can be trivially solved. Even in the case of complex temperature dependence of viscosity, approximate analytical solutions can be constructed for this simple ODE. However, numerical methods for (5.1.1) and (5.1.2) can come in very handy, for complicated $\mu(y)$.

5.1.1 Shooting Method

Here we solve a related initial value problem

$$\frac{\mathrm{d}^2 u_0}{\mathrm{d}y^2} = \frac{G}{\mu(y)} \tag{5.1.3}$$

along with the initial conditions

$$u_0(y = -h) = 0 \qquad \text{and} \qquad \frac{\mathrm{d}u_0}{\mathrm{d}y}(y = -h) = 0\,. \tag{5.1.4}$$

The second condition in (5.1.4) is an arbitrary guess for the velocity gradient at the bottom plate ($y = -h$) and has replaced the velocity boundary condition at the other end ($y = h$) to convert the boundary value ODE into an initial value problem. Equation (5.1.3) can be written as a system of two first-order ODEs, and time integration schemes, such as the Runge–Kutta and predictor–corrector schemes of Chapter 3, can be used to integrate from the bottom plate ($y = -h$) to the top plate ($y = h$).

In general $u_0(y = h)$ will not be U and the velocity condition at the upper plate is not satisfied. In this sense our very first shot has missed the target, but for linear boundary value problems it is a simple matter to correct the guess for the velocity gradient $\frac{\mathrm{d}u_0}{\mathrm{d}y}(y = -h)$ and integrate again. In order to obtain the correction, we need to establish the linear relation between the velocity at the top plate and the velocity gradient at the bottom plate in the absence of a pressure gradient. This can be accomplished by solving the following initial value problem:

$$\frac{\mathrm{d}^2 u_1}{\mathrm{d}y^2} = 0 \tag{5.1.5}$$

[2] The actual equation for streamwise velocity that results from the Navier–Stokes equation is

$$0 = -\frac{\mathrm{d}p}{\mathrm{d}x} + \frac{\mathrm{d}}{\mathrm{d}y}\left(\mu(y)\frac{\mathrm{d}u}{\mathrm{d}y}\right),$$

which can be solved by the shooting method just as easily. Later in this chapter we will consider the shooting method for a general linear second-order ODE.

with starting conditions

$$u_1(y=-h)=0 \qquad \text{and} \qquad \frac{\mathrm{d}u_1}{\mathrm{d}y}(y=-h)=1\,. \tag{5.1.6}$$

The above ODE can be trivially solved to obtain $u_1(y)=h+y$, from which it is clear that a unit velocity gradient at the bottom plate corresponds to a velocity of $2h$ at the top plate. Thanks to the linearity of the problem, the velocity field can be written as a sum

$$u(y)=u_0(y)+\alpha\, u_1(y)\,, \tag{5.1.7}$$

where α is the yet to be determined velocity gradient at the bottom plate. It can be easily verified that (5.1.7) automatically satisfies (5.1.1) and the no-slip boundary condition at the bottom plate. From (5.1.7) the velocity at the top plate becomes

$$u(y=h)=u_0(y=h)+2\,\alpha\, h\,. \tag{5.1.8}$$

The above equation provides the mapping for the velocity at the top plate in terms of the velocity gradient at the bottom plate, $\frac{\mathrm{d}u}{\mathrm{d}y}(y=-h)=\alpha$ (see Figure 5.2 for a graphical illustration of this mapping). Here the mapping is linear and therefore can be solved for the actual velocity condition at the top plate (second of (5.1.2)) to obtain

$$\alpha=\frac{1}{2h}\left(U-u_0(y=h)\right)\,. \tag{5.1.9}$$

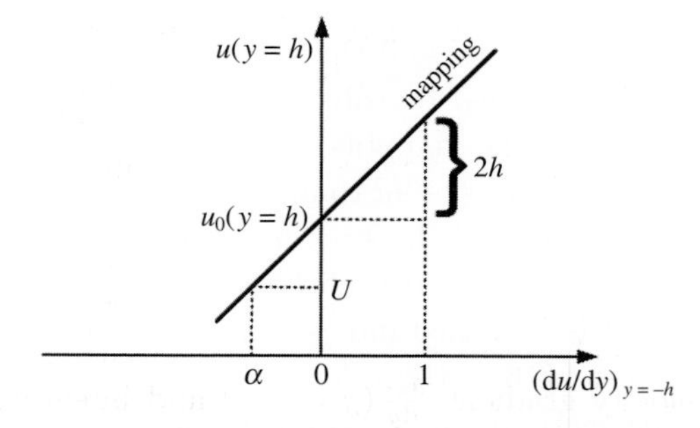

Figure 5.2 Graphical illustration of the linear mapping. An initial guess of zero velocity gradient at the bottom boundary yielded a velocity of u_0 at the top boundary. A second guess of unit velocity gradient at the bottom boundary yielded a velocity of u_0+2h at the top boundary. Thus, the value of velocity gradient at the bottom boundary that will yield a velocity of U at the top boundary can be found.

A general linear second-order boundary value ODE can be written as

$$\frac{\mathrm{d}^2u}{\mathrm{d}y^2}+p(y)\frac{\mathrm{d}u}{\mathrm{d}y}+q(y)\,u+r(y)=0 \quad \text{for} \quad 0\le y\le L \tag{5.1.10}$$

along with two Dirichlet boundary conditions

$$u(y=0)=U_{\mathrm{B}} \qquad \text{and} \qquad u(y=L)=U_{\mathrm{T}}\,. \tag{5.1.11}$$

The shooting method for the above problem involves solving the following two related initial value problems for u_0 and u_1 along with their starting conditions at $y = 0$:

$$\left.\begin{array}{ll} \dfrac{d^2u_0}{dy^2} + p(y)\dfrac{du_0}{dy} + q(y)\,u_0 + r(y) = 0\,; & \dfrac{d^2u_1}{dy^2} + p(y)\dfrac{du_1}{dy} + q(y)\,u_1 = 0\,; \\ u_0(0) = U_B\,; & u_1(0) = 0\,; \\ \dfrac{du_0}{dy}(0) = 0\,; & \dfrac{du_1}{dy}(0) = 1\,. \end{array}\right\} \tag{5.1.12}$$

It can be easily shown that the superposition (5.1.7) satisfies (5.1.10) and the boundary condition at $y = 0$. Furthermore, u at the other boundary ($y = L$) is related to the velocity gradient, $\frac{du}{dy}(y = 0) = \alpha$, by

$$u(y = L) = u_0(u = L) + \alpha\, u_1(y = L)\,. \tag{5.1.13}$$

The second velocity boundary condition in (5.1.11) can now be used to obtain

$$\alpha = \frac{U_T - u_0(y = L)}{u_1(y = L)}\,. \tag{5.1.14}$$

For this more general linear boundary value problem, the implementation of the shooting method requires the numerical solution of both the initial value problems in (5.1.12). From the value of u_0 and u_1 at $y = L$, the unknown constant α can be evaluated from (5.1.14). Solution to the boundary value problem (5.1.10) can then be obtained from the superposition (5.1.7). In essence, since the mapping between α and the velocity boundary condition at $y = L$ is linear, two points, as obtained by solving the two initial value problems, are sufficient to fully determine the mapping and hence the solution.

Boundary conditions other than Dirichlet type may be associated with (5.1.10). In the shooting method the starting conditions of the two initial value problems (5.1.12) must be modified such that on superposition the boundary condition at $y = 0$ must be satisfied independent of α, and secondly, the two initial value problems must be linearly independent. Provided these conditions are satisfied the shooting method can be applied with Neumann and mixed type boundary conditions as well. The proper value of α must of course be evaluated to satisfy the boundary condition at $y = L$.

5.1.2 Relaxation Method

Here we begin with the discretization of the domain $0 \le y \le L$ by $N + 1$ points with $y_0 = 0$, $y_N = L$ and all other y_j in between. As outlined in the previous chapter, the vector of derivatives at the grid points can be expressed in terms of the grid point values of u as

$$u'_j = [D_1]_{jk} u_k \qquad \text{and} \qquad u''_j = [D_2]_{jk} u_k\,, \tag{5.1.15}$$

where primes denote differentiation with respect to y, and the matrices $\mathbf{D}_1$ and $\mathbf{D}_2$ are the first- and second-derivative matrices. Here sum over the repeated index, i.e., sum over $k = 0, \ldots, N$, is implied. Henceforth, unless otherwise stated, this convention will be followed. The discussion to follow is independent of which finite or compact difference scheme is used, both in the interior and near the boundaries, in defining the first- and second-derivative matrices.

Equation (5.1.10) without the boundary conditions can now be expressed in terms of the following linear system:

$$[L]_{jk} u_k = -r_k \,, \tag{5.1.16}$$

where the matrix $\mathbf{L}$ takes the following form:

$$[L]_{jk} = [D_2]_{jk} + [p]_{jl}\,[D_1]_{lk} + [q]_{jl}\,[I]_{lk} \,. \tag{5.1.17}$$

In (5.1.17), $\mathbf{I}$ is the identity matrix of size $(N+1) \times (N+1)$; also

$$\mathbf{p} = \begin{bmatrix} p(y_0) & & & & \\ & p(y_1) & & & \\ & & \ddots & & \\ & & & p(y_{N-1}) & \\ & & & & p(y_N) \end{bmatrix} \tag{5.1.18}$$

and

$$\mathbf{q} = \begin{bmatrix} q(y_0) & & & & \\ & q(y_1) & & & \\ & & \ddots & & \\ & & & q(y_{N-1}) & \\ & & & & q(y_N) \end{bmatrix} \tag{5.1.19}$$

are diagonal matrices with the $N+1$ grid point values of p and q along the diagonal. The boundary conditions (5.1.11) can be easily incorporated by modifying the matrix operator as

$$\mathbf{L}^* = \begin{bmatrix} -1 \quad 0 \quad 0 & \cdots & \\ & \text{Rows 2 to } N \text{ same as } \mathbf{L} & \\ & \cdots & 0 \quad 0 \quad -1 \end{bmatrix} \tag{5.1.20}$$

and the right-hand side vector as

$$-\vec{r}^{\,*} = -\begin{bmatrix} U_\mathrm{B} \\ \text{Elements 2 to } N \text{ same as } \vec{r} \\ U_\mathrm{T} \end{bmatrix} . \tag{5.1.21}$$

The resulting linear system for (5.1.10) which has been modified appropriately for the boundary conditions (5.1.11) becomes

$$[L]^*_{jk} u_k = -r^*_j \,. \tag{5.1.22}$$

In the case of linear boundary value problems, contrary to what the name suggests, the relaxation method requires no initial guess and subsequent relaxation towards the correct answer. The grid point values of u can straightaway be obtained by solving the system of $(N+1)$ linear equations represented by equation (5.1.22). Numerical methodologies for the solution of such linear systems are a very rich field in itself. There are several direct and iterative methods to choose from, each with their own list of advantages and disadvantages. In this book we will not go into the details of numerical methods for linear systems. We simply refer the reader to excellent texts on this subject by Golub and van Loan (2012), Varga (1962), Strang (2016), and Axelsson (1996), among others.

5.1.3 Neumann Boundary Conditions

If boundary conditions other than Dirichlet type are associated with (5.1.10), then in the relaxation method, it is a simple matter to modify the first and the last rows of L appropriately to define the corresponding boundary modified matrix operator L^*. The top and bottom elements of the right-hand side are also appropriately modified to discretely implement the boundary conditions. Once these changes are made, a solution can be obtained, provided the resulting linear system of the form (5.1.22) can be solved. In particular, the problem of Neumann boundary conditions at both ends poses an interesting challenge, since the operator in this case is often not invertible. Meaningful solutions can still be obtained with careful attention. Here we will address this problem in the context of a very simple physical problem.

Consider heat conduction in a one-dimensional rod with distributed heat source along its length. Let the two ends of the rod be maintained adiabatic. The governing equation for steady state temperature distribution within the rod is then given by

$$\frac{\mathrm{d}^2 T}{\mathrm{d}y^2} = S(y) \,, \tag{5.1.23}$$

where $T(y)$ is the non-dimensional temperature and $S(y)$ is the non-dimensional heat source/sink. Scaling according to the length of the rod, we have $0 \le y \le 1$. The adiabatic condition at the two ends of the rod translates to the following Neumann boundary conditions for $T(y)$:

$$\left.\frac{\mathrm{d}T}{\mathrm{d}y}\right|_{y=0} = 0 \quad \text{and} \quad \left.\frac{\mathrm{d}T}{\mathrm{d}y}\right|_{y=1} = 0 \,. \tag{5.1.24}$$

Integrating (5.1.23) once and applying (5.1.24) results in the following integral condition that the net source/sink of heat be zero:

$$\int_0^1 S(y)\,\mathrm{d}y = 0\,. \tag{5.1.25}$$

The above condition translates to the familiar constraint that steady state heat conduction in an adiabatic system is possible only when the net source of heat is balanced by net sink. It is also clear that solution to (5.1.23) is arbitrary to an additive constant, since a constant value when added to a solution will still satisfy the equation and the Neumann boundary conditions. It should be an interesting question to investigate how this non-uniqueness of solution translates numerically.

The discrete version of the governing equation with boundary conditions can be written as

$$[L^*]_{jk}\,T_k = [S^*]_j\,, \tag{5.1.26}$$

where the matrix operator on the left and the right-hand side vector are

$$\mathbf{L}^* = \begin{bmatrix} \text{Row 1 of } \mathbf{D_1} \\ \vdots \\ \text{Rows 2 to } N \text{ of } \mathbf{D_2} \\ \vdots \\ \text{Row } (N+1) \text{ of } \mathbf{D_1} \end{bmatrix} \quad \text{and} \quad \vec{S}^* = \begin{bmatrix} 0 \\ S_1 \\ \vdots \\ S_{N-1} \\ 0 \end{bmatrix}. \tag{5.1.27}$$

By construction the inner product of the first row of $\mathbf{L}^*$ with T_k (i.e., $[L^*]_{1k}\,T_k$) will yield the first derivative at $y = 0$. The inner product of the last row of $\mathbf{L}^*$ with T_k (i.e., $[L^*]_{Nk}\,T_k$) will yield the first derivative at $y = 1$. Correspondingly the boundary values of the temperature gradient, which are zero according to (5.1.24), are applied as the first and last elements of the right-hand side vector. Equation (5.1.26) cannot be readily solved since $\mathbf{L}^*$ is not invertible. One of the eigenvalues of $\mathbf{L}^*$ is zero; in other words there exists a null mode. Remember, back in Section 4.2.1 we discussed the null mode of the second-derivative operator with Neumann boundary conditions. In contrast, the second-derivative operator with at least one Dirichlet or mixed boundary condition has all non-zero eigenvalues.

A solution to the above singular problem can still be obtained by **singular value decomposition**, which will be explained below. First, consider the eigen-decomposition of $\mathbf{L}^*$ as follows:

$$\mathbf{L}^* = \mathbf{E}\boldsymbol{\lambda}\mathbf{E}^{-1}\,, \tag{5.1.28}$$

where $\mathbf{E}$ is the eigenvector matrix whose $(N+1)$ columns are the $(N+1)$ eigenvectors of $\mathbf{L}^*$ (i.e., $\vec{E}_0, \ldots, \vec{E}_N$):

$$\mathbf{E} = \begin{bmatrix} \uparrow & & \uparrow \\ \vec{E}_0 & \cdots & \vec{E}_N \\ \downarrow & & \downarrow \end{bmatrix}. \tag{5.1.29}$$

Here $\mathbf{E}^{-1}$ is the inverse of the eigenvector matrix $\mathbf{E}$ and $\boldsymbol{\lambda}$ is a diagonal matrix consisting of the corresponding $(N+1)$ eigenvalues of $\mathbf{L}^*$:

$$\boldsymbol{\lambda} = \begin{bmatrix} \lambda_0 & & \\ & \ddots & \\ & & \lambda_N \end{bmatrix}. \tag{5.1.30}$$

Since $\mathbf{L}^* = \mathbf{E}\,\boldsymbol{\lambda}^{-1}\mathbf{E}^{-1}$, the solution in terms of eigenvectors and eigenvalues can be expressed as

$$T_k = [E]_{kl}\,[\lambda^{-1}]_{lm}\,[E^{-1}]_{mj}\,S_j^*\,, \tag{5.1.31}$$

where λ^{-1} is still a diagonal matrix consisting of the inverse of the $(N+1)$ eigenvalues of $\mathbf{L}^*$. The difficulty when one of the eigenvalues is zero becomes clear in the above representation (the inverse of the zero eigenvalue blows up). To be specific, assume the zeroth eigenvalue to be zero, i.e., $\lambda_0 = 0$. The corresponding eigenvector, $\vec{E}_0$, corresponds to the null mode of the operator $\mathbf{L}^*$. For the solution to exist we require that the projection of the right-hand side on the null (zeroth) mode vanish, i.e.,

$$\sum_{j=0}^{N}[E^{-1}]_{0j}\,S_j^* = 0\,. \tag{5.1.32}$$

This condition is the discrete analog of the integral requirement (5.1.25) that the net heat source/sink be zero.

It is quite amazing indeed to note that we have stumbled upon a discrete integration scheme. With the interior grid points, $y_1, \ldots, y_{N-1}$, as the nodes and $\left[E^{-1}\right]_{0j}$ for $j = 1, \ldots, N-1$ as the weights, the integral $\int_0^1 (.)\,dx$ can be approximated. Thus, from any numerical approximation to the second derivative, with Neumann boundary condition at both ends, we can obtain a discrete approximation to integration over the domain from the null eigenvector. After all, differentiation and integration are mathematically related and it is no surprise that their numerical approximations are related as well. The above integration operator can be compared against those defined in Section 4.5. Clearly they are different since the above approximation involves only the interior grid points. The above is the result of the discrete system (5.1.26) applying boundary conditions at the boundary points and using the second derivative only at the interior points.

Now the ratio $\left([E^{-1}]_{0j}S_j^*\right)/\lambda_0$ corresponds to zero divided by zero and the resulting numerical indeterminacy accurately reflects the non-uniqueness of the equation (5.1.23) with the Neumann boundary condition (5.1.24) that we started with. One simple option is to arbitrarily set this ratio $\left([E^{-1}]_{0j}S_j^*\right)/\lambda_0$, which in this case is equivalent to choosing the mean non-dimensional temperature, to be zero. Other choices for the ratio simply translate to a different mean non-dimensional temperature. This arbitrariness that the solution is known only to an additive constant is indeed the expected non-unique behavior.

In essence the strategy for solving the singular linear system (5.1.26) is to identify and isolate the null mode and set the projection of the solution on the null mode to be equal to zero. This strategy can be pursued even in other linear systems with multiple zero eigenvalues and associated null modes. We will revisit the singularity associated with the above problem of elliptic (second-derivative) operator with Neumann boundary conditions in the context of pressure Poisson equation in the numerical solution of an incompressible fluid flow, later in Chapter 8.

5.2 Nonlinear Boundary Value Problems

We will describe the shooting and relaxation methods for nonlinear problems through an example. What better example than the Blasius equation

$$f''' + \tfrac{1}{2} f\, f'' = 0\,, \tag{5.2.1}$$

with boundary conditions

$$f(\eta = 0) = 0\,, \quad f'(\eta = 0) = 0 \quad \text{and} \quad f'(\eta \to \infty) \to 1\,. \tag{5.2.2}$$

Of course in this problem η plays the role of the independent variable y of the previous examples.

5.2.1 Shooting Method

Fortunately a simple feature of the boundary conditions for the Blasius equation makes the shooting method essentially trivial. We now turn to a description of how this comes about. Let $f_0(\eta)$ be a solution to the Blasius equation (5.2.1) with boundary conditions

$$f_0(\eta = 0) = 0\,, \quad f_0'(\eta = 0) = 0 \quad \text{and} \quad f_0''(\eta = 0) = 1\,. \tag{5.2.3}$$

The boundary condition at the other end is not satisfied, i.e., $f_0'(\eta \to \infty) \neq 1$. Fortunately, for any value of α, the quantity $f(\eta) = \alpha\, f_0(\alpha\, \eta)$ is also a solution to (5.2.1) with boundary conditions (5.2.3). Now, as $\eta \to \infty$, the limit of $f'(\eta)$ is α^2 times the limit of $f_0'(\eta)$. Thus all we need to do in order to solve the Blasius problem is to determine the behavior of f_0 reliably for large value of its argument η, and then choose α as the limiting value of $[f_0'(\eta)]^{-1/2}$ as $\eta \to \infty$. The solution f_0 can be conveniently obtained by writing the Blasius equation as a system of three first-order ordinary differential equations – see (3.1.3) – and solving as an initial value problem with starting conditions (5.2.3) using any one of the methods outlined in Chapter 3. Figure 5.3 shows a plot of the boundary layer profile $f'(\eta)$ (the curve labeled $m = 0$; the other curves are discussed later).

Computer Exercise 5.2.1 Write a program to integrate the Blasius equation and produce a table of the solution accurate to four digits. Use the ideas just developed to generate a solution $f_0(\eta)$ and then get the desired $f(\eta)$ from the scaling. Start with $f_0''(\eta = 0) = 1$, do the computation accurately to get α. Although theoretically this single integration from $\eta = 0$ to $\eta \to \infty$ is sufficient, in order improve accuracy you may have to start over again and solve (5.2.1) once more with an improved starting condition $f_0''(\eta = 0) = \alpha^3$ instead of the arbitrary condition $f_0''(\eta = 0) = 1$. If necessary, repeat this step a few times till the desired level of accuracy is attained.

Although you are computing a function that has been computed many times before, and tables thus exist for comparison, it is generally useful for debugging purposes to have some analytic information about the solution that you are computing numerically. This kind of "background information" is provided by the following problems. Use some or all of this information in verifying your numerical work.

Problem 5.2.2 Find all solutions of (5.2.1) of the form $f(\eta) = A\,\eta^n$, where A and n are constants. (These will satisfy boundary conditions different from the required ones for the boundary layer case.)

Problem 5.2.3 Consider the function $f_0(\eta)$ that satisfies (5.2.1) and (5.2.3) (this is really the function you are computing according to the developments above). Assume f_0 is expanded in a power series:

$$f_0(\eta) = \sum_{k=0}^{\infty} F_k\,\eta^{3k+2}\,. \tag{5.2.4}$$

You can show that only every *third* power is represented starting with η^2. By substituting the series into (5.2.1) derive *Weyl's recursion formula* (see Rosenhead, 1963)

$$(3n+2)(3n+1)3n\,F_n = -\frac{1}{2}\sum_{k=0}^{n-1}(3k+2)(3k+1)\,F_k\,F_{n-k-1}\,. \tag{5.2.5}$$

With the assumption that $F_0 = 1/2$ (such that $f_0''(\eta = 0) = 1$) these can be solved to give $F_k = A_k/(-2)^k(3k+2)!$ where

$$A_0 = A_1 = 1\,;\; A_2 = 11\,;\; A_3 = 375\,;\; A_4 = 2789\,;\; A_5 = 3817137\,;\quad \text{etc.} \tag{5.2.6}$$

Hence, a power series solution valid for small η can be constructed.

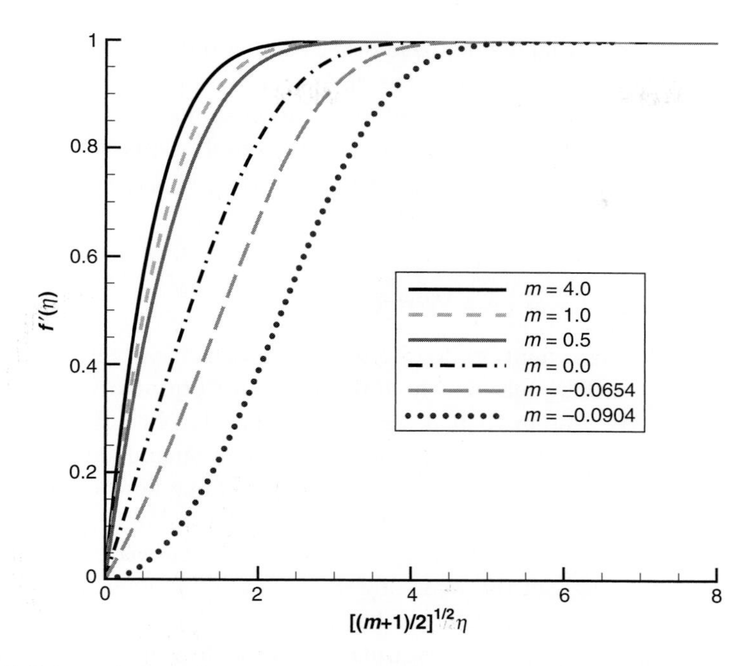

Figure 5.3 Boundary layer solutions of flow over a wedge as obtained by solving the Falkner–Skan equation for varying values of the parameter m, which is related to the half angle of the wedge by $m = \alpha/(\pi - \alpha)$. The case of $m = 0$ corresponds to the Blasius solution of zero-pressure gradient flow over a flat plate and the solution for $m = 1$ is the stagnation point flow. Positive values of m correspond to favorable pressure gradient and the velocity profile can be seen to get fuller. Negative values of m correspond to a boundary layer development under adverse pressure gradient. For the limiting case of $m = -0.0904$ the velocity gradient at the wall becomes zero, implying zero frictional force on the wall. For values of m lower than this limit flow separation, with reverse flow close to the wall, would result and the boundary layer approximation breaks down.

We already considered mappings, and so it may be worthwhile to mention how H. Weyl turned the boundary layer problem (5.2.1) and (5.2.2) into a mapping. If $g = f'$, (5.2.1) says that $(\log g)' = -\frac{1}{2}f$. Thus by (5.2.2)

$$\log g(\eta) = -\tfrac{1}{2}\int_0^\eta f(x)\,\mathrm{d}x\,. \tag{5.2.7}$$

Repeated integration by parts and use of the boundary conditions turns this into the relation

$$g(\eta) = \exp\left\{-\tfrac{1}{4}\int_0^\eta (\xi-\eta)^2\, g(\xi)\,\mathrm{d}\xi\right\}. \tag{5.2.8}$$

Problem 5.2.4 Verify equation (5.2.8).

We may think of (5.2.8) as defining a mapping from one function, g, entered on the right-hand side, to another coming out on the left, i.e., $g = \Phi(g)$, where $\Phi(g)$ is the expression on the right-hand side of (5.2.8). Weyl's result then says that *the second derivative of the solution to the boundary layer equation that we seek is a fixed point of the transformation* Φ. Equation (5.2.8) suggests that f'' can be constructed via an iteration $g_{n+1} = \Phi(g_n)$ with the starting function g_0 chosen in some appropriate way.

5.2.2 Boundary Layer on a Wedge

The Blasius equation describes the boundary layer on a flat plate. A close "relative" of this equation, the **Falkner–Skan equation**, studied by V.M. Falkner and S.W. Skan in 1930, describes boundary layers on wedges of different opening angles. The potential flow about a wedge of opening angle 2α, placed symmetrically with apex pointing towards the oncoming flow as shown in Figure 5.4, has the form $U_0(x/l)^m$ along the wedge, where U_0 and l are scales of velocity and length, x is measured from the apex along the wedge, and m is related to the opening angle of the wedge by $m = \alpha/(\pi - \alpha)$. Thus, $m = 0$ (or $\alpha = 0$) reduces the problem to the Blasius, flat-plate boundary layer, and $m = 1$ (or $\alpha = \pi/2$) produces the case of **stagnation point flow** first studied by K. Hiemenz in 1911 and L. Howarth in 1935 (see Batchelor, 1967, §5.5).

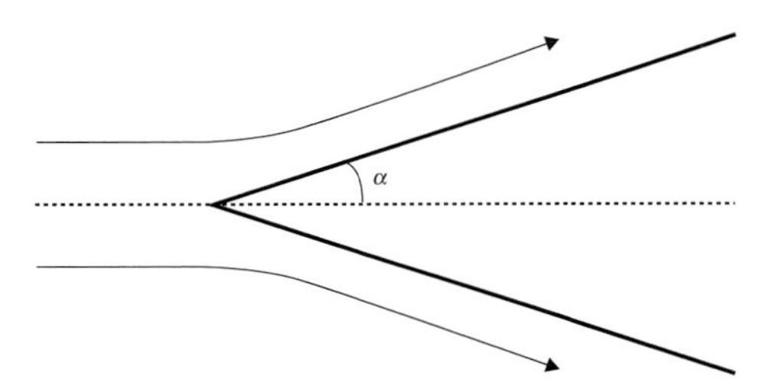

Figure 5.4 Schematic of the Falkner–Skan flow over a wedge of angle 2α. For a uniform flow of velocity U_0 approaching the wedge, the potential flow solution over the wedge is given by $U_0(x/l)^m$, where x is along the wedge.

For all these cases the boundary layer approximation admits a similarity solution for the streamfunction

$$\psi(\eta) = \sqrt{\nu\, x\, U(x)}\, f(\eta)\,; \qquad \text{where} \qquad \eta = y\sqrt{U(x)/\nu\, x}\,. \tag{5.2.9}$$

Substituting (5.2.9) into the boundary layer equations shows that f must satisfy

$$f''' + \tfrac{1}{2}(m+1)\, f\, f'' + m\,(1 - f'^2) = 0\,. \tag{5.2.10}$$

For $m = 0$ this is again the Blasius equation, (5.2.1). For $m = 1$ the similarity solution is an exact solution of the 2D Navier–Stokes equations without any boundary layer assumption! We retain the boundary conditions (5.2.2).

Just as for the Blasius problem, we first solve an equation identical to (5.2.10)

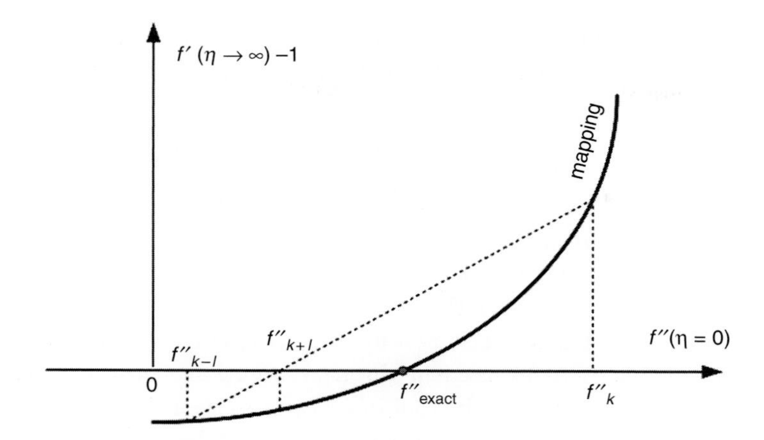

Figure 5.5 Schematic of the secant algorithm for converging towards the correct value of $f''(\eta = 0)$, which will satisfy the far field boundary condition of $f'(\eta \to \infty) = 1$.

for f_0

$$f_0''' + \tfrac{1}{2}(m+1)\, f_0\, f_0'' + m\,(1 - f_0'^2) = 0\,, \tag{5.2.11}$$

with (5.2.3) as the boundary conditions instead of (5.2.2). Unfortunately, $f(\eta) = \alpha\, f_0(\alpha\,\eta)$ is not a solution to (5.2.10) for any value of α. Thus a simple scaling will not give us the answer f. This time a real shooting method is required. Let us use $f_{0,0}$ and $f_{0,\infty}$ as shorthand notations for $f_0(\eta = 0)$ and $f_0(\eta \to \infty)$, and similarly for the derivatives. In general, the starting guess of $f''_{0,0} = 1$ is not perfect and therefore $f'_{0,\infty}$ will miss the target and will not be equal to 1. The second derivative at $\eta = 0$ must be corrected appropriately in order to satisfy $f'_{0,\infty} = 1$. In other words, $f'(\eta \to \infty)$ can be considered as a function of $f''(\eta = 0)$ and for the solution we require the root of $\{f'(\eta \to \infty) - 1 = 0\}$.

In a linear problem, the mapping between $f'(\eta \to \infty)$ and $f''(\eta = 0)$ would be linear and therefore in addition to f_0 it is sufficient to solve one other initial value problem similar to (5.2.11) for a linearly independent f_1 with starting conditions

$$f_{1,0} = 0\,, \quad f'_{1,0} = 0 \quad \text{and} \quad f''_{1,0} = \alpha \neq 1\,. \tag{5.2.12}$$

The functions f_0 and f_1 will uniquely determine the value of $f''(\eta = 0)$ that will satisfy $f'(\eta \to \infty) = 1$ as

$$f''_{2,0} = f''_{1,0} - \frac{\{f'_{1,\infty} - 1\}}{\{f'_{1,\infty} - f'_{0,\infty}\}}\,\{f''_{1,0} - f''_{0,0}\}\,. \tag{5.2.13}$$

For a nonlinear boundary value problem as given by (5.2.10), the expression (5.2.13) does not provide the exact $f''(\eta = 0)$ required for satisfying the far field condition at $\eta \to \infty$. Instead (5.2.13) can be used as the third boundary condition for the next iteration. That is, the value from (5.2.13) can be used as $f''_{2,0}$. This, along with the other two boundary conditions, $f_{2,0} = 0$ and $f'_{2,0} = 0$, can be used to start the next iteration and shoot towards $\eta \to \infty$. In this manner, with appropriate generalization, (5.2.13) becomes

$$f''_{k+1,0} = f''_{k,0} - \frac{\left\{f'_{k,\infty} - 1\right\}}{\left\{f'_{k,\infty} - f'_{k-1,\infty}\right\}} \left\{f''_{k,0} - f''_{k-1,0}\right\}, \tag{5.2.14}$$

which uses results from iterations k and $k-1$ to obtain improved estimation of $f''(\eta = 0)$ for the next, $(k+1)$th, iteration. Hopefully this sort of iteration will converge towards the exact answer. Equation (5.2.14) is simply a straight line fit through the last two iterates as illustrated in Figure 5.5 for estimating $f''_{k+1}(\eta = 0)$ to be used in the next iteration. This is nothing but the *secant algorithm* for finding the root of $\{f'(\eta \to \infty) - 1 = 0\}$. The order of convergence of the secant algorithm to the exact root can be shown to be the "golden ratio," 1.618....

Iteration can be sped up with Newton's method, where $f''(\eta = 0)$ for the next iteration can be obtained as

$$f''_{k+1,0} = f''_{k,0} - \frac{\left\{f'_{k,\infty} - 1\right\}}{\left\{\frac{\mathrm{d}f'_{k,\infty}}{\mathrm{d}f''_{k,0}}\right\}}, \tag{5.2.15}$$

where the derivative in the denominator can be obtained either analytically or with a finite difference approximation. As seen in the chapter on mappings, the convergence of Newton's method is quadratic. Of course, other methods such as bisection method, the false position method, or Ridder's method are available to continue the iteration.

Figure 5.3 shows several profiles of this more general variety, first obtained numerically by Hartree (1937). The boundary layer problem has also been investigated for flows of the form $-U_0(x/l)^m$. Now the physical interpretation is that of a boundary layer on the *converging* surfaces of intersecting planes. For example, the case $m = -1$ leads to the flow investigated by K. Pohlhausen in 1921.

A generalized form of equation (5.2.10),

$$f''' + \alpha f f'' + \beta (1 - f'^2) + \gamma f'' = 0, \tag{5.2.16}$$

which also allows suction at the wall represented by the term $\gamma f''$, has been explored numerically for various combinations of the parameters. When $\beta < 0$ but $\alpha > 0$ (which is possible for slightly negative values of m) the solution is not unique. The solutions for negative values of m shown in Figure 5.3 are those for which $f''(\eta = 0) > 0$. They join smoothly with the solutions for positive values of m. For $m = -0.0904\ldots$ the second derivative $f''(\eta = 0)$ falls to zero, implying that the frictional force on the wall vanishes everywhere for such a boundary layer. Stewartson (1954) has discussed these problems in detail, and also investigated the second set of solutions for $m < 0$. These join smoothly onto the ones shown in Figure 5.3 at $m = -0.0904$, but *do not* yield the Blasius solution for the boundary layer on a flat plate as $m \to 0$. All of these solutions show *a region of reversed flow near the wall.* Batchelor (1967) remarks that "their physical significance is not clear." Inspection of Figure 5.3 reveals that the solutions for $m \leq 0$ have a *point of inflection.* For $m = 0$ this is right at the wall (i.e., at $\eta = 0$). For $m < 0$ it is within the flow (see also the discussion in Rosenhead (1963), §V.21).

Computer Exercise 5.2.5 Modify the program written in Computer Exercise 5.2.1 so as to be able to solve the Falkner–Skan equations for $m \neq 0$. Find, for example, the Hiemenz–Howarth solution for $m = 1$, the "uniform wall-stress solution" for $m = 1/3$, and explore the $m < 0$ interval. Also try out the wall suction effect with a non-zero γ. Consult the literature for guidance and comparisons.

5.2.3 Relaxation Method

We now go back to the simpler Blasius problem given by equation (5.2.1) and boundary conditions (5.2.2) to illustrate this method. In the implementation it is convenient to reduce the higher-order ODE into a system of second- and first-order ODEs (it is not necessary to reduce to a system of only first-order ODEs as in the integration of initial value problems). The Blasius equation can then be written as

$$g = f' \qquad \text{and} \qquad g'' + \tfrac{1}{2} f\, g' = 0\,, \tag{5.2.17}$$

and accordingly the boundary conditions become

$$f(\eta = 0) = 0\,, \quad g(\eta = 0) = 0 \quad \text{and} \quad g(\eta \to \infty) \to 1\,. \tag{5.2.18}$$

In the relaxation method we start with an initial guess, f_0, from which $g_0 = f_0'$ can be obtained. In general, f_0 and g_0 will not satisfy (5.2.17) – otherwise the problem is solved in the first place. Now we are set to determine the perturbations ξ and ζ, such that

$$f = f_0 + \xi \qquad \text{and} \qquad g = g_0 + \zeta \tag{5.2.19}$$

will satisfy (5.2.17). Substituting (5.2.19) into (5.2.17) and eliminating quadratic and higher nonlinear terms in the perturbation, we obtain the following coupled linear equations

$$\xi' - \zeta = -\left(f_0' - g_0\right)\,, \tag{5.2.20a}$$

$$\zeta'' + \tfrac{1}{2}\left(f_0\zeta' + g_0'\xi\right) = -\left(g_0'' + \tfrac{1}{2} f_0 g_0'\right)\,. \tag{5.2.20b}$$

The right-hand sides of the above equations can be identified as the negative of the residue when f_0 and g_0 are used as solution in (5.2.17). The perturbations ξ

and ζ as evaluated from (5.2.20a, 5.2.20b) will not completely nullify the residue, due to the neglect of the quadratic term $\frac{1}{2}\xi\,\zeta'$ on the left-hand side of (5.2.20b). Therefore (5.2.19) must be reinterpreted as

$$f_1 = f_0 + \xi \qquad \text{and} \qquad g_1 = g_0 + \zeta\,. \tag{5.2.21}$$

They only provide the next iterate of f and g. Had the problem been linear, (5.2.20a, 5.2.20b) would have involved no approximation and the problem would then be solved with the correction step (5.2.19). Because of linearization, (5.2.20a, 5.2.20b) are not exact, and thus the solution must be obtained iteratively, with f_1 and g_1 replacing f_0 and g_0 in those equations, and so on. So the term *relaxation method* is more appropriate in the solution of nonlinear ODEs.

If we choose f_0 and g_0 to be such that they satisfy the boundary conditions (5.2.18), then the appropriate boundary conditions for the perturbation at all levels of iteration are homogeneous:

$$\xi(\eta = 0) = 0\,, \quad \zeta(\eta = 0) = 0 \quad \text{and} \quad \zeta(\eta \to \infty) \to 0\,. \tag{5.2.22}$$

The advantage of linearization is that during each iteration we need to solve only a system of two linear boundary value ODEs, (5.2.20a, 5.2.20b) with boundary conditions (5.2.22), about which we should now be quite familiar after our discussion in Section 5.1. This linearization and iteration is nothing but Newton's method applied to the solution of the Blasius equation. Convergence towards the answer is extremely rapid for most reasonable choices of the initial guess f_0 and g_0.

We will now discuss further the numerical solution of (5.2.20a, 5.2.20b). Typically the outer edge of the boundary layer will be truncated at a large but finite value, L, so that for computational purposes $0 \le \eta \le L$.[3] Also consider a suitable distribution of $N+1$ grid points ranging from $\eta_0 = 0$ to $\eta_N = L$. It may be advantageous to consider a non-uniform distribution of grid points with clustering of points close to $\eta = 0$. The details of the grid distribution and the choice of finite or compact difference schemes will certainly affect the definition of the discrete first- and second-derivative matrix operators. Here if we consider these matrices to be appropriately defined (following the discussion in Chapter 4), then further investigation can proceed independent of these details. The discretized

[3] Truncation to a finite domain is not always needed. One could map the infinite domain onto a finite domain through appropriate change of the independent variable from η to say $\tilde{\eta}$. The problem could now be solved in $\tilde{\eta}$ after appropriately modifying the ODEs and the boundary conditions. For the Blasius problem, both approaches give essentially the same answer provided L is greater than few times the boundary layer thickness.

version of (5.2.20a, 5.2.20b) for the kth iteration takes the following form:

$$\begin{bmatrix} \mathbf{L_{11}} & \mathbf{L_{12}} \\ \\ \mathbf{L_{21}} & \mathbf{L_{22}} \end{bmatrix} \begin{bmatrix} \vec{\xi} \\ \\ \vec{\zeta} \end{bmatrix} = - \begin{bmatrix} \vec{f_k'} - \vec{g_k} \\ \\ \vec{g_k''} + \frac{1}{2} \vec{f_k}\, \vec{g_k'} \end{bmatrix}, \tag{5.2.23}$$

where the block matrices on the left-hand side are of size $(N+1) \times (N+1)$ and are given by

$$\mathbf{L_{11}} = \mathbf{D_1}\,; \quad \mathbf{L_{12}} = -\mathbf{I}\,; \quad \mathbf{L_{21}} = \frac{\mathbf{1}}{\mathbf{2}}\mathbf{g_k'}\,; \quad \mathbf{L_{22}} = \mathbf{D_2} + \frac{\mathbf{1}}{\mathbf{2}}\,\mathbf{f_k}\,\mathbf{D_1}\,. \tag{5.2.24}$$

The first- and second-derivative matrices $\mathbf{D_1}$ and $\mathbf{D_2}$ are of size $(N+1)\times(N+1)$. Here $\mathbf{I}$ is the identity matrix of size $(N+1) \times (N+1)$. The perturbations $\vec{\xi}$ and $\vec{\zeta}$, now defined at the grid points, are vectors of length $(N+1)$. Current iterates $\vec{f_k}$, $\vec{g_k}$ and their derivatives are also known at the grid points and therefore on the right-hand side they are vectors of length $(N+1)$, whereas on the left-hand side $\mathbf{f_k}$ and $\mathbf{g_k'}$ need to be properly interpreted as diagonal matrices of size $(N+1) \times (N+1)$. Thus (5.2.23) is a linear system of $2(N+1)$ equations, of which the first $(N+1)$ enforce (5.2.20a) at the grid points, while the next $(N+1)$ enforce (5.2.20b) at the grid points.

Boundary conditions (5.2.22) need to be incorporated into the discretized matrix equation (5.2.23). To do this we first recognize that the first row of (5.2.23) is the discretized version of equation (5.2.20a) at $\eta = 0$. The $(N+2)$th row of (5.2.23) is the discretized version of equation (5.2.20b) at $\eta = 0$. Similarly the $2(N+1)$th row of (5.2.23) is the discretized version of equation (5.2.20b) at $\eta = L$. We must forgo enforcing these in order to satisfy the boundary conditions (5.2.22). After all, only $2(N+1)$ conditions can be satisfied to determine the $2(N+1)$ grid point values of ξ and ζ uniquely.

In principle, any three rows of (5.2.23) can be sacrificed for the enforcement of the boundary conditions, but the connection between rows 1, $N+2$ and $2(N+1)$ with the boundary conditions (5.2.22) is obvious. Boundary corrections to (5.2.23) can be listed as:

(1) Condition (5.2.22) (left) can be easily enforced by replacing the first row of $\mathbf{L_{11}}$ and $\mathbf{L_{12}}$ by $\begin{bmatrix} 1 & 0 & 0 & \cdots & \end{bmatrix}$ and $\begin{bmatrix} 0 & 0 & 0 & \cdots & \end{bmatrix}$. Correspondingly the first element of the right-hand side vector is set to 0.

(2) Condition (5.2.22) (middle) can be enforced by replacing the first row of $\mathbf{L_{21}}$ and $\mathbf{L_{22}}$ by $\begin{bmatrix} 0 & 0 & 0 & \cdots & \end{bmatrix}$ and $\begin{bmatrix} 1 & 0 & 0 & \cdots & \end{bmatrix}$. Correspondingly the $(N+2)$th element on the right-hand side vector is set to 0.

(3) Condition (5.2.22) (right) can be enforced by replacing the last row of $\mathbf{L_{21}}$ and $\mathbf{L_{22}}$ by $\begin{bmatrix} 0 & 0 & 0 & \cdots & \end{bmatrix}$ and $\begin{bmatrix} & \cdots & 0 & 0 & 1 \end{bmatrix}$. Correspondingly the last element of the right-hand side vector is set to 0.

The resulting boundary corrected linear system can be readily solved by any one of the linear system solvers available. There are several commercial (such as MATLAB, IMSL) and freeware (such as LAPACK) packages that contain suitable linear algebra routines. For theory on linear system solvers the reader is referred to Golub and van Loan (2012), Varga (1962), Strang (2016) and Axelsson (1996). It is important to realize that in (5.2.23) as well as in its boundary modified version, both the matrix on the left and the vector on the right will change from iteration to iteration. The choice of linear system solver must take into account this aspect of the problem.

5.3 Boundary Value Problems in Viscous Flow

There are a number of cases in which the full Navier–Stokes equations can be reduced to a system of ODEs because of the extreme symmetry of the boundaries or through the introduction of a similarity *ansatz* for the solution. Such solutions give numerous analytical insights that are suggestive of general trends. Furthermore, since these solutions can typically be computed to high accuracy, they are very useful as benchmark cases for numerical methods aimed at solving the full PDEs.

It would be inappropriate to go through the rather detailed analysis that is required for each of these solutions here, so we shall limit ourselves to a brief review of some of the simplest cases. Some of these, such as plane Poiseuille and plane Couette flow, have already been considered.

5.3.1 Poiseuille Flow

The Poiseuille flow problem to be addressed here is the steady, fully-developed flow through a pipe of arbitrary but uniform cross-section D (of any shape). Gravity is ignored. Since each fluid particle is moving at a constant velocity, the Navier–Stokes equations reduce to the statement that the viscous force is balanced by the difference in pressure forces across the particle, i.e., that the viscous term in the Navier–Stokes equation balances a constant pressure gradient. Since the only variation of importance is in the cross-stream direction (taken as the yz-plane), and the only non-zero component of velocity, u, is along the pipe axis, we obtain the 2D equation in the cross-section D:

$$\left.\begin{array}{rcll} \nabla^2 u & = & -1 & \text{within } D, \\ u & = & 0 & \text{on the boundary of } D. \end{array}\right\} \tag{5.3.1}$$

Here we have scaled u by the constant $\Delta p/\mu L$, where μ is the dynamic viscosity, and Δp is the magnitude of the drop in pressure along a length of pipe L. We shall discuss how to solve a PDE such as (5.3.1) numerically in Chapter 8.

For particular symmetries the PDE in (5.3.1) reduces to an ODE. For example, if the pipe has a cross-section with circular symmetry, we can use polar coordinates r, φ and cast (5.3.1) in the form:

$$\frac{1}{r}\frac{\mathrm{d}}{\mathrm{d}r}\left(r\frac{du}{\mathrm{d}r}\right)+\frac{1}{r^2}\frac{\mathrm{d}^2u}{\mathrm{d}\phi^2}=-1\,. \tag{5.3.2}$$

Then, assuming the solution depends only on r, we have a simple second-order ODE:

$$\frac{\mathrm{d}}{\mathrm{d}r}\left(r\frac{du}{\mathrm{d}r}\right)=-r \tag{5.3.3}$$

with the general solution

$$u(r)=-\frac{1}{4}r^2+C\,\log r+C'\,, \tag{5.3.4}$$

where C and C' are two arbitrary constants. If $r=0$ is within the domain, C must vanish. If the domain is bounded internally by a circle of radius R_1 and externally by a circle of radius R_2, where $R_1<R_2$, the solution is

$$u=\frac{1}{2}\left\{R_2^2-r^2+\frac{R_2^2-R_1^2}{\log\left(R_2/R_1\right)}\log\left(r/R_2\right)\right\}\,, \tag{5.3.5}$$

where the logarithmic term is to be read as 0 for $R_1=0$.

The solution (5.3.4) can also be found by superposition of the elemental solutions r^2 and $\log r$ and "adjustment" of constants in much the same way as one finds solutions of Laplace's equation for potential flow about bodies in the elementary theory of inviscid flows. This procedure gives for a pipe of elliptical cross-section,

$$\frac{x^2}{a^2}+\frac{z^2}{b^2}=1\,, \tag{5.3.6}$$

the solution

$$u=\frac{a^2b^2}{2\left(a^2+b^2\right)}\left(1-\frac{x^2}{a^2}-\frac{z^2}{b^2}\right)\,. \tag{5.3.7}$$

As a final example of a simple closed form solution of the Poiseuille problem we give the solution for flow in a pipe with equilateral triangle cross-section:

$$u=\frac{2}{s\sqrt{3}}h_1\,h_2\,h_3\,, \tag{5.3.8}$$

where h_1, h_2, h_3 are the distances of the point in question from the sides of the equilateral triangle of side s. All these solutions are given in Landau and Lifshitz (1987), Article 17, Problems 1–3. The last two, of course, only relate to our current theme of ODEs if written in elliptical or triangular coordinates, respectively, but may be very useful later for checking a finite difference program aimed at solving the PDE (5.3.1).

Problem 5.3.1 Find the volume discharge for each of the flows (5.3.5)–(5.3.8). Compare the results for pipes of the same cross-sectional area (and the same pressure drop per unit length and fluid viscosity).

5.3.2 Couette Flow

The *circular Couette flow* problem (for a historical review see Donnelly, 1991) addresses the flow between concentric cylinders both of which can be rotated independently. In the simplest case, before the onset of any of the interesting instabilities, the velocity is purely circumferential, and depends only on the radial distance, r, from the common rotation axis of the cylinders. The obvious equation in this situation is Newton's second law, which says that the centripetal acceleration of a fluid element is provided by the pressure gradient

$$\frac{\rho\, v_\phi^2}{r} = \frac{\mathrm{d}p}{\mathrm{d}r}\,. \tag{5.3.9}$$

This equation allows us to determine the pressure field once we know the velocity. The ODE that we seek for v_ϕ is basically the torque equation, which says that the rate of change of the angular momentum of a cylindrical shell is equal to the torque of the tangential frictional forces acting on its faces. This equation is (see Batchelor, 1967, §4.5)

$$\frac{\mathrm{d}^2 v_\phi}{\mathrm{d}r^2} + \frac{1}{r}\frac{\mathrm{d}v_\phi}{\mathrm{d}r} - \frac{v_\phi}{r^2} = 0\,. \tag{5.3.10}$$

The solution of this ODE with the boundary conditions $v_\phi = \Omega_1 R_1$ for $r = R_1$, $v_\phi = \Omega_2 R_2$ for $r = R_2$ is

$$v_\phi(r) = \frac{\Omega_2 R_2^2 - \Omega_1 R_1^2}{R_2^2 - R_1^2}\, r + \frac{(\Omega_1 - \Omega_2)\, R_1^2 R_2^2}{R_2^2 - R_1^2}\frac{1}{r}\,. \tag{5.3.11}$$

Here $R_1 < R_2$ are the cylinder radii, and Ω_1, Ω_2 their angular velocities. It is noteworthy that the kinematic viscosity does not appear in (5.3.11).

We remark that the steady Stokes flow problem can be solved for the case when the cylinders are no longer concentric by using a transformation to bipolar coordinates. This solution is quite useful for numerical verification of PDE solvers since it displays considerable richness of structure. (We have already mentioned it in Example 1.6.2 and in connection with Figure 2.9.) The full analytic solution for this case is not known, but in the limit of a narrow gap an approximate analysis, known as *lubrication theory*, may be invoked.

5.3.3 von Kármán Flow over a Rotating Disk

Consider the flow in a halfspace bounded by an infinite disk rotating uniformly with angular velocity Ω. In steady state an axisymmetric, swirling flow will be set up. Because of the nature of the boundary conditions, the only length scale in the problem is the viscous penetration depth $\sqrt{\nu/\Omega}$. Hence, in place of the vertical coordinate y, along the rotation axis of the disk, one introduces the dimensionless variable $\eta = y\sqrt{\Omega/\nu}$. In cylindrical coordinates the radial and circumferential velocity components, v_r and v_ϕ, are assumed to have the forms

$$v_r = \Omega\, r\, F(\eta)\,; \qquad v_\phi = \Omega\, r\, G(\eta)\,. \tag{5.3.12}$$

The axial velocity, on the other hand, is expected to go to a constant value as $y \to \infty$, and is assumed to have the form

$$v_y = \sqrt{\Omega \nu}\, H(\eta)\,. \tag{5.3.13}$$

A similar *ansatz* is assumed for the pressure, as required for consistency in the equations of motion.

We substitute the above velocity expressions into the Navier–Stokes equations, written in cylindrical coordinates, and into the condition of vanishing divergence of the velocity field. This results in the following system of ODEs:

$$\left.\begin{aligned} F^2 - G^2 + F'\,H &= F''\,, \\ 2F\,G + G'\,H &= G''\,, \\ 2F + H' &= 0\,. \end{aligned}\right\} \tag{5.3.14}$$

The boundary conditions on the system (5.3.14) are:

$$\left.\begin{array}{lll} F(0) = 0\,; & G(0) = 1\,; & H(0) = 0\,; \\ F(\eta \to \infty) \to 0\,; & G(\eta \to \infty) \to 0\,, & \end{array}\right\} \tag{5.3.15}$$

where the first three are no-slip and no-penetration conditions on the disk, while the last two are far-field conditions. Unlike the Poiseuille and Couette flow problems, (5.3.14) forms a system of nonlinear boundary value problem. The reduction of the full viscous flow problem to the system of ODEs (5.3.14) was given by T. von Kármán in 1921. The first numerical integrations of (5.3.14), done essentially by hand using a calculating machine, were reported by Cochran (1934) in one of the early works of the CFD literature. Of particular interest is the limiting value of $H(\eta \to \infty)$. Numerically one finds this value to be approximately -0.886.

Problem 5.3.2 Write a system of first-order ODEs, equivalent to (5.3.14), that is suitable for numerical integration.

Computer Exercise 5.3.3 Using the results of Problem 5.3.2 set up an integration procedure for the system (5.3.14). Use the shooting method and write a program to determine the solution.

Problem 5.3.4 Formulate the relaxation method to solve the system of ODEs (5.3.14) for the von Kármán rotating disk problem. Write down the system of linear perturbation equations that results from the relaxation method. Now consider a discretization of $(N+1)$ points spanning $0 \le \eta \le \eta_{\text{large}}$ and let $\mathbf{D_1}$ and $\mathbf{D_2}$ be the corresponding first- and second-derivative matrices. Express the linear system that you obtained above in the form of a matrix equation, similar to that in (5.2.23). What are the appropriate boundary conditions for this problem?

Computer Exercise 5.3.5 Make an appropriate choice for the grid and choose a finite or compact difference scheme for the first- and second-derivative operators. Then write a program to implement the relaxation method and compare the accuracy and efficiency of computation with the shooting method.

5.4 Eigenvalue Problems

We continue with our philosophy and illustrate the eigenvalue problem with the example of thermal development in a pipe flow called the **Graetz problem**. We consider the flow to be hydrodynamically fully developed with a velocity profile

$$u(r) = 2\bar{u}\left(1 - \frac{r^2}{R^2}\right), \tag{5.4.1}$$

where R is the radius of the pipe and $\bar{u}$ is the average velocity through the pipe. However, the flow will be considered to be thermally developing. This could be the result of a sudden change in temperature of the pipe wall from T_0 to T_w in a fully developed flow, or can be considered as the limiting case of high Prandtl number when the thermal entrance length is much larger than the development length of the velocity profile.

If we neglect heating due to viscous dissipation and axial conduction of heat, the energy equation for the temperature, $T(r, x)$, which is now a function of the radial (r) and the axial (x) direction, can be written as

$$u\frac{\partial T}{\partial x} = \frac{\alpha}{r}\frac{\partial}{\partial r}\left(r\frac{\partial T}{\partial r}\right), \tag{5.4.2}$$

where α is thermal diffusivity of the fluid. Equation (5.4.2) is an initial-boundary value PDE, with the thermal development along the axial (x) direction playing an analogous role to temporal development in an initial value problem. So we specify the starting temperature profile along the axial direction as the uniform inlet temperature, $T(r, x = 0) = T_0$. The appropriate boundary condition along the radial direction is given by

$$T(r = R, x) = \begin{cases} T_0 & \text{for} \quad x < 0 \\ T_w & \text{for} \quad x > 0 \end{cases} \tag{5.4.3}$$

where without loss of generality $x = 0$ is taken to be the point at which the wall temperature changes.

Introducing the following non-dimensional variables (see White, 1974, §3.3.7)

$$\theta = \frac{T_w - T}{T_w - T_0}\,; \qquad \eta = \frac{r}{R}\,; \qquad \xi = \frac{x}{2R\,\mathrm{Re}\,\mathrm{Pr}}\,, \tag{5.4.4}$$

where $\mathrm{Re} = 2R\,\bar{u}/\nu$ is the Reynolds number based on pipe diameter and $\mathrm{Pr} = \nu/\alpha$ is the Prandtl number of the fluid, the energy equation (5.4.2) can be rewritten as

$$\frac{\partial \theta}{\partial \xi} = \frac{2}{\eta(1-\eta^2)} \frac{\partial}{\partial \eta}\left(\eta \frac{\partial \theta}{\partial \eta}\right). \tag{5.4.5}$$

The initial and boundary conditions transform to $\theta(\eta, \xi = 0) = 1$ and $\theta(\eta = 1, \xi > 0) = 0$. The linearity of (5.4.5) allows separation of variables and we look for a solution of the separable form

$$\theta(\eta, \xi) = \Psi(\eta)\,\Xi(\xi)\,. \tag{5.4.6}$$

On substitution into (5.4.5) it is easy to see that the axial solution exhibits exponential decay

$$\Xi(\xi) = A\,\exp\left[-2\lambda\xi\right], \tag{5.4.7}$$

where A is the constant of integration and λ is a positive constant arising from the separation of variables. The characteristic equation for the radial behavior

$$\eta\,\Psi'' + \Psi' = \lambda\,\eta\,\left(\eta^2 - 1\right)\Psi \tag{5.4.8}$$

needs to be solved with the following temperature boundary condition at the pipe wall:

$$\Psi(\eta = 1) = 0\,, \tag{5.4.9}$$

supplemented by a symmetry condition for temperature profile at the pipe centerline:

$$\Psi'(\eta = 0) = 0\,. \tag{5.4.10}$$

As usual the prime denotes derivative with respect to the argument. Note η is the non-dimensional radius.

We are finally at the eigenvalue problem. Equation (5.4.8) along with boundary conditions (5.4.9) and (5.4.10) is a second-order eigenvalue problem for $\Psi(\eta)$. Note that the equation and the boundary conditions are both linear and homogeneous. Non-trivial solutions, known as the **eigenfunctions**, to the above problem are possible for only certain values of λ, known as the **eigenvalues**. These eigenvalues and corresponding eigenfunctions can be ordered and enumerated with a subscript n as λ_n and Ψ_n. The overall solution to (5.4.5) can then be written as a superposition of all the different eigensolutions as

$$\theta(\eta, \xi) = \sum_n A_n\,\Psi_n(\eta)\,\exp\left[-2\lambda_n \xi\right]. \tag{5.4.11}$$

The weighting constants A_n are obtained from the condition $\theta(\eta, \xi = 0) = 1$ as

$$\sum_n A_n \Psi_n(\eta) = 1 . \tag{5.4.12}$$

From the theory of second-order linear boundary value problems it can be established that the eigenfunctions, Ψ_n, are orthogonal with respect to weight $\eta(1-\eta)$. The orthogonality condition leads to the estimation of weighting constants as

$$A_n = \frac{\int_0^1 \eta(1-\eta)\, \Psi_n \, d\eta}{\int_0^1 \eta(1-\eta)\, \Psi_n^2 \, d\eta} . \tag{5.4.13}$$

Table 5.1 lists the first few of the eigenvalues along with the weighting constants A_n. The above problem was first solved more than a century ago in 1885 by Graetz. Two years earlier in 1883 he solved the related problem of thermal development in a slug flow with a constant velocity across the pipe (instead of the parabolic one). Graetz's solutions were in fact numerical. Since Graetz, many famous researchers, including Nusselt, have refined the numerical approach and significantly improved the accuracy of the solution.

Table 5.1 First few eigenvalues of (5.4.8) along with the weighting constants A_n. In the evaluation of (5.4.13), the eigenfunctions have been normalized such that $\Psi_n(\eta = 0) = 1$.

n	$(\lambda_n)^{1/2}$	A_n
0	2.70436	1.46622
1	6.67903	−0.80247
2	10.67338	0.58709
3	14.67108	−0.47490
4	18.66987	0.40440

A numerical solution of the eigenvalue problem (5.4.8) can be sought using the shooting method. Equation (5.4.8) can be written as a system of two linear ODEs and integrated from $\eta = 0$ to $\eta = 1$, starting with condition (5.4.10) supplemented by an arbitrary condition at the centerline that $\Psi(\eta = 0) = 1$. Since the eigenvalue λ is not known *a priori*, the shooting method begins with a guess value for λ and the corresponding solution is checked at the other end to see if condition (5.4.9) is satisfied. The integration of (5.4.8) must be continued till a suitable value of λ is found such that condition (5.4.9) is satisfied. In principle, it is possible to repeat this procedure several times to extract the different eigenvalues and corresponding eigenfunctions that satisfy (5.4.8) along with the boundary conditions. In practice, it is difficult to extract more than a few eigensolutions. One must resort to periodic removal of already extracted eigensolutions in order to obtain new solutions. This procedure is known as *ortho-renormalization.* Here we will not pursue the shooting method further since the relaxation method provides a direct and efficient way to get the eigensolutions.

We begin with a discretization of the domain $0 \leq \eta \leq 1$ into $N+1$ points: η_j for $j = 0$ to N. In the present problem it may be advantageous to cluster the grid points close to $\eta = 1$, since the strongest variation in temperature can be expected close to the pipe wall. Nevertheless, the first- and second-derivative matrices can be appropriately defined once the grid points and the finite or compact difference scheme are chosen. The discretized version of (5.4.8) becomes

$$\begin{bmatrix} \eta_0 & & & & \\ & \eta_1 & & & \\ & & \ddots & & \\ & & & \eta_{N-1} & \\ & & & & \eta_N \end{bmatrix} \mathbf{D_2} \begin{bmatrix} \Psi_0 \\ \Psi_1 \\ \vdots \\ \Psi_{N-1} \\ \Psi_N \end{bmatrix} + \mathbf{D_1} \begin{bmatrix} \Psi_0 \\ \Psi_1 \\ \vdots \\ \Psi_{N-1} \\ \Psi_N \end{bmatrix}$$

$$= \lambda \begin{bmatrix} R_0 & & & & \\ & R_1 & & & \\ & & \ddots & & \\ & & & R_{N-1} & \\ & & & & R_N \end{bmatrix} \begin{bmatrix} \Psi_0 \\ \Psi_1 \\ \vdots \\ \Psi_{N-1} \\ \Psi_N \end{bmatrix}, \tag{5.4.14}$$

where $\mathbf{D_1}$ and $\mathbf{D_2}$ are the first- and second-derivative matrices and the diagonal entries in the $\mathbf{R}$ matrix on the right-hand side are $R_j = \eta_j\left(\eta_j^2 - 1\right)$. The terms on the left-hand side can be combined to represent the above into a generalized matrix eigenvalue problem:

$$\mathbf{L}\,\boldsymbol{\Psi} = \lambda\,\mathbf{R}\,\boldsymbol{\Psi}\,. \tag{5.4.15}$$

Before solving (5.4.15), the left- and right-hand side matrices must be modified to account for the boundary conditions. This can be accomplished with the following steps:

(1) The Dirichlet boundary condition (5.4.9) can be incorporated into (5.4.15) by replacing the last row of $\mathbf{L}$ by $[0\ \ldots\ 0\ 0\ 1]$ and correspondingly the last row of $\mathbf{R}$ by all zeros.

(2) The Neumann boundary condition (5.4.10) at the centerline can be implemented by replacing the first row of $\mathbf{L}$ by the first row of the first-derivative matrix, $\mathbf{D_1}$, and correspondingly the first row of the right-hand side matrix $\mathbf{R}$ by all zeros.

Thus the final generalized matrix eigenvalue problem to be solved numerically is identical to (5.4.15) with the left- and right-hand side matrices replaced by their corresponding boundary corrected version, $\mathbf{L}^*$ and $\mathbf{R}^*$.

There are several library routines that are part of MATLAB and other packages such as LAPACK that can solve generalized matrix eigenvalue problem.

These routines typically take matrices $\mathbf{L}^*$ and $\mathbf{R}^*$ as input and provide all eigenvalues, λ_n, and if appropriate switches are turned on, all the corresponding eigenvectors, $\boldsymbol{\Psi}_n$, as output. The actual numerical computation of the eigenvalues and eigenvectors from the left- and right-hand side matrices is quite complex and falls under the general topic of numerical linear algebra. Once again we refer the reader to books by Golub and van Loan (2012), Axelsson (1996) and Wilkinson (1965) and to the user guides for MATLAB and LAPACK packages. Before leaving this topic, we just note that the generalized eigenvalue problem can be converted into a regular eigenvalue problem of the form

$$(\mathbf{L}^{*-1}\,\mathbf{R}^*)\,\boldsymbol{\Psi} = \frac{1}{\lambda}\,\boldsymbol{\Psi}\,, \tag{5.4.16}$$

where λ^{-1} is the eigenvalue of $(\mathbf{L}^{*-1}\mathbf{R}^*)$.

Computer Exercise 5.4.1 Use your favorite finite or compact difference scheme and construct the boundary corrected left- and right-hand side matrices $\mathbf{L}^*$ and $\mathbf{R}^*$. Use library routines to solve either the generalized (5.4.15) or the standard eigenvalue (5.4.16) problem and compare the computed eigenvalues with those provided in Table 5.1. See how many grid points are needed to obtain the first eigenvalue to three decimal accuracy and verify that the required number of points decrease with higher-order differencing. Repeat the above for the first few eigenvalues.

5.5 Hydrodynamic Instability

Investigation of the stability of fluid flows has played an important role in fluid dynamics research. In linear stability analysis one considers only the evolution of infinitesimal disturbances and in this case the stability problem often reduces to solving a system of linear eigenvalue ODEs. The above outlined matrix-based direct method for eigenvalue calculation is a power tool and you may be surprised to know that you are now in a position to solve most of the classic linear stability problems of the past and even some stability problems of current interest. Of course, the challenge is not just in numerically solving the eigenvalue ODEs, but also in mathematically formulating the problem and in the analysis of the ensuing results.

The linear stability analysis begins with a **base flow**, whose stability to infinitesimal disturbances is usually the subject matter of the investigation. The base flow can be as simple as a Poiseuille or a Couette flow, or can be more complex as in the case of flow between two eccentric cylinders considered in Example 1.6.2, or can be even more complex that only a numerical solution of the base flow is possible. Infinitesimal perturbations are then added to the base flow and the governing equations for the time evolution of the perturbations is obtained by demanding that the base flow plus the perturbations satisfy the continuity, Navier–Stokes and if necessary the energy equations. For further details read the texts on hydrodynamic stability by Drazin and Reid (2004) and Criminale *et al.* (2003).

In relatively simpler problems the base flow is steady and one-dimensional (i.e., the velocity components show strong variation along only one coordinate direction). In which case, the stability of the base flow greatly simplifies and the governing equations for the infinitesimal perturbation reduce to an eigenvalue system of ODEs. Linear stability of both time-periodic and multi-dimensional base flows have also been considered in the past, but the resulting eigenvalue problem involves PDEs. Here we limit our attention to simpler base flows and the resulting eigenvalue ODEs.

5.5.1 Thermal Instability

We first consider the problem of thermal instability of a layer of fluid of infinite horizontal extent, heated at the bottom and cooled at the top. This problem was originally considered by Lord Kelvin, Bénard and Lord Rayleigh and the resulting buoyancy induced convective flow now goes by the name **Rayleigh–Bénard convection**. In order to characterize this instability one needs to consider both the flow and thermal fields. The base state is steady and one dimensional, given by no fluid motion and a linear temperature drop from the bottom of the fluid layer to the top.

The governing equations in this problem are the Boussinesq equations, which consider the fluid to be incompressible except in the buoyancy term where the temperature-dependence of density is considered to be important and linear. Furthermore, heating due to internal friction and heating/cooling due to adiabatic compression/expansion are considered unimportant in the energy equation. Infinitesimal velocity and temperature perturbations are added to the base state and substituted into the Boussinesq equations. Non-dimensionalizing with the fluid layer height, H, as the length scale, H^2/κ as the time scale and temperature drop across the layer, ΔT, as the temperature scale, we get the following linearized equations for the non-dimensional velocity, $\mathbf{u}'$, and non-dimensional temperature, θ', perturbations:

$$\left.\begin{aligned} \nabla \cdot \mathbf{u}' &= 0\,, \\ \frac{\partial \mathbf{u}'}{\partial t} &= -\nabla p' + \mathrm{Ra}\ \mathrm{Pr}\ \theta'\,\mathbf{j} + \mathrm{Pr}\ \nabla^2 \mathbf{u}'\,, \\ \frac{\partial \theta'}{\partial t} &= v' + \nabla^2 \theta'\,, \end{aligned}\right\} \tag{5.5.1}$$

where v' is the vertical velocity component and $\mathbf{j}$ is the unit vector along the vertical direction. The two non-dimensional parameters – the Rayleigh number,

Ra, and the Prandtl number, Pr – are given by

$$\mathrm{Ra} = \frac{g\,\alpha \Delta T H^3}{\kappa\,\nu} \qquad \text{and} \qquad \mathrm{Pr} = \frac{\nu}{\kappa}\,, \tag{5.5.2}$$

where g is acceleration due to gravity, α is the thermal expansivity of the fluid, and κ and ν are the thermal diffusivity and kinematic viscosity of the fluid.

The appropriate iso-thermal and no-penetration boundary conditions at the top and bottom of the convective layer lead to

$$\theta' = v' = 0\,. \tag{5.5.3}$$

Two different types of boundary conditions are usually considered for the horizontal velocity components. The first is a no-slip condition appropriate for a solid boundary:

$$u' = w' = 0\,. \tag{5.5.4}$$

It is also referred to as the *rigid* boundary condition. The alternative is a stress-free (or *free*) boundary condition appropriate for a freely slipping boundary (as in the case of an air–water interface):

$$\frac{\partial u'}{\partial y} = \frac{\partial w'}{\partial y} = 0\,. \tag{5.5.5}$$

The rigid and free boundary conditions are used in different combinations at the top and the bottom of the fluid layer. The case of a *free–free* boundary condition at both the top and bottom, although unrealistic, is often considered for its theoretical simplicity. With a *rigid–rigid* boundary condition the problem of convection between two parallel bounding plates can be explored. In the case of a free surface at the top, a *rigid–free* boundary condition may be more appropriate.

The next step in the linear stability analysis is to make the normal mode *ansatz* for the velocity and temperature perturbation. The perturbation equations (5.5.1) and the boundary conditions (5.5.3)–(5.5.5) are both linear and homogeneous along the horizontal (x and z) directions and therefore admit normal mode solutions of the form

$$\left.\begin{aligned} \mathbf{u}' &= \tilde{\mathbf{u}}(y)\,\exp\left[\iota a x\right]\,\exp\left[\lambda t\right]\,, \\ \theta' &= \tilde{\theta}(y)\,\exp\left[\iota a x\right]\,\exp\left[\lambda t\right]\,. \end{aligned}\right\} \tag{5.5.6}$$

In an infinite layer there is no preferred horizontal direction and therefore without loss of generality we have chosen the horizontal waveform to be along the x-direction, independent of z. Substitution of (5.5.6) into (5.5.1) yields a system of ODEs that is an eigenvalue problem for the velocity and temperature eigenfunctions, $\tilde{\mathbf{u}}(y)$ and $\tilde{\theta}(y)$. In (5.5.6) a is the horizontal wavenumber and λ is the eigenvalue of the normal mode. The ODEs can be manipulated to obtain the following sixth-order eigenvalue system for the vertical velocity and temperature eigenfunctions:

$$\left(\frac{\mathrm{d}^2}{\mathrm{dy}^2} - a^2\right) \tilde{v} - \tilde{\phi} = \frac{\lambda}{\mathrm{Pr}} \tilde{v}\,, \tag{5.5.7a}$$

$$\left(\frac{\mathrm{d}^2}{\mathrm{dy}^2} - a^2\right) \tilde{\phi} - a^2\,\mathrm{Ra}\,\tilde{\theta} = 0\,, \tag{5.5.7b}$$

$$\left(\frac{\mathrm{d}^2}{\mathrm{dy}^2} - a^2\right) \tilde{\theta} + \tilde{v} = \lambda\,\tilde{\theta}\,, \tag{5.5.7c}$$

where $\tilde{\phi}$ is an intermediate function. For additional details see Drazin and Reid (2004).

The problem of a *free–free* boundary condition was considered by Rayleigh in 1916 and for this case the appropriate boundary conditions are

$$\tilde{v} = \tilde{\phi} = \tilde{\theta} = 0 \quad \text{at } y = 0,\, 1\,. \tag{5.5.8}$$

It can be easily verified by substitution that (5.5.7a–5.5.7c) and (5.5.8) admit eigensolutions of the form

$$\tilde{v}_n(y) = \sin\left(n\,\pi\,y\right) \qquad \text{for} \quad n = 1,\, 2,\, \ldots\,, \tag{5.5.9}$$

and the corresponding $\tilde{\phi}$ and $\tilde{\theta}$ can be evaluated from (5.5.7a) and (5.5.7b) as

$$\left.\begin{aligned} \tilde{\phi} &= -\left(n^2\,\pi^2 + a^2 + \frac{\lambda}{\mathrm{Pr}}\right) \sin\left(n\,\pi\,y\right) \\ \tilde{\theta} &= \frac{1}{a^2\,\mathrm{Ra}} \left(n^2\,\pi^2 + a^2\right) \left(n^2\,\pi^2 + a^2 + \frac{\lambda}{\mathrm{Pr}}\right) \sin\left(n\,\pi\,y\right)\,. \end{aligned}\right\} \tag{5.5.10}$$

The **dispersion relation** for the eigenvalue can finally be obtained by substituting the above into (5.5.7c)

$$\left(n^2\,\pi^2 + a^2\right) \left(n^2\,\pi^2 + a^2 + \lambda\right) \left(n^2\,\pi^2 + a^2 + \frac{\lambda}{\mathrm{Pr}}\right) = a^2\,\mathrm{Ra}\,. \tag{5.5.11}$$

For the different modes (i.e., for different n) the Rayleigh number criterion for **marginal stability** (i.e., $\lambda = 0$) is given by

$$\mathrm{Ra} = \frac{\left(n^2\,\pi^2 + a^2\right)^3}{a^2}\,. \tag{5.5.12}$$

For any given n and horizontal wavenumber a, as long as the Rayleigh number is below the value given in (5.5.12), the growth rate $\lambda < 0$ and the infinitesimal perturbation will decay. As a result, if $\mathrm{Ra} > (\pi^2 + a^2)^3/a^2$ then the first mode ($n = 1$) will grow; if $\mathrm{Ra} > (4\pi^2 + a^2)^3/a^2$ then the second mode ($n = 2$) will also grow and so on. Therefore, as the Rayleigh number increases from zero, $n = 1$ is the first mode to become unstable. The critical Rayleigh number and corresponding critical wavenumber of this first mode are then obtained by minimizing $(\pi^2 + a^2)^3/a^2$ and we obtain

$$\mathrm{Ra_c} = \frac{27}{4}\,\pi^4 \qquad \text{and} \qquad a_\mathrm{c} = \frac{\pi}{\sqrt{2}}\,. \tag{5.5.13}$$

The implication is that below this $\mathrm{Ra_c}$ *all* infinitesimal disturbances decay in time and the base state of pure conduction is stable. For Rayleigh number above $\mathrm{Ra_c}$ infinitesimal disturbances of horizontal wavenumber around a_c will grow exponentially in time.

With *rigid–rigid* top and bottom boundaries, (5.5.8) must be replaced by

$$\tilde{v} = \frac{\mathrm{d}\tilde{\mathrm{v}}}{\mathrm{dy}} = \tilde{\theta} = 0 \quad \text{at} \quad y = 0, 1\,. \tag{5.5.14}$$

An explicit dispersion relation like (5.5.11) is not possible for this case. Nevertheless, (5.5.7a–5.5.7c) can be solved numerically along with (5.5.14) for the eigenvalues and eigenfunctions with Ra, Pr and a as input parameters. The critical Rayleigh number now increases to $\mathrm{Ra_c} = 1708$ and the corresponding critical wavenumber is $a_c = 3.117$. For the case of a *rigid–free* boundary condition, (5.5.14) is applied at $y = 0$ and (5.5.8) is applied at $y = 1$. The critical Rayleigh number is $\mathrm{Ra_c} = 1101$ and the corresponding critical wavenumber is $a_c = 2.682$.

Computer Exercise 5.5.1 In this exercise we will numerically solve the eigenvalue problem (5.5.7a–5.5.7c). Discretize the domain, $0 \le y \le 1$, with $N + 1$ suitable grid points and choose your favorite finite or compact difference scheme to obtain the second-derivative matrix operator (as can be seen from (5.5.7a–5.5.7c) the first-derivative operator is not needed except perhaps for the boundary correction). Discretize the equations into a generalized 3×3 block matrix eigenvalue problem, with each block of size $(N+1)\times(N+1)$. Incorporate boundary conditions (5.5.8) by replacing the first and last rows of each of the blocks appropriately. Solve the resulting generalized matrix eigenvalue problem for different values of the parameter Ra, Pr and a. Verify the theoretical results (5.5.9) and (5.5.11).

Computer Exercise 5.5.2 Repeat for the *rigid–rigid* and *rigid–free* boundary conditions and verify the critical Rayleigh number and critical wavenumber for these cases.

5.5.2 Instability of Wall-Bounded Shear Flows

Laminar solutions to Poiseuille, Couette and von Kármán flows obtained in Section 5.3 are stable only below a certain critical Reynolds number, where the Reynolds number, $\mathrm{Re} = UL/\nu$, is defined with appropriate velocity (U) and length (L) scales in each flow. Above the critical Reynolds number, the laminar flow becomes unstable to exponentially growing disturbances. The theoretical analysis of the instability becomes particularly simple in the case of one-dimensional laminar shear flows, the simplest example of which are the plane

Poiseuille or plane Couette flows considered in Section 5.3 and which, under constant viscosity, take the following non-dimensional form:

$$\left.\begin{array}{llll}\textbf{Poiseuille flow:} & u(y) = (1-y^2) & \text{for} & -1 \le y \le 1\,; \\ \textbf{Couette flow:} & u(y) = (1+y) & \text{for} & -1 \le y \le 1\,,\end{array}\right\} \tag{5.5.15}$$

where the half height between the two parallel plates (h) is the length scale, the velocity at the centerline (U_{cl}) is the velocity scale, and the corresponding Reynolds number is $\mathrm{Re} = U_{\mathrm{cl}} h/\nu$.

The stability of the above one-dimensional base flows to infinitesimal disturbances is governed by the following linearized continuity and momentum equations:

$$\left.\begin{array}{rcl}\nabla\cdot\mathbf{u}' & = & 0\,, \\ \left(\dfrac{\partial}{\partial t} + u\dfrac{\partial}{\partial x}\right)\mathbf{u}' + v'\dfrac{\mathrm{du}}{\mathrm{dy}}\mathbf{i} & = & -\nabla p' + \dfrac{1}{\mathrm{Re}}\nabla^2\mathbf{u}'\,,\end{array}\right\} \tag{5.5.16}$$

where $\mathbf{i}$ is the unit vector along the base flow direction and v' is the perturbation velocity component along y. One then makes the normal mode *ansatz* and assumes velocity perturbations to be given by

$$\mathbf{u}' = \tilde{\mathbf{u}}(y)\,\exp\left[\iota\,(ax+bz)\right]\,\exp\left[\lambda t\right]\,, \tag{5.5.17}$$

where a and b are the wavenumbers of the perturbation along x and z. Substitution of (5.5.17) into (5.5.16) yields

$$\left.\begin{array}{r}\iota(a\,\tilde{u} + b\,\tilde{w}) + \dfrac{\mathrm{d}\tilde{v}}{dy} = 0\,, \\ \left\{\dfrac{\mathrm{d}^2}{\mathrm{d}y^2} - (a^2+b^2) - \iota\,a\,\mathrm{Re}\,u\right\}\tilde{u} - \mathrm{Re}\,\dfrac{du}{dy}\tilde{v} - \iota\,a\,\mathrm{Re}\,\tilde{p} = \lambda\,\mathrm{Re}\,\tilde{u}\,, \\ \left\{\dfrac{\mathrm{d}^2}{\mathrm{d}y^2} - (a^2+b^2) - \iota\,a\,\mathrm{Re}\,u\right\}\tilde{v} - \mathrm{Re}\,\dfrac{\mathrm{d}\tilde{p}}{\mathrm{d}y} = \lambda\,\mathrm{Re}\,\tilde{v}\,, \\ \left\{\dfrac{\mathrm{d}^2}{\mathrm{d}y^2} - (a^2+b^2) - \iota\,a\,\mathrm{Re}\,u\right\}\tilde{w} - \iota\,b\,\mathrm{Re}\,\tilde{p} = \lambda\,\mathrm{Re}\,\tilde{w}\,.\end{array}\right\} \tag{5.5.18}$$

Squire (1933) devised an ingenious way to reduce the above three-dimensional problem into an equivalent two-dimensional one. This requires the following transformation of variables:

$$\left.\begin{array}{c}\hat{a} = \left(a^2+b^2\right)^{1/2}\,, \quad \hat{a}\,\hat{u} = a\,\tilde{u} + b\,\tilde{w}\,, \quad \hat{p} = \dfrac{\hat{a}}{a}\tilde{p}\,, \\ \hat{v} = \tilde{v}\,, \quad \hat{\lambda} = \dfrac{\hat{a}}{a}\lambda\,, \quad \widehat{\mathrm{Re}} = \dfrac{a}{\hat{a}}\,\mathrm{Re}\,,\end{array}\right\} \tag{5.5.19}$$

where $\hat{a}$ is the horizontal wavenumber of the effective two-dimensional disturbance, $\widehat{\mathrm{Re}}$ is the equivalent Reynolds number and $\hat{\lambda}$ is the eigenvalue of the two-dimensional disturbance. The resulting equations for the effective two-dimensional disturbance are

$$\iota\,\hat{a}\,\hat{u} + \frac{d\hat{v}}{dy} = 0\,, \qquad (5.5.20a)$$

$$\left\{\frac{d^2}{dy^2} - \hat{a}^2 - \iota\,\hat{a}\widehat{\mathrm{Re}}u\right\}\hat{u} - \widehat{\mathrm{Re}}\frac{du}{dy}\hat{v} - \iota\,\hat{a}\,\widehat{\mathrm{Re}}\,\hat{p} = \hat{\lambda}\,\widehat{\mathrm{Re}}\,\hat{u}\,, \qquad (5.5.20b)$$

$$\left\{\frac{d^2}{dy^2} - \hat{a}^2 - \iota\,\hat{a}\,\mathrm{Re}\,u\right\}\hat{v} - \widehat{\mathrm{Re}}\frac{d\hat{p}}{dy} = \hat{\lambda}\,\widehat{\mathrm{Re}}\,\hat{v}\,. \qquad (5.5.20c)$$

From the transformation (5.5.19) it is clear that $\hat{a} \geq a$ and as a result $\hat{\lambda} \geq \lambda$ and $\widehat{\mathrm{Re}} \leq \mathrm{Re}$. **Squire's theorem** for the instability of one-dimensional shear flows can then be stated as:

Squire's Theorem *There exists for each three-dimensional disturbance a corresponding more unstable two-dimensional disturbance with a lower effective Reynolds number.*

With increasing Reynolds number, the two-dimensional disturbances begin to grow before three-dimensional disturbances. Thus, it is sufficient to investigate the stability of one-dimensional base flows to two-dimensional disturbances.

The advantage of the two-dimensional problem is that (5.5.20a–5.5.20c) can be reduced to a single equation with the introduction of streamfunction. We introduce a perturbation streamfunction, ψ', of the following form:

$$\psi' = \hat{\psi}(y)\,\exp\left[\iota\,\hat{a}\,\hat{x}\right]\,\exp\left[\lambda t\right]\,, \qquad (5.5.21)$$

where $\hat{\psi}$ is related to the two components of velocity by

$$\hat{u} = \frac{d\hat{\psi}}{dy} \qquad \text{and} \qquad \hat{v} = -\iota\,\hat{a}\,\hat{\psi}\,. \qquad (5.5.22)$$

With the above definition of streamfunction (5.5.20a) is automatically satisfied. Equations (5.5.20b) and (5.5.20c) can be combined to obtain the following equation for the streamfunction:

$$\frac{1}{\widehat{\mathrm{Re}}}\left(\frac{d^2}{dy^2} - \hat{a}^2\right)^2\hat{\psi} - \iota\,\hat{a}\,u\left(\frac{d^2}{dy^2} - \hat{a}^2\right)\hat{\psi} + \iota\,\hat{a}\,\frac{d^2u}{dy^2}\hat{\psi} = \hat{\lambda}\left(\frac{d^2}{dy^2} - \hat{a}^2\right)\hat{\psi}\,. \qquad (5.5.23)$$

The above fourth-order ODE is the celebrated **Orr–Sommerfeld** equation, which has been intensely studied by numerous researchers.

From (5.5.22) the no-slip and no-penetration boundary conditions for the velocity components at the two parallel plates transform to the following boundary conditions:

$$\hat{\psi} = \frac{d\hat{\psi}}{dy} = 0 \quad \text{at} \quad y = \pm 1\,. \qquad (5.5.24)$$

Equation (5.5.23) along with (5.5.24) forms an eigenvalue problem and by appropriately choosing one of the base flows given in (5.5.15), the stability of either the Poiseuille or the Couette flow can be investigated.

In **temporal stability analysis**, the effective two-dimensional wavenumber, $\hat{a}$, is assumed to be real, and equation (5.5.23) is an eigenvalue problem for the eigenvalue, $\hat{\lambda}$, and the eigenfunction, $\hat{\psi}$. Here $\hat{a}$ and $\widehat{\mathrm{Re}}$ are parameters. The eigenvalue and the corresponding eigenfunctions are in general complex. From (5.5.21) it is easy to see that the real part of the eigenvalue, $\hat{\lambda}_R$, determines the growth rate of the disturbance. Positive (or negative) values correspond to growth (or decay) of the disturbance. The imaginary part of the eigenvalue, $\hat{\lambda}_I$, corresponds to the temporal frequency of the disturbance. The eigenvalue problem then results in a dispersion relation of the following functional form:

$$\hat{\lambda}_R = F(\hat{a}, \widehat{\mathrm{Re}}) \quad \text{and} \quad \hat{\lambda}_{\mathrm{I}} = G(\hat{a}, \widehat{\mathrm{Re}}) \,. \tag{5.5.25}$$

For numerical solution, it is convenient to write (5.5.23) as a system of two second-order ODEs:

$$\left.\begin{aligned} \hat{\phi} - \left(\frac{\mathrm{d}^2}{\mathrm{d}y^2} - \hat{a}^2\right)\hat{\psi} &= 0 \\ \frac{1}{\widehat{\mathrm{Re}}}\left(\frac{\mathrm{d}^2}{\mathrm{d}y^2} - \hat{a}^2\right)\hat{\phi} - \iota\,\hat{a}\,u\,\hat{\phi} + \iota\,\hat{a}\,\frac{\mathrm{d}^2 u}{\mathrm{d}y^2}\hat{\psi} &= \hat{\lambda}\,\hat{\phi}\,. \end{aligned}\right\} \tag{5.5.26}$$

Upon discretization of the domain ($-1 \leq y \leq 1$) with $N+1$ points, (5.5.26) can be written in the following generalized matrix eigenvalue problem

$$\begin{bmatrix} \mathbf{L_{11}} & \mathbf{L_{12}} \\ \mathbf{L_{21}} & \mathbf{L_{22}} \end{bmatrix} \begin{bmatrix} \vec{\hat{\phi}} \\ \vec{\hat{\psi}} \end{bmatrix} = \hat{\lambda} \begin{bmatrix} \mathbf{0} & \mathbf{0} \\ \mathbf{R_{21}} & \mathbf{0} \end{bmatrix} \begin{bmatrix} \vec{\hat{\phi}} \\ \vec{\hat{\psi}} \end{bmatrix}, \tag{5.5.27}$$

where the block matrix operators on the left and right-hand sides are given by

$$\mathbf{L_{11}} = \mathbf{R_{21}} = \mathbf{I}\,; \mathbf{L_{12}} = \hat{a}^2\,\mathbf{I} - \mathbf{D_2}\,; \mathbf{L_{21}} = -\frac{1}{\widehat{\mathrm{Re}}}\mathbf{L_{12}} - \iota\,\hat{a}\,\mathbf{u}\,; \mathbf{L_{22}} = \iota\,\hat{a}\,\frac{\mathrm{d}^2\mathbf{u}}{\mathrm{d}y^2}\,, \tag{5.5.28}$$

where $\mathbf{D_2}$ and $\mathbf{I}$ are the second-derivative and identity matrices of size $(N+1) \times (N+1)$. In (5.5.28) $\mathbf{u}$ and $\mathrm{d}^2\mathbf{u}/\mathrm{d}y^2$ must be regarded as diagonal matrices of size $(N+1) \times (N+1)$ with the value of the base flow, $u(y)$, and its second derivative evaluated at the $N+1$ grid points as the entries along the diagonals.

The four boundary conditions given in (5.5.24) can be incorporated into (5.5.27) in the following manner:

(1) The first row of the matrix equation (5.5.27) can be used to implement $\hat{\psi} = 0$ at $y = -1$. The first row of $\mathbf{L_{11}}$ and $\mathbf{L_{12}}$ will be replaced by $[0,\ 0,\ \ldots\ ]$ and $[1,\ 0,\ 0,\ \ldots\ ]$, respectively.
(2) The boundary condition, $\hat{\psi} = 0$ at $y = 1$, can be enforced by replacing the last row of $\mathbf{L_{11}}$ and $\mathbf{L_{12}}$ by $[0,\ 0,\ \ldots\ ]$ and $[\ldots,\ 0,\ 0,\ 1\ ]$.

(3) The $(N+2)$th row of (5.5.27) can be used to implement $d\hat{\psi}/dy = 0$ at $y = -1$. This can be achieved by replacing the first row of $\mathbf{L_{21}}$ by all zeros, $[0,\ 0,\ \ldots\ ]$, and the first row of $\mathbf{L_{22}}$ by the first row of the first-derivative matrix $\mathbf{D_1}$. Correspondingly the first row of the right-hand side matrix, $\mathbf{R_{21}}$, must also be replaced by all zeros, $[0,\ 0,\ \ldots\ ]$.

(4) The last row of (5.5.27) can be used to implement $d\hat{\psi}/dy = 0$ at $y = 1$. This can be achieved by replacing the last row of $\mathbf{L_{21}}$ and $\mathbf{R_{21}}$ by $[0,\ 0,\ \ldots\ ]$ and the last row of $\mathbf{L_{22}}$ by the last row of the first-derivative matrix.

After incorporating all the boundary conditions, the generalized matrix eigenvalue problem, (5.5.27), can be numerically solved by calling appropriate MATLAB routines or from packages such as IMSL and LAPACK. The solution will result in $2(N+1)$ eigenvalues, $\hat{\lambda}_n$, and the corresponding $2(N+1)$ eigenfunctions, $\left(\vec{\hat{\phi}}_n, \vec{\hat{\psi}}_n\right)$. A dispersion relation of the form (5.5.25) can be obtained for each of the eigenmodes. Attention is usually focused only on eigenmodes that become unstable (with positive real part for the eigenvalue). In principle, the eigenvalue problem can result in many modes of instability; in which case attention can be limited to the *most unstable mode* with the largest disturbance growth rate. However, for the Poiseuille and Couette base flows considered in (5.5.15), there exists only one instability mode.

For the plane Poiseuille flow given in (5.5.15), the 60 eigenvalues with the largest growth rate $\hat{\lambda}_R$ are plotted in Figure 5.6(a) for $\widehat{\mathrm{Re}} = 10000$ and $\hat{a} = 1$. The eigenvalues in red correspond to symmetric modes and the eigenvalues in green correspond to antisymmetric modes. Following Mack (1976), we classify the eigenvalues as belonging to one of three families of modes: A, P and S which are plotted as square, circle and gradient symbols, respectively. The A and P family are finite in number, while the S family has infinite number of stable modes. From the figure it is clear that only one of the symmetric A modes is unstable (i.e., its $\hat{\lambda}_R > 0$).

The real and imaginary parts of the eigenfunction $\hat{\psi}$ of the unstable mode are shown in Figure 5.6(b). It can be seen that the eigenfunction is symmetric about the centerline of the channel ($y = 0$), hence the mode is termed symmetric. A contour plot of the growth rate, $\hat{\lambda}_R$, is shown in Figure 5.6(c) as a function of $\hat{a}$ and $\widehat{\mathrm{Re}}$. The contour corresponding to $\hat{\lambda}_R = 0$ is the **neutral curve** and it demarcates the growing disturbance ($\hat{\lambda}_R > 0$) from the decaying ones ($\hat{\lambda}_R < 0$). The lowest Reynolds number below which all infinitesimal disturbances are stable is the **critical Reynolds number** ($\mathrm{Re_c}$) and for the plane Poiseuille flow $\mathrm{Re_c} = 5772.2$. The corresponding critical wavenumber $\hat{a}_c = 1.021$. Above the critical Reynolds number, a group of infinitesimal disturbance of wavenumber within a narrow wavenumber band begins to grow exponentially. For $\widehat{\mathrm{Re}} = 10000$ Figure 5.6(d) shows the growth rate and frequency of the unstable symmetric mode as a function of $\hat{a}$. We can see that the maximum growth rate of about 0.00416 occurs at a streamwise wavenumber of about 0.95. Also the frequency of the growing mode appears to linearly increase with $\hat{a}$ and this suggests a nearly constant group velocity.

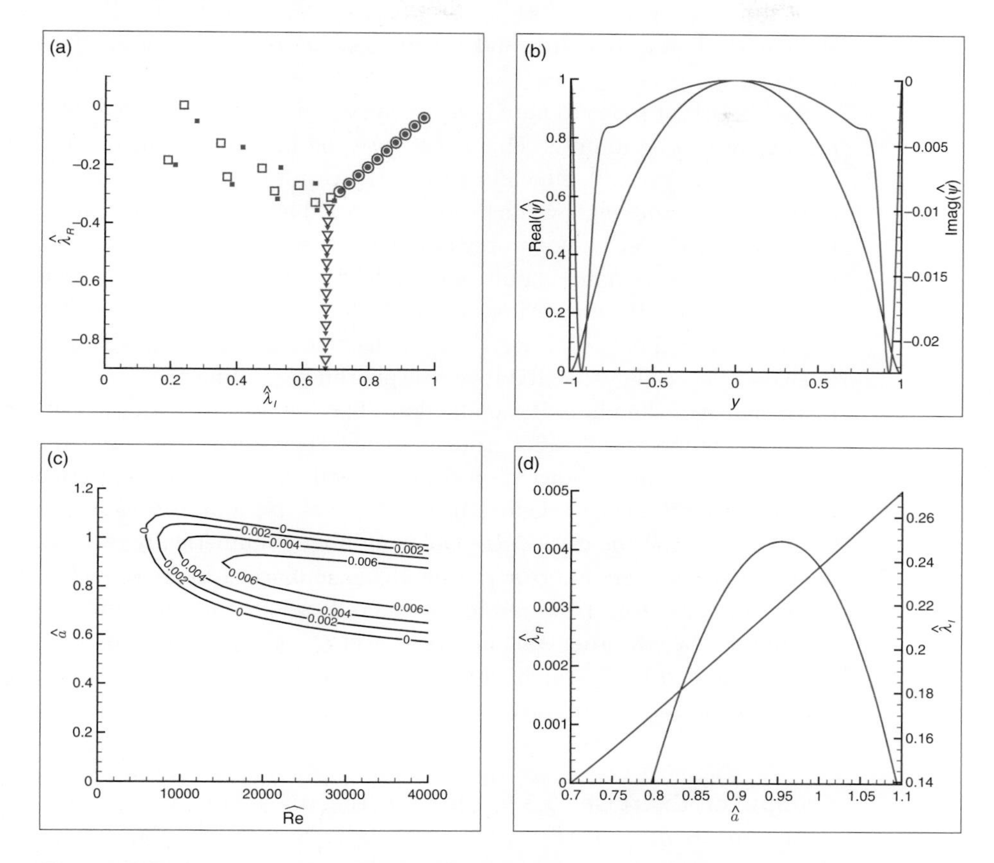

Figure 5.6 Hydrodynamic stability results for the plane Poiseuille flow given in (5.5.15). (a) A plot of the top 60 eigenvalues with the largest growth rate $\hat{\lambda}_R$ for $\widehat{\mathrm{Re}} = 10000$ and $\hat{a} = 1$. The eigenvalues in red correspond to symmetric mode and the eigenvalues in green correspond to antisymmetric mode. Following Mack (1976) we classify the eigenvalues as belonging to one of three families of modes: A, P and S, which are plotted as square, circle and gradient symbols. (b) The real and imaginary parts of the eigenfunction $\hat{\psi}$ of the unstable mode. (c) A contour plot of the growth rate, $\hat{\lambda}_R$, as a function of $\hat{a}$ and $\widehat{\mathrm{Re}}$. (d) For $\widehat{\mathrm{Re}} = 10000$ the growth rate and frequency of the unstable symmetric mode are plotted as a function of $\hat{a}$.

Computer Exercise 5.5.3 Using the derivative matrices, construct the block matrices that appear on the left- and right-hand sides of (5.5.27). Modify their first and last rows to incorporate the appropriate boundary conditions. Assemble the blocks to form the boundary corrected left- and right-hand side matrices. Use a generalized eigenvalue routine to compute the eigenvalues. Hold the wavenumber to be 1.021 and repeat the eigenvalue calculation for varying $\widehat{\text{Re}}$, initially with a modest grid resolution of say $N = 100$. Start with a Reynolds number sufficiently lower than $\widehat{\text{Re}}_c$ and verify that the real parts of all the eigenvalues are negative.

An important point to note here is that you may find about half of the $2(N+1)$ eigenvalues to be infinite! This is because the first block row on the right-hand side of (5.5.27) is null, which in turn is the result of somewhat artificial splitting of the fourth-order eigenvalue problem (5.5.23) into a system of two second-order ODEs (5.5.26). Let us not worry about these spurious eigenvalues. It is only the real parts of the remaining finite eigenvalues that need to be negative for stability.

Now increase $\widehat{\text{Re}}$ above $\widehat{\text{Re}}_c$ in small steps and determine at what Reynolds number the real part of at least one of the eigenvalues changes sign and becomes positive. This eigenvalue and its corresponding eigenfunction signify the unstable disturbance of the plane Poiseuille flow. The numerical estimation of the critical Reynolds number will differ from its exact theoretical value and the difference will depend on your choice of grid points and the order of accuracy of the finite or compact difference scheme. Increase (or if appropriate decrease) resolution and see how well the critical Reynolds number is numerically estimated. In fact, a plot of the numerical error ($|$exact $\widehat{\text{Re}}_c$ – numerical $\widehat{\text{Re}}_c|$) versus N on log–log scale is a great way to evaluate the overall order of accuracy of the numerical scheme. For a pth-order scheme, we expect the error to decrease as N^{-p}, which will correspond to a straight line of slope $-p$ in a log–log plot.

Computer Exercise 5.5.4 Repeat the above for varying values of the wavenumber, $\hat{a}$, at a resolution deemed sufficient from the previous exercise. Trace the neutral curve near the critical Reynolds number. For larger Reynolds number, it may be more advantageous to hold the Reynolds number fixed and increase (or decrease) $\hat{a}$ in order to trace the rest of the neutral curve.

In a similar manner it can easily be verified that the plane Couette flow in (5.5.15) is unconditionally stable to infinitesimal disturbances at all Reynolds numbers. In other words, $\hat{\lambda}_R$ for all the eigenvalues over the entire range of $\hat{a}$ and $\widehat{\text{Re}}$ is negative, indicating that all infinitesimal disturbances decay in time.

In fact, we can extend the stability analysis to the combination plane Poiseuille–Couette flow considered in Section 5.3. For constant viscosity the base flow becomes

$$\textbf{Poiseuille–Couette Flow: } u(y) = A(1-y^2) + B(1+y) \text{ for } -1 \le y \le 1\,, \tag{5.5.29}$$

where A and B are constants that must satisfy $A + B = 1$ in order for the non-dimensional velocity to be unity along the channel centerline. In the limit of $A \to 1$ and $B \to 0$, we recover the plane Poiseuille flow and the corresponding critical Reynolds number of 5772.2. The critical Reynolds number increases with the addition of Couette flow (i.e., as B increases). It can easily be shown that in the limit of $A \to 0.74$ and $B \to 0.26$, the critical Reynolds number approaches infinity and the corresponding critical wavenumber approaches zero.

Computer Exercise 5.5.5 Repeat Computer Exercises 5.5.3 and 5.5.4 for the Poiseuille–Couette flow with increasing Couette flow component. Track the critical Reynolds number as B increases from zero. Also keep track of the decreasing critical wavenumber.

The above stability analysis and in particular the Orr–Sommerfeld equation (5.5.23) or (5.5.26) are often used even in the context of nearly parallel flows such as the Blasius and Falkner–Skan boundary layers. Unlike the one-dimensional parallel flows considered in (5.5.15) and (5.5.29), the boundary layers have a slow streamwise (x-directional) development and as a consequence in addition to the order-one streamwise velocity component there is also an order-$(\mathrm{Re})^{-1/2}$ wall-normal (v) velocity component, where Reynolds number is based on free stream velocity and thickness of the boundary layer. The self-similar Blasius and Falkner–Skan velocity profiles given by (5.2.1) and (5.2.16) account for this slow streamwise development of the velocity profile, within the limits of the boundary layer approximation. However, in the stability analysis one assumes the flow to be locally parallel at any given streamwise location and investigates the stability of the base flow at that streamwise location. If we use the local free stream velocity $U(x)$ as the velocity scale and $\sqrt{\nu\, x/U(x)}$ as the length scale, then the function f', as obtained from (5.2.1) or (5.2.16), provides the local non-dimensional streamwise velocity, $u(y)$ for $0 \le y \le \infty$. Here y is nothing but the non-dimensional wall-normal coordinate; the same as η. The appropriate local Reynolds number now becomes $\mathrm{Re} = \sqrt{U(x)\, x/\nu}$. The discretized Orr–Sommerfeld equation can now be used to investigate the local stability of Blasius and Falkner–Skan boundary layers, the only difference being that the domain is now $0 \le y \le \infty$ and correspondingly the boundary conditions (5.5.24) are applied at $y = 0$ and at $y \to \infty$.

Computer Exercise 5.5.6 Use the Blasius solution obtained in Computer Exercise 5.2.1 and investigate its stability at varying $\widehat{\mathrm{Re}}$ and wavenumber $\hat{a}$. For computational purposes, you can truncate the computational domain at some large but finite value of y, say $y_{\max} \approx 25$.

Figure 5.7 displays the dispersion relation of the Blasius boundary layer, in terms of contours of constant growth rate plotted as a function of the Reynolds number and disturbance wavenumber in frame (a) and as a function of the

Reynolds number and disturbance frequency in frame (b). Following standard convention, for plotting purposes the length scale has been chosen to be the displacement thickness, $\delta^* = 1.72\sqrt{\nu x/U(x)}$, and therefore in the plot $\widehat{\mathrm{Re}}^* = 1.72\widehat{\mathrm{Re}}$, $\hat{a}^* = \hat{a}$ and $\hat{\lambda}^* = \hat{\lambda}$. From the neutral curve, marked as the thick solid line, the critical Reynolds number and critical wavenumber can be determined as $\widehat{\mathrm{Re}}_c^* = 519.06$ and $\hat{a}_c^* = 0.3038$.

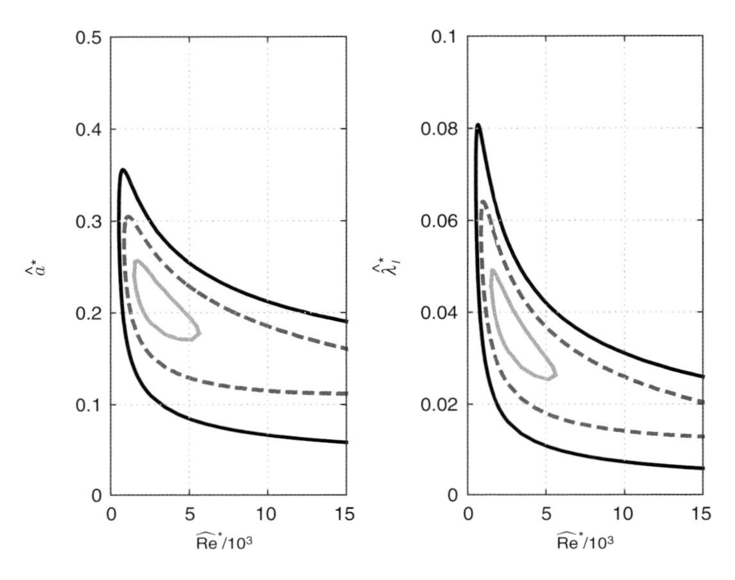

Figure 5.7 Contours of growth rate $\hat{\lambda}_R = 0.0, 0.0015, 0.0023$ for the Blasius flat-plate boundary layer. (a) Plotted as a function of $\widehat{\mathrm{Re}}$ and $\hat{a}$. (b) Plotted as a function of $\widehat{\mathrm{Re}}$ and $\hat{\lambda}_I$.

An important aspect of instability that can be observed in both Figures 5.6 and 5.7 is that the unstable wavenumber band broadens as Reynolds number increases above the critical value, reaches a maximum, and with further increase in Reynolds number the unstable wavenumber band slowly decreases. In other words, the **upper** and **lower branches** of the neutral curve approach each other as Reynolds number is increased. In the limit of infinite Reynolds number, it can be shown that the Poiseuille–Couette and Blasius boundary layer flows are stable to all infinitesimal disturbances. It may sound a bit odd, but wall bounded shear flows such as the Poiseuille–Couette and Blasius flows require viscous effects to become unstable!

In the case of thermal instability considered in Section 5.5.1, buoyancy is destabilizing, while viscous effects are stabilizing. Above the critical Rayleigh number buoyancy wins over viscosity and the pure conduction state is destabilized. The same viscous effects are however destabilizing in wall bounded shear flows through the critical layer (for a thorough discussion on the very intriguing nature of viscous effects see Drazin and Reid, 2004). Therefore, in the limit of infinite Reynolds number, when viscous effects are absent, these flows are stabilized. Dispersion relations shown in Figures 5.6 and 5.7 are typical of such **viscous instability**. This viscous instability mode now goes by the name of **Tollmien–Schlichting instability**, after Tollmien and Schlichting, who made pioneering contributions to this problem.

The stability of Falkner–Skan profiles as defined by (5.2.16) has been studied in detail by Wazzan *et al.* (1968) in the limit of $\alpha = 0$ for varying pressure gradient parameter β. They also computed and presented neutral curves for β in the range $-0.1988 \leq \beta \leq 1.0$. The value $\beta = 1$ corresponds to the plane stagnation point flow and $\beta = 0$ corresponds to the Blasius boundary layer. The critical Reynolds numbers for the onset of instability for different values of β obtained from the neutral curves are presented in Table 5.2. It is evident from the neutral curves that the boundary layer is stabilized with a favorable pressure gradient ($\beta > 0$). With an adverse pressure gradient ($\beta < 0$), instability is enhanced and in fact the nature of the upper branch of the neutral curve changes dramatically. Under an adverse pressure gradient, instability persists over a finite wavenumber band even in the limit of infinite Reynolds number. The Falkner–Skan profiles for $\beta < 0$ are marked by the presence of a point of inflection. This point moves away from the wall as β decreases from zero, and below $\beta < -0.1988$ the boundary layer flow separates. The relation between the inflectional nature of the one-dimensional velocity profile and the nature of the instability will be explored below in the context of free shear flows.

5.5.3 Instability of Free Shear Flows

The stability of free shear flows such as jets, wakes and mixing layers can be analyzed in a similar manner. Like the boundary layer, these boundary-free shear flows are also not strictly parallel, but their downstream development can be considered to take a self-similar form. A two-dimensional laminar jet sufficiently away from the jet source has the following self-similar form

$$u(y) = \mathrm{sech}^2(y) \quad \text{for} \quad -\infty \leq y \leq \infty\,, \tag{5.5.30}$$

where the streamwise velocity has been normalized by the jet centerline velocity, $U(x)$, and the transverse coordinate, y, has been scaled by $3\left(\frac{16}{9}\frac{x^2\,\mu^2}{J\,\rho}\right)^{1/3}$, which is proportional to the local jet thickness and therefore increases as the $2/3$ power of the streamwise coordinate, x. The centerline velocity of the jet correspondingly decreases as the $1/3$ power of x and is given by

$$U(x) = \frac{2}{3}\left(\frac{9}{16}\right)^{2/3}\left(\frac{J^2}{\rho\,\mu\,x}\right)^{1/3}\,; \tag{5.5.31}$$

here J is the net streamwise momentum flux of the jet integrated across y which remains a constant independent of x. The terms ρ and μ are the density and dynamic viscosity of the fluid, respectively.

Table 5.2 Critical Reynolds number $\widehat{\mathrm{Re}}_c^*$ for varying values of the parameter β for the Falkner–Skan boundary-layer profiles. After Wazzan *et al.* (1968).

β	$\widehat{\mathrm{Re}}_c^*$
−0.1988	67
−0.1	199
−0.05	318
0.0	520
0.1	1380
0.2	2830
0.4	6230
0.8	10920
∞	21675

The laminar wake flow behind a two-dimensional object at sufficient distance downstream of that object is closely related to the two-dimensional jet flow, since the streamwise momentum lost as drag can be considered as the opposite of momentum flux imparted in a jet. The velocity deficit in a developing wake has been observed to fit the sech^2 profile reasonably well. Whereas, far downstream the wake deficit takes a Gaussian form and these two self-similar wake profiles can be expressed as

$$u(y) = \begin{cases} 1 - u_d \exp(-y^2) \\ \quad\text{or} \\ 1 - u_d \,\mathrm{sech}^2(y) \end{cases} \quad \text{for} \quad -\infty \le y \le \infty\,. \tag{5.5.32}$$

Here the transverse coordinate, y, has been scaled by the wake thickness and velocity has been scaled by the free stream velocity, U. In (5.5.32) u_d is the non-dimensional velocity deficit at the wake centerline.

The laminar mixing layer that forms at the interface of two parallel streams of velocity takes a self-similar form in a coordinate that moves downstream at the average velocity of the two streams, $(U_1 + U_2)/2$. In this frame of reference only the velocity difference, $(U_1 - U_2)/2$, is of importance and is usually chosen as the reference velocity scale. The self-similar mixing layer profile then takes the following form:

$$u(y) = \tanh(y) \quad \text{for} \quad -\infty \le y \le \infty\,, \tag{5.5.33}$$

where the transverse coordinate, y, again has been scaled by appropriate mixing layer thickness. For a thorough discussion of the boundary layer theory used in obtaining the above laminar jet (5.5.30), wake (5.5.32) and mixing layer (5.5.33) profiles, the reader should consult the classic book on boundary layers by Schlichting (1968).

The linear stability of the above self-similar laminar profiles to infinitesimal disturbances can be investigated using the Orr–Sommerfeld equation, if one makes the locally parallel flow assumption and considers the streamwise development to be unimportant in the stability analysis. It is then a simple matter to solve the generalized matrix eigenvalue problem given in (5.5.27) with the non-dimensional base flow and its derivative appropriately evaluated from the self-similar profiles (5.5.30), (5.5.32) or (5.5.33). The Reynolds number is based on the velocity scale, U, and the transverse length scale of the jet, wake or mixing layer.

The instability of the sech^2 profile for the laminar jet flow has been studied by Curle (1957), Howard (1958) and Silcock (1975), among many others. Two modes of instabilities are observed: the **sinuous mode** whose streamfunction is symmetric about the jet centerline; and the **varicose mode** whose streamfunction is anti-symmetric about the jet centerline. Stability analysis shows that the sinuous mode is the first one to become unstable as the Reynolds number is increased and its critical Reynolds number is about 4 (a very small value indeed)!

Stability analysis for both forms of the wake profile given in (5.5.32) places the critical Reynolds number for the laminar wake flow to be about 4 again (see Sato and Kuriki, 1961 and Taneda, 1963). The tanh mixing layer profile is in fact unstable at all Reynolds numbers (the critical Reynolds number is zero)!

The extremely low critical Reynolds number for the free shear flows, in comparison with those of wall bounded flows, suggests that the instability in jets, wakes and mixing layers is much more potent than in wall bounded boundary layers. The growth rate of unstable disturbance in free shear layers can be more than an order of magnitude larger than that of the wall bounded boundary layers. Furthermore, in the case of free shear layers, the instability persists even in the limit of infinite Reynolds number. In fact, the peak disturbance growth rate is realized in the limit of infinite Reynolds number. Unlike the plane Poiseuille–Couette and Blasius flows, for free shear flows viscosity is not the destabilizing agent. The role of viscosity is simply to stabilize the flow and it completely succeeds in doing so below the critical Reynolds number. Thus, the nature of instability in the free shear flows appears to be different from that of the plane Poiseuille–Couette and Blasius flows.

A better understanding of the instability mechanism in free shear flows can be obtained by investigating the infinite Reynolds number limit. The advantage is that in this limit the Orr–Sommerfeld equation reduces to the following second-order problem:

$$-\iota \hat{a} u \left(\frac{\mathrm{d}^2}{\mathrm{d}y^2} - \hat{a}^2 \right) \hat{\psi} + \iota \hat{a} \frac{\mathrm{d}^2 u}{\mathrm{d}y^2} \hat{\psi} = \hat{\lambda} \left(\frac{\mathrm{d}^2}{\mathrm{d}y^2} - \hat{a}^2 \right) \hat{\psi} . \tag{5.5.34}$$

The above equation investigates the inviscid instability of the base flow, $u(y)$, and is called the **Rayleigh stability equation**. In the inviscid limit, the appropriate boundary conditions for the eigenfunction of the streamfunction are

$$\hat{\psi} = 0 \quad \text{at} \quad y = \pm\infty . \tag{5.5.35}$$

Based on (5.5.34), Rayleigh (1880) proposed the following **inflection-point theorem** for the inviscid instability of parallel shear flows:

Theorem *A necessary condition for inviscid instability is that the base flow velocity profile should have an inflection point.*

In other words, $\mathrm{d}^2 u/\mathrm{d}y^2$ must become zero somewhere in the domain. A stronger theorem was later obtained by Fjørtoft (1950), which states:

Theorem *A necessary condition for inviscid instability is that*

$$\frac{\mathrm{d}^2 u}{\mathrm{d}y^2}(u - u_s) < 0$$

somewhere in the flow, where u_s is the base flow velocity at the point of inflection.

The free shear layer velocity profiles given in (5.5.30), (5.5.32) and (5.5.33) are all marked by the presence of an inflection point. For example, the $\mathrm{sech}^2(y)$ velocity profile of the jet has two inflection points at $y = \pm\tanh^{-1}(1/3)$. In contrast, the plane Poiseuille–Couette and favorable pressure gradient boundary layer profiles have none. Only under an adverse pressure gradient will a boundary layer develop an inflection point, which progressively moves away from the wall as the adverse pressure gradient increases. The effect of the inflection point is to change the character of the neutral curve at large values of Reynolds number. In the case of a favorable streamwise pressure gradient ($\beta > 0$) there is no inflection point in the boundary layer profile and the instability is viscous in origin. As $\widehat{\mathrm{Re}}$ increases the neutral curve closes on itself (the upper and lower branches merge) and there is no instability at large enough Reynolds number. In the case of an adverse streamwise pressure gradient, the instability is inviscid in nature and the neutral curve remains open (the upper and lower branches do not merge) as $\widehat{\mathrm{Re}} \to \infty$.

The numerical investigation of inviscid instability with Rayleigh's stability equation is in general simpler than the full finite Reynolds number stability analysis represented by the Orr–Sommerfeld equation. Upon discretization (5.5.34) leads to a generalized eigenvalue system of size $(N+1) \times (N+1)$:

$$\mathbf{L} \vec{\hat{\psi}} = \hat{\lambda} \mathbf{R} \vec{\hat{\psi}} , \tag{5.5.36}$$

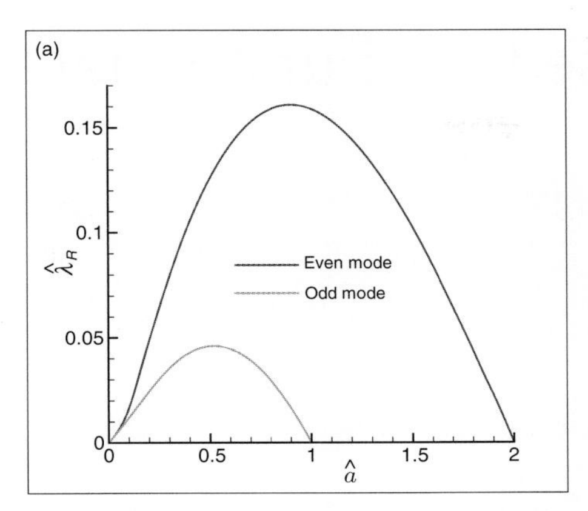

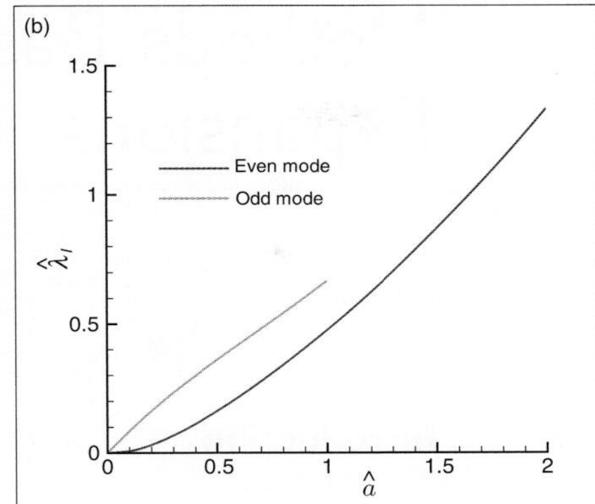

Figure 5.8 Inviscid instability of the laminar jet profile (5.5.30) to sinuous (even) and varicose (odd) modes. (a) Plot of growth rate vs streamwise wavenumber. (b) Plot of frequency versus streamwise wavenumber.

where

$$\mathbf{R} = \mathbf{D_2} - \mathbf{\hat{a}^2\,I} \quad \text{and} \quad \mathbf{L} = -\iota\,\mathbf{\hat{a}}\left[\mathbf{u\,R} - \frac{\mathrm{d}^2\mathbf{u}}{\mathrm{dy}^2}\right]. \tag{5.5.37}$$

Owing to the symmetry of the base flow profiles about the centerline ($y = 0$), it is sufficient to consider only one half of the free shear layer, say the region $0 \le y \le \infty$. While the boundary condition at infinity remains the same as (5.5.35), the boundary condition at the centerline depends on the nature of the disturbance mode we are seeking. For the sinuous mode, the even boundary condition can be applied as $d\hat{\psi}/dy = 0$ at $y = 0$; and for the varicose mode, the odd boundary condition at the centerline is applied as $\hat{\psi} = 0$ at $y = 0$.

Computer Exercise 5.5.7 Investigate the inviscid instability of the laminar jet profile (5.5.30) to sinuous (even) and varicose (odd) modes. Figure 5.8 shows the growth rate and frequency of the two modes as a function of the streamwise wavenumber. Using your numerical code reproduce these plots.

6 Methods Based on Functional Expansions

6.1 Introduction

The finite difference schemes for derivatives of continuous functions are based on *local* approximations. For example, a three-point scheme approximates the derivative based on a parabolic fit through the three local points. In contrast, methods based on functional expansions, also known as **spectral methods**, are *global* in nature. Spectral methods belong to the general class of method of weighted residuals (for details on method of weighted residuals see the review article by Finlayson, 1966. One of the earliest applications of spectral methods to ordinary differential equations was by Lanczos (1938). However, spectral methods have found widespread application in scientific computations in fluid mechanics only since the pioneering works of Kreiss and Oliger (1972), Orszag (1969, 1970) and Orszag and Patterson (1972). The last few decades have seen significant advances in the theory and application of spectral methods which can be found in a number of books, for example Gottlieb and Orszag (1983), Canuto *et al.* (2006), Boyd (2001), Mercier (1989), Funaro (1992) and Fornberg (1998). In this chapter we will introduce spectral methods. For a more detailed look the reader is referred to the above texts.

As the name suggests, in methods based on functional expansions the solution, $u(x)$, to the ODE (or PDE) is first approximately expanded as a weighted sum of basis functions as follows:

$$u(x) \approx \sum_k \hat{u}_k \, \phi_k(x), \qquad (6.1.1)$$

where $\hat{u}_k$ are the expansion coefficients and $\phi_k(x)$ are the sequence of basis functions, also known as the *trial functions*. Unlike finite difference and finite element methods, in spectral methods the trial functions are infinitely differentiable smooth global functions that span the entire computational domain. Typically they are eigenfunctions of the singular Sturm–Liouville problem and thus form a complete basis for the representation of continuous solutions. The proper choice of trial functions is dictated by the rapid (or exponential) convergence of the expansion. With appropriate sequence of trial functions the expansion coefficients, $\hat{u}_k$, decay most rapidly with increasing k, so that you need only a limited number of these coefficients.

6.2 Fourier Approximation

In periodic problems the choice of trial functions is straightforward. They are the trigonometric functions, which in a periodic domain $0 \leq x \leq L_x$ take the form

$$\phi_k(x) = \exp\left[i\,k\frac{2\pi}{L_x}x\right], \tag{6.2.1}$$

in which case (6.1.1) is nothing but the Fourier expansion given by

$$u(x) = \sum_{k=-\infty}^{\infty} \hat{u}_k \exp\left[ik\frac{2\pi}{L_x}x\right]. \tag{6.2.2}$$

The Fourier coefficients $\hat{u}_k$ are complex and are given by the following forward Fourier transform:

$$\hat{u}_k = \frac{1}{L_x}\int_0^{L_x} u(x)\exp\left[-ik\frac{2\pi}{L_x}x\right]\,\mathrm{d}x. \tag{6.2.3}$$

The corresponding backward Fourier transform is given by the expansion (6.2.2).

The theory of Fourier expansion shows that provided $u(x)$ is infinitely differentiable and is periodic over the domain $0 \leq x \leq L_x$,[1] the expansion given in (6.2.2) converges exponentially. In other words, the amplitude of the kth Fourier coefficient, $|\hat{u}_k|$, decays exponentially (i.e., as e^{-k}). Note that exponential decay is faster than any negative power of k. This is important since in practical applications the Fourier summation (6.2.2) will be truncated to a finite number of terms. We can define the **truncation error**, $E(x)$, as

$$E(x) = u(x) - \sum_{k=-N/2}^{N/2} \hat{u}_k \exp\left[ik\frac{2\pi}{L_x}x\right], \tag{6.2.4}$$

and from (6.2.2) it can be seen that this is the error due to the neglect of terms in the expansion with $|k| > N/2$. It can be shown that the truncation error decays exponentially, faster than any negative power of N. The **exponential convergence** is commonly termed as **spectral accuracy** or **infinite-order accuracy** (see Gottlieb and Orszag, 1983 for further details).

In the case of non-periodic problems the Fourier expansion will be inappropriate since the convergence of the expansion (6.2.2) will not be exponential; convergence will only be algebraic. The effect of non-periodicity or discontinuities within the domain can be severe. In a simplistic sense the effect on convergence can be stated as follows: any discontinuity in the jth derivative (i.e., in $\mathrm{d}^j u/\mathrm{d}x^j$) will result in the Fourier coefficients converging at $O(k^{-j-1})$. The discontinuity is not limited to the interior of the domain; it can occur at the boundary through violation of the periodic boundary condition as $\mathrm{d}^j u/\mathrm{d}x^j\big|_{x=0} \neq \mathrm{d}^j u/\mathrm{d}x^j\big|_{x=L_x}$.

[1] If the function is continuous and periodic then the function's derivatives to all order will also be continuous and periodic.

Example 6.2.1 Consider the function $u(x) = x$ over the periodic domain $0 \leq x \leq 2\pi$. The corresponding Fourier coefficients

$$\hat{u}_k = \begin{cases} \pi, & k = 0\,, \\ \iota/k, & \text{otherwise}\,, \end{cases} \tag{6.2.5}$$

exhibit a slow algebraic decay of only $O(k^{-1})$. This linear decay is consistent with the non-periodicity of the function (i.e., $u(x = 0) \neq u(x = 2\pi)$).

Example 6.2.2 Consider the following triangular function defined over the periodic domain $0 \leq x \leq 2\pi$:

$$u(x) = \begin{cases} x, & 0 \leq x < \pi\,, \\ 2\pi - x, & \pi \leq x < 2\pi\,. \end{cases} \tag{6.2.6}$$

The corresponding Fourier coefficients

$$\hat{u}_k = \begin{cases} \pi/2, & k = 0\,, \\ (-1)^{k+1}/\pi k^2 & \text{otherwise}\,, \end{cases} \tag{6.2.7}$$

exhibit a slow algebraic decay of $O(k^{-2})$. In this example the function is continuous and periodic; it is only the first derivative that is discontinuous and non-periodic. So in this case $j = 1$ and explains the quadratic decay of the Fourier coefficients.

Problem 6.2.3 Obtain the Fourier coefficients of the following step-function defined over the periodic domain $0 \leq x \leq 2\pi$:

$$u(x) = \begin{cases} 0, & 0 \leq x < \pi\,, \\ 1, & \pi \leq x < 2\pi\,, \end{cases} \tag{6.2.8}$$

and explain the rate of decay.

Problem 6.2.4 Obtain the Fourier coefficients of $u(x) = 1 - [(x - \pi)/\pi]^2$ defined over the periodic domain $0 \leq x \leq 2\pi$ and explain the rate of decay.

Problem 6.2.5 Show that the Fourier coefficients of $u(x) = 1/[3 - \sin x]$ defined over the periodic domain $0 \leq x \leq 2\pi$ have exponential decay.

The use of periodic trial functions (6.2.1) for a non-periodic problem will

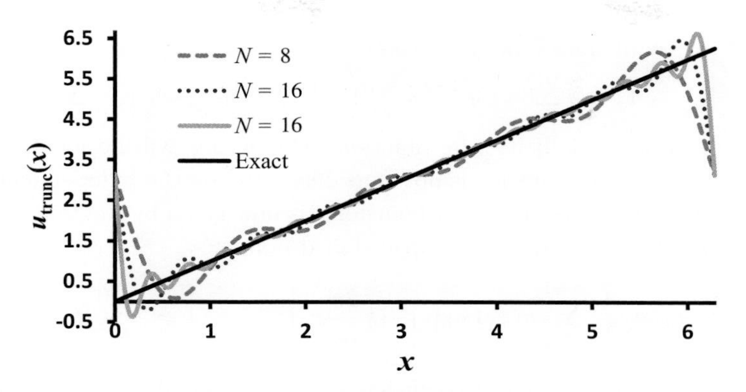

Figure 6.1 The function $u(x) = x$ over the domain $0 \le x \le 2\pi$ is approximated with truncated Fourier series of $N = 8, 16$ and 32 terms. Note that with increasing resolution the oscillations decrease in the interior, but not close to the boundaries, where periodicity is violated. With higher resolution the oscillations grow and they are pushed closer to the boundaries. This is known as Gibbs oscillation.

result in **Gibbs' phenomenon**, which is related to the slow algebraic decay of the Fourier coefficients. This phenomenon for Example 6.2.1 is vividly illustrated in Figure 6.1 where the truncated expansion

$$u_{\mathrm{tr}}(x) \approx \sum_{k=-N/2}^{N/2} \hat{u}_k \exp\left[ik\frac{2\pi}{L_x}x\right], \tag{6.2.9}$$

is plotted for $N = 8$, 16 and 32. Contrary to expectation, with increasing N the truncated expansion (6.2.9) does not provide improved approximation to $u(x) = x$. For large N the amplitude of the oscillation decreases in the interior, away from the boundaries. However, the undershoot and the overshoot that occur near $x = 0$ and $x = 2\pi$ persist even for very large N. Increasing N will only push the location of undershoot to approach $x = 0$ and overshoot to approach $x = 2\pi$ as $O(1/N)$; but the amplitude of undershoot and overshoot reaches an asymptotic value. For further discussion on Gibbs' phenomenon the reader can refer to mathematical textbooks on functional expansion or to any one of the above referred books on spectral methods. Nevertheless it is quite clear that Fourier expansion is inadequate for non-periodic functions.

6.2.1 Fourier Interpolation

Often the data to be Fourier transformed is not known as an analytic function, but is available only through the function values at selected grid points. In which case, the computation of Fourier coefficients as given by the integral in (6.2.3) is not feasible. Discrete versions of the forward and backward Fourier transforms must be employed. The appropriate set of grid points for the discrete transforms are equi-spaced points given by

$$x_j = j\,\Delta x \quad \text{for} \quad j = 0, 1, \ldots, N-1, \quad \text{where} \quad \Delta x = L_x/N. \tag{6.2.10}$$

Due to periodicity, the function value at x_N will be identically the same as at x_0 and therefore x_N is not considered among the independent set of points. The discrete forward Fourier transform is now given by the following summation over all N independent equi-spaced grid points:

$$\tilde{u}_k = \frac{1}{N}\sum_{j=0}^{N-1} u(x_j)\, \exp\left[-ik\frac{2\pi}{L_x}x_j\right] \quad \text{for} \quad k = -N/2, \ldots, (N/2)-1. \tag{6.2.11}$$

The discrete Fourier coefficients, $\tilde{u}_k$, as given above will not be equal in general to the actual Fourier coefficients, $\hat{u}_k$. Combining (6.2.11) and (6.2.3) it can be shown that $\tilde{u}_k$ and $\hat{u}_k$ are related through the relation

$$\tilde{u}_k = \hat{u}_k + \sum_{\substack{l=-\infty \\ l\neq 0}}^{\infty} \hat{u}_{(k+Nl)} \quad \text{for} \quad k = -N/2, \ldots, (N/2)-1\,. \tag{6.2.12}$$

As a result the interpolated trigonometric function

$$u_{\text{int}}(x) \approx \sum_{k=-N/2}^{(N/2)-1} \tilde{u}_k \, \exp\left[ik\frac{2\pi}{L_x}x\right] \tag{6.2.13}$$

will in general be different from the truncated approximation $u_{\text{tr}}(x)$ to the actual function $u(x)$. While the difference between $u_{\text{tr}}(x)$ and $u(x)$ is termed the truncation error, the difference between $u_{\text{int}}(x)$ and $u_{\text{tr}}(x)$ is called the **aliasing error**. Thus, interpolation through a set of N equi-spaced points has both the truncation and aliasing errors.

Note the subtle difference in the definition of the truncated (6.2.9) and the interpolated (6.2.13) trigonometric polynomials. The summation in the truncated expansions extends to include the $k = N/2$ mode. For $k \neq 0$ the Fourier coefficients, $\hat{u}_k$, are in general complex. For the truncated expansion $u_{tr}(x)$ to be a real valued function, we have the condition: $\hat{u}_{-k} = \hat{u}_k^*$. Thus each positive $(k > 0)$ mode combines with the corresponding negative $(k < 0)$ mode, whose amplitudes are complex conjugates of each other, to form a pair. Thus, in (6.2.9) the $k = N/2$ mode is required to pair with the $k = -N/2$ mode. Note that these two Fourier modes at the grid points are given by

$$\exp\left[-\iota\left(\pm\frac{N}{2}\right)\frac{2\pi}{L_x}x_j\right] = \exp[-\iota\, j\,\pi] = (-1)^j\,. \tag{6.2.14}$$

Thus it can also be seen from (6.2.11) that a real valued $u(x_j)$ will result in a real valued $\tilde{u}_{-N/2}$. It can also be verified that $\tilde{u}_{N/2}$, if computed, will be real and the same as $\tilde{u}_{-N/2}$. Thus, the $k = -N/2$ mode has been included in the Fourier interpolation (6.2.13) without the $k = N/2$ mode.

Let us investigate the nature of aliasing error a bit more. Equation (6.2.12)

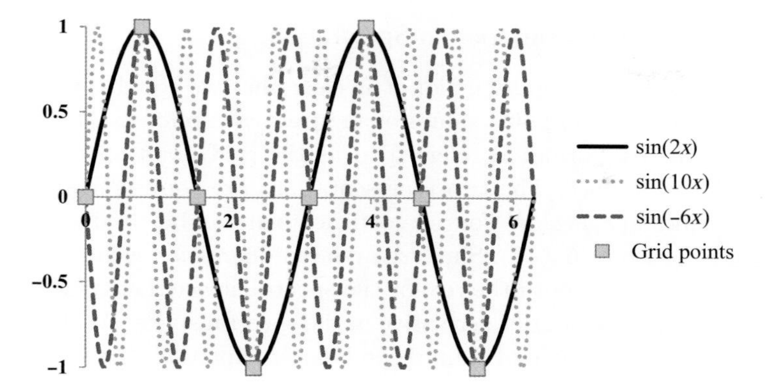

Figure 6.2 Plotted are three sinusoidal functions which alias on a grid of $N = 8$. Although the three functions are quite different within the periodic domain $0 \leq x \leq 2\pi$, they cannot be distinguished at the eight grid points.

shows that the kth mode of the interpolated function is not just $\hat{u}_k$; infinite number of other modes $\hat{u}_{k\pm N}, \hat{u}_{k\pm 2N}, \ldots$, are also spuriously aliased in $\tilde{u}_k$. In essence, on a set of N equi-spaced points given by (6.2.10), there is no difference between $\exp\left[i\, k\, 2\pi x_j/L_x\right]$ and the spurious higher modes $\exp\left[i\,(k \pm N)\, 2\pi x_j/L_x\right]$, etc. From the function value at the grid points alone it is impossible to distinguish between these aliased Fourier modes. This aliasing behavior of the Fourier modes is illustrated in Figure 6.2 for $N = 8$. However, it is clear from (6.2.12) that the aliasing modes, represented by the second term on the right, are precisely those neglected in the truncated representation (6.2.9). It can therefore be argued that the aliasing error is of the same order as the truncation error. Specifically, for a continuously differentiable periodic function, the aliasing error decays exponentially with increasing N. Thus for such smooth periodic functions, the number of grid points, N, can be chosen sufficiently large enough to contain both the truncation and aliasing errors below a desired level.

In spite of the truncation and aliasing errors, the interpolated trigonometric function (6.2.13) collocates to the exact value at the grid points. The discrete backward Fourier transform corresponding to (6.2.11) is given by

$$u(x_j) = \sum_{k=-N/2}^{(N/2)-1} \tilde{u}_k \exp\left[ik\frac{2\pi}{L_x}x_j\right]. \tag{6.2.15}$$

Using the following orthogonality relation of the Fourier modes:

$$\frac{1}{N}\sum_{j=0}^{N-1} \exp\left[i(k-m)\frac{2\pi}{L_x}x_j\right] = \begin{cases} 1 & k = m \pm lN,\ l = 0, 1, \ldots, \\ 0 & \text{otherwise}, \end{cases} \tag{6.2.16}$$

it is a simple matter to see that (6.2.11) followed by (6.2.15) gives back the exact value of the function at the grid points.

6.2.2 Fourier Differentiation in Spectral Space

From the above discussion it is clear that we could consider a function in **real space** as grid point values or in **spectral space** as Fourier coefficients. Sometimes the former is referred to as the grid point space, and the latter as the Fourier space. The two representations are equivalent. From the N grid point values $u(x_j)$ the N Fourier coefficients $\tilde{u}_k$ can be obtained from (6.2.11) and vice versa from (6.2.15). Similarly, related quantities such as derivatives can also be evaluated in both the real or the spectral space. Let us start by considering how to compute Fourier derivative in the spectral space. We define the first derivative $u'(x)$ as a weighted sum of the Fourier expansion basis as

$$u'(x) = \sum_{k=-\infty}^{\infty} \hat{u}'_k \exp\left[ik\frac{2\pi}{L_x}x\right]. \tag{6.2.17}$$

By differentiating both sides of (6.2.2) it can be readily seen that the Fourier coefficients of the derivative are simply related to the Fourier coefficients of the function as

$$\hat{u}'_k = \left(ik\frac{2\pi}{L_x}\right)\hat{u}_k, \tag{6.2.18}$$

where the Fourier coefficients $\hat{u}_k$ are given in (6.2.3). As with the function, truncating the summation (6.2.17) to only N terms will result in truncation error for the derivative. But provided the function being differentiated is periodic and infinitely differentiable, the truncation error will decrease exponentially with increasing N.

The Fourier coefficients of the second derivative can be similarly expressed as

$$\hat{u}''_k = \left(ik\frac{2\pi}{L_x}\right)^2\hat{u}_k = -\left(k\frac{2\pi}{L_x}\right)^2\hat{u}_k. \tag{6.2.19}$$

Coefficients of higher-order derivatives can be similarly expressed. For later consideration we rewrite (6.2.18) and (6.2.19) in terms of Fourier first- and second-derivative operators as

$$\hat{u}'_k = [\mathcal{D}_{\mathcal{F}1}]_{kl}\,\hat{u}_l \quad \text{and} \quad \hat{u}''_k = [\mathcal{D}_{\mathcal{F}2}]_{kl}\,\hat{u}_l. \tag{6.2.20}$$

Unlike the finite difference derivative matrices defined in Chapter 4, which yield derivative values at the grid points given the functions values at the grid points, the above derivative operators are in the spectral space and yield Fourier coefficients of the derivative in terms of Fourier coefficients of the function. To distinguish the derivative operation in spectral space from that in the grid point space, we denote the derivative operators in (6.2.20) with a script font. The advantage of the derivative operator in spectral space is clear. The derivative operators are diagonal and as a result each Fourier mode of the derivative depends only on the corresponding mode of the function. Such a statement cannot be made in the grid point space, since the derivative at any grid point will depend on function values at two or more neighboring grid points.

6.2.3 Fourier Differentiation in Grid Point Space

From Fourier interpolation, Fourier derivative matrices for the grid points (6.2.10) can be defined. Let us first consider the discretized version of the first-derivative operator. The Fourier interpolant (6.2.13) can be differentiated to obtain the first derivative, u', at the grid points as

$$u'(x_j) = \sum_{k=-N/2}^{(N/2)-1} \tilde{u}_k \left(\iota k \frac{2\pi}{L_x}\right) \exp\left[\iota k \frac{2\pi}{L_x} x_j\right]. \tag{6.2.21}$$

Upon substituting for the Fourier coefficients from (6.2.11) we get

$$\begin{aligned} u'(x_j) &= \sum_{l=0}^{N-1} \left\{ \frac{1}{N} \sum_{k=-N/2}^{N/2-1} \left(\iota k \frac{2\pi}{L_x}\right) \exp\left[\iota k \frac{2\pi}{L_x}(x_j - x_l)\right] \right\} u(x_l) \\ &= \sum_{l=0}^{N-1} [D_{F1}]_{jl}\, u(x_l). \end{aligned} \tag{6.2.22}$$

The **Fourier first-derivative matrix, $\mathbf{D}_{F1}$**, is defined as the summation within the curly parentheses. Much like the finite difference first-derivative operators defined in Chapter 4, $\mathbf{D}_{F1}$ is a matrix of size $N \times N$, which when operates on the vector of grid point values, $u(x_l)$ for $l = 0, 1, \ldots, N-1$, will yield the vector of grid point derivative values, $u'(x_l)$ for $l = 0, 1, \ldots, N-1$.

Amazingly, the summation within the curly parentheses can be carried out and the Fourier first-derivative matrix can be explicitly written as

$$[D_{F1}]_{jl} = \frac{2\pi}{L_x} \begin{cases} \frac{1}{2}(-1)^{j+l} \cot\left[\frac{(j-l)\pi}{N}\right], & j \neq l, \\ 0, & j = l. \end{cases} \tag{6.2.23}$$

Fourier higher-order derivative matrix operators can be similarly defined by taking higher-order derivatives of the Fourier interpolant (6.2.13). The Fourier pth derivative matrix can be written as

$$[D_{Fp}]_{jl} = \left\{ \frac{1}{N} \sum_{k=-N/2}^{(N/2)-1} \left(\iota k \frac{2\pi}{L_x}\right)^p \exp\left[\iota k \frac{2\pi}{L_x}(x_j - x_l)\right] \right\}. \tag{6.2.24}$$

In particular, the **Fourier second-derivative matrix $\mathbf{D}_{F2}$** can be explicitly written as

$$[D_{F2}]_{jl} = \left(\frac{2\pi}{L_x}\right)^2 \begin{cases} \frac{1}{4}(-1)^{j+1} N + \dfrac{(-1)^{j+l+1}}{2\sin^2\left(\frac{(j-l)\pi}{N}\right)}, & j \neq l, \\ 0, & j = l. \end{cases} \tag{6.2.25}$$

The accuracy of the Fourier first- and second-derivative operators can be evaluated in terms of their eigenvalues as outlined in Section 4.2.1. The eigenvalues of the Fourier first-derivative matrix are

$$\iota \frac{2\pi\, k}{L_x} \quad \text{for} \quad k = -(N/2), \ldots, -1, 0, 1, \ldots, (N/2) - 1, \tag{6.2.26}$$

and the eigenvalues of the Fourier second-derivative matrix are

$$-\left(\frac{2\pi k}{L_x}\right)^2 \quad \text{for} \quad k = -(N/2), \ldots, -1, 0, 1, \ldots, (N/2) - 1. \tag{6.2.27}$$

The above eigenvalues of the first- and second-derivative operators are identical to the first N exact eigenvalues of the continuous first- and second-derivative operators of periodic functions given in (4.2.6) and (4.2.12). Thus, the spectral approximation is 100% accurate in differentiating the first N Fourier modes. The only error is in truncating the complete expansion (6.2.2) to the first N terms. Thus, for derivative operation in both Fourier space using (6.2.20) and in grid point space using (6.2.22) the error arises only from truncation and possibly aliasing.

6.3 Polynomial Approximation

For non-periodic problems, exponential convergence can be obtained with orthogonal polynomials, which are solutions to the following singular Sturm–Liouville problem (Canuto *et al.*, 2006, §9.2):

$$\left.\begin{aligned} -\frac{\mathrm{d}}{\mathrm{d}x}\left(p(x)\frac{\mathrm{d}\phi_k}{\mathrm{d}x}\right) + q(x)\,\phi_k = \lambda_k\, w(x)\,\phi_k \quad &\text{over} \quad -1 \le x \le 1\,, \\ \text{Singularity:} \quad p(1) = p(-1) = 0\,, \\ \text{Boundary conditions:} \quad p(x)\frac{\mathrm{d}\phi_k}{\mathrm{d}x} \to 0 \quad &\text{as} \quad x \to \pm 1\,. \end{aligned}\right\} \tag{6.3.1}$$

The variables coefficients, $p(x)$, $q(x)$ and $w(x)$, are all continuous real functions with $q(x)$ and $w(x)$ non-negative, and $p(x)$ strictly positive in the interval $-1 < x < 1$. The orthogonal system of polynomials, $\phi_k(x)$ for $k = 0, 1, \ldots$, are the eigenfunctions of (6.3.1) and the corresponding eigenvalues are λ_k.

A two-parameter family of Jacobi polynomials results with the following choice of coefficients:

$$p(x) = (1-x)^{1+\alpha}\;(1+x)^{1+\beta}\;; \quad q(x) = 0\,; \quad w(x) = (1-x)^{\alpha}\,(1+x)^{\beta}\,, \tag{6.3.2}$$

where the two parameters, α and β, are constrained to be greater than -1. For a detailed discussion on the theory and properties of Jacobi polynomials the reader is referred to mathematical texts on orthogonal polynomials (Sansone, 1959; Luke, 1969). Of the entire family of Jacobi polynomials the two most commonly used are the special cases of Chebyshev ($\alpha = \beta = -\frac{1}{2}$) and Legendre ($\alpha = \beta = 0$) polynomials. The popularity of Chebyshev and Legendre polynomials arises from the relative theoretical simplicity in their analysis. The basic classification of spectral methods is based on the trial functions in use; therefore the most popular spectral methods in use are the **Fourier spectral**, **Chebyshev spectral** and **Legendre spectral** methods. In the following subsections we will present some basic properties and recurrence relations associated with Chebyshev and Legendre polynomials. For additional details consult the texts on orthogonal polynomials and spectral methods.

6.3.1 Chebyshev Polynomials

Chebyshev polynomials arise with the following coefficients in the singular Sturm–Liouville problem (6.3.1):

$$p(x) = \sqrt{(1-x^2)}\,; \quad q(x) = 0\,; \quad w(x) = 1/\sqrt{(1-x^2)}\,. \tag{6.3.3}$$

Conventionally the kth Chebyshev polynomial is denoted by $T_k(x)$ which can be expanded in power series as

$$T_k(x) = \frac{k}{2}\sum_{j=0}^{\lfloor k/2 \rfloor} (-1)^k \,\frac{(k-j-1)!}{j!\,(k-2j)!}\,(2x)^{k-2j}\,, \tag{6.3.4}$$

where $\lfloor k/2 \rfloor$ is the integer part of $k/2$. A simpler representation of Chebyshev polynomials in terms of cosines is possible with a change of variable as

$$T_k(x) = \cos(k\theta) \quad \text{where} \quad \theta = \cos^{-1}(x). \tag{6.3.5}$$

The first few Chebyshev polynomials are

$$T_0(x) = 1\,; \quad T_1(x) = x\,; \quad T_2(x) = 2x^2 - 1\,, \tag{6.3.6}$$

and higher-order ones are given by the recurrence

$$T_{k+1}(x) = 2x\,T_k(x) - T_{k-1}(x). \tag{6.3.7}$$

Chebyshev polynomials satisfy the following orthogonality relation:

$$\int_{-1}^{1} \frac{T_k(x)T_l(x)}{\sqrt{1-x^2}}\,\mathrm{d}x = \begin{cases} 0 & \text{if} \quad k \neq l\,, \\ \pi & \text{if} \quad k = l = 0\,, \\ \pi/2 & \text{if} \quad k = l \geq 1\,. \end{cases} \tag{6.3.8}$$

Other useful properties of the Chebyshev polynomials are:

$$T'_{k+1}(x) = 2\,(k+1)\,T_k(x) + \frac{(k+1)}{(k-1)}T'_{k-1}(x), \tag{6.3.9a}$$

$$|T_k(x)| \leq 1 \quad \text{and} \quad |T'_k(x)| \leq k^2 \quad \text{for} \quad -1 \leq x \leq 1, \tag{6.3.9b}$$

$$T_k(\pm 1) = (\pm 1)^k, \quad T'_k(\pm 1) = (\pm 1)^{k+1}k^2, \tag{6.3.9c}$$

$$T_k(x) = 0 \quad \text{for} \quad x = \cos\left(\frac{2j-1}{2k}\pi\right) \quad \text{with} \quad j = 1,\, 2, \ldots, k, \tag{6.3.9d}$$

$$T'_k(x) = 0 \quad \text{for} \quad x = \cos\left(\frac{j\pi}{k}\right) \quad \text{with} \quad j = 0,\, 1, \ldots, k. \tag{6.3.9e}$$

The first one is a recurrence relation for the first derivative of the Chebyshev polynomials (prime indicates derivative with respect to x). Equation (6.3.9d) provides an explicit expression for the zeros of the Chebyshev polynomials and it can be seen that the kth Chebyshev polynomial has k zero crossings. Equation (6.3.9e) gives the extrema of the Chebyshev polynomials, which are $(k+1)$ in number and include the end points $x = \pm 1$.

Forward Chebyshev transforms of a function defined over the range $-1 \leq x \leq 1$ are given by the following integral:

$$\hat{u}_k = \frac{2}{\pi\, c_k} \int_{-1}^{1} \frac{u(x)\, T_k(x)}{\sqrt{1-x^2}} \mathrm{d}x, \tag{6.3.10}$$

where $c_k = 2$ if $k = 0$ and $c_k = 1$ otherwise. The corresponding Chebyshev expansion is

$$u(x) = \sum_{k=0}^{\infty} \hat{u}_k\, T_k(x). \tag{6.3.11}$$

As with the Fourier expansion, a truncated representation can be obtained by limiting the above summation to only N terms.

6.3.2 Legendre Polynomials

Legendre polynomials arise with the following coefficients in the singular Sturm–Liouville problem (6.3.1):

$$p(x) = 1 - x^2\,; \quad q(x) = 0\,; \quad w(x) = 1\,. \tag{6.3.12}$$

The kth Legendre polynomial is usually denoted by $L_k(x)$ which can be expanded in power series as

$$L_k(x) = \frac{1}{2^k} \sum_{j=0}^{\lfloor k/2 \rfloor} (-1)^j \begin{pmatrix} k \\ j \end{pmatrix} \begin{pmatrix} 2(k-j) \\ k \end{pmatrix} (x)^{k-2j}\,. \tag{6.3.13}$$

Unlike Chebyshev polynomials, here a representation in terms of cosines is not possible and therefore the cosine transform cannot be used in the case of Legendre expansion. The first few Legendre polynomials are

$$L_0(x) = 1\,; \quad L_1(x) = x\,; \quad L_2(x) = \frac{3}{2}x^2 - \frac{1}{2}, \tag{6.3.14}$$

and higher-order ones are given by the recurrence

$$L_{k+1}(x) = \frac{2k+1}{k+1} x\, L_k(x) - \frac{k}{k+1} L_{k-1}(x). \tag{6.3.15}$$

Legendre polynomials satisfy the following orthogonality relation:

$$\int_{-1}^{1} L_k(x) L_l(x)\, \mathrm{d}x = \begin{cases} 0 & \text{if} \quad k \neq l\,, \\ \left(k + \frac{1}{2}\right)^{-1} & \text{if} \quad k = l\,. \end{cases} \tag{6.3.16}$$

Other useful properties of the Chebyshev polynomials are:

$$L'_{k+1}(x) = (2k+1)\, L_k(x) + L'_{k-1}(x), \tag{6.3.17a}$$

$$|L_k(x)| \le 1 \quad \text{and} \quad |L'_k(x)| \le \tfrac{1}{2}k(k+1) \quad \text{for} \quad -1 \le x \le 1, \tag{6.3.17b}$$

$$L_k(\pm 1) = (\pm 1)^k, \quad L'_k(\pm 1) = (\pm 1)^{k+1}\tfrac{1}{2}k(k+1). \tag{6.3.17c}$$

Unlike the Chebyshev polynomials, there are no simple expressions for the zeros and extrema of the Legendre polynomials. Nevertheless, the kth Legendre polynomial has k zero crossings and $(k+1)$ extrema. Again the locations of the extrema include the end points $x = \pm 1$. The zeros and extrema can be evaluated using an iterative procedure.

Forward Legendre transform of a function defined over the range $-1 \le x \le 1$ is given by the following integral:

$$\hat{u}_k = (k + \frac{1}{2}) \int_{-1}^{1} u(x)\, L_k(x)\, \mathrm{d}x. \tag{6.3.18}$$

The corresponding Legendre expansion is then

$$u(x) = \sum_{k=0}^{\infty} \hat{u}_k\, L_k(x). \tag{6.3.19}$$

6.3.3 Polynomial Interpolation

The equi-spaced distribution of grid points (6.2.10) used in Fourier interpolation is optimal for periodic functions. The optimality is in the sense that the discrete summation (6.2.11) provides the best approximation to the integral in (6.2.3). However, an equi-spaced distribution of grid points is inappropriate in approximating the integrals (6.3.10) and (6.3.18) in the case of Chebyshev and Legendre transforms. The question then is: what distribution of grid points, x_j for $j = 0, 1, \ldots, N$, will provide the best approximation to the integrals given in (6.3.10) and (6.3.18)? We have already seen the answer in Section 4.6.4 as given by the appropriate quadrature rules. There, in the context of numerical approximation to integration, we obtained the optimal distribution of points and weights for evaluating integrals as weighted sum of the function values given at the optimal grid points. Here we will use that information for approximating the integrals (6.3.10) and (6.3.18).

For the Chebyshev polynomials the optimal distribution of grid points is given by the **Chebyshev–Gauss points**

$$x_j = \cos\left(\frac{\pi(2j+1)}{2N+2}\right) \qquad j = 0, 1, \ldots, N\,, \tag{6.3.20}$$

and the discrete version of the forward Chebyshev transform (6.3.10) becomes

$$\tilde{u}_k = \frac{2}{\pi\, c_k} \sum_{j=0}^{N} u(x_j)\, T_k(x_j)\, w_j \quad \text{for} \quad k = 0, 1, \ldots, N, \tag{6.3.21}$$

where the Chebyshev Gauss weights are $w_j = \pi/(N+1)$. As with Fourier interpolation, in general $\tilde{u}_k$ will not be equal to $\hat{u}_k$, with the difference being the aliasing error. However, the choice of Gauss points and the corresponding Gauss weights guarantee that the approximation (6.3.21) becomes exact (i.e., $\tilde{u}_k=\hat{u}_k$) provided $u(x)$ is a polynomial of degree $2N+1-k$ or lower. Thus with a distribution of $N+1$ Gauss points, all the Chebyshev coefficients for $k=0,1,\ldots,N$, can be accurately obtained provided $u(x)$ is a polynomial of degree utmost $N+1$. If $u(x)$ is a polynomial of degree greater than $N+1$, then aliasing error will be present.

The problem with Gauss points is that they do not include the boundary points; all the $N+1$ points defined by (6.3.20) are in the interior. Often it is desirable to include the boundaries as first and last points, in order to enforce appropriate boundary conditions. By restricting $x_0 = 1$ and $x_N = -1$, one obtains the **Chebyshev–Gauss–Lobatto** distribution of points

$$x_j = \cos\frac{\pi j}{N} \qquad j=0,1,\ldots,N \tag{6.3.22}$$

and the corresponding discrete version of the forward Chebyshev transform is

$$\tilde{u}_k = \frac{1}{d_k}\sum_{j=0}^{N} u(x_j)\,T_k(x_j)\,w_j \quad \text{for} \quad k=0,1,\ldots,N, \tag{6.3.23}$$

where the Chebyshev–Gauss–Lobatto weights and prefactors are given by

$$w_j = \begin{cases} \pi/2N & j=0,\, N\,, \\ \pi/N & \text{otherwise} \end{cases} \qquad \text{and} \qquad d_j = \begin{cases} \pi & j=N\,, \\ \pi\, c_k/2 & \text{otherwise.} \end{cases} \tag{6.3.24}$$

Again, due to aliasing error, $\tilde{u}_k$ will not be equal to $\hat{u}_k$. But for the Gauss–Lobatto points and weights the approximation (6.3.23) is exact (i.e., $\tilde{u}_k=\hat{u}_k$) provided $u(x)$ is a polynomial of degree $2N-1-k$ or lower. Thus the accuracy of the Gauss–Lobatto points is two polynomial degrees lower than that of Gauss points. Clearly this is the penalty for fixing two of the grid points to be the boundary points.

For both the Gauss and Gauss–Lobatto distribution of points, the inverse transform is given by

$$u_{\text{int}}(x) = \sum_{k=0}^{N} \tilde{u}_k\, T_k(x). \tag{6.3.25}$$

Upon substitution of either (6.3.21) or (6.3.23) it can be easily verified that the interpolated polynomial collocates to the exact function values at the Gauss or Gauss–Lobatto points, respectively.

The **Legendre–Gauss points** are the zeros of $L_{N+1}(x)$ and the corresponding weights are

$$w_j = \frac{2}{(1-x_j^2)\left[L'_{N+1}(x_j)\right]^2} \quad \text{for} \quad j=0,\,1,\,\ldots,\,N\,. \tag{6.3.26}$$

The **Legendre–Gauss–Lobatto points** include the end points $x_0 = 1$, $x_N = -1$, along with the zeros of $L'_N(x)$. The corresponding weights are

$$w_j = \frac{2}{N(N+1)\,[L_N(x_j)]^2} \quad \text{for} \quad j = 0,\, 1,\, \ldots,\, N. \tag{6.3.27}$$

For both sets of points the forward Legendre transform is given by

$$\tilde{u}_k = \frac{1}{d_k} \sum_{j=0}^{N} u(x_j)\, L_k(x_j)\, w_j \quad \text{for} \quad k = 0, 1, \ldots, N, \tag{6.3.28}$$

where the prefactors are $d_k = 1/(k + \frac{1}{2})$ for both the Gauss or Gauss–Lobatto distribution of points, except for the Gauss–Lobatto points $d_N = 2/N$. The inverse transform is correspondingly

$$u_{\text{int}}(x) = \sum_{k=0}^{N} \tilde{u}_k\, L_k(x). \tag{6.3.29}$$

Again it can be shown that the interpolation collocates to the exact function values at the Gauss or Gauss–Lobatto points.

6.3.4 Chebyshev Differentiation in Spectral Space

We first expand the first derivative of $u(x)$ in terms of Chebyshev polynomials as

$$u'(x) = \sum_{k=0}^{N} \hat{u}'_k\, T_k(x) = \sum_{k=0}^{N} \hat{u}_k\, T'_k(x). \tag{6.3.30}$$

In the above the prime in $\hat{u}'_k$ indicates that the Chebyshev coefficient is that of the first derivative, while the prime in T'_k indicates the first derivative of the kth Chebyshev polynomial. Using the recurrence relation (6.3.9a), we obtain

$$\begin{bmatrix} 1 & 0 & -\frac{1}{2} & & & \\ & \frac{1}{4} & 0 & -\frac{1}{4} & & \\ & & \frac{1}{6} & 0 & -\frac{1}{6} & \\ & & & \frac{1}{8} & 0 & -\frac{1}{8} \\ & & & & \ddots & \ddots \\ & & & & & \frac{1}{2N} \end{bmatrix} \begin{bmatrix} \hat{u}'_0 \\ \hat{u}'_1 \\ \hat{u}'_2 \\ \hat{u}'_3 \\ \vdots \\ \hat{u}'_{N-1} \end{bmatrix} = \begin{bmatrix} \hat{u}_1 \\ \hat{u}_2 \\ \hat{u}_3 \\ \hat{u}_4 \\ \vdots \\ \hat{u}_N \end{bmatrix}. \tag{6.3.31}$$

From the above, for an N-mode truncation starting from $\hat{u}'_N = 0$, $\hat{u}'_{N-1} = 2N\,\hat{u}_N$ and $\hat{u}'_{N-2} = 2(N-1)\,\hat{u}_{N-1}$, we can work our way up and find all $\hat{u}'_k$ in terms of $\hat{u}_k$. This process can be explicitly written as

$$\hat{u}'_k = \frac{2}{c_k} \sum_{\substack{p=k+1 \\ p+k \text{ odd}}}^{\infty} p\,\hat{u}_p\,, \quad k \geq 0, \tag{6.3.32}$$

where $c_k = 2$ if $k = 0$ and $c_k = 1$ otherwise. We can also write it in matrix form as

$$\begin{bmatrix} \hat{u}'_0 \\ \hat{u}'_1 \\ \hat{u}'_2 \\ \hat{u}'_3 \\ \vdots \\ \hat{u}'_N \end{bmatrix} = \underbrace{\begin{bmatrix} 0 & 1 & 0 & 3 & 0 & \cdots \\ & 0 & 4 & 0 & 8 & \ddots \\ & & 0 & 6 & 0 & \ddots \\ & & & 0 & 8 & \ddots \\ & & & & \ddots & \ddots \\ & & & & & 0 \end{bmatrix}}_{\mathcal{D}_{C1}} \begin{bmatrix} \hat{u}_0 \\ \hat{u}_1 \\ \hat{u}_2 \\ \hat{u}_3 \\ \vdots \\ \hat{u}_N \end{bmatrix}. \tag{6.3.33}$$

Note $\mathcal{D}_{C1}$ is a strictly upper triangular matrix, which is simply a reflection of the fact that the derivative of an Nth-order polynomial is of order $(N-1)$. The corresponding second derivative is of order $(N-2)$ and the Chebyshev coefficients $\hat{u}''_k$ can be expressed as

$$\hat{u}''_k = \frac{1}{c_k} \sum_{\substack{p=k+2 \\ p+k \text{ even}}}^{\infty} p(p^2 - k^2)\, \hat{u}_p, k \geq 0\,. \tag{6.3.34}$$

Correspondingly, the second-derivative operator in spectral space is

$$\begin{bmatrix} \hat{u}''_0 \\ \hat{u}''_1 \\ \hat{u}''_2 \\ \hat{u}''_3 \\ \vdots \\ \hat{u}''_N \end{bmatrix} = \underbrace{\begin{bmatrix} 0 & 0 & 4 & 0 & 32 & \cdots \\ & 0 & 0 & 24 & 0 & \ddots \\ & & 0 & 0 & 48 & \ddots \\ & & & 0 & 0 & \ddots \\ & & & & \ddots & 0 \\ & & & & & 0 \end{bmatrix}}_{\mathcal{D}_{C2}} \begin{bmatrix} \hat{u}_0 \\ \hat{u}_1 \\ \hat{u}_2 \\ \hat{u}_3 \\ \vdots \\ \hat{u}_N \end{bmatrix}. \tag{6.3.35}$$

Contrast these matrices with the Fourier derivative operators $\mathcal{D}_{\mathcal{F}1}$ and $\mathcal{D}_{\mathcal{F}2}$, which are diagonal.

Problem 6.3.1 Derive the expressions given in (6.3.32) and (6.3.34). Then obtain the operators $\mathcal{D}_{C1}$ and $\mathcal{D}_{C2}$.

6.3.5 Chebyshev Differentiation in Grid Point Space

We now proceed to form the Chebyshev derivative matrices in real space. These derivative matrices are based on Chebyshev interpolation operation and can be obtained for both Gauss and Gauss–Lobatto grid points. The N-mode Chebyshev interpolant (6.3.25) can be differentiated and evaluated at the grid points to yield

$$u'(x_j) = \sum_{k=0}^{N} \tilde{u}_k \, T_k'(x_j). \tag{6.3.36}$$

By substituting the appropriate Chebyshev coefficients for the Gauss or the Gauss–Lobatto points the corresponding first-derivative matrix operators can be defined for Gauss or Gauss–Lobatto points. For example, for the Gauss–Lobatto points (6.3.23) can be substituted in (6.3.36) to yield

$$u'(x_j) = \sum_{l=0}^{N} \left\{ \sum_{k=0}^{N} \frac{1}{d_k} T_k'(x_j)\, T_k(x_l)\, w_l \right\} u(x_l), \tag{6.3.37}$$

where the term within the curly parentheses defines the **Chebyshev first-derivative matrix** on the Gauss–Lobatto points. This Chebyshev first-derivative matrix can be evaluated and explicitly written as

$$[D_{\mathrm{C1}}]_{jl} = \begin{cases} \dfrac{e_j(-1)^{j+l}}{e_l(x_j - x_l)} & j \neq l\,, \\[2ex] \dfrac{-x_l}{2(1-x_l^2)} & 1 \leq j = l \leq N-1\,, \\[2ex] \dfrac{2N^2+1}{6} & j = l = 0\,, \\[2ex] -\dfrac{2N^2+1}{6} & j = l = N\,, \end{cases} \tag{6.3.38}$$

where $e_k = 2$ if $k = 0$ or N and $e_k = 1$ otherwise. Higher-order Chebyshev derivatives at the Gauss–Lobatto points can be defined as powers of the first-derivative matrix:

$$\mathbf{D}_{\mathrm{C}p} = [\mathbf{D}_{\mathrm{C1}}]^p\,. \tag{6.3.39}$$

These Chebyshev derivative operators can be used in place of the finite difference operators defined earlier in Chapter 4. The superior accuracy of the spectral methods can be examined in the following computer exercises.

Computer Exercise 6.3.2 The accuracy of the Chebyshev first-derivative matrix can be established by repeating the Computer Exercise 4.2.4 with $\mathbf{D}_{\mathrm{C1}}$. Show that the error for the first N test functions is identically zero.

Computer Exercise 6.3.3 Repeat the above analysis for the Chebyshev second-derivative operator.

We now leave to the reader the task of obtaining the first- and second-derivative matrix operators in the spectral space (i.e., $\mathcal{D}_{\mathrm{L1}}$ and $\mathcal{D}_{\mathrm{L2}}$) for Legendre polynomials. Legendre first- and second-derivative matrix operators in the grid point space (i.e., $\mathbf{D}_{\mathrm{L1}}$ and $\mathbf{D}_{\mathrm{L2}}$) can also be numerically constructed.

6.4 Galerkin, Tau, Collocation and Pseudo-spectral Methods

In the numerical implementation of spectral methods, the summation (6.1.1) will be truncated to a finite number of terms. In the case of Fourier expansion the summation will extend from $k = -\frac{N}{2}$ to $k = \frac{N}{2}$ and in the case of polynomial expansions the summation will range from $k = 0$ to $k = N$. The task of numerically solving a given ODE for $u(x)$ boils down to evaluating the $N+1$ expansion coefficients $\hat{u}_k$. Once the expansion coefficients are evaluated, the solution is available over the entire domain from (6.1.1). The requirement of appropriate choice of trial functions for rapid convergence of (6.1.1) is obvious. The spectral expansion can then be truncated to fewer terms and the number of expansion coefficients to be computed for accurate representation of the solution can be minimized.

The next step in the development of spectral methods is to identify the $N+1$ equations that must be solved to obtain the expansion coefficients. These equations must be obtained from the governing equations and the boundary conditions. Depending on how these $N+1$ equations are chosen and enforced, four different types of spectral methods: **Galerkin**, **tau**, **collocation**, and **pseudo-spectral** methods arise.

We first illustrate these methods for a linear constant coefficient ODE

$$\mathbf{L}\left(u(x)\right) = \left[c_2 \frac{\mathrm{d}^2}{\mathrm{d}\mathrm{x}^2} + c_1 \frac{\mathrm{d}}{\mathrm{d}\mathrm{x}} + c_0\right] u(x) = f(x) \tag{6.4.1}$$

where $\mathbf{L}$ is a linear constant coefficient differential operator. The above equation is supplemented with appropriate boundary conditions. The orthogonal expansion (6.1.1) can be substituted into the left-hand side of the above equation and rearranged to obtain

$$\sum_k \hat{b}_k \, \phi_k(x). \tag{6.4.2}$$

Here $\hat{b}_k$ are the expansion coefficients, which are related to $\hat{u}_k$ through

$$\hat{b}_k = \sum_l \mathcal{L}_{kl} \hat{u}_l = \sum_l \left[c_2 \left(\mathcal{D}_2\right)_{kl} + c_1 \left(\mathcal{D}_1\right)_{kl} + c_0 \delta_{kl}\right] \hat{u}_l. \tag{6.4.3}$$

The matrix operator $\mathcal{L}_{kl}$ is the discretized version of $\mathbf{L}$ in spectral space. The exact form of this matrix operator depends both on the operator $\mathbf{L}$ and on the trial functions. For example, $\mathcal{D}_1$ and $\mathcal{D}_2$ in (6.4.3) are $\mathcal{D}_{\mathcal{F}1}$ and $\mathcal{D}_{\mathcal{F}2}$ in the case of Fourier trial functions and are $\mathcal{D}_{\mathrm{C1}}$ and $\mathcal{D}_{\mathrm{C2}}$ in the case of Chebyshev trial functions. The right-hand side, $f(x)$, can also be expressed as a weighted sum of the trial functions

$$f(x) = \sum_k \hat{f}_k \, \phi_k(x), \tag{6.4.4}$$

where the expansion coefficients of the forcing function, $\hat{f}_k$, can be evaluated as

$$\begin{aligned}
&\text{Fourier:} \quad && \hat{f}_k = \frac{1}{2\pi} \int_0^{L_x} f(x) \exp\left(-\frac{\iota k \, 2\pi}{L_x} x\right) \mathrm{d}x \\
&\text{Chebyshev:} && \hat{f}_k = \frac{2}{\pi \, c_k} \int_{-1}^{1} \frac{f(x) \, T_k(x)}{\sqrt{1-x^2}} \, \mathrm{d}x \, , \\
&\text{Legendre:} && \hat{f}_k = (k + \frac{1}{2}) \int_{-1}^{1} f(x) \, L_k(x) \, \mathrm{d}x \, ,
\end{aligned} \tag{6.4.5}$$

where

$$c_k = \begin{cases} 2 & \text{if} \quad k = 0 \, , \\ 1 & \text{if} \quad k \geq 1 \, . \end{cases}$$

Upon substitution of (6.4.2) and (6.4.4) into the ODE (6.4.1) one obtains

$$\sum_k (\hat{b}_k - \hat{f}_k) \, \phi_k(x) = R(x), \tag{6.4.6}$$

where the ODE appears on the left-hand side and $R(x)$ is the residue. The objective is then to minimize the residue in some appropriate way.

6.4.1 Galerkin Methods

In the Galerkin method we first require that each of the trial functions individually satisfy all the boundary conditions of the ODE and as a result any finite truncation of (6.1.1) will also automatically satisfy the boundary conditions. It is then demanded that the residue be orthogonal to the first $N+1$ trial functions. That is,

$$\int R(x) \, \phi_j(x) \, w(x) \, \mathrm{d}x = 0, \tag{6.4.7}$$

for j in the range $-\frac{N}{2}$ to $\frac{N}{2} - 1$ in the case of Fourier spectral method and in the range 0 to N in the case of polynomial expansions. From the orthogonality relations (6.3.8) and (6.3.16) we obtain[2] the simple relation

$$\hat{b}_k = \hat{f}_k . \tag{6.4.8}$$

Substituting for $\hat{b}_k$ from (6.4.3) we obtain the following linear system that must be solved:

$$\sum_{l=-N/2}^{N/2-1} \mathcal{L}_{kl} \hat{u}_l = f_k . \tag{6.4.9}$$

[2] For the Fourier expansion we have the following orthonormality condition, $\int_0^{L_x} \frac{1}{2\pi} \exp\left(\frac{\iota \, k \, 2\pi \, x}{L_x}\right) \exp\left(\frac{\iota \, l \, 2\pi \, x}{L_x}\right) \, \mathrm{d}x = \delta_{kl}$, where the weighting function, $w(x) = 1/2\pi$.

From the coefficients $\hat{u}_k$ the solution $u(x)$ can be evaluated using the truncated expansion.

In the case of periodic problems, each Fourier mode (6.2.1) of the Fourier expansion automatically satisfies the periodic boundary conditions and therefore implementation of the Fourier–Galerkin spectral method is relatively straightforward. Furthermore, since the first- and second-derivative operators in the Fourier space are diagonal, the matrix operator $\mathcal{L}_{kl}$ is also diagonal and (6.4.9) can be solved independently for each mode.

For non-periodic problems, as can be seen from (6.3.9c) and (6.3.17c), both the Chebyshev and Legendre polynomials do not individually satisfy any particular boundary condition. However, by defining a modified sequence of trial functions as linear combinations of Chebyshev or Legendre polynomials, Chebyshev Galerkin and Legendre Galerkin methods can be developed. For example, the following set of trial functions:

$$\phi_k(x) = \begin{cases} T_k(x) - T_0(x) \qquad \text{or} \qquad L_k(x) - L_0(x) \text{ for even } k \\ T_k(x) - T_1(x) \qquad\qquad\; L_k(x) - L_1(x) \text{ for odd } k \end{cases} \tag{6.4.10}$$

can be used to implement Dirichlet boundary condition. Similar linear combinations can be worked out to implement other boundary conditions as well. However, as we will see below the tau method is better suited for non-periodic problems and therefore we proceed to it now.

6.4.2 Tau Methods

The primary difference between the Galerkin and tau methods is that in the latter each individual trial function is not required to satisfy the boundary conditions. Instead, boundary conditions will be explicitly enforced in the evaluation of the expansion coefficients. Since the Fourier modes automatically satisfy the periodicity condition, Fourier–Galerkin and Fourier–tau methods are one and the same. The real use of the tau method is for non-periodic problems, since it avoids the need for a special expansion basis that automatically satisfies the boundary conditions. Thus, Chebyshev–tau and Legendre–tau methods can be used with a variety of non-periodic boundary conditions.

Consider boundary conditions of the form

$$\mathbf{B}_1(u) = \alpha \quad \text{at} \quad x = -1 \qquad \text{and} \qquad \mathbf{B}_2(u) = \beta \quad \text{at} \quad x = 1, \tag{6.4.11}$$

where $\mathbf{B}_1$ and $\mathbf{B}_2$ are linear operators. In the tau method these boundary conditions are enforced as

$$\mathbf{B}_1\left(\sum_k \hat{u}_k\, \phi_k(x)\right) = \alpha \text{ at } x = -1 \quad \text{and} \quad \mathbf{B}_2\left(\sum_k \hat{u}_k\, \phi_k(x)\right) = \beta \text{ at } x = 1, \tag{6.4.12}$$

which provide two relations to be satisfied by the expansion coefficients $\hat{u}_k$. For example, Dirichlet–Dirichlet boundary conditions, where $\mathbf{B}_1$ and $\mathbf{B}_2$ are identity operators, yield the following two relations (see (6.3.9c) and (6.3.17c)):

$$\sum_k \hat{u}_k = \alpha \quad \text{and} \quad \sum_k (-1)^k \hat{u}_k = \beta\,. \tag{6.4.13}$$

As in the Galerkin method, the residue in the discretized ODE (6.4.6) is required to be orthogonal to the trial functions. However, since there are two equations (6.4.13) that arise from the boundary conditions already relating the $N+1$ expansion coefficients, the residue need be orthogonal only to the first $N-1$ trial functions. Thus (6.4.12) (or (6.4.13) in the case of Dirichlet–Dirichlet boundary conditions) will be supplemented with

$$\hat{b}_k = \hat{f}_k \quad \text{for} \quad k = 0, 1, \ldots, N-2 \tag{6.4.14}$$

for the evaluation of the expansion coefficients. From (6.4.3), for Dirichlet–Dirichlet boundary conditions we can explicitly write

$$\begin{bmatrix} \mathcal{L}_{00} & \mathcal{L}_{01} & \mathcal{L}_{02} & \cdots & \mathcal{L}_{0N} \\ \mathcal{L}_{10} & \mathcal{L}_{11} & \mathcal{L}_{12} & \cdots & \mathcal{L}_{1N} \\ \vdots & \vdots & \vdots & & \vdots \\ \mathcal{L}_{N-2,0} & \mathcal{L}_{N-2,1} & \mathcal{L}_{N-2,2} & \cdots & \mathcal{L}_{N-2,N} \\ 1 & 1 & 1 & \cdots & 1 \\ 1 & -1 & 1 & \cdots & (-1)^N \end{bmatrix} \begin{bmatrix} \hat{u}_0 \\ \hat{u}_1 \\ \hat{u}_2 \\ \hat{u}_3 \\ \vdots \\ \hat{u}_N \end{bmatrix} = \begin{bmatrix} \hat{f}_1 \\ \hat{f}_2 \\ \vdots \\ \hat{f}_{N-2} \\ \alpha \\ \beta \end{bmatrix}. \tag{6.4.15}$$

From (6.4.3) and from the definitions of $\mathcal{D}_{C1}$ and $\mathcal{D}_{C2}$ in (6.3.33) and (6.3.35) the linear operator $\mathcal{L}_{kl}$ is upper tridiagonal. However, the last two lines that arise from the boundary conditions make the solution of the above linear system more involved.

6.4.3 Collocation Methods

By requiring that the residue be orthogonal to the trial functions, in some sense, the Galerkin and tau methods enforce the governing ODE in the spectral space. In contrast, in the collocation method the ODE is satisfied at a set of grid points. This is accomplished by requiring that the residue be zero at the grid points, which can be expressed similar to (6.4.7) as

$$\int R(x)\,\delta(x - x_j)\,\mathrm{d}x = 0, \tag{6.4.16}$$

where x_j are the grid points. The optimum distribution of grid points for Fourier collocation is equi-spaced as given in (6.2.10). In the case of non-periodic problems the choice can be either Gauss or Gauss–Lobatto points. However, Gauss–Lobatto points are the usual choice, since boundary conditions are to be enforced.

Equation (6.4.16) simply states that the governing equation (6.4.1) is satisfied in a discrete sense at the grid points. This can be accomplished by obtaining a discretized version of the continuous operator $\mathbf{L}$, which when operates on the discretized function $u(x_j)$ will yield $\mathbf{L}u$ at the grid points. The grid point values of $\mathbf{L}u$ can then be equated to the right-hand side f evaluated at the grid points. In other words,

$$\sum_{j=0}^{N} [L]_{ij}\, u_j = f_j. \tag{6.4.17}$$

For the linear constant coefficient operator in (6.4.1) the corresponding discrete approximation can be written as

$$[L]_{ij} = c_2\,[D_2]_{kl} + c_1\,[D_1]_{kl} + c_0\delta_{kl}. \tag{6.4.18}$$

Here $\mathbf{D}_1$ and $\mathbf{D}_2$ are the first- and second-derivative operators. In the case of Fourier trial functions they are $\mathbf{D}_{\mathrm{F}1}$ and $\mathbf{D}_{\mathrm{F}2}$ defined in (6.2.23) and (6.2.25), and in the case of Chebyshev trial functions they are $\mathbf{D}_{\mathrm{C}1}$ and $\mathbf{D}_{\mathrm{C}2}$ defined in (6.3.38) and (6.3.39). Thus, the collocation method is similar to the boundary value and eigenvalue approaches outlined in Chapter 5. The only difference is the Fourier or Chebyshev first- and second-derivative matrices will be used in place of finite difference operators. In the case of periodic problems no further consideration regarding boundary conditions is necessary. In the case of non-periodic problems, Dirichlet and Neumann boundary conditions will be enforced very similarly to how they were incorporated for the finite difference approach explained in Chapter 5.

6.4.4 Galerkin and Tau Methods for Nonlinear Equations

Now that we have learnt three different methods for solving linear constant coefficient ODEs using spectral methods, let us proceed to consider nonlinear equations. We add to the linear terms on the left-hand side of (6.4.1) a simple quadratic nonlinear term $u(x)v(x)$. Here $v(x)$ can be $u(x)$ or (du/dx) and thus correspond to quadratic nonlinearity, or can be a variable coefficient dependent on x. Irrespective, we will see here how to handle such products of functions of x, first in the context of the Fourier–Galerkin method for periodic problems. Similar to (6.2.2) $v(x)$ can be expanded in a Fourier series as

$$v(x) = \sum_{k=-\infty}^{\infty} \hat{v}_k \exp\left[ik\frac{2\pi}{L_x}x\right]. \tag{6.4.19}$$

Fourier coefficients of the product,

$$\hat{a}_k = \frac{1}{L_x}\int_0^{L_x} u(x)\,v(x)\,\exp\left[-ik\frac{2\pi}{L_x}x\right]\,\mathrm{dx}, \tag{6.4.20}$$

can be expressed as the **convolution sum** of the Fourier coefficients of $u(x)$ and $v(x)$ as

$$\hat{a}_k = \sum_{l+m=k} \hat{u}_l \, \hat{v}_m \quad \text{for} \quad k = 0, \pm 1, \pm 2, \ldots, \tag{6.4.21}$$

where summation is over all possible values of l and m.

The numerical implementation of the Galerkin method will involve an N-mode truncation for $k = -N/2, \ldots, N/2 - 1$. The discretized version of (6.4.1) with the nonlinear term on the left-hand side becomes

$$\sum_{l=-N/2}^{N/2-1} \mathcal{N}_{kl} \hat{u} = \sum_{l=-N/2}^{N/2+1} \left[c_2 \left(\mathcal{D}_2 \right)_{kl} + c_1 \left(\mathcal{D}_1 \right)_{kl} + c_0 \delta_{kl} + \hat{v}_{k-l} \right] \hat{u}_l = \hat{f}_k. \tag{6.4.22}$$

Here $\mathcal{N}_{kl}$ represents the discretized nonlinear operator in spectral space. It can be readily recognized that the nonlinear term complicates the computational task. The simplicity of the Fourier–Galerkin method for linear problems was due to the diagonal nature of the operators $\mathcal{D}_1$ and $\mathcal{D}_2$. Unfortunately for any non-constant periodic function $v(x)$, the term $\hat{v}_{k-l}$ in (6.4.22) will not be diagonal. In fact, for an arbitrary function $v(x)$ the matrix operator $\mathcal{N}_{kl}$ will be full and a dense matrix solver is required.

If $v(x)$ is not a variable coefficient specified in the problem, but is a linear function of $u(x)$ or its derivative, then the term $\sum_{l=-N/2}^{N/2+1} \hat{v}_{k-l} \, \hat{u}_l$ in (6.4.22) is quadratic in the unknown Fourier coefficients. Thus, a nonlinear system must be solved for the solution.

There is a more fundamental difference in the accuracy of results obtained for linear and nonlinear problems. In a linear problem the solution $\hat{u}_l$ obtained by solving (6.4.9) is *exact* for the N modes ($k = -N/2, \ldots, N/2 - 1$). The only error is in the neglected higher modes. This is true even in the case where $v(x)$ in the nonlinear term is just a variable coefficient, provided we use $2N$ Fourier coefficients ($\hat{v}_k$ for $k = -N, \ldots, N - 1$) in evaluating the term $\sum_{l=-N/2}^{N/2+1} \hat{v}_{k-l} \, \hat{u}_l$ in (6.4.22). This way all N Fourier coefficients of $u(x)$ are accurately computed without any truncation or approximation.

In a nonlinear problem, if the Fourier expansion of $u(x)$ is truncated to N modes, the exact evaluation of the nonlinear term will require higher modes that are not being computed. Thus, the Galerkin solution of a nonlinear equation will involve unavoidable error even in the modes being computed. However, it can be shown that this error is of the same order as the truncation error and as a result if we include sufficient number of modes in the expansion, this error can be kept below any desired value. The details of the proof can be found for example in Canuto *et al.* (2006).

Evaluation of the N Fourier coefficients of the quadratic term by the convolution sum (6.4.21) requires $O\left(N^2\right)$ floating point operations as represented by the summation. The convolution can be extended to higher nonlinearities. For example, an expansion similar to (6.4.21) for the Fourier coefficients of a cubic term will involve double summation and require $O\left(N^3\right)$ floating point operations and thus will be computationally even more involved. As the nonlinearity of the equation increases, its Galerkin representation becomes more and more complicated.

For nonlinear equations in non-periodic domains, Chebyshev or Legendre expansions must be used. Furthermore, the boundary conditions can be more easily implemented with the tau methodology. In the tau method, as presented in (6.4.15), two of the relations satisfy the boundary conditions. The other relations are used to eliminate the residue along the $N-2$ modes, where the complexity of handling the nonlinear terms remains the same as in the Galerkin method. As a result, in general, it is not practical to consider Galerkin or tau methodology for problems higher than quadratic nonlinearity.

6.4.5 Pseudo-spectral Methods

In the Galerkin and tau methods the governing equations are solved in spectral space, while in the collocation method the equations are solved at grid points (i.e., in real space). Both these contrasting approaches have their definite advantages. As discussed in Sections 6.2.2 and 6.3.4, computing derivatives in spectral space involves multiplication with a diagonal or an upper diagonal matrix, in the case of Fourier or polynomial expansions, respectively. In contrast, evaluating derivatives at the grid points in real space requires multiplication with a dense matrix. Thus, it is computationally easier to evaluate derivatives in spectral space than in real space.

However, as discussed above in Section 6.4.4, evaluating the nonlinear terms in spectral space becomes increasingly cumbersome with increasing degree of nonlinearity. In contrast, evaluating nonlinear terms such as $u(x)v(x)$ is straightforward in real space, as it only requires multiplication of u_j and v_j at each grid point. Even exponential nonlinearities such as e^u can be easily computed as e^{u_j} locally at each grid point, while it is extremely difficult to represent, let alone evaluate, such nonlinear terms in spectral space.

By combining the advantages of real and spectral space representations, Orszag proposed an ingenious idea that simplifies the application of Galerkin and tau approaches to nonlinear problems. This new approach is carried out still in spectral space, except when a nonlinear product is to be evaluated the individual components are first transformed to the grid points, where the nonlinear product is evaluated easily in real space, which is then transformed to the spectral space and the computation proceeds. Orszag coined the term **pseudo-spectral method** for this approach. Pseudo-spectral implementation of the quadratic nonlinear product $u(x)v(x)$, in place of (6.4.21), is shown below:

$$
\begin{array}{lll}
\hat{u}_l \xrightarrow[\text{Transform}]{\text{Inverse}} u(x_j) & & \\
 & \searrow & \\
 & & u(x_j)v(x_j) \xrightarrow[\text{Transform}]{\text{Forward}} \tilde{a}_k . \\
 & \nearrow & \\
\hat{v}_m \xrightarrow[\text{Transform}]{\text{Inverse}} v(x_j) & &
\end{array}
\tag{6.4.23}
$$

A similar approach can be followed for even higher nonlinear terms.

Since grid points are defined and used in the evaluation of nonlinear terms, the pseudo-spectral method is in some sense related to the collocation method. In fact it can be rigorously shown that for most fluid mechanical problems involving periodic boundaries, Fourier pseudo-spectral and Fourier collocation methods are algebraically equivalent. However, in non-periodic problems polynomial pseudo-spectral and polynomial collocation methods are not equivalent, but are closely related. For details the reader should refer to Canuto *et al.* (2006).

6.4.6 Aliasing Error and De-aliasing Techniques

Note that in (6.4.23) the tilde in the Fourier coefficients $\tilde{a}_k$ distinguish them from those defined in (6.4.20). They are different because of aliasing errors. To see this, let us write out the steps presented in (6.4.23) and compare the outcome with (6.4.21). First, the inverse transforms can be written as

$$u_j = \sum_{l=-N/2}^{N/2-1} \hat{u}_l \exp\left[il\frac{2\pi}{L_x}x_j\right] \quad \text{and} \quad v_j = \sum_{m=-N/2}^{N/2-1} \hat{v}_m \exp\left[im\frac{2\pi}{L_x}x_j\right]. \tag{6.4.24}$$

Then the forward transform of the nonlinear product can be expressed as

$$\tilde{a}_k = \frac{1}{N}\sum_{j=0}^{N-1} (u_j v_j) \exp\left[-ik\frac{2\pi}{L_x}x_j\right]. \tag{6.4.25}$$

Substituting (6.4.24) into (6.4.25) and using the orthogonality property of the exponentials it can be shown that

$$\tilde{a}_k = \sum_{l+m=k} \hat{u}_l\,\hat{v}_m + \sum_{l+m=k\pm N} \hat{u}_l\,\hat{v}_m = \hat{a}_k + \sum_{l+m=k\pm N} \hat{u}_l\,\hat{v}_m. \tag{6.4.26}$$

When compared with (6.4.21) the second term on the right can be identified as the aliasing error. The similarity between this and the aliasing error in the context of Fourier interpolation is clear: see (6.2.12).

Problem 6.4.1 Follow the steps presented above and obtain (6.4.26).

Since Fourier collocation and Fourier pseudo-spectral methods are equivalent, Fourier collocation method for nonlinear equations also suffers from the aliasing error in the evaluation of the nonlinear terms. Again, the aliasing error can be shown to be of the same order as the truncation error. Thus, provided sufficient number of modes are used in the pseudo-spectral method or sufficient number of grid points are used in the collocation method, the aliasing error will remain small. However, if the aliasing error can be eliminated then the more efficient pseudo-spectral and collocation methods can be used to solve nonlinear problems,

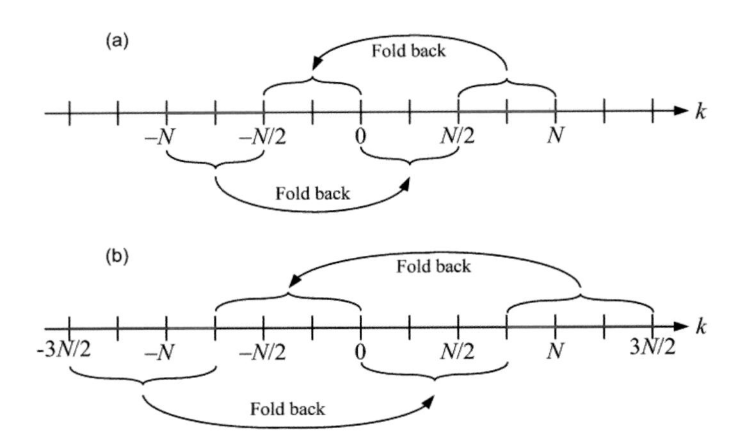

Figure 6.3 Schematic illustrating aliasing error due to quadratic nonlinear products. (a) Higher modes fold back to corrupt the main computed N modes ($k = -N/2, \ldots, N/2 - 1$). (b) 3/2-rule de-aliasing technique, where the modes that fold back to the computed N modes ($k = -N/2, \ldots, N/2 - 1$) are identically zero (marked in blue).

without the penalty of aliasing error. In particular, in the context of Navier–Stokes equations, to address the quadratic nonlinearity, few different **de-aliasing techniques** have been developed.

To devise an effective de-aliasing strategy let us first look at the source of aliasing. The quadratic product can be followed on the wavenumber line shown in Figure 6.3. Starting with $\hat{u}_l$ and $\hat{v}_m$ for the N modes ($l, m = -N/2, \ldots, N/2-1$), the coefficients of the quadratic product $\hat{a}_k$ given by the convolution sum (6.4.21) expand the wavenumber spectrum to $2N$ modes ($k = -N, \ldots, N - 1$). As given in (6.4.26), when the quadratic product is evaluated using the pseudo-spectral method, the modes that are "outside" the main N modes, i.e., ($k = -N, \ldots, -N/2 - 1$) and ($k = N/2, \ldots, N - 1$) "fold back" to corrupt the main modes with aliasing error: see Figure 6.3.1(a). This understanding leads to a simple de-aliasing strategy where the fold back of the nonlinearly expanded modes is pushed outside the main N modes of interest. This strategy is called the **3/2 de-aliasing rule**, and it can be written in terms of the following steps:

$$
\begin{array}{l}
\hat{u}_l \xrightarrow{\text{Expand}} \bar{\hat{u}}_l \xrightarrow[\text{Transform}]{\text{Inverse}} \bar{u}(x_j) \\
\qquad\qquad\qquad\qquad\qquad \searrow \\
\qquad\qquad\qquad\qquad\qquad\qquad \bar{u}(x_j)\,\bar{v}(x_j) \xrightarrow[\text{Transform}]{\text{Forward}} \bar{\hat{a}}_k. \\
\qquad\qquad\qquad\qquad\qquad \nearrow \\
\hat{v}_m \xrightarrow{\text{Expand}} \bar{\hat{v}}_m \xrightarrow[\text{Transform}]{\text{Inverse}} \bar{v}(x_j)
\end{array}
\tag{6.4.27}
$$

Here the expanded Fourier coefficients $\bar{\hat{u}}_l$ and $\bar{\hat{v}}_m$ are defined for $M = 3N/2$ modes i.e., ($l, m = -3N/4, \ldots, 3N/4 - 1$) by setting them equal to $\hat{u}_l$ and $\hat{v}_m$

for the inner N modes and the outside $N/2$ modes are set to zero. That is,

$$\bar{\hat{u}}_l = \begin{cases} \hat{u}_l & l = -\frac{N}{2}, \ldots, \frac{N}{2} - 1, \\ 0 & l < -\frac{N}{2}, \quad l \geq \frac{N}{2} \end{cases} \quad \text{and } \bar{\hat{v}}_m = \begin{cases} \hat{v}_m & m = -\frac{N}{2}, \ldots, \frac{N}{2} - 1, \\ 0 & m < -\frac{N}{2}, \ m \geq \frac{N}{2}. \end{cases} \tag{6.4.28}$$

The expanded Fourier coefficients are inverse Fourier transformed to M equispaced grid points $\overline{x}_j = jL_x/M$ for $j = 0, 1, \ldots, M-1$:

$$\overline{u}_j = \sum_{l=-M/2}^{M/2-1} \bar{\hat{u}}_l \exp\left[il\frac{2\pi}{L_x}\overline{x}_j\right] \quad \text{and} \quad \overline{v}_j = \sum_{m=-M/2}^{M/2-1} \bar{\hat{v}}_m \exp\left[im\frac{2\pi}{L_x}\overline{x}_j\right]. \tag{6.4.29}$$

The nonlinear product is then carried out at each of the dense grid points $\overline{x}_j$, which then can be forward transformed to obtain the M Fourier coefficients of the quadratic product as

$$\bar{\hat{a}}_k = \frac{1}{M}\sum_{j=0}^{M-1} (\overline{u}_j\overline{v}_j) \exp\left[-ik\frac{2\pi}{L_x}\overline{x}_j\right]. \tag{6.4.30}$$

Substituting (6.4.29) into (6.4.30) and using the orthogonality property of the exponentials we obtain

$$\bar{\hat{a}}_k = \sum_{l+m=k} \bar{\hat{u}}_l \, \bar{\hat{v}}_m + \sum_{l+m=k\pm M} \bar{\hat{u}}_l \, \bar{\hat{v}}_m. \tag{6.4.31}$$

The second term on the right is the aliasing error, but for the M modes. Comparing frames (a) and (b) in Figure 6.3, it is clear that the aliasing error for the expanded M modes follows the same pattern as the original N modes. In other words, modes $(k = -M, \ldots, -M/2)$ fold back to modes $(k = 0, \ldots, M/2)$; and modes $(k = M/2, \ldots, M)$ fold back to $(k = -M/2, \ldots, 0)$. However, since we started with only N mode expansion for u and v their quadratic product is nonzero only for the inner $2N$ modes $(k = -N, \ldots, N-1)$. Thus, for $M = 3N/2$ a large fraction of modes that fold back is zero and they are marked blue in Figure 6.3(b). In particular, when compared to (6.4.21), it can be readily seen that

$$\bar{\hat{a}}_k = \hat{a}_k \quad \text{for } (k = -N/2, \ldots, N/2-1). \tag{6.4.32}$$

Thus, when (6.4.30) is used for the N inner modes of interest, namely $k = -N/2, \ldots, N/2-1$, the 3/2-rule eliminates the aliasing error in pseudo-spectral evaluation of quadratic nonlinearities.

The 3/2-rule can also be used in the collocation scheme for evaluating quadratic products without aliasing error. Here instead of multiplying u_j and v_j at each of the N grid points, they are first individually transformed to spectral space, followed by the de-aliasing steps outlined in (6.4.27), the resulting $\bar{\hat{a}}_k$ is truncated to N modes and finally transformed back to the real space. The resulting product u_jv_j will be free of aliasing error at the N grid points.

There are other techniques for de-aliasing, such as method of phase shifts and $\frac{1}{2}$-mode truncation. Also the de-aliasing steps presented in (6.4.27) can be applied to polynomial expansions. De-aliased Chebyshev and Legendre pseudo-spectral methods can thus be developed. The de-aliasing ideas presented above can also be extended to cubic and higher-order nonlinearities, although at additional computational cost. The reader is again referred to textbooks devoted to spectral methods.

6.4.7 Fast Fourier Transforms (FFT)

With the pseudo-spectral scheme we see the need to go from the spectral coefficients to the grid point values and from the grid point values to the Fourier coefficients. In other words we need to carry out the forward and backward transforms given in (6.2.11) and (6.2.15). Therefore speed at which these transforms are performed becomes important. We can greatly accelerate the computation by pre-computing and storing the weights

$$W^{kj} = \exp\left[-ik\frac{2\pi j}{N}\right]. \tag{6.4.33}$$

Even with this simplification, the forward and backward Fourier transforms involve matrix–vector multiplications. In other words, $O(N^2)$ floating point operations are required. The key ingredient that initially made pseudo-spectral methodology very attractive is the discovery that a very efficient technique exists for the calculation of the transforms (6.2.11) and (6.2.15). It is known as the **Fast Fourier Transform**, abbreviated FFT.

There are many texts and articles that treat this neat algorithm. It has a curious history. The algorithm is frequently associated with the names of J.W. Cooley and J.W. Tukey who published about it in the 1960s. Later it turned out that others had in fact discovered and implemented the method much earlier without bringing it to general attention. One such early discovery leads to what Press *et al.* (1986) call the Danielson–Lanczos lemma. This is the basic result that a discrete Fourier transform of length N can be replaced by two transforms of length $N/2$. The FFT algorithm results by doing this kind of reduction recursively.

The task at hand is to discover an efficient way of computing (6.2.11) and (6.2.15). The FFT allows one to reduce the number of floating point operations from $O(N^2)$ to $O(N \log N)$. Let us see how this comes about for (6.2.11) by rewriting it as follows:

$$\begin{aligned} N\tilde{u}_k &= \sum_{j=0}^{N-1} u(x_j)\, W^{kj} \\ &= \sum_{\substack{j=0 \\ j=\text{ even}}}^{N-1} u(x_j)\, W^{kj} + \sum_{\substack{j=0 \\ j=\text{ odd}}}^{N-1} u(x_j)\, W^{kj} \\ &= \sum_{j'=0}^{N/2-1} u(x_{2j'})\, W^{k2j'} + \sum_{j'=0}^{N/2-1} u(x_{2j'+1})\, W^{k(2j'+1)} \end{aligned}$$

$$= \left[\sum_{j'=0}^{N/2-1} u(x_{2j'}) \, W^{2kj'} \right] + W^k \left[\sum_{j'=0}^{N/2-1} u(x_{2j'+1}) \, W^{2kj'} \right] . \quad (6.4.34)$$

For simplicity here we have assumed N to be even. The last line contains two discrete Fourier transforms of the data at the odd and even grid points. The W-factors needed in a transform of length $N/2$ is just the square of the corresponding factor in a transform of length N. Changing N to $N/2$ in $W = \exp(-2\pi\iota kj/N)$ produces W^2. As a result the terms within the square parentheses in the last equation are Fourier transforms of length $N/2$ (except for a scaling prefactor).

Now we address a subtle point in the implementation of the above strategy. Although k ranged from $-N/2$ to $(N/2) - 1$ when we started with the first step in (6.4.34), as we get to the two DFTs of length $N/2$ in the final step, the corresponding $k = -N/4, \ldots, (N/4) - 1$. So how do we obtain $\tilde{u}_k$ for the remaining k? If we set $k = m + N/2$, and use the fact that $W^{kj} = -W^{mj}$ we obtain

$$N\tilde{u}_{m+N/2} = \left[\sum_{j'=0}^{N/2-1} u(x_{2j'}) \, W^{2mj'} \right] - W^m \left[\sum_{j'=0}^{N/2-1} u(x_{2j'+1}) \, W^{2mj'} \right] \quad (6.4.35)$$

and similarly for $k = m - N/2$. Thus, one long Fourier transform has been divided into two smaller transforms of half the length. If we take the number of floating point operations for the matrix–vector multiplication of the undivided Fourier transform to be cN^2, where C is an $O(1)$ constant, the corresponding cost of the two smaller transforms of half the length is $cN^2/2$. In (6.4.34) the cost of additional tasks of scaling and recombining the two smaller ones to make the desired transform is of $O(N)$. But we can be more effective.

Assume, in particular, that N is a power of 2, say $N = 2^h$. Then we can iterate the procedure above in the following way: we do the transform on an array of length 2^h by splitting it into two transforms on arrays of length 2^{h-1}. Now, instead of actually doing the two transforms of length 2^{h-1}, we split each of them into two transforms of length 2^{h-2}. There are two pairs to recombine and each recombination will take $O(2^{h-2})$ operations. That is, the total number of floating point operations for recombination is again of $O(N)$. You keep proceeding in this fashion. At step p you have 2^p transforms of length 2^{h-p}. Recombining terms from two of these takes $O(2^{h-p})$ operations, and there are 2^{p-1} pairs to recombine. Thus, at each step you use $O(N)$ operations for recombination. You keep this up until you reach the step $(\log_2 N)$, where you have to do 2^h transforms of length 1. The transforms themselves are trivial; only the recombination of terms incurs an operation count, and again it is $O(N)$. Finally, you have reduced the calculation to $(\log_2 N)$ steps each consisting of $O(N)$ operations. Hence, the total operation count is $O(N \log N)$ as advertised.

To fix the ideas consider Table 6.1 where the schematics of the repeated splitting in an FFT on eight points has been written out. The leftmost column gives the index in binary; the second column gives the original data. After the first

Table 6.1 Schematic of the FFT algorithm on points.

Index (binary)	Original data	First split	Second split	Third split	Index (binary)
000	u_0	u_0	u_0	u_0	000
001	u_1	u_2	u_4	u_4	100
010	u_2	u_4	u_2	u_2	010
011	u_3	u_6	u_6	u_6	110
100	u_4	u_1	u_1	u_1	001
101	u_5	u_3	u_5	u_5	101
110	u_6	u_5	u_3	u_3	011
111	u_7	u_7	u_7	u_7	111

split we have the even indices in a smaller array at the top, the odd numbered data in a smaller array at the bottom. The two arrays are separated by a horizontal line. The second split produces four arrays of length 2. The final split produces eight arrays of length 1. By consistently writing the "even arrays" first we see that the net result from the leftmost column to the rightmost is to shuffle the data. A little bit of scrutiny shows that the rule for this shuffle is that the index in binary has been bit reversed. So, as described here, the FFT algorithm consists in a reordering of the data by bit reversal followed by recombination. However, you could also do the converse. You could do the recombination first, and then reorder later. Or, in problems where the input data and the output data are both wanted in real space, and two transforms are to be done, you can omit the reordering entirely!

There are many additional technical details regarding the FFT for which you are referred to the literature. Modified algorithms exist for providing fast transforms for data that is not a power of 2 in length. Very efficient FFT algorithms exist for arrays of length $N = 2^p 3^q 5^r$. The overall divide and conquer strategy remains the same. The data vector to be Fourier transformed is divided into p smaller transforms of length 2, q transforms of length 3 and r transforms of length 5. The overall cost of the transform still scales as $N \log N$. Variations exist for treating real or complex data.

Finally, Chebyshev expansion enjoys a great advantage among all polynomial expansions. Chebyshev transforms can be rewritten in terms of Fourier cosine transforms. This allows the use of FFT for Chebyshev transforms. As a result, the operations count of Chebyshev transforms also scale as $N \log N$. What we have said may suffice to give you the gist of the method. You can usually find a good implementation of the FFT in whatever computer language you are using.

6.4.8 Resolution and Computational Efficiency

The resolution of the Galerkin, tau and pseudo-spectral methods depends on how many modes we keep, while the resolution of the collocation method depends on the number of grid points. Of course if we need to transform back and forth between the grid points and the modes, their counts must be equal. Let us now inquire about the computational price of these methods.

In the case of linear equations, the Galerkin and tau methods can be implemented entirely in the spectral space, without the need for convolution sum or pseudo-spectral methods. The number of floating point operations scale only as $O(N)$ and therefore Galerkin and tau methods are highly accurate and extremely fast for linear problems.

For nonlinear problems the option is between pseudo-spectral and collocation methods. First, let us consider these methods without the de-aliasing step. In the pseudo-spectral method the forward and backward transforms shown of (6.4.23) are computationally the most demanding steps. With the use of FFT the number of floating point operations for these transforms scale as $O(N \log N)$. The operation counts of all other steps scale as $O(N)$ and therefore can be ignored in comparison.

In the collocation method the first- and second- derivative operators are dense (full) matrices. Due to this global nature of spectral approximation in real space, derivative at any grid point is dependent on the function value at all other points. Thus, the operation counts of computing derivatives scales as $O(N^2)$. If we additionally consider steps involved in de-aliasing of the nonlinear terms, the operation count of the forward and backward transforms of the expanded data scale as $O(M \log M)$. Since $M = 3N/2$, it can be seen that the overall scaling of floating point operations is not altered by de-aliasing.

It can thus be concluded that in terms of total number of floating point operations the pseudo-spectral method is more economical than the collocation approach. However, this advantage does not always materialize in terms of computational time. While the shuffling and recombination steps of FFT operate on more complex patterns of data, the matrix–vector multiplication of the collocation method operates on contiguous data set. As a result, for modest values of N, matrix–vector multiplication is often computationally faster than an FFT. Thus, the collocation method, for its simplicity in implementation, is often used in problems where N is modest.

The computational cost of collocation methods can be compared with that of finite difference schemes. The bandedness of the finite difference derivative matrices arises from the finite size stencil used in defining the derivatives. This difference between dense versus banded operators becomes important when we come to solve the resulting matrix equations. The computational cost of solving banded tridiagonal and pentadiagonal linear systems of size N scales as $O(b^2N)$, where b is the bandwidth. The operation count for solving corresponding dense systems scales as $O(N^3)$ (since bandwidth becomes N). Thus for the same number of points the collocation method will involve far more floating point operations than finite difference methods. Two factors greatly alleviate this disparity. First, since spectral approximations are far more accurate than finite difference approximations, for the same level of accuracy, far fewer points are required with the spectral approach. Second, the dense matrix–vector and matrix–matrix multiplication operations involved in the spectral method are often very highly optimized in most computers.

6.5 Some Examples

Here we consider examples in fluid mechanics where the spectral method is used to reduce the problem to a system of ODEs which can be solved easily using the time marching techniques of Chapter 3, or using the boundary value and eigenvalues techniques of Chapter 5. In fact all the computer exercises of Chapter 5 can be repeated with spectral methods.

You may be surprised to learn that we have already seen an application of spectral methods in Chapter 5. In the investigation of Rayleigh–Bénard convection in Section 5.5.1, we considered Fourier modes along the horizontal direction ($\exp(iax)$) in (5.5.6) and sinusoidal function in the vertical direction ($\sin(n\pi y)$ in (5.5.9)). For $-\infty < a < \infty$ and $n = 1, 2, \ldots$, these are the trial functions and they satisfy the stress-free boundary conditions at the top and bottom boundaries. Since the governing equations (5.5.1) are linear, each Fourier mode can be solved alone, independent of all other modes. Furthermore, due to the autonomous nature of the problem (the coefficients in (5.5.1) are independent of time) the time dependence of each Fourier mode reduces to simple exponent growth or decay. Thus, this is nothing but an application of the Fourier–Galerkin method for solving (5.5.1). In this case, the solution is the dispersion relation (5.5.11).

We now consider the Galerkin method applied to nonlinear Rayleigh–Bénard convection. Owing to nonlinearity, the time evolution of the solution will no longer be simple exponential growth or decay. As we will see below, the solution can be chaotic. The governing equations are again the Boussinesq equations and we follow the non-dimensionalization outlined in Section 5.5.1. One important difference is that the perturbation to the conduction state is not assumed to be small and thus we will not neglect the quadratic terms here. The resulting nonlinear equations for the perturbation are

$$
\begin{aligned}
\nabla \cdot \mathbf{u}' &= 0 \,, \\
\frac{\partial \mathbf{u}'}{\partial t} + \mathbf{u}' \cdot \nabla \mathbf{u}' &= -\nabla p' + \mathrm{Ra}\,\mathrm{Pr}\,\theta' \mathbf{j} + \mathrm{Pr}\ \nabla^2 \mathbf{u}' \,, \\
\frac{\partial \theta'}{\partial t} + \mathbf{u}' \cot \nabla \theta' &= v' + \nabla^2 \theta' \,.
\end{aligned}
\tag{6.5.1}
$$

The problem is first simplified by considering only a two-dimensional flow. As a result the z-velocity and z-variation of variables are neglected; i.e., $w' = 0$ and $\partial/\partial z = 0$. Again we choose the trial functions to be $\exp(i\,a\,x)$ along the horizontal direction and $\sin(n\,\pi\,y)$ or $\cos(n\,\pi\,y)$ along the vertical direction and the flow variables are expanded as follows:

$$\begin{aligned}
u'(x,z,t) &= \frac{1}{2\pi}\int_{-\infty}^{\infty}\left[\sum_{n=1}^{\infty}\hat{u}_n(t;a)\,\cos(n\pi y)\right]\exp(iax)\,\mathrm{d}a\,,\\
v'(x,z,t) &= \frac{1}{2\pi}\int_{-\infty}^{\infty}\left[\sum_{n=1}^{\infty}\hat{v}_n(t;a)\,\sin(n\pi y)\right]\exp(iax)\,\mathrm{d}a\,,\\
\theta'(x,z,t) &= \frac{1}{2\pi}\int_{-\infty}^{\infty}\left[\sum_{n=1}^{\infty}\hat{\theta}_n(t;a)\,\sin(n\pi y)\right]\exp(iax)\,\mathrm{d}a\,.
\end{aligned} \tag{6.5.2}$$

In order to obtain a numerical solution using the Galerkin method we need to limit attention to a finite number of test functions. In other words, we need to truncate the above expansions to only a few modes (each mode will be represented by the pair a and n).

For the linear problem (5.5.1) the time evolution of each mode is independent. Here the governing equation (6.5.1) is nonlinear and the time evolutions of the different modes are coupled. As a result the solution depends on how many modes are considered and their values of the pair a and n. A natural question then arises: what is the minimal truncation that retains the fewest number of trial functions and still allows us to explore the rich physics of thermal convection? Such a truncation was obtained in 1963 by E.N. Lorenz, who was a meteorologist and mathematician at MIT. In his model the Galerkin truncation retained only one velocity mode and two thermal modes, which can be written as

$$\left.\begin{aligned}
u'(x,z,t) &= -\frac{\sqrt{2}(a^2+\pi^2)}{a}A(t)\cos(\pi\,y)\sin(ax)\,,\\
v'(x,z,t) &= \frac{\sqrt{2}(a^2+\pi^2)}{\pi}A(t)\sin(\pi\,y)\cos(ax)\,,\\
\theta'(x,z,t) &= \frac{\sqrt{2}}{\pi\,\mathrm{Ra_c}}B(t)\sin(\pi\,y)\cos(ax) - \frac{\sqrt{2}}{\pi\,\mathrm{Ra_c}}C(t)\sin(2\pi y)\,,
\end{aligned}\right\} \tag{6.5.3}$$

where $A(t)$, $B(t)$ and $C(t)$ are the time-dependent amplitudes of the modes. Note that the full exponentials are not needed along the x-direction; sine and cosine functions are sufficient. The scaling factors multiplying the different modes are just to make the final ODEs simple.

It can be readily noted that the above truncated expansion satisfies the continuity equation. The above expansion is substituted into the momentum and energy equations (6.5.1). The nonlinear terms are carefully written out and coefficients of the $\sin(\pi\,y)$ and $\sin(2\,\pi\,y)$ modes are equated and upon further simplification we obtain the **Lorenz equations**

$$\left.\begin{aligned}\frac{\mathrm{d}A}{\mathrm{d}t} &= \mathrm{Pr}(B-A)\\ \frac{\mathrm{d}B}{\mathrm{d}t} &= rA - AC - B\\ \frac{\mathrm{d}C}{\mathrm{d}t} &= AB - bC\,,\end{aligned}\right\} \tag{6.5.4}$$

where Pr is the Prandtl number, $r = \mathrm{Ra}/\mathrm{Ra_c}$ is the ratio of Rayleigh number to the critical value for onset of convection and the constant $b = 4\pi^2/(\pi^2 + a^2)$. For stress-free top and bottom boundaries in Section 5.5.1 it was shown that $\mathrm{Ra_c} = 27\pi^4/4$ and if we use the critical value of $a_c = \pi/\sqrt{2}$ for the wavenumber, we obtain $b = 8/3$.

Problem 6.5.1 Derive the above ODEs by substituting (6.5.3) into (6.5.1).

The above system of three coupled nonlinear ODEs, along with the initial values $A(0)$, $B(0)$ and $C(0)$, can be solved using any of the time marching techniques presented in Chapter 3. But it is far more instructive to study the dynamical properties of the above equations. The fixed points of (6.5.4) can be obtained by setting $\mathrm{d}A/\mathrm{d}t = \mathrm{d}B/\mathrm{d}t = \mathrm{d}C/\mathrm{d}t = 0$. The three fixed points are

$$\mathrm{I}: A = B = C = 0, \quad \mathrm{II}\ \&\ \mathrm{III}: A = B = \pm\sqrt{b(r-1)},\ C = r - 1. \tag{6.5.5}$$

The stability of the fixed points can also be investigated using the methods presented in Chapter 2. For a given fluid, the Prandtl number and b are fixed and the only control variable is r. With increasing r three different regimes can be identified.

For $r < 1$, it is observed that only the fixed point I is stable, the other two fixed points are unstable. The fixed point I corresponds to conduction with no convective fluid motion. The fixed point stability analysis of Chapter 2 pertains only to small perturbations from the fixed point. However it can be shown that for Rayleigh numbers below the critical value (i.e., $r < 1$), the fixed point I is globally stable and that all perturbations will decay and the pure conduction state is stable.

Fixed points II and III correspond to steady convective roll cells and they differ only by half a wavelength shift in the location of the convective roll cells. It can be shown that both these fixed points are stable in the interval

$$1 < r < r_2 = \frac{\mathrm{Pr}\,(\mathrm{Pr} + b + 3)}{\mathrm{Pr} - b - 1}. \tag{6.5.6}$$

Thus, at $r = 1$ the flow undergoes a **pitchfork bifurcation** where one steady state gives way to another steady state. Over the Rayleigh number range given in

(6.5.6) steady convection will prevail and the amplitude of thermal perturbation increases as $(r-1)$, while the velocity perturbation goes as $\sqrt{r-1}$. Depending on the value of initial perturbation one of these two steady states will be reached.

For $r > r_2$ none of the three fixed points are stable. The solution exhibits a chaotic behavior, shown in Figure 1.1 back at the start of the book. That figure showed the time evolution of the three variables for $\mathrm{Pr} = 10$ and $r = 28$. It can be seen that the flow revolves around one of the two fixed points II and III and randomly jumps to revolve around the other and back again and so on. This picture is often referred to as the **Lorenz butterfly**. The Lorenz model and this chaotic behavior have profound implication on our understanding of chaotic systems and important problems such as weather prediction.

Computer Exercise 6.5.2 Solve (6.5.4) for $b = 8/3$, $\mathrm{Pr} = 10$ and for varying values of r. Observe the pitchfork bifurcation at $r = 1$ and for large values of r plot the chaotic trajectory of the solution.

Computer Exercise 6.5.3 Use the Chebyshev first- and second-derivative operators to obtain the solution of von Kármán flow over a rotating disk. You can use the same code as that you developed for Computer Exercise 5.3.5. The only difference is that you need to choose the Chebyshev Gauss–Lobatto distribution of points between the disk and the outer boundary (which can be chosen to be several boundary layer thickness) and use the Chebyshev first- and second-derivatives for this grid. Compare the numerical solution with that obtained with the finite difference approximation. Show that convergence can be obtained with far fewer points with the spectral methodology.

Computer Exercise 6.5.4 Repeat Computer Exercise 5.4.1 for the Graetz problem with the Chebyshev spectral methodology. Construct the boundary corrected left- and right-hand side matrices $\mathbf{L}^*$ and $\mathbf{R}^*$. Use library routines to solve either the generalized (5.4.15) or the standard eigenvalue (5.4.16) problem and compare the computed eigenvalues with those provided in Table 5.1. Plot the error in the first eigenvalue as a function of number of grid points and show that exponential convergence is achieved (i.e., error decreases as e^{-N}). Show that far fewer points are needed with the spectral methodology compared to the finite difference approximation. Repeat the above for the first few eigenvalues.

Computer Exercise 6.5.5 Repeat the hydrodynamic stability Computer Exercises 5.5.1–5.5.7 with the Chebyshev spectral methodology.

7 Partial Differential Equations

In this chapter we turn, finally, to subjects that are considered more mainstream CFD, viz., the variety of methods in use for discretizing partial differential equations (PDEs) on uniform grids in space and time, thus making them amenable to numerical computations. This is a very wide field on which many volumes have been (and will be) written. There are many different methods that have been advanced and promoted over the years. Many of the more well-established methods and schemes are named for their originators, although substantial leeway exists for improvements in detail and implementation. It is also not difficult to come away with feelings of helplessness and confusion concerning which method to choose and why. Our goal in this chapter will be to shed light on some of the guiding principles in constructing these methods.

A digression on semantics is in order. The words *method* and *scheme* are both in use for numerical procedures. It may be of interest to recall their precise meaning as given, for example, by *The Merriam–Webster Dictionary*:

Method: from the Greek *meta* (with) & *hodos* (way).
(1) a procedure or process for achieving an end;
(2) orderly arrangement; plan.

Scheme
(1) a plan for doing something; esp. a crafty plot.
(2) a systematic design.

Although both these terms seem to apply for numerical procedures, it would seem from these definitions that the term *method* is somewhat more desirable than scheme! But we have used these terms interchangeably, perhaps with a slight emphasis on methods, and we will continue to do so.

7.1 Definitions and Preliminaries

Before we begin to consider numerical methods for PDEs, here we start by defining and classifying the PDEs that we wish to solve. In a PDE the quantity being solved, such as fluid velocity or temperature, is a function of multiple independent variables, such as time and space (t and x) or multiple space directions (x and y). As mathematical entities, PDEs come in many different forms with varying properties, and they are a very rich topic in their own right. Here we will stay focused on only those classes of PDEs that we generally encounter in fluid mechanical applications.

If the problem being solved involves only one unknown quantity (or only one dependent variable) then we need to solve a **scalar PDE** for the unknown quantity. If the problem involves multiple quantities (for example, different components of velocity, pressure and temperature) then we need to solve a **system of PDEs**. The number of PDEs in the system must be equal to the number of quantities being solved. A system of PDEs is **linear** if in each term of each PDE all the unknown quantities and their spatial and temporal derivatives appear utmost at power 1. In other words, quadratic and higher products of the unknown quantities cannot appear in a linear system. Otherwise the system of PDEs will be termed **nonlinear**. Clearly, the Navier–Stokes equations represent a nonlinear system of PDEs for fluid velocity, pressure and temperature. However, before we plunge into numerical methods for Navier–Stokes equations in Chapter 8, we will start this chapter by considering numerical methods for a simple scalar linear PDE. The advantage of a linear PDE is that its solution can be defined precisely in an appropriate function space, and there is very well-developed theory of linear PDEs.

If all the terms in a PDE involve the unknown quantity being solved, i.e., if none of the terms are free of the dependent variable, then the PDE is said to be unforced or **homogeneous**. The solution of a homogeneous PDE depends only on the initial and boundary conditions. In other words, there are no external source terms.

A **constant coefficient** linear PDE is one in which the coefficients that multiply the unknown quantity or its derivatives, in each term of the PDE, are not a function of the independent variables. The coefficients cannot be a function of the dependent variable as well; otherwise the PDE will be nonlinear. If the coefficients are functions of the independent variables then we have a **variable coefficient** PDE. If the coefficients are independent of the time variable (t) then the PDE is termed **autonomous**. Thus an autonomous PDE is one that is constant coefficient in terms of the time variable.

It will come as no surprise that most progress in terms of numerical analysis is possible for constant coefficient linear PDEs. A second-order PDE can be classified according to the coefficients that multiply its highest derivatives into the three standard types: **elliptic**, **hyperbolic** and **parabolic**. Some PDEs do not fall into any of these categories, as they may change character between elliptic, hyperbolic and parabolic over the domain of the independent variables, and are therefore termed **mixed type**. However, we will not be concerned with mixed type PDEs. Certain features inherent in the different physics of these three types will reappear as we discuss the numerical methods, but the first impression is that numerical procedures are surprisingly insensitive to this important mathematical difference. Elliptic equations usually lead to boundary value problems.

Hyperbolic and parabolic equations lead to initial-boundary value problems. Hyperbolic problems usually have conservation properties.

To illustrate these points let us consider a linear homogeneous scalar PDE for one dependent variable, $u(x,t)$, in terms of the independent space (x) and time (t) variables. A class of PDEs can be expressed as

$$\frac{\partial u}{\partial t} = \mathbf{P}\left(x, t, \frac{\partial}{\partial x}\right) u, \tag{7.1.1}$$

where the operator $\mathbf{P}$ is a polynomial in the spatial derivative and thus has the form

$$\mathbf{P}\left(x, t, \frac{\partial}{\partial x}\right) = \sum_{l=0}^{p} a_l(x,t) \frac{\partial^l}{\partial x^l}. \tag{7.1.2}$$

In general, the order of the PDE depends on the highest spatial derivative and thus the above PDE is pth order. If the coefficients a_l are independent of t then the PDE is autonomous and if the coefficients a_l are independent of both x and t then the PDE is constant coefficient. Since u appears in every term only once (at power 1) the PDE is linear and homogeneous. Let us now look at some simple examples that we shall consider in greater detail later in this chapter.

Example 7.1.1 The **advection equation** is an example of a hyperbolic PDE and is given by

$$\mathbf{P}\left(x, t, \frac{\partial}{\partial x}\right) = -c\frac{\partial}{\partial x}, \tag{7.1.3}$$

yielding

$$\frac{\partial u}{\partial t} = -c\frac{\partial u}{\partial x}, \tag{7.1.4}$$

$$u(x, t=0) = f(x), \tag{7.1.5}$$

where the second equation is the initial condition. Here c is the speed of advection and the exact solution can be expressed as $u(x,t) = f(x-ct)$. If the speed of advection c is positive then the solution is advected to the right, i.e., along the positive x-direction. If the spatial domain is bounded, then for a right-traveling wave the appropriate boundary condition for u must be specified at the left boundary, i.e., at the inlet. A very important property of the advection equation is that it conserves energy and we shall see more about it in Section 7.1.1.

Example 7.1.2 The **diffusion equation** is an example of a parabolic PDE and is given by

$$\mathbf{P}\left(x, t, \frac{\partial}{\partial x}\right) = \nu\frac{\partial^2}{\partial x^2}, \tag{7.1.6}$$

yielding

$$\frac{\partial u}{\partial t} = \nu \frac{\partial^2 u}{\partial x^2}, \tag{7.1.7}$$

$$u(x, t = 0) = f(x). \tag{7.1.8}$$

Here ν is the diffusion coefficient, which is required to be non-negative. If u is fluid velocity or vorticity then ν is kinematic viscosity of the fluid. If u is temperature then ν is thermal diffusivity of the material, and so on. The diffusion equation is second-order and thus requires two boundary conditions both at the left and the right ends of the domain. A very important property of the diffusion equation is that it is dissipative and the total energy of the system monotonically decays over time and we shall see more about it in Section 7.1.1.

Example 7.1.3 The **advection–diffusion equation** is given by

$$\frac{\partial u}{\partial t} = -c \frac{\partial u}{\partial x} + \nu \frac{\partial^2 u}{\partial x^2}, \tag{7.1.9}$$

where the time evolution of u is governed by the combination of advection and diffusion effects. Here again an initial condition similar to (7.1.5) is required, along with boundary conditions at both the left and right ends of the spatial domain. Due to the presence of the diffusion term the above equation is dissipative.

Example 7.1.4 An important variant of the advection–diffusion equation is **Burgers' equation** given by

$$\frac{\partial u}{\partial t} = -u \frac{\partial u}{\partial x} + \nu \frac{\partial^2 u}{\partial x^2}. \tag{7.1.10}$$

Burgers' equation is a big favorite in elementary fluid mechanics because it has the right form and it looks somewhat like a one-dimensional version of the Navier–Stokes equation. Furthermore it can be solved analytically using Cole–Hopf transformation. It was introduced by J.M. Burgers in the 1940s. The key difference from the simple advection–diffusion equation is that instead of the constant advection speed c, in Burgers' equation the advection of momentum is at the local fluid velocity u. As a result Burgers' equation is nonlinear and gives rise to interesting new physics. If the diffusion effect is absent (i.e., $\nu = 0$) we obtain the **inviscid Burgers equation** whose important feature is that the solution develops shocks, which can be studied using kinematic wave theory.

As we can see from the above examples the general form given in equations (7.1.1) and (7.1.2) covers a number of PDEs of physical significance. However, this form is not all inclusive. There are other PDEs that cannot be readily cast in the form of equations (7.1.1) and (7.1.2). For example, consider the **wave equation**:

$$\frac{\partial^2 u}{\partial t^2} = c^2 \frac{\partial^2 u}{\partial x^2}, \tag{7.1.11}$$

where the second derivative in time makes it appear different from the ones considered above. The above wave equation can be written as two separate first-order equations as

$$\frac{\partial u}{\partial t} = \pm c \frac{\partial v}{\partial x} \quad \text{and} \quad \frac{\partial v}{\partial t} = \pm c \frac{\partial u}{\partial x}, \tag{7.1.12}$$

which together reduce to the wave equation. If we now consider u and v to be the components of a vector-valued function $\mathbf{u} = (u, v)$ then the above system of PDEs can be written in the form (7.1.1) for $\mathbf{u}$. Note that the wave equation resembles the advection equation, but with characteristics traveling both left and right at speed $\pm c$. As a result the wave equation has the same conservation properties as the advection equation.

Finally we consider the **Helmholtz equation**, which is an example of an elliptic PDE. The Helmholtz equation in two spatial dimensions is given by

$$\frac{\partial^2 u}{\partial x^2} + \frac{\partial^2 u}{\partial y^2} - \lambda u = g. \tag{7.1.13}$$

The above is a boundary value problem, as boundary conditions for u must be specified all around the domain within which the equation is being solved in the (x–y)-plane. Note that g on the right-hand side is the forcing and as a result the Helmholtz equation is inhomogeneous. As we will see later in the context of the time-splitting technique for the Navier–Stokes equation, Helmholtz' equation arises for the velocity components. In the above equation, if $\lambda = 0$ the resulting equation is commonly known as the **Poisson equation**, which again arises in the time-splitting technique as the equation that governs pressure.

7.1.1 Well-posedness

Of course, one has to understand the nature of what constitutes a well-posed problem for the different types of equations. Asking any method, numerical or otherwise, to solve a problem that is not well-posed, will surely lead to disaster regardless of the method. Before we seek numerical methods we should ask the following questions regarding properties of the PDE and the nature of the solution that we expect to obtain:

(1) Does a solution exist?
(2) If it exists, is it uniquely dependent on initial and boundary conditions?
(3) Will the solution remain smooth (smooth means continuous and continuously differentiable) if started from a smooth initial condition? Or will the solution or its derivatives develop discontinuities over time?

(4) Will the solution remain non-smooth if started from a non-smooth initial condition? Or will the solution become continuous and continuously differentiable?
(5) Will the solution remain bounded (or well defined)?

Now we shall discuss these questions a bit more. If a solution does not exist there is no point in seeking a numerical approximation. So we can move to the second question. In Section 5.1.3 we considered the steady one-dimensional heat conduction with adiabatic boundary conditions and observed the solution to be non-unique. To any solution of the equation, an arbitrary constant can be added and still be a solution. This is an example of an ODE whose solution cannot be uniquely determined. A multi-dimensional extension of this problem is steady two-dimensional or three-dimensional heat conduction with adiabatic boundary condition on all the sides. This problem yields a PDE with a non-unique solution.

In fact, in some sense the solution of the incompressible Navier–Stokes equation itself is non-unique. As you may recall, pressure appears only in the momentum equation as a spatial gradient. Thus, pressure can be obtained only up to an arbitrary additive constant. That is, once you obtain pressure you can add a constant to the pressure field everywhere and still have a good solution. However, such simple non-uniqueness has not stopped us from numerically solving the incompressible Navier–Stokes equation. The non-uniqueness can be removed by simply anchoring the pressure at any one point within the domain.

As we will see in the example of the inviscid Burgers equation, even when we start with a smooth initial condition, shocks, which are discontinuities, can develop in the solution. These discontinuities by themselves do not make the solution bad. In fact, these discontinuities are physically correct solutions of the governing equation. Any numerical method for solving such equations must be able to handle discontinuities. The PDE becomes inapplicable at the discontinuity (since $\partial/\partial x$ is not well defined) and an equivalent conservation law must be satisfied across the discontinuity. Such discontinuities will not arise in a viscous Burgers equation as the discontinuities will be smoothed by the diffusive effect of the viscous term. In fact, even if we start with a discontinuous initial condition the effect of the diffusion term will be to immediately smooth the solution. In this sense it is easier to simulate viscous flows than inviscid flows.

Finally, we come to the last question on boundedness. An important quantity that determines whether a solution remains bounded (well defined) or not is the energy of the system, appropriately defined in a suitable norm. The definitions of norm and energy depend on the physical problem being solved. Here we define energy of the solution as the L^2 norm of u, which then takes the form

$$E(t) = \int_{\Omega} |u(x,t)|^2 \, \mathrm{d}x \,, \tag{7.1.14}$$

where the integration is over the entire spatial domain. If the energy of the system $E(t)$ remains constant or decays then the PDE can be considered to be **strongly well-posed**. If the energy grows exponentially then the PDE is **ill-posed**. In between, if the energy grows, but only algebraically, the PDE can be considered to be **weakly well-posed**.

Now we can discuss the well-posedness of the examples considered above. In the advection and wave equations, energy is conserved and thus $E(t)$ remains a constant over time. The diffusion and Burgers equations are dissipative and as a result $E(t)$ strictly decays over time.

Note that well-posedness is a property of the PDE and is independent of the numerical methodology. In fact, the solution of the PDE being bounded is similar to the solution of a numerical method being stable. If we consider a well-posed PDE then it makes natural sense to require a stable numerical methodology to solve that PDE. Otherwise, the unstable numerical methodology will yield an erroneous solution, whose energy grows with time. The numerical solution thus will differ from the bounded behavior of the analytic solution.

Note that there are fluid mechanical problems, such as the hydrodynamic stability problem discussed in Section 5.5, where the energy of the perturbation can grow exponentially in time. In this sense the PDE that governs the growing perturbation is ill-posed. However, in Section 5.5 this was not an issue since the initial-value PDE was converted to an eigenvalue ODE system, before resorting to numerical methodology. An alternate approach will be to solve the linear PDE system directly without the normal mode *ansatz*. In this initial value approach the solution of the PDE will grow exponentially in time, when the perturbation is unstable. Thus, on occasions we may want to solve ill-posed PDEs, to learn something about the instability of a physical system. Even in such cases we should use stable numerical schemes. Clearly we only want to let the unstable disturbance grow in time, but not the noise that is present in the system. However, the methodology must not be too numerically dissipative and destroy the unstable nature of the physical instability. In any case, we will not pursue this line of reasoning further in this chapter. We will mainly focus on well-posed problems, as most of the CFD applications and simulations are concerned with stable systems.

7.2 The Advection Equation

Then, let us start with something that is supposedly quite simple, the equation for advection in one dimension given in (7.1.4) and (7.1.5). Here u is the advected field, and the speed of advection c will be taken as a constant. We have already studied much more complicated versions of this problem in three dimensions from a Lagrangian perspective, and we know that many intricate things can happen (see Section 2.7.1). But in one dimension the solution is simple and is given by $u(x,t) = f(x - ct)$ for any function f. If f has a discontinuity, we have a *shock wave*, and we say that $f(x - ct)$ is a weak solution. Not much of a challenge you may say, but let us proceed to discretize it anyway.

7.2.1 The FTCS Method

For simplicity let us consider a bounded domain of $0 \le x \le 2\pi$ with periodic boundary condition (i.e., $[\partial^p u/\partial x^p]_{x=0} = [\partial^p u/\partial x^p]_{x=2\pi}$ for all non-negative p). Let us fully discretize the advection equation, where time is represented by time intervals and space is discretized into grid points as

$$\begin{aligned} x_j &= j\Delta x \quad \text{for } j = 0, 1, \ldots, N-1, \\ t_n &= n\Delta t \quad \text{for } n = 0, 1, \ldots, \end{aligned} \tag{7.2.1}$$

where $\Delta x = 2\pi/N$. We will distinguish the numerical approximation from the exact solution and denote the numerical solution at the jth grid point at the nth time by $V_{j,n}$. First let us try the simplest approach that uses forward integration in time and three-point central difference in space, hence its name, the FTCS method:

$$\left(\frac{V_{j,n+1} - V_{j,n}}{\Delta t}\right) = -c\left(\frac{V_{j+1,n} - V_{j-1,n}}{2\Delta x}\right), \tag{7.2.2}$$

where the term within the parentheses on the left-hand side is the forward difference approximation to $\partial u/\partial t$ and the term within the parentheses on the right-hand side is the central difference approximation to $\partial u/\partial x$.

The space-time diagram, sometimes called the **stencil** of the FTCS method, is shown in Figure 7.1. The points (j, n) and $(j, n+1)$ are connected by a dashed line because they are connected via the time derivative. The points $(j-1, n)$ and $(j+1, n)$ are connected by a solid line because they are connected via the spatial derivative. As can be seen from the figure this is a three-point two-level method.

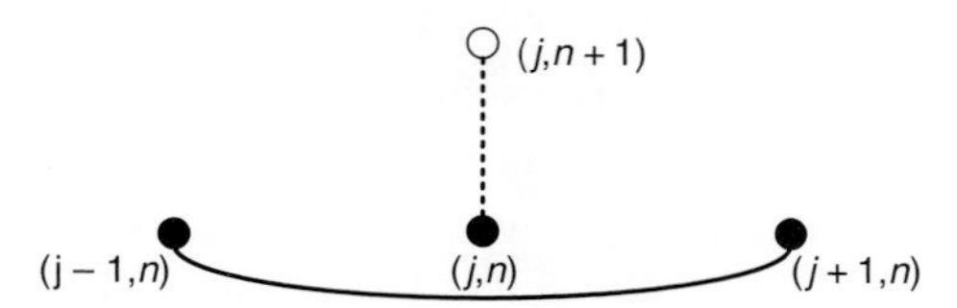

Figure 7.1 Space-time diagram of the FTCS method. The dashed line connects grid values related by the discretization of the time derivative. The solid line connects grid points related by the discretization of the spatial derivative.

The accuracy of the FTCS method can be established by investigating how well the exact solution satisfies the difference equation (7.2.2). The **truncation error** of the FTCS scheme is given by

$$\begin{aligned} \varepsilon_{j,n} = \; & \frac{u(j\Delta x, (n+1)\Delta t) - u(j\Delta x, n\Delta t)}{\Delta t} \\ & + c\frac{u((j+1)\Delta x, n\Delta t) - u((j-1)\Delta x, n\Delta t)}{2\Delta x}. \end{aligned} \tag{7.2.3}$$

Taylor expanding each term on the right-hand side about $j\Delta x$ and $n\Delta t$ we obtain the following expression for the truncation error:

$$\varepsilon_{j,n} = \left(\frac{\partial u}{\partial t}\right)_{j,n} + \frac{\Delta t}{2!}\left(\frac{\partial^2 u}{\partial t^2}\right)_{j,n} + \cdots + c\left(\frac{\partial u}{\partial x}\right)_{j,n} + \frac{c\Delta x^2}{3!}\left(\frac{\partial^3 u}{\partial x^3}\right)_{j,n} + \cdots . \tag{7.2.4}$$

The first and the third terms on the right cancel by virtue of the advection equation and we are left with the second and the fourth terms as the leading order errors of the approximations to the time and space derivatives. Therefore we have

$$\varepsilon_{j,n} = O(\Delta t) + O(\Delta x)^2. \tag{7.2.5}$$

Thus, the FTCS method is first-order accurate in time and second-order accurate in space. Since $\varepsilon_{j,n} \to 0$ as $\Delta x, \Delta t \to 0$, we say that the FTCS method is **consistent**.

The stability of the FTCS method can be investigated using **von Neumann stability analysis**. We notice that (7.2.2) admits solutions of the form

$$V_{j,n} = \xi^n \exp(i\,k\,j\Delta x)\,, \tag{7.2.6}$$

where k is the wavenumber and ξ is the amplitude, both of which are constants (i.e., independent of grid point j and time step n). We may think of (7.2.6) as a typical mode in a Fourier series representation of the numerical solution.

Substitution of (7.2.6) into (7.2.2) and simplification gives the relation

$$\xi = \frac{\xi^{n+1}}{\xi^n} = 1 - i\left(\frac{c\Delta t}{\Delta x}\right)\sin(k\,\Delta x). \tag{7.2.7}$$

Since the numerical solution increases in amplitude by ξ each time step, for stability we require that $|\xi| \leq 1$. Unfortunately, relation (7.2.7) shows that the FTCS method is not very useful. Since $|\xi|$ is always larger than 1, any solution of the form (7.2.6) will grow in time. The form (7.2.6) is rather general. Since it is a Fourier mode, and because the equation is linear, finding the solution for every Fourier mode essentially solves the problem. This shows that (7.2.2) is unstable for virtually any initial condition.

In spite of the negative result that the FTCS method is unconditionally unstable we have made some useful progress. We have seen how PDEs are discretized in the discrete time–discrete space framework. We have introduced the graphical representation of such discretizations. We have seen how to get a condition for instability by trying solutions of the form (7.2.6). We have also established a methodology for checking consistency and for obtaining the spatial and temporal orders of accuracy.

7.2.2 Stability Diagram Approach

Could we have predicted this instability? After all we have studied the properties of forward integration in time (Explicit Euler) in Section 3.2 and central difference in Section 4.1. The answer is *yes*, we could predict the stability properties of the combined space-time FTCS method. Note that the stability region of explicit Euler integration is a circle of unit radius centered at $(-1, 0)$ in the complex plane (see Figure 3.1). Stability of the FTCS method can now be restated in the following way. Only when *all* the eigenvalues of the discrete spatial operator represented by the right-hand side of (7.2.2) fall within the stable region of the time integration scheme is the overall space-time method stable. As we saw in Section 4.1.1 on modified wavenumbers, the eigenvalues of the central difference first-derivative operator are purely imaginary. Thus, all the eigenvalues fall outside the stable region of the explicit Euler integration, and as a result FTCS is unconditionally unstable.

This latter way of investigating the stability of space-time discretization of a PDE will be termed the **stability diagram approach**. In Section 3.5 we already saw, in the context of stability analysis of a system of ODEs, the theoretical justification for this approach. To extend it to the stability analysis of a PDE, let us consider semi-discretization of the PDE in space alone. With central difference in space, as in the FTCS method, or with the upwinding method, which we will soon discuss, or with any other spatial discretization for that matter, the semi-discretized advection equation can be written out as

$$\frac{\mathrm{d}\vec{V}}{\mathrm{d}t} = \mathbf{L}\vec{V}\,, \tag{7.2.8}$$

where $\mathbf{L}$ is the matrix operator that arises from spatial discretization of the right-hand side of (7.1.4). Since the advection equation (7.1.4) is being solved with periodic boundary conditions, we follow the discussion on finite difference matrix operators presented in Section 4.2.1 and deduce that $\mathbf{L}$ is a circulant matrix. In the above semi-discrete equation $\vec{V}(t) = V_j(t)$ represents the vector of grid point values of the solution.

We now follow the discussion of Section 5.1.3 and eigendecompose $\mathbf{L}$ as

$$\mathbf{L} = \mathbf{E}\boldsymbol{\lambda}\mathbf{E}^{-1}\,, \tag{7.2.9}$$

where $\mathbf{E}$ is the eigenvector matrix and $\boldsymbol{\lambda}$ is the diagonal eigenvalue matrix. We now change variables from $\vec{V}$ to $\vec{W}$ with the linear transformation

$$\vec{V} = \mathbf{E}\vec{W}\,. \tag{7.2.10}$$

This operation is nothing but transforming the grid point values of the solution to the eigencoordinate. Substituting the transformation into (7.2.8) we obtain

$$\mathbf{E}\frac{\mathrm{d}\vec{W}}{\mathrm{d}t} = \mathbf{E}\boldsymbol{\lambda}\mathbf{E}^{-1}\mathbf{E}\vec{W}\,, \tag{7.2.11}$$

where we have used the fact that the matrix operator $\mathbf{L}$ and its eigenvalue matrix $\mathbf{E}$ are independent of time. The above can be simplified and, using index notation, can be written as

$$\frac{\mathrm{d}W_j}{\mathrm{d}t} = \lambda_j W_j \quad \text{for} \quad j = 0, 1, \ldots, N-1\,. \tag{7.2.12}$$

Note there is no summation over j. Thus, the original coupled system of ODEs given in (7.2.8) has been transformed to N decoupled ODEs along the eigendirections. This greatly simplifies the stability analysis, since time integration of each ODE is on its own. For stability we now require $\lambda_j \Delta t$, for all the eigenvalues, to lie within the stability region of whatever time integration scheme we choose to solve (7.2.12). This is the essence of the stability diagram approach. See Chapter 32 of Trefethen and Embree (2005) for some examples of the stability diagram approach.

If we already know the stability region of the time integration scheme and the eigenvalues of the discrete space operator, then this is a quick way of finding the stability properties of the overall method. However, this works well only if space and time discretizations are cleanly separable. That is, in (7.2.2) the discrete time derivative on the left-hand side is entirely at the jth grid point, and the discrete space derivative on the right is entirely at the nth time level. As we will see later we can come up with space-time discretizations where the space and time parts are interconnected and cannot be cleanly separated. In which case, one must resort to von Neumann analysis to establish the stability properties, which does not use any boundary information explicitly. However, by employing Fourier expansion, von Neumann analysis implicitly assumes periodic boundary condition. In contrast, as discussed in Section 4.2, eigenvalues of the space operator can be calculated for both periodic and non-periodic boundary conditions. Thus, the stability diagram approach can be applied to non-periodic problems as well. But as we will see in Section 7.6 von Neumann stability analysis is very useful even for non-periodic problems.

7.2.3 Upwind Methods

The central difference approximation to $\partial u/\partial x$ in FTCS has the desired property that it is non-dissipative, which corresponds to the eigenvalues being purely imaginary. But with the explicit Euler scheme, this proves to be fatal because its stability region does not include any part of the imaginary axis. There are two ways of stabilizing the FTCS method. The first is to choose a different time integration method, such as third-order Runge–Kutta or third-order Adams–Bashforth, whose stability diagram includes a portion of the imaginary axis, and with a proper choice of Δt all the eigenvalues can be contained within the stable region. We will explore this further in Problem 7.2.1.

The second approach to stabilization is to use an upwind approximation to the space derivative, whose eigenvalues are away from the imaginary axis and have a negative real part. However, such upwind methods are dissipative (while the true equation is not) and while numerical dissipation helps to stabilize the method, it is at the cost of losing some accuracy. The simplest two-point upwind approximation to the advection equation can be expressed as

$$\left(\frac{V_{j,n+1} - V_{j,n}}{\Delta t}\right) = -c\left(\frac{V_{j,n} - V_{j-1,n}}{\Delta x}\right). \tag{7.2.13}$$

Following the steps of (7.2.3) it can be readily shown that the forward-in-time and two-point upwind-in-space method is first-order accurate in both time and space. That is,

$$\varepsilon_{j,n} = O(\Delta t) + O(\Delta x). \tag{7.2.14}$$

Although less accurate than the FTCS method, this second approach has the very valuable property of stability. To see this we can apply von Neumann analysis and show that the amplification factor now becomes

$$\xi = \frac{\xi^{n+1}}{\xi^n} = 1 - \frac{c\Delta t}{\Delta x}\left(1 - \cos(k\Delta x) + \sin(k\Delta x)\right) . \tag{7.2.15}$$

The requirement for stability is that the amplification factor be less than 1 in amplitude (i.e., $|\xi| < 1$). This leads to the condition

$$\left|\frac{c\Delta t}{\Delta x}\right| < 1. \tag{7.2.16}$$

The non-dimensional grouping $\left|\frac{c\Delta t}{\Delta x}\right|$ is the well-known **Courant–Friedrichs–Lewy (CFL) number**, also sometimes referred to as the **Courant number**. Inequality (7.2.16), which limits the CFL number for stability, is called the **CFL criterion**. It has a simple physical origin, which we shall see now.

In Figure 7.2 we show a space-time grid and a particular point (j, n) on it. Now, a hyperbolic equation such as (7.1.4) has characteristics, i.e., curves along which information is propagated. For the simple advection equation we just have information propagating to the right with speed c, viz., the initial condition $u(x, t = 0) = f(x)$ is propagated with speed c to become the solution: $u(x, t) = f(x - ct)$. In Figure 7.2 we show the **cone of influence** or **domain of dependency** spreading backwards bounded by the characteristics $x = \pm ct$. (Only $x = +ct$ of these is relevant for the advection equation. Each would be relevant for the wave equation, where two sets of characteristics go both left and right.) Analytically the solution at (j, n) needs access to previous solution values inside this cone. Thus, if we make a consistent discretization using the grid of space-time points shown, it is intuitively clear that the domain of dependency on this discrete grid must respect the analytical constraint. If the numerical method has a cone of influence that lies within the analytically given one, as in Figure 7.2(a), information will propagate more rapidly in the numerical method than in the continuum equation, and needed information will be missing in the numerical representation. This is likely to lead to some kind of disaster. On the other hand, it seems acceptable to have the numerical method take a smaller time step than the theoretically possible optimum as illustrated in Figure 7.2(b).

In essence, for stability the numerical domain of dependence must be larger than the analytical domain of dependence, and embed the analytical domain of dependence within it. Since the slope of the analytical cone of influence is the wave speed c, and for the upwind scheme (7.2.13) the slope of numerical cone of influence is $\Delta t/\Delta x$, the condition for stability reduces to (7.2.16) obtained

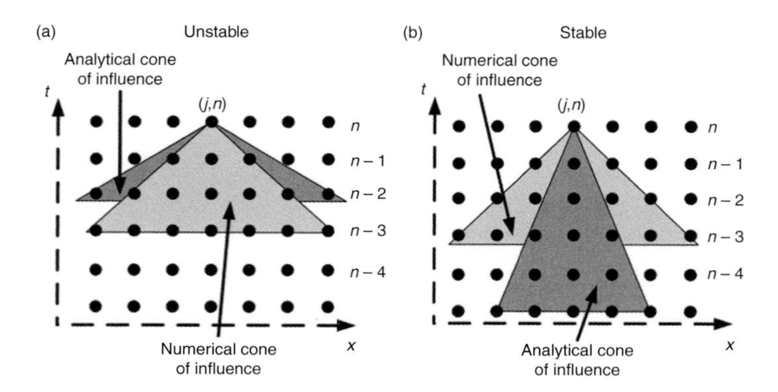

Figure 7.2 Relation between analytical and numerical *cones of influence* in the integration of a hyperbolic PDE. In (a) the numerical cone misses part of the actually required domain. This leads to instability. The situation is corrected in (b). The diagrams are drawn as if Δx and Δt are held fixed and c is varied in our model problem.

from von Neumann analysis. It must be emphasized that the above requirement on domain of dependence should be treated carefully. For example, with proper choice of Δt and Δx the numerical domain of dependence of the FTCS scheme can be made to include the analytical domain of dependence. Nevertheless FTCS is unstable. Thus, loosely speaking, the condition on domain of dependence is not sufficient. Furthermore, the numerical domain of dependence is hard to establish in multi-stage and multi-step schemes. Nevertheless, this line of argument shows that any explicit scheme for advection equation will face a CFL limitation of the kind given in (7.2.16), but the constant on the right-hand side may be different from 1, depending on the time and space discretization.

Problem 7.2.1 Consider the advection equation in a periodic domain $0 \leq x \leq 2\pi$. Let the domain be discretized with N equi-spaced grid points ($\Delta x = 2\pi/N$) and let the spatial derivative be approximated by three-point central difference. (a) Express the eigenvalues of the spatial operator in terms of c and Δx. (b) Let the third-order Runge–Kutta scheme be used for time integration whose stability diagram is given in Figure 3.3. Use the requirement that all the distinct eigenvalues of the spatial operator fall within the stability diagram and obtain the stability restriction.

Problem 7.2.2 Consider the advection equation in a periodic domain $0 \leq x \leq 2\pi$. Let the domain be discretized with N equi-spaced grid points ($\Delta x = 2\pi/N$) and consider the upwind scheme given in (7.2.13). (a) Obtain the stability condition (7.2.16) using von Neumann stability analysis. (b) Express the eigenvalues of the spatial operator in terms of c and Δx and obtain the stability condition (7.2.16) using the stability diagram.

7.2.4 Lax's Theorem on Convergence

An important property that we desire of any numerical method is that the numerical solution should approach the exact solution as we decrease the grid spacing and the time step. Mathematically speaking this property can be stated as

$$\lim_{\substack{\Delta x \to 0 \\ \Delta t \to 0}} |V_j(t) - u(x_j, t)| \to 0, \tag{7.2.17}$$

where $u(x_j, t)$ is the analytic solution at the grid points and the $V_j(t)$ are numerical approximations. A natural question that one might ask is what about accuracy? Does accuracy imply the above requirement? Not really. After all, the FTCS method was first-order accurate in time and second-order accurate in space. In other words, each time step the truncation error can be made zero by taking the limit $\Delta x, \Delta t \to 0$. But the FTCS method was unconditionally unstable no matter how small a time step is used. How can this be? How can a method be good for a single time step (accurate), but yet be unstable?

The important point to recognize is that accuracy is a statement on truncation error for a single time step. In contrast, stability is a statement on how the error accumulates over long time. Note that (7.2.17) compares the analytic solution and the numerical approximation at a finite time t. As $\Delta t \to 0$ this requires infinite number of time steps to reach the finite time. Thus, accuracy alone does not guarantee (7.2.17). We also need the method to be stable. A method that satisfies the important property (7.2.17) will be termed **convergent**.

Convergence of a numerical method is often hard to establish in practical problems, since it requires knowledge of the exact analytic solution. It is in this regard that **Lax's theorem** is an important and a powerful result, which can be stated as follows, *for a well-posed linear initial value PDE a numerical discretization that is consistent will converge to the exact solution if and only if it is stable.* This is a powerful theorem since it guarantees convergence when both accuracy and stability are achieved. Since we know how to establish the stability and accuracy of a numerical method, with the Lax theorem we also have a way of establishing convergence. The proof of the theorem is somewhat involved, and we refer the reader to standard textbooks such as Richtmyer and Morton (1994).

7.2.5 Lax's Method

Lax's method for the advection equation can be expressed as:

$$\left(\frac{V_{j,n+1} - \frac{1}{2}\left(V_{j+1,n} + V_{j-1,n}\right)}{\Delta t} \right) = -c \left(\frac{V_{j+1,n} - V_{j-1,n}}{2\Delta x} \right). \tag{7.2.18}$$

Thus, compared to FCTS, in the time derivative the values of u at (j,n) has been replaced by the average of the values at $(j+1,n)$ and $(j-1,n)$. The stencil for Lax's method is shown in Figure 7.3. The truncation error of Lax's method is obtained by Taylor expanding each term of (7.2.18):

$$\varepsilon_{j,n} = \frac{\Delta t}{2!}\left(\frac{\partial^2 u}{\partial t^2}\right)_{j,n} + \frac{\Delta x^2}{2\Delta t}\left(\frac{\partial^2 u}{\partial x^2}\right)_{j,n} + \frac{c\Delta x^2}{3!}\left(\frac{\partial^3 u}{\partial x^3}\right)_{j,n} + \cdots . \tag{7.2.19}$$

The second term on the right is new, being absent in the FTCS method. The order of accuracy of Lax's method is $O(\Delta t) + O(\Delta x)^2 + O(\Delta x^2/\Delta t)$. Thus, for consistency we not only require $\Delta x, \Delta t \to 0$, but also $\left(\Delta x^2/\Delta t\right) \to 0$. This complexity is due to the fact that in (7.2.18) the numerical approximation to the time derivative is not entirely at the jth grid point, and involves the neighboring points as well. Here, the discrete time and space operators are not completely separable, and as a result the time- and space-dependence of error do not decouple.

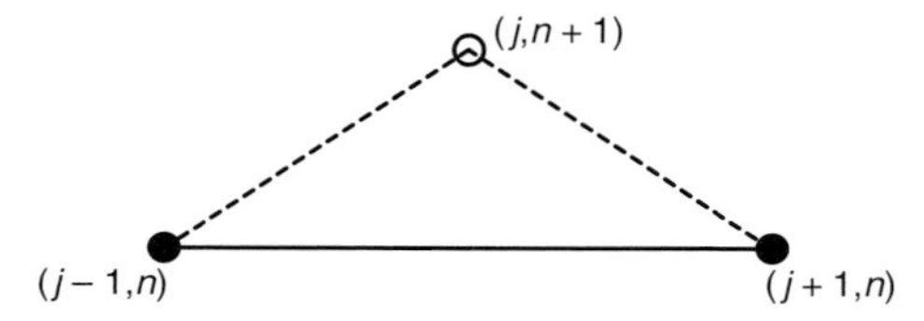

Figure 7.3 Space-time stencil for Lax's method. Dashed lines connect grid values related by the discretization of the time derivative. The solid line connects grid points related by the discretization of the spatial derivative.

To establish stability we use von Neumann analysis leading to the condition

$$\xi = \cos(k\,\Delta x) - i\left(\frac{c\Delta t}{\Delta x}\right)\sin(k\,\Delta x)\,, \tag{7.2.20}$$

in place of (7.2.7), and from which we obtain the CFL condition for stability as

$$\left|\frac{c\Delta t}{\Delta x}\right| \le 1. \tag{7.2.21}$$

How did Lax's modification of the FTCS method from (7.2.2) to (7.2.18) achieve stability? In the case of the upwind scheme, stability can be expected due to numerical diffusion (amplitude error) of the upwind scheme. Lax's method seems to maintain the symmetry about the jth point. Nevertheless, the modification to the time derivative seems to have done the trick by inducing numerical diffusion. To see this consider the following equation:

$$\frac{\partial u}{\partial t} = -c\frac{\partial u}{\partial x} + \nu\frac{\partial^2 u}{\partial x^2}, \tag{7.2.22}$$

where $\nu = (\Delta x^2/2\Delta t)$ is the numerical diffusivity. Discretize this in the manner of FTCS to get

$$\left(\frac{V_{j,n+1}-V_{j,n}}{\Delta t}\right)=-c\left(\frac{V_{j+1,n}-V_{j-1,n}}{2\Delta x}\right)+\nu\left(\frac{V_{j+1,n}-2V_{j,n}+V_{j,n}}{\Delta x^2}\right). \tag{7.2.23}$$

By rearranging it is easy to verify that the above discretization is identical to Lax's method (7.2.18). In other words, Lax's method, which started with the objective of approximating the advection equation, is the three-point central difference method for the advection–diffusion equation with a specific value of diffusivity equal to $(\Delta x^2/2\Delta t)$. Although the stability diagram approach cannot be directly applied to (7.2.18), since space and time discretizations are coupled, in this case by rewriting as (7.2.23) we were able to apply the stability diagram approach and come to the CFL condition for stability given in (7.2.21).

In fact, from the truncation error (7.2.19) it can be stated that Lax's method exactly solves the following **modified equation**:

$$\frac{\partial u}{\partial t}=-c\frac{\partial u}{\partial x}+\frac{\Delta t}{2!}\left(\frac{\partial^2 u}{\partial t^2}\right)_{j,n}+\frac{\Delta x^2}{2\Delta t}\left(\frac{\partial^2 u}{\partial x^2}\right)_{j,n}+\frac{c\Delta x^2}{3!}\left(\frac{\partial^3 u}{\partial x^3}\right)_{j,n}+\cdots. \tag{7.2.24}$$

However, since we are solving the advection equation, from (7.2.24) it can be seen that Lax's method is first-order accurate in time and second-order accurate in space. Because of the third term on the right, for Lax's method we need $(\Delta x^2/\Delta t)\to 0$ for consistency. This approach to evaluating the accuracy of a numerical method is often referred to as *accuracy via the modified equation.*

7.3 The Diffusion Equation

We will now consider the one-dimensional diffusion equation given in (7.1.8). Here u is the field being diffused over time and ν is the diffusion coefficient, which will be taken as a constant. Let us proceed to discretize the diffusion equation.

7.3.1 The FTCS Method

Again we consider the bounded domain $0\le x\le 2\pi$ with periodic boundary condition. We discretize time by equi-spaced time intervals, and space by equi-spaced grid points (see Section 7.2.1). Let us first try the simplest FTCS approach that uses explicit Euler in time and three-point central difference in space:

$$\left(\frac{V_{j,n+1}-V_{j,n}}{\Delta t}\right)=\nu\left(\frac{V_{j+1,n}-2V_{j,n}+V_{j-1,n}}{\Delta x^2}\right), \tag{7.3.1}$$

where the term within the parentheses on the right is the central difference approximation to $\partial^2 u/\partial x^2$. The **stencil** of the FTCS method for the diffusion equation is the same as that shown in Figure 7.1. So it is still a three-point two-level method.

The accuracy of the FTCS method is established by investigating how well the exact solution satisfies the difference equation (7.3.1). The truncation error of the FTCS scheme can be obtained by Taylor expanding each term of (7.3.1). Or more readily from the explicit Euler and three-point central difference we note the leading order error of the method as $O(\Delta t) + O(\Delta x)^2$. Accordingly, the FTCS method for the diffusion equation is consistent.

The stability of the FTCS method can be investigated using von Neumann stability analysis. Substitution of (7.2.6) into (7.3.1) and simplification gives the relation

$$\xi = 1 + 2\left(\frac{\nu \Delta t}{\Delta x^2}\right)[\cos(k\,\Delta x) - 1]\,. \tag{7.3.2}$$

For stability we require $|\xi| < 1$, from which we obtain the following **diffusion condition**:

$$\frac{\nu \Delta t}{\Delta x^2} < \frac{1}{2}. \tag{7.3.3}$$

The CFL criterion (7.2.16) can be thought of as time step limitation given the values of advection speed c and grid spacing Δx. Similarly, the above diffusion condition limits the time step for given values of diffusion coefficient ν and Δx. The above stability condition can also be obtained from the stability diagram. Recall that the stable region of the explicit Euler scheme extends from -2 to 0 along the real axis (see Figure 3.1(b)). Also from Section 4.2.1 it can be seen that for the three-point central difference approximation to $\partial^2/\partial x^2$ with periodic boundary conditions, the eigenvalue of largest magnitude is $-4/\Delta x^2$; see (4.2.13). Thus, the requirement that all the eigenvalues fall within the stability diagram leads to exactly the same condition (7.3.3).

Any explicit time integration of the diffusion equation will lead to a diffusion condition similar to (7.3.3). Only the constant on the right-hand side will depend on the actual time integration method. However, there is a big difference between CFL and diffusion conditions. While the CFL condition requires $\Delta t \propto \Delta x$, the diffusion condition is more constraining and requires $\Delta t \propto \Delta x^2$. As a result, with increasing refinement of the spatial grid, one is forced to choose a much smaller time step from stability restriction of the diffusion equation. The question is do we need to take such small time steps?

Broadly speaking, the time step Δt and the grid size Δx must be chosen such that the numerical solution at any specified later time t_n is close to the actual solution within a desired error bound. For a well-posed PDE, Lax's theorem guarantees that with a stable and consistent numerical method we can converge to the actual solution and reduce the error to below a desired threshold, with a proper choice of Δt and Δx. But it is not possible to determine a priori the values of Δt and Δx one must choose, as they are problem dependent. However, it is generally observed that the time step chosen to satisfy the CFL condition is sufficiently small enough to provide accurate numerical solution that stays converged to the actual solution over long time integration. As we discussed before, this CFL condition can be understood to arise from the physical requirement that the numerical domain of dependence must include the physical domain of dependence (see Figure 7.2). The even smaller time step demanded by the diffusion stability condition in (7.3.3) is not generally needed from the accuracy point of view. This line of argument leads to the following important practical conclusion. *An explicit scheme is acceptable for the advection term, while an implicit scheme is desired for the diffusion term.* We shall discus this further under implicit methods in Section 7.8.

7.3.2 The Du Fort–Frankel Method

Here we will discuss an explicit method for the diffusion equation with interesting properties. The Du Fort–Frankel method is a three-point three-step method written as

$$\left(\frac{V_{j,n+1} - V_{j,n-1}}{2\Delta t}\right) = \nu \left(\frac{V_{j+1,n} - (V_{j,n+1} + V_{j,n-1}) + V_{j-1,n}}{\Delta x^2}\right). \tag{7.3.4}$$

Thus, compared to FCTS, the leap-frog scheme is used for time integration.

Problem 7.3.1 Consider the Du Fort–Frankel method for the diffusion equation. Obtain its leading order truncation error as

$$\varepsilon_{j,n} = -\frac{\Delta t^2}{3!}\left(\frac{\partial^3 u}{\partial t^3}\right)_{j,n} + \frac{\nu \Delta x^2}{12}\left(\frac{\partial^4 u}{\partial x^4}\right)_{j,n} + \nu\left(\frac{\Delta t}{\Delta x}\right)^2 \left(\frac{\partial^2 u}{\partial t^2}\right)_{j,n} + \cdots.$$

Also, show that leading-order modified equation of the Du Fort–Frankel scheme has a wave-equation-like character.

Problem 7.3.2 Using von Neumann analysis show that the Du Fort–Frankel method is unconditionally stable.

From Problem 7.3.1 it is clear that, for consistency, the Du Fort–Frankel method requires, in addition to $\Delta x \to 0$ and $\Delta t \to 0$, an extra condition that $(\Delta t/\Delta x) \to 0$. In this sense the Du Fort–Frankel method is like Lax's method for the advection equation. Note that the spatial discretization on the right-hand side of (7.3.4) is not entirely at the nth time level and this is the cause of the additional consistency requirement. In both the Lax and the Du Fort–Frankel methods, since the truncation error is not guaranteed to become zero as $\Delta x \to 0$ and $\Delta t \to 0$, some authors call these methods inconsistent. Here we use the terminology that these methods are consistent provided the additional consistency requirement is satisfied.

Due to leap-frog-like discretization of time, the accuracy of the Du Fort–Frankel method is second-order both in time and space. Also, most amazingly, it is unconditionally stable, despite being an explicit method. But is it really so? Not really. In fact, the additional requirement of consistency that $(\Delta t/\Delta x) \to 0$ is more restrictive than the CFL condition and is satisfied by the diffusion condition. Thus, even though the Du Fort–Frankel method is unconditionally stable, for consistency requirements one is forced to choose small time steps in accordance with the diffusion condition.

Problem 7.3.3 Consider diffusion equation in a periodic domain $0 \le x \le 2\pi$. Let the domain be discretized with N equi-spaced grid points ($\Delta x = 2\pi/N$) and let the spatial second derivative be approximated by three-point central difference. (a) Express the eigenvalues of the spatial operator in terms of ν and Δx. (b) If a third-order Runge–Kutta scheme is used for time integration, using the stability diagram obtain the stability restriction on Δt.

7.4 The Advection–Diffusion Equation

We are now ready to put the advection and diffusion effects together and consider numerical approximations. First we will consider centered schemes for the advection–diffusion equation and then consider upwind schemes. In each case we will investigate the accuracy of the method and its stability properties. It may be natural to expect the diffusion effect to be stabilizing; after all the energy of the analytic solution decreases monotonically. So, it will be revealing to see if and how the diffusion effect helps to stabilize the numerical method.

7.4.1 Analysis of a General Explicit Three-Point Method

We want to start a bit more general. Consider any explicit method using three points to discretize (7.1.9). It looks like this:

$$a_0 V_{j,n+1} = A_1 V_{j+1,n} + A_0 V_{j,n} + A_{-1} V_{j-1,n}. \tag{7.4.1}$$

By expanding (7.4.1) in Taylor series and collecting terms we arrive at certain conditions on the coefficients a_0 and As. The first of these is

$$a_0 = A_1 + A_0 + A_{-1} = \frac{1}{\Delta t}. \tag{7.4.2}$$

The first equality in (7.4.2) comes about since there is no term with just u itself in the advection–diffusion equation. The second comes from reproducing the time derivative of u. To produce the spatial derivative, we require

$$-A_1 + A_{-1} = \frac{c}{\Delta x}. \tag{7.4.3}$$

The idea is that a_0 and each As can be written as series in inverse powers of Δt and Δx. When these are multiplied out in the Taylor series expansion of (7.4.1), a number of terms arise. Some of these represent the quantities we want in our advection–diffusion equation, (7.1.9). Others are lumped into the truncation error. One of the solutions of (7.4.2) and (7.4.3) is

$$a_0 = \frac{1}{\Delta t}; \; A_1 = -\frac{c}{2\Delta x} + \frac{\nu}{\Delta x^2}; \; A_0 = \frac{1}{\Delta t} - \frac{2\nu}{\Delta x^2}; \; A_{-1} = \frac{c}{2\Delta x} + \frac{\nu}{\Delta x^2}, \tag{7.4.4}$$

which yields the Lax method written out in (7.2.23).

A von Neumann analysis allows us to investigate the stability of the discretization (7.4.1) in a general way. We shall do this first, and then substitute in coefficient values for particular methods subsequently. This will let us read off stability results for other methods, with essentially no additional work. We again invoke the form (7.2.6) for the *ansatz*. Substitution in (7.4.1) gives

$$\xi = \frac{1}{a_0}\left(A_1 \exp(ik\Delta x) + A_0 + A_{-1}\exp(-ik\Delta x)\right). \tag{7.4.5}$$

Our stability criterion is that $|\xi| < 1$ for all k. Imposing this on (7.4.5) we get

$$a_0^2 > \left[(A_1 + A_{-1})\cos(k\Delta x) + A_0\right]^2 + (A_1 - A_{-1})^2 \sin^2(k\Delta x). \tag{7.4.6}$$

This inequality must be satisfied for all values of $X = \cos(k\Delta x)$ between -1 and 1. The inequality says that a certain quadratic in X is to be non-negative; $X = 1$ is a zero of this quadratic. Hence, the inequality (7.4.6) has the form

$$\alpha(X-1)(X-\rho) > 0, \tag{7.4.7}$$

where α and ρ can be obtained from (7.4.6) as

$$\alpha = -4A_1A_{-1}; \qquad \alpha\rho = a_0^2 - A_0^2 - (A_1 - A_{-1})^2. \tag{7.4.8}$$

Since $X \le 1$, this implies that $\alpha(X-\rho) \le 0$. But $\alpha(X-\rho)$ is the equation of a line, and the only way it can be below the x-axis for all x in the interval from -1 to 1 is for both $\alpha(1-\rho)$ and $\alpha(\rho-1)$ to be negative. In other words, we must have $|\alpha| < \alpha\rho$. Following this line of reasoning we arrive at the following two conditions for stability:

$$A_0\left(A_1 + A_{-1}\right) > 0; \quad A_0\left(A_1 + A_{-1}\right) + 4A_1A_{-1} > 0. \tag{7.4.9}$$

With the coefficients in (7.4.4) the condition for stability reduces to

$$\frac{\nu\Delta t}{\Delta x^2} < \frac{1}{2}; \qquad c^2 < \frac{2\nu}{\Delta t}. \tag{7.4.10}$$

We introduce the abbreviation

$$C = \frac{c\Delta t}{\Delta x} \tag{7.4.11}$$

for the Courant number of the numerical method. It measures the distance over which a fluid particle is advected in a unit time step in units of grid spacing. We also introduce the **diffusion number**

$$D = \frac{\nu\Delta t}{\Delta x^2} \tag{7.4.12}$$

as a dimensionless measure of the diffusion length (in units of grid spacing). In terms of these two quantities our stability requirements are

$$D < \frac{1}{2}; \qquad |C| < \sqrt{2D}. \tag{7.4.13}$$

Figure 7.4 shows the stability domain of the central difference scheme given by (7.4.1) and (7.4.4) in the (C–D)-plane. For $D = 0.5$ we get the stability condition of Lax's method, i.e., $|C| < 1$. For $D = 0$ we get the instability of the FTCS method for any finite C.

The Courant and diffusion numbers can be combined to define a **cell Reynolds number**

$$\mathrm{Re}_\Delta = \frac{c\,\Delta x}{\nu} = \frac{C}{D}. \tag{7.4.14}$$

Sometimes Re_Δ is referred to as the **cell Péclet number**. In terms of Courant and Péclet numbers the stability requirement of the central difference scheme becomes

$$2C \leq \mathrm{Re}_\Delta \leq \frac{2}{C}. \tag{7.4.15}$$

Note that the above two conditions can be satisfied only for $C \leq 1$ and the stability domain in the (C–Re_Δ)-plane is shown in Figure 7.4(b).

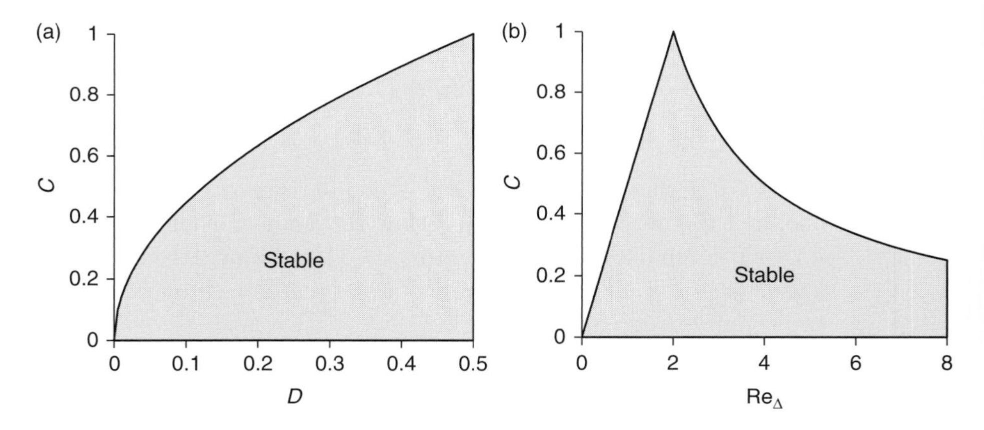

Figure 7.4 (a) Stable region in the C–D plane for the (7.4.17) method. For $D = 0$ this reduces to the FTCS scheme, that is always unstable. For $D = 0.5$ we reproduce Lax's method, which is stable for $0 \leq C \leq 1$. (b) The same stable region for (7.4.17) plotted in the C–Re_Δ-plane.

7.4.2 Upwind Differencing

There are various ways in which the stability boundary of Figure 7.4 can be extended, in particular, in the regime where advection is "strong" and diffusion is "weak," i.e., the $D = 0$ portion of Figure 7.4. The method we considered in the previous subsection was symmetric. Physically we would only expect the site that is *downstream* to feel the disturbance, i.e., for $c > 0$ we would expect the point $(j+1, n+1)$ to feel the perturbation from the point (j, n); and for $c < 0$, the point $(j-1, n+1)$. The notion of **upwind differencing**, or, as it is more simply known **upwinding**, arises from these considerations.

To keep things reasonably general consider the method

$$\left(\frac{V_{j,n+1} - V_{j,n}}{\Delta t}\right) = -\frac{c}{2}\left((1-w)\frac{V_{j+1,n} - V_{j,n}}{\Delta x} + (1+w)\frac{V_{j,n} - V_{j-1,n}}{\Delta x}\right)$$
$$+\nu\left(\frac{V_{j+1,n} - 2V_{j,n} + V_{j-1,n}}{\Delta x^2}\right), \tag{7.4.16}$$

where w is a fixed number between -1 and $+1$. We want to investigate the effects of having a bias like this in the evaluation of the spatial derivative. For $w = 0$ we get the symmetric FTCS method studied in Section 7.2.1.

The method (7.4.16) has the general format of (7.4.1). The coefficients are

$$\begin{aligned} a_0 &= \frac{1}{\Delta t}; \quad A_1 = -\frac{c(1-w)}{2\Delta x} + \frac{\nu}{\Delta x^2}; \\ A_0 &= \frac{1}{\Delta t} - \frac{cw}{\Delta x} - \frac{2\nu}{\Delta x^2}; \quad A_{-1} = \frac{c(1+w)}{2\Delta x} + \frac{\nu}{\Delta x^2}. \end{aligned} \tag{7.4.17}$$

Substituting these into the stability requirements, namely equations (7.4.9), we find the conditions:

$$0 < (1 - w\,C - 2D)(w\,C + 2D); \qquad 0 < w\,C + 2D - C^2, \tag{7.4.18}$$

where the abbreviations C and D have been used. Combining these two inequalities we obtain the stability condition:

$$C^2 < w\,C + 2D < 1. \tag{7.4.19}$$

Ignore diffusion for the moment, i.e., set $D = 0$. Then (7.4.19) reduces to $C < w$. Thus, introducing the upwind bias in the method can lead to a finite range of stable values of C even for $D = 0$. Stability for positive C implies that w must be chosen positive. Thus, we must "mix in" a bit more of the derivative in the upstream direction (forward differencing) than in the downstream direction (backward differencing). Conversely, if C is negative, we find that w must be negative, and we must use more of the downstream derivative in (7.4.16) for stability. In the case $C = 1$ from (7.4.19) we see that for stability we must have $w = 1$. The only spatial derivative to be included is the one in the upstream direction. Similar remarks pertain to the case $C = -1$.

Assume that w is chosen according to this idea of upwinding, i.e., we set $w = \text{sign}\,(C)$. Then the left inequality in (7.4.19) is always satisfied for $|C| < 1$, and the right inequality gives

$$|C| + 2D < 1. \tag{7.4.20}$$

The stability region in the $(C\text{–}D)$-plane is thus within a triangle bounded by $D = 0$, and the two lines $1 - 2D = \pm C$. This region is indicated in Figure 7.5

along with the previously determined region of stability of the FTCS method. We see that the stability interval in C is now finite for small values of D. Thus, upwinding allows us to compute stably for a finite range of C values at small D. (For $D = 0$, in fact, for all $|C| \leq 1$). This mechanism can help considerably in stabilizing diffusion-free problems (or, in general, dissipation-free problems).

Problem 7.4.1 Find the exact values of D and C at which the two stability boundaries in Figure 7.5 intersect.

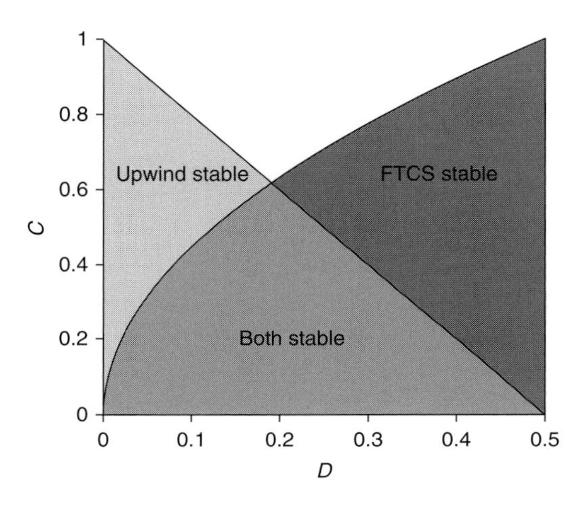

Figure 7.5 Stable region for upwind differencing superimposed on the earlier results on stability of the FTCS method. Upwinding gives enhanced stability for small values of D although the method is only first-order accurate for the spatial derivative.

Problem 7.4.2 Use three-point central difference and semi-discretize the advection–diffusion equation to the form (7.2.8). Using Theorem 4.2.1, evaluate the eigenvalues of the resulting circulant matrix $\mathbf{L}_*$. With these eigenvalues and the stability diagram of the explicit Euler scheme obtain the result 7.4.13, and thereby demonstrate the equivalence of von Neumann stability analysis and the stability diagram approach for periodic boundary conditions.

Problem 7.4.3 Repeat the analysis of Problem 7.4.2 for the upwinding-included scheme outlined in Section 7.4.2 and obtain the result 7.4.19.

Problem 7.4.4 Consider the following two-step Lax–Wendroff method for the advection equation:

$$\frac{V_{j,n+1} - V_{j,n}}{\Delta t} = -c \frac{V_{j+\frac{1}{2},n+\frac{1}{2}} - V_{j-\frac{1}{2},n+\frac{1}{2}}}{\Delta x},$$

where

$$\frac{V_{j+\frac{1}{2},n+\frac{1}{2}} - \frac{1}{2}\left[V_{j+1,n} + V_{j,n}\right]}{\Delta t/2} = -c \frac{V_{j+1,n} - V_{j,n}}{\Delta x},$$

and a similar equation for $V_{j-\frac{1}{2},n+\frac{1}{2}}$. Note that the second equation is just Lax's method at the half points for a half-time step. This is followed by the first equation which is the leap-frog method for the remaining half-time step.

(a) Combine the two steps to show that they reduce to the familiar FTCS scheme but with an effective diffusivity of $c^2 \Delta t/2$.
(b) Show that the method is second-order in both time and space.
(c) Obtain the CFL condition for stability for the above method.

7.5 Godunov's Theorem

In the preceding section we considered a relatively general procedure for generating finite difference approximations to the advection–diffusion equation. We were primarily concerned with the issue of stability. In this section we want to return to accuracy. How are we doing in terms of truncation error in these methods, and what might the constraints be?

Let us write a general explicit scheme in the form

$$V_{j,n+1} = \sum_{l=-\infty}^{\infty} A_l \, V_{j+l,n}. \tag{7.5.1}$$

The coefficients A_l are not quite the same as in Section 7.4.1 – we have factored out a_0. Even though the sum formally extends over all l, we assume, of course, that only a finite number of the coefficients are non-zero. We will just worry about the advection equation for now. This time our interest is in establishing the order of accuracy of the approximation in the sense of a truncation error. Substituting the exact solution in (7.5.1) and expanding in Taylor series we get

$$\begin{aligned} u_{j,n} &+ \Delta t \left(\frac{\partial u}{\partial t}\right)_{j,n} + \frac{\Delta t^2}{2!}\left(\frac{\partial^2 u}{\partial t^2}\right)_{j,n} + \cdots \\ &= \sum_{l=-\infty}^{\infty} A_l \left[u_{j,n} + l \Delta x \left(\frac{\partial u}{\partial x}\right)_{j,n} + \frac{l^2 \Delta x^2}{2!}\left(\frac{\partial^2 u}{\partial x^2}\right)_{j,n} + \cdots \right]. \end{aligned} \tag{7.5.2}$$

If we want a method that is second-order accurate, the terms written explicitly in (7.5.2) must balance exactly when we use the PDE to write $\partial u/\partial t$ as $-c(\partial u/\partial x)$ and $\partial^2 u/\partial t^2$ as $c^2(\partial^2 u/\partial x^2)$. This leads to the following constraints on the coefficients:

$$\sum_{l=-\infty}^{\infty} A_l = 1; \qquad \sum_{l=-\infty}^{\infty} l A_l = -C; \qquad \sum_{l=-\infty}^{\infty} l^2 A_l = C^2, \tag{7.5.3}$$

where C is again the Courant number. So far so good. But now Godunov (1959) noticed that these conditions are really quite restrictive. Suppose, for example, that the initial state $f(x)$ is a monotone function of the spatial coordinate x. Then, the analytical solution has the property that it will remain monotone for ever, since analytically $u(x,t) = f(x - ct)$. It is natural to impose this condition on our numerical method (7.5.1). Thus, we are led to the following theorem.

Theorem *The scheme* (7.5.1) *is both second-order accurate and monotonicity preserving if all the* A_l *are non-negative.*

Proof The condition is certainly *sufficient*. To see this consider from (7.5.1) the difference between one grid point value and the next at time step $n+1$, i.e.,

$$V_{j+1,n+1} - V_{j,n+1} = \sum_{l=-\infty}^{\infty} A_l \, (V_{j+l+1,n} - V_{j+l,n}) \, . \tag{7.5.4}$$

Certainly if all the differences on the right have the same sign at step n, and if all the A_l are non-negative, the differences at step $n+1$ will have the same uniform sign, and the method will be monotonicity preserving.

The condition is also *necessary*. Assume one of the A_l is negative, say $A_r < 0$. Then we could consider the following initial condition: $V_{l,0} = 0$ for $l \le r$ and $V_{l,0} = 1$ for $l > r$ (i.e., a step change in V). This initial state is certainly monotone. In (7.5.4) set $n = 0$ and $j = 0$. In the sum on the right only the term for $l = r$ does not vanish. Indeed, it equals A_r; but $A_r < 0$. Thus, at time step 1 we find $V_{1,1} - V_{0,1} < 0$, and the monotonicity has been broken. □

Still all appears well. We found our conditions for a second-order method of the type desired, and we have resolved that all the coefficients need to be non-negative for the method to have the desired feature of monotonicity preservation. But now disaster strikes: if all the A_l are non-negative, (7.5.3) have only trivial solutions. This follows from the Cauchy–Schwarz inequality which gives

$$C^2 = \left| \sum_{l=-\infty}^{\infty} l A_l \right|^2 = \left| \sum_{l=-\infty}^{\infty} l \sqrt{A_l} \sqrt{A_l} \right|^2 \le \left(\sum_{l=-\infty}^{\infty} l^2 A_l \right) \left(\sum_{l=-\infty}^{\infty} A_l \right) = C^2. \tag{7.5.5}$$

We have used (7.5.3) and the result that the A_l are non-negative (to take the square roots). Recall that for the Cauchy–Schwarz inequality to hold as an equality, as we see it must here, $l\sqrt{A_l}$ has to be proportional to $\sqrt{A_l}$ for each l. It is not difficult to see that there are no generally acceptable solutions. (If $l\sqrt{A_l} = \lambda\sqrt{A_l}$

for all l, conditions (7.5.3) give $\lambda = -C$. Thus, for $l = 0$ we see that $A_0 = 0$, and C is restricted to being an integer, say $C = -m$, where A_m is the only coefficient in the method that can be non-zero, and then by (7.5.3) must equal 1. Equation (7.5.1) then degenerates to $V_{j,n+1} = V_{j+m,n}$. This is, of course, true for $C = -m$, but can hardly pass as a general numerical method for solving the advection equation.)

In conclusion Godunov's theorem can be stated as follows.

Theorem (Godunov's Theorem) *A linear two-step numerical method for solving a partial differential equation can at best be first-order accurate if it is monotonicity preserving.*

Clearly the linear difference equation (7.5.1) that we started with can be an approximation to a variety of PDEs, not just limited to advection or diffusion equation. In the following section we will consider implicit time-stepping. Here we hasten to point out that the above theorem is equally valid for implicit two-step methods as well. This can be verified by writing the implicit version of (7.5.1) as

$$V_{j,n+1} + \sum_{l=1}^{L} \alpha_l \left(V_{j+l,n+1} + V_{j-l,n+1}\right) = \sum_{l=-\infty}^{\infty} B_l \, V_{j+l,n} \,, \tag{7.5.6}$$

which can be linearly transformed to the form (7.5.1) with the coefficients A now expressed in terms of Bs and αs.

What a difference from our discussion of ODEs, where we had all kinds of options for improving the accuracy! Here we are, faced with one of the simplest conceivable linear PDEs, an equation that we can solve analytically, for which we cannot even construct a method that is second-order accurate without putting spurious wiggles into our solution!

This is indeed a big challenge, but it must be correctly seen as research opportunity. As suggested in the proof for the necessary condition of Godunov's theorem, the spurious wiggles appeared in the case of a step jump in the solution. It was shown by van Leer (1974) that by considering second-order methods that are nonlinear (even for a linear PDE!) one can circumvent the restriction imposed by Godunov's theorem. He introduced the **MUSCL** scheme which stands for **monotonic upstream-centered scheme for conservation laws**. There has been outstanding numerical research under the titles **total variation diminishing (TVD)**, **essentially non-oscillatory (ENO)**, and **weighted ENO (WENO)** methods, where a controlled amount of nonlinearity is carefully introduced near locations where the solution shows rapid variation. The net desired effect is to avoid spurious oscillations, perhaps at the expense of slightly smoothening any sharp jump.

7.6 More on Stability: Non-periodic Boundary Conditions

So far we have only considered periodic domains. It is time we consider other boundary conditions, albeit in the context of the linear advection–diffusion equation. To fix the idea, let us consider the advection–diffusion equation (7.1.9) within the domain $0 \le x \le 2\pi$ with the following Dirichlet boundary conditions:

$$u(0,t) = U_{\mathrm{L}}(t) \quad \text{and} \quad u(2\pi,t) = U_{\mathrm{R}}(t)\,. \tag{7.6.1}$$

Unlike periodic boundary conditions, the discretization will involve $N+1$ grid points starting from $x_0 = 0$ to $x_N = 2\pi$. The explicit Euler three-point method for the advection–diffusion equation given in (7.4.1) along with the Dirichlet boundary conditions can be expressed as

$$\begin{bmatrix} V_{0,n+1} \\ V_{1,n+1} \\ \vdots \\ V_{N,n+1} \end{bmatrix} = \begin{bmatrix} 1 & 0 & 0 & \cdots & 0 \\ A_{-1}/a_0 & A_0/a_0 & A_1/a_0 & 0 & \cdots \\ 0 & \ddots & \ddots & \ddots & \ddots \\ 0 & \cdots & 0 & 0 & 1 \end{bmatrix} \begin{bmatrix} U_{\mathrm{L}}(t) \\ V_{0,n} \\ \vdots \\ U_{\mathrm{R}}(t) \end{bmatrix}. \tag{7.6.2}$$

We now choose the three-point method to be the central difference approximation; the coefficients a_0, A_1, A_0 and A_{-1} are then given by (7.4.4). Let $\vec{V}_{\mathrm{I},n+1}$ and $\vec{V}_{\mathrm{I},n}$ be column vectors of length $(N-1)$ containing the interior grid point values of V at time levels $(n+1)$ and n respectively. If we define the boundary correction vector to be

$$\vec{f}_{\mathrm{I}} = \begin{bmatrix} A_{-1}U_{\mathrm{L}}(t)/a_0 \\ 0 \\ \vdots \\ A_1 U_{\mathrm{R}}(t)/a_0 \end{bmatrix} = \begin{bmatrix} \left(D + \frac{C}{2}\right) U_{\mathrm{L}}(t) \\ 0 \\ \vdots \\ \left(D - \frac{C}{2}\right) U_{\mathrm{R}}(t) \end{bmatrix}, \tag{7.6.3}$$

then the mapping from the nth to the $(n+1)$th time level can be expressed as

$$\vec{V}_{\mathrm{I},n+1} = \mathbf{A}\,\vec{V}_{\mathrm{I},n} + \vec{f}_{\mathrm{I}}\,. \tag{7.6.4}$$

The matrix operator $\mathbf{A}$ is a tridiagonal matrix with sub-diagonal, diagonal and super-diagonal elements given by $\{D + C/2,\ 1-2D,\ D - C/2\}$.

We now ask the question, under what condition will $\vec{V}_{\mathrm{I}}$ remain bounded and not grow uncontrolledly? In other words, we want to investigate the stability of the explicit Euler time integration scheme for advection–diffusion equation in conjunction with the three-point central difference approximation. But now the stability analysis is with Dirichlet boundary conditions. We already have the result (7.4.13) for periodic boundary conditions from von Neumann stability analysis. The results of the periodic and Dirichlet boundary conditions can then be compared.

In order to perform the stability analysis with the non-periodic boundary conditions we follow the **matrix method**. We first note that the mapping (7.6.4) has a solution

$$\vec{V}_{\mathrm{I},n+1} = \mathbf{A}^{n+1}\,\vec{V}_{\mathrm{I},0} + \left(\sum_{l=1}^{n} \mathbf{A}^{l}\right)\vec{f}_{\mathrm{I}}\,. \tag{7.6.5}$$

From this we can obtain the condition for stability to be

$$||\mathbf{A}^{n}|| < 1 \quad \text{for all } \; n\,, \tag{7.6.6}$$

where the matrix norm is defined in terms of the vector norm by

$$||\mathbf{A}|| = \max_{||\vec{x}|| \neq 0} \frac{||\mathbf{A}\vec{x}||}{||\vec{x}||}\,, \tag{7.6.7}$$

where $\vec{x}$ is any vector of length $(N-1)$ and $||\vec{x}||$ is the L_2 norm of the vector. We refer the reader to textbooks such as Smith (1985) or Mitchell and Griffiths (1980) for the definitions and properties of matrix norms in the context of numerical methods for PDEs, and for a detailed discussion of stability analysis by the matrix method. From the definition of the matrix norm it can be shown $||\mathbf{A}^{n}|| \leq ||\mathbf{A}||^{n}$. Therefore, instability growth over a single time step is a necessary condition (although not sufficient) for long term instability. If we demand stability for all times, then the necessary and sufficient condition for stability can be restated as

$$||\mathbf{A}|| \leq 1\,. \tag{7.6.8}$$

The L_2 norm of a matrix can be bounded from below by its spectral radius as

$$\sigma(\mathbf{A}) \leq ||\mathbf{A}||\,, \tag{7.6.9}$$

where the spectral radius $\sigma(\mathbf{A})$ is defined as the largest eigenvalue in magnitude. If $\mathbf{A}$ is a normal matrix then its L_2 norm is equal to the spectral radius and the necessary and sufficient condition for stability becomes

$$\sigma(\mathbf{A}) \leq 1\,. \tag{7.6.10}$$

It can be verified that the above condition of containing all the eigenvalues of $\mathbf{A}$ within a unit circle is the same as requiring all the eigenvalues of $\mathbf{L}$ (see (7.2.9)) to be contained within the stability diagram of the time-integration scheme. After all, if we discretize (7.2.8) further in time, $\mathbf{A}$ can be expressed in terms of $\mathbf{L}$. Thus, when time and space discretizations are separable it can be shown that the matrix method and the stability diagram are the same. Even in cases where time and space discretizations are not separable, by augmenting the vector of unknowns to multiple stages or time steps, the numerical method can be reduced to a mapping of the form (7.6.4) and thus the matrix method can be employed to investigate stability.

With central difference, in the limit of pure advection (i.e., when $D = 0$) $\mathbf{A}$ is skew-symmetric and therefore normal. In the limit of pure diffusion (i.e., when $C = 0$), $\mathbf{A}$ is symmetric and therefore normal. Thus, in these limits stability can be established in terms of the largest eigenvalue of $\mathbf{A}$ by magnitude. Unfortunately in the context of advection–diffusion equation $\mathbf{A}$ is non-normal. See Trefethen and Embree (2005), Chapter 12, for an excellent discussion on the non-normality of the advection–diffusion operator.[1]

Nevertheless, let us proceed to calculate the eigenvalues of $\mathbf{A}$ for the explicit Euler three-point method. We use Theorem 4.2.2 and obtain

$$\lambda_j = (1-2D) + 2\sqrt{D^2 - \frac{C^2}{4}}\cos\left[\frac{\pi j}{N}\right] \quad \text{for} \quad j = 1, 2, \ldots, N-1\,. \tag{7.6.11}$$

According to the matrix method the question of stability reduces to finding the largest eigenvalue in magnitude. Two different cases can be identified:

$$\text{if } D^2 \geq \frac{C^2}{4} \quad \lambda_{\min} = 1 - 2D - 2\sqrt{D^2 - \frac{C^2}{4}}\,, \tag{7.6.12}$$

and

$$\text{if } D^2 < \frac{C^2}{4} \quad |\lambda|_{\max} = \sqrt{1 - 4D + C^2}\,. \tag{7.6.13}$$

In the former case all the λ are real and are constrained to be below 1.0 in value. So we are concerned with the possibility of $\lambda_{\min} < -1$. In the latter case the eigenvalues are complex and we are concerned with their amplitude. From the above we can obtain the following conditions for stability:

$$\begin{aligned} &0 < D < \frac{1}{1 + \sqrt{1 - \left[\dfrac{C}{2D}\right]^2}} && \text{for} \quad \left[\frac{C}{2D}\right]^2 \leq 1\,, \\ &0 < D < \left[\frac{2D}{C}\right]^2 && \text{for} \quad \left[\frac{C}{2D}\right]^2 \geq 1\,. \end{aligned} \tag{7.6.14}$$

The above stability region of the explicit Euler three-point central difference method is plotted in Figure 7.6. Compared to Figure 7.4(b) the above stability region is much larger. The only difference between the two is the boundary condition. Figure 7.4(b) was for periodic boundary conditions and therefore the stability region obtained from von Neumann analysis and the matrix method are the same. Whereas Figure 7.6 was obtained for Dirichlet boundary conditions with the matrix method. What a difference boundary conditions seem to have caused! Changing from periodic to Dirichlet boundary conditions has greatly expanded the stable region. But how can this be? There can be hundreds to thousands of interior points and we have altered only the two boundary points. Could this affect stability so dramatically?

This situation may look discouraging – but there is hope. As we will see below,

[1] Our discussion has so far been on stability properties of the discrete matrix operator $\mathbf{A}$. The question of normality and non-normality is more general and extends to the continuous advection, diffusion and advection–diffusion operators defined in Examples 7.1.1–7.1.3. Trefethen and Embree discussed the non-normality of the continuous advection–diffusion equation with Dirichlet boundary conditions.

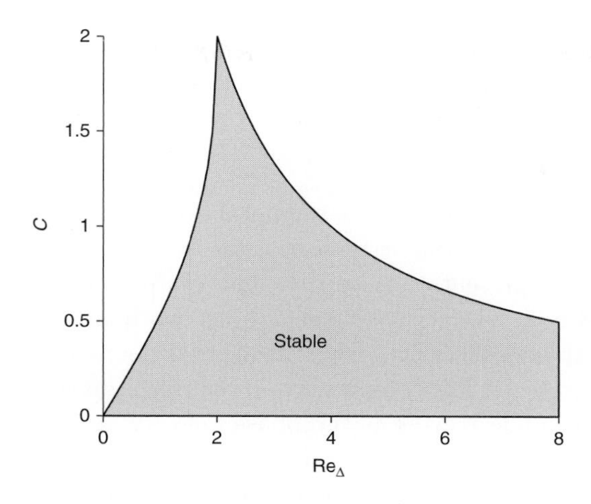

Figure 7.6 Stable region of the explicit Euler central difference method for the advection–diffusion equation with Dirichlet boundary conditions obtained with the matrix method. The above stable region plotted in the C–Re_Δ is for absolute instability, which can be compared with that shown in Figure 7.4(b), corresponding to a region of temporal instability. The region between the two boundaries then corresponds to convective instability.

the results of both von Neumann analysis and the matrix method, despite their limitations, are quite useful and they capture different stability properties of the numerical method. For now we point out that the larger region of stable behavior in Figure 7.6 is due to non-normality of $\mathbf{A}$. In this situation, $\sigma(\mathbf{A}) \leq 1$ is only a necessary condition and as a result the true stable region given by $\|\mathbf{A}\| \leq 1$ will be smaller than that given by the spectral radius (significantly smaller in the present case).

The stable region shown in Figure 7.6 goes back to the work of Siemieniuch and Gladwell (1978), who noted the curious fact that the numerical solution exhibits instability even when D and C are chosen to be within the stable region. Later workers, particularly Morton (1980), have clarified the role of non-normality and emphasized the relevance of von Neumann stability analysis even in the context of non-periodic boundary conditions.

7.6.1 Convective and Absolute Instabilities

We now proceed to the question: is there any physical significance to the stability region presented in Figure 7.6? The paper by Cossu and Loiseleux (1998) is of particular relevance. Let us first look at the behavior of the exact advection–diffusion equation, without the boundary influence. It admits normal modes of the form

$$\exp\left[\iota(kx - \omega t)\right] . \tag{7.6.15}$$

Substituting the above into the advection–diffusion equation we obtain the exact **dispersion relation**

$$\omega = ck - \iota\nu k^2 \,. \tag{7.6.16}$$

This relation connects the temporal evolution of the solution (i.e., the frequency ω) with the spatial structure of the solution (i.e., the wavenumber k). In this general representation both ω and k are complex and the dispersion relation is a complex relation between these two quantities. The phase velocity of the normal mode is ω_R/k_R, where R denotes the real part (the imaginary part will be denoted by I). The complex group velocity is given by $d\omega/dk$. While an individual normal mode travels at the phase velocity, a wave packet that forms from the superposition of normal modes travels at the group velocity (see Trefethen, 1982 for an interesting discussion on group velocity in finite difference schemes).

Different kinds of instabilities can be defined based on the dispersion relation. First, the condition for **temporal instability** is $\omega_I > 0$ for some real k. Temporal instability ($\omega_I > 0$) is the necessary and sufficient condition for the existence of any instability. If the flow is temporally unstable then one can proceed to examine if this instability is convective or absolute.

To investigate **convective instability** ω is taken to be real and the dispersion relation is considered as an expression for the complex wavenumber in terms of the real frequency. That is, (7.6.16) will be considered as equations for $k_R(\omega_R)$ and $k_I(\omega_R)$. From (7.6.15) each normal mode can be expressed as $\exp\left[\iota(k_R x - \omega_R t)\right] \exp\left[-k_I x\right]$. For $k_I > 0$ the disturbance grows upstream ($x < 0$) and for $k_I < 0$ the disturbance grows downstream ($x > 0$). Thus, it may appear that for any value ω_R the disturbance either grows along the upstream or downstream directions depending on the sign of k_I. However, this is not so. Only those convective modes that can be related to a temporal instability qualify as a genuine convective instability.

An absolute normal mode of the dispersion relation is defined as one with zero group velocity. In the complex k-plane, each value of k at which $d\omega/dk = 0$ is satisfied is an absolute mode with zero group velocity. Let us denote the complex wavenumber of such an absolute mode by k_a and the corresponding complex frequency by ω_a. The condition for **absolute instability** is that the imaginary part of the complex frequency be positive (i.e., the temporal growth rate of the absolute mode ω_{aI} must be positive). In the case of absolute instability, the perturbation grows and spreads in place without being convected upstream or downstream.

Absolute and convective instabilities have been much studied as part of hydrodynamic stability theory. In many flows the temporal instability appears first as convective instability above a critical Reynolds number Re_1. Only above a second, higher, critical Reynolds number Re_2 does convective instability give way to absolute instability. In other words, for $Re_1 > Re$ the flow remains stable, for $Re_2 > Re > Re_1$ the flow is convectively unstable and for $Re > Re_2$ the flow becomes absolutely unstable. As a result the stable region of absolute instability ($\mathrm{Re}_2 > \mathrm{Re}$) is larger than that of convective instability ($\mathrm{Re}_1 > \mathrm{Re}$).

From the exact dispersion relation (7.6.16) it can be easily verified that the advection–diffusion equation has no physical temporal instability (a result that is to be expected). Since the equation is temporally stable, there is no convective or absolute instability.

To study the corresponding behavior of the explicit Euler three-point central method, we obtain the following dispersion relation by substituting (7.6.15) into the difference equation (7.4.1) with (7.4.4) to obtain

$$\exp(-\iota\omega\Delta t) = 1 - \iota C \sin(k\Delta x) + 2D(\cos(k\Delta x) - 1)\,. \tag{7.6.17}$$

The temporal instability of the above dispersion relation yields conditions given in (7.4.13), which we previously obtained with von Neumann stability analysis. Most interestingly, Cossu and Loiseleux (1998) showed that the absolute instability of the above dispersion relation yields conditions given in (7.6.14), which were obtained from the matrix method.

We are now ready to interpret the two different stability boundaries that we have obtained so far. Figure 7.4(b) shows the boundary between stability and temporal instability, while Figure 7.6 corresponds to the region of absolute stability. In the region between these two stability boundaries the numerical solution will be convectively unstable. These different regions of stability and instability are shown in Figure 7.7. In the region below the black line the numerical solution is stable, while above the gray line the solution is absolutely unstable.

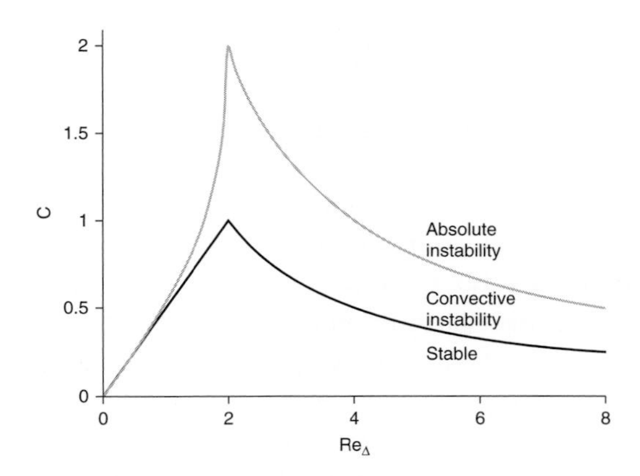

Figure 7.7 A composite picture of the stable and unstable regions of the explicit Euler central difference method for the advection–diffusion equation. The gray line is the stability boundary obtained with Dirichlet boundary conditions using the matrix method and was shown in Figure 7.6. The black line is the result from von Neumann stability analysis presented in Figure 7.4(b). The region below the black line is the region of stability. In the region between the black and gray lines the numerical solution will be convectively unstable. Above the gray line will result in absolute instability.

Problem 7.6.1 Consider the dispersion relation (7.6.17). Obtain (7.4.13) as the condition for temporal instability, and (7.6.14) as the condition for absolute instability.

With periodic boundary conditions any disturbance that convects upstream or downstream will exit the domain, but re-enter through the periodic boundary. In essence, with periodic boundary conditions a locally convective instability becomes globally absolute, whereas with non-periodic boundary conditions, convective instabilities will leave the domain if you wait long enough. Only absolute instability will continue to grow in place. Thanks to the assumption of periodic boundary conditions, von Neumann analysis captures both locally absolute and convective instabilities, whereas for non-periodic boundary conditions, the matrix method identifies only the region of absolute instability.

It must be amply clear that, even in case of linear constant coefficient PDEs, the topic of the stability of a numerical method is not a simple matter. Issues such as nonlinearity, variable coefficients and lack of translational invariance can further complicate the situation. There has been significant advancement in our understanding of properties of non-normal operators and matrices and their role in the stability analysis of numerical methods. We recommend the book by Trefethen and Embree (2005) and the numerous references given there. For our purposes of an introduction to CFD, here we summarize the following main points:

(1) As argued by Morton (1980), von Neumann analysis (using Fourier modes) is relevant even in case of non-periodic boundary conditions.
(2) Instead of von Neumann analysis, the stability diagram approach can be used with the eigenvalues of the circulant spatial operator (i.e., assuming periodic boundary condition). Though conclusions will be the same, the stability diagram approach allows separation of the time-integration and spatial discretization methods.
(3) If time and space discretizations are coupled, as in the Lax or Du Fort–Frankel methods, then von Neumann analysis or the matrix method can be pursued.
(4) If the mapping operator $\mathbf{A}$ is normal then the matrix method will yield stability conditions identical to the von Neumann and stability diagram approaches.
(5) If the mapping operator $\mathbf{A}$ is non-normal, the condition $\sigma(\mathbf{A}) \leq 1$ is only a necessary condition. The eigenvalue spectrum in this case is not very useful. Following the ideas pioneered by Trefethen one can define a pseudospectrum. The condition for stability can then be stated as: the pseudospectrum of $\mathbf{A}$ must lie within the unit circle. For details we ask the reader to consult Trefethen and Embree (2005).

A few more comments are in order. We discussed the scenario of convective instabilities leaving the domain in case of non-periodic boundary conditions. In a numerical method this requires careful implementation of the boundary conditions. Chapter 34 of Trefethen and Embree (2005) presents an example where an improper implementation of the boundary condition reflects the outgoing convective disturbance and converts it to a globally absolute instability. Furthermore, the boundary condition itself could be a source of non-modal instability, whose analysis goes by the name **GKS stability**, after the authors of the classic paper Gustafsson *et al.* (1972). This theory is complicated and beyond the scope of this book. In the following section we will present a simple example that illustrates what boundary conditions can do.

7.6.2 Boundary-induced Oscillations

There is more to boundary conditions. Let us consider a numerical solution of the advection–diffusion equation starting from an initial condition $u(x, t = 0) = 0$, with Dirichlet boundary conditions:

$$u(0,t) = 0 \quad \text{and} \quad u(2\pi, t) = 1\,. \tag{7.6.18}$$

Just one time step after the start (i.e., at $t = \Delta t$), from (7.6.2) the solution at the first grid point to the left of the $x = 2\pi$ boundary (i.e., at x_{N-1}) can be written as

$$V_{N-1,1} = \left[D + \frac{C}{2}\right] V_{N-2,0} + (1 - 2D)V_{N-1,0} + \left[D - \frac{C}{2}\right], \tag{7.6.19}$$

where we have used $V_{N,0} = 1$. Using the fact that u is initially zero everywhere we obtain the following result:

$$V_{N-1,1} = \left(D - \frac{C}{2}\right) = \frac{D}{2}(2 - Re_\Delta)\,, \tag{7.6.20}$$

where in the second expression we use the definition of cell Reynolds number. For $\mathrm{Re}_\Delta > 2$, we make the very interesting observation that $V_{N-1,1}$ is negative! The numerical solution at all other points to the left of x_{N-1} will remain zero after the first step. Thus, for $\mathrm{Re}_\Delta > 2$ the numerical solution develops an unphysical oscillation. Contrast this with the exact solution, which is always positive and monotonically decreases to the left of the right boundary for all time.

In fact, if we continue to solve the advection–diffusion equation using the explicit Euler three-point central difference scheme for $\mathrm{Re}_\Delta > 2$, the oscillations will persist and propagate into the domain. We can choose Δx and Δt carefully so that $\mathrm{Re}_\Delta < 2/C$. Then, as per (7.4.15) and Figure 7.6, we are within the stable region. Even then the solution will exhibit grid-point-to-grid-point oscillation if $\mathrm{Re}_\Delta > 2$ but the oscillation will remain bounded. In this sense the numerical solution is stable – but it is useless due to unphysical oscillations. These oscillations, or wiggles, are known as **two-delta waves**. You may recall that in

Section 2.2 on explicit Euler advancement of a mapping, we saw such stable but spuriously oscillatory solutions, where it was called overstability.

These numerical oscillations are trying to tell us some physics. To understand this let us look at the exact steady-state solution that will be approached after very long time:

$$u(x, t \to \infty) = \frac{\exp\left(\mathrm{Re}\,\dfrac{x}{2\pi}\right) - 1}{\exp(\mathrm{Re}) - 1}\,, \tag{7.6.21}$$

where $\mathrm{Re} = 2\pi c/\nu$ is, as usual, the Reynolds number. This solution exhibits a boundary layer structure; for large Re, the solution falls off rapidly over a thin region close to the boundary $x = 2\pi$, and over the rest of the domain the solution remains close to zero. The thickness of the boundary layer can be evaluated from the exact steady-state solution to be $\delta = 2\pi/\mathrm{Re}$. Now the condition $\mathrm{Re}_\Delta < 2$ for avoiding oscillation can be restated in terms of the boundary layer thickness as

$$\Delta x \leq 2\delta\,. \tag{7.6.22}$$

In other words, to avoid oscillation there must be at least one grid point within the first two boundary layer thickness.

Physically, with $c > 0$, advection tends to move the solution to the right and steepen the gradient near the boundary $x = 2\pi$, while diffusion tends to diffuse the solution and spread it away. The relative strength of these two mechanisms in a numerical implementation is given by Re_Δ. When Re_Δ increases above 2, advection dominates over diffusion and the numerical solution exhibits unphysical oscillation.

The boundary layer that we are concerned with here is one where there is normal velocity approaching a boundary. This is the kind of boundary layer you would encounter in a stagnation point flow, for example, at the leading edge of an airfoil. Here wall-normal advection and diffusion compete, giving rise to the possibility of two-delta waves. In the case of Blasius-type boundary layers, advection is predominantly in the streamwise direction, while diffusion is along the wall-normal direction and there is no direct competition. Nevertheless, what we have observed is a robust and a powerful result. It provides a very useful practical guideline even when we go on to solve more complex nonlinear equations, such as the Navier–Stokes equation. In general, it is a good idea to have sufficient number of grid points within the momentum and thermal boundary layers in order to avoid two-delta waves.

If one insists on using a larger cell Reynolds number, one way to avoid the two-delta waves of the central difference is to upwind the advection operator as outlined in Section 7.4.2. For the upwind scheme given in (7.4.16), just one time step after the start, the solution at the first grid point to the left of the $x = 2\pi$

boundary can be written as

$$V_{N-1,1} = \left[D + \frac{C}{2}(1-w)\right] V_{N-2,0} + (1-2D)V_{N-1,0} + \left[D - \frac{C}{2}(1-w)\right] . \tag{7.6.23}$$

The condition for avoiding two-delta waves becomes

$$\mathrm{Re}_\Delta < \frac{2}{1-w} . \tag{7.6.24}$$

Thus, in the limit of pure upwinding (i.e., $w = 1$) there is no limit on cell Reynolds number. Unfortunately this comes at a significant loss of accuracy. As the saying goes "there is no free lunch." For $w = 1$ the modified equation that corresponds to the discretization (7.4.16) is

$$\frac{\partial u}{\partial t} + c\frac{\partial u}{\partial x} + \nu\left[1 + \frac{\mathrm{Re}_\Delta}{2}\right]\frac{\partial^2 u}{\partial x^2} = -c^2\frac{\Delta t}{2}\frac{\partial^2 u}{\partial x^2}\cdots . \tag{7.6.25}$$

It can be seen that upwinding stabilizes the two-delta oscillation with an added numerical viscosity of $\nu\,\mathrm{Re}_\Delta/2$. So, if we insist on $\mathrm{Re}_\Delta \gg 2$, an upwind discretization will be stable and free of oscillation, but the numerical viscosity will be much larger than the physical viscosity and as a result we will obtain a physically meaningless numerical solution with a very thick boundary layer. This lack of accuracy with upwinding can be partially addressed with higher-order upwinding, with methods such as **quadratic upwind interpolation for convective kinematics**. This method was originally proposed by Leonard (1979) and is popularly known as the **QUICK scheme**. For further information you can consult CFD textbooks (for example Versteeg and Malalasekera, 2007).

7.7 Burgers' Equation

Much of what we have done in formulating methods for linear PDEs can be carried over to nonlinear PDEs that are written in flux conservation form. What we mean by this is probably most easily seen by stating an example. Take Burgers' equation,

$$\frac{\partial u}{\partial t} + u\frac{\partial u}{\partial x} = \nu\frac{\partial^2 u}{\partial x^2} , \tag{7.7.1}$$

which is nothing but the advection–diffusion equation, with the important difference that advection is now at velocity u as opposed to the constant velocity c. We write the Burgers equation in the general form

$$\frac{\partial u}{\partial t} + \frac{\partial F}{\partial x} = 0 , \tag{7.7.2}$$

where the flux $F = (u^2/2 - \nu\frac{\partial u}{\partial x})$ and it depends on u and its spatial derivative. The above is also called the **conservation form**.

When $\nu = 0$ we get the inviscid Burgers equation. An exact solution to this equation can be constructed using the method of characteristics. For an initial condition $u(x,t=0)=f(x)$, the solution remains constant along the characteristic $x=x_0+f(x_0)t$ and takes the form

$$u(x,t)=f(x-f(x_0)t)\,, \tag{7.7.3}$$

where x_0 is the initial location of the characteristic. If $f(x_0)$ is not a constant then all the characteristics will not be of the same slope and therefore a key feature of the inviscid Burgers equation is the formation of shocks due to the intersection of the characteristics. For a right-moving flow ($u>0$), the earliest time when the characteristics start to intersect and form a shock can be estimated as

$$t_{\mathrm{sh}}=-\frac{1}{\min\left(\dfrac{\mathrm{d}f}{\mathrm{d}x}\right)}\,. \tag{7.7.4}$$

Thus, a shock will form only when velocity decreases with x (i.e., when $\mathrm{d}f/\mathrm{d}x<0$). Regions of velocity increase will result in an expansion fan. Corresponding results can be obtained for a left-moving ($u<0$) flow as well. The viscous Burgers equation can also be analytically treated using the Cole–Hopf transformation, which converts Burgers' equation to a diffusion equation. The main effect of viscosity is to regularize the shock and give it a tanh shape of finite thickness.

We can now proceed to discretize the conservation form of Burgers' equation in much the same way we discretized the special cases of advection equation ($F=cu$) and the advection–diffusion equation ($F=cu-\nu\partial u/\partial x$) in the earlier sections. Hence, you will see the terms "Lax method" etc., applied in this wider context also. The stability of these methods for the Burgers equation, however, is more complicated since (7.7.2), in general, will not have plane wave solutions of the form used in von Neumann stability analysis.

Under rather general conditions the solution $u(x,t)$ will be expected to die out at infinity. Thus,

$$\int_{-\infty}^{\infty}\frac{\partial u}{\partial t}\mathrm{d}x=\frac{\mathrm{d}}{\mathrm{d}t}\left[\int_{-\infty}^{\infty}u\mathrm{d}x\right]=-F(\infty)+F(-\infty)=0\,, \tag{7.7.5}$$

and we expect $\int_{-\infty}^{\infty}u\mathrm{d}x$ to be conserved in time. The question arises as to when the numerical method will respect this important feature of the analytical problem. The issue of what integrals of motion of an analytical problem will be retained by a numerical method is an important one. Usually only a few (if there are many) can be accommodated numerically. Sometimes spurious ones arise. We can show examples of both phenomena. Here we simply remark that if a numerical method of the form

$$V_{j,n+1}-V_{j,n}=\frac{\Delta t}{\Delta x}\left[G(V_{j+k-1,n},\ldots,V_{j-k,n})-G(V_{j+k,n},\ldots,V_{j-k+1,n})\right] \tag{7.7.6}$$

can be found, then the numerical solution will have a counterpart of the analytical integral. The sum over all nodal values will be invariant in time. Methods of the form (7.7.6) are referred to as **conservative**. For the inviscid case, where shocks form, the treatment represented by (7.7.6) is called **shock capturing**.

A variety of numerical options are available for solving the Burgers equation in conservation form. The simplest among them is upwinding for the spatial derivative in (7.7.2). If higher-order accuracy is desired the multi-step Lax–Wendroff method of Problem 7.4.4 can be adopted as

$$\frac{V_{j,n+1} - V_{j,n}}{\Delta t} = \frac{F_{j+\frac{1}{2},n+\frac{1}{2}} - F_{j-\frac{1}{2},n+\frac{1}{2}}}{\Delta x}, \tag{7.7.7}$$

where

$$\frac{V_{j+\frac{1}{2},n+\frac{1}{2}} - \frac{1}{2}\left[V_{j+1,n} + V_{j,n}\right]}{\Delta t/2} = \frac{F_{j+1,n} - F_{j,n}}{\Delta x}, \tag{7.7.8}$$

with a similar equation for $V_{j+\frac{1}{2},n+\frac{1}{2}}$.

We can also consider other important ideas, somewhat reminiscent of the second-order Runge–Kutta method for ODEs, contributed by MacCormack (1969) and widely known as the **MacCormack method**. It looks like this:

$$\hat{V}_{j,n+1} = V_{j,n} - \frac{\Delta t}{\Delta x}\left[F_{j+1,n} - F_{j,n}\right], \tag{7.7.9}$$

$$V_{j,n+1} = V_{j,n} - \frac{\Delta t}{2\Delta x}\left[F_{j+1,n} - F_{j,n} + \hat{F}_{j,n+1} - \hat{F}_{j-1,n+1}\right], \tag{7.7.10}$$

where the $\hat{F}$s are evaluated using the intermediate values $\hat{u}$.

Problem 7.7.1 Show that if $F = cu$, the MacCormack method reduces to the FTCS method with $2D = C^2$.

7.7.1 Finite Volume Approach

So far the focus of the book has been from the point of view of grid points and finite difference approximations. Here we will venture into the cell- (or element-) based finite volume method, but only very briefly. This is an important approach in CFD and there is a large body of literature with many good textbooks. It is particularly well suited for equations written in conservation form and so this is a good point for its introduction.

In the finite volume approach we associate a cell E_j with the point x_j and the spatial extent of this cell is given as $E_j \in \left[x_{j-1/2}, x_{j+1/2}\right]$. The typical working variable is the average value of the solution within this cell

$$\bar{u}_j = \int_{x_{j-1/2}}^{x_{j+1/2}} u(x,t)\,dx\,. \tag{7.7.11}$$

Integrate (7.7.2) across the jth cell and use Gauss' theorem to obtain

$$\frac{\mathrm{d}\bar{u}_j}{\mathrm{d}t} = -\frac{1}{\Delta x}\left[F(x_{j+1/2},t) - F(x_{j-1/2},t)\right] . \tag{7.7.12}$$

We then integrate over one time step to obtain

$$\bar{u}_j(t_{n+1}) = \bar{u}_j(t_n) - \frac{\Delta t}{\Delta x}\left[\hat{F}(x_{j+1/2}) - \hat{F}(x_{j+1/2})\right] , \tag{7.7.13}$$

where $\hat{F}(x_{j-1/2})$ represents the flux at $x_{j-1/2}$ integrated over the nth time step. So far there is no approximation: equation (7.7.13) is exactly Burgers' equation integrated over a finite cell volume and over one time step. So the problem reduces to one of calculating $\hat{F}$ at all the cell interfaces given all the cell average values $\bar{u}_j$. Different numerical approximations to $\hat{F}$ give rise to different finite volume methods.

Let the numerical representation of $\bar{u}$ at time level t_n be denoted simply as $V_{j,n}$ (let us drop the overbar, since we know V is numerical approximation, which here is the cell average). The numerical problem can now be stated as follows: at the $(j-1/2)$ cell interface between the $(j-1)$th and jth elements we need to approximate the time-averaged flux as

$$\hat{F}(x_{j-1/2}) \approx G(V_{j-1,n}, V_{j,n}) , \tag{7.7.14}$$

in terms of the left and right cell-averaged values. For the inviscid Burgers equation ($F = u^2/2$) an explicit central difference approximation can be written as

$$G(V_{j-1,n}, V_{j,n}) = \frac{1}{4}\left[(V_{j-1,n})^2 + (V_{j,n})^2\right] . \tag{7.7.15}$$

As we have seen in earlier sections, it is not a good idea to use central difference for the advection problem. Therefore, an upwind scheme may be a more stable choice. In which case an explicit approximation can be written as

$$G(V_{j-1,n}, V_{j,n}) = \begin{cases} \frac{1}{2}\,(V_{j-1,n})^2 & \text{if } V_{j-1,n}, V_{j,n} > 0\,, \\ \frac{1}{2}\,(V_{j,n})^2 & \text{if } V_{j-1,n}, V_{j,n} < 0\,. \end{cases} \tag{7.7.16}$$

Of course, the explicit upwind scheme will be only first-order accurate in both space and time. Furthermore, the CFL condition will apply for stability. For higher-order accuracy there is an extensive body of research and it is mainly divided between two related issues.

The first relates to the **Riemann problem**, which for the Burgers equation will correspond to the initial condition

$$u(x,0) = \begin{cases} u_L & \text{for } x < 0\,, \\ u_\mathrm{R} & \text{for } x > 0\,. \end{cases} \tag{7.7.17}$$

Solving the Riemann problem is straightforward for the inviscid Burgers equation. Since the finite volume approach approximates the solution with a jump across each interface (for example, $V_{j-1,n} \neq V_{j,n}$ at $x_{j-1/2}$), the flux through the interface can be calculated with the solution of the corresponding Riemann problem and integrating the solution between t_n and t_{n+1} to obtain an approximation for $\hat{F}(x_{j-1/2})$.

Exact solutions to the Riemann problem are available for the Euler equation and a number of other hyperbolic equations in fluid mechanics. The book by Toro (1999) contains an excellent summary of these exact solutions and, more importantly, how you can numerically approximate the solution to the Riemann problem.

For a higher-order finite volume approach, the second issue relates to knowing the left and right states $(u_{\mathrm{L}}, u_{\mathrm{R}})$ of an interface from the cell-averaged values. These left and right states are needed in order to solve the Riemann problem at each interface. The simplest approach is to take u_{L} to be the cell average of the cell to the left of the interface and u_{R} to be the cell average of the cell to the right of the interface. But you can do better by using cell averages of neighboring cells to obtain improved estimates for u_{L} and u_{R} at each interface. This goes by the name **reconstruction**. But reconstruction is not a straightforward task: remember Godunov's theorem – any linear higher-order method will result in oscillations. MUSCL and WENO type schemes, along with exact or numerical approximation to the Riemann problem is what is needed. We will not pursue this topic further, since there are excellent texts on finite volume methods for conservation equations such as Toro (1999).

7.8 Implicit Time-differencing

We saw already in Chapters 2 and 3, admittedly in a quite preliminary way, that implicit time-differencing can lead to enhanced stability properties. In this section we explore how this works in the present context of the advection–diffusion equation. For example, it is possible to enhance the stability of the FTCS method described in earlier sections by using an implicit method for time integration.

Consider then the following discretization:

$$\left(\frac{V_{j,n+1} - V_{j,n}}{\Delta t}\right) = -c\left(\frac{V_{j+1,n+1} - V_{j-1,n+1}}{2\Delta x}\right) + \nu\left(\frac{V_{j+1,n+1} - 2V_{j,n+1} + V_{j-1,n+1}}{\Delta x^2}\right), \quad (7.8.1)$$

aimed, once again, at integrating the advection–diffusion equation. Since the spatial derivatives are entirely computed at the next time level, this method is called **fully implicit** (FI).

Problem 7.8.1 Using von Neumann analysis show that the fully implicit method is unconditionally stable for all values of C and D.

7.8.1 Stability Versus Accuracy

Some general remarks on implicit methods are in order at this point. As we have just seen, one great advantage of such methods is their stability properties. But we have also found explicit methods that are stable, so what is the real issue? In the preceding chapter we identified, for any given spatial discretization, two general criteria for the maximum allowable time step of an explicit scheme. On one hand we have that the CFL limit of the form $c\Delta t/\Delta x$ must be smaller than a constant. On the other hand, we have that the diffusion limit of the form $\nu\Delta t/\Delta x^2$ must be smaller than a constant. It turns out that the first of these is usually not the main obstacle. Although the CFL limit may restrict Δt to be less than $\Delta x/c$, a method that is first-order accurate in time and second-order accurate in space will in general not give sufficient accuracy if time-integrated at the CFL stability limit. Methods that are second or higher order in time can be constructed, and such methods can then be run at their CFL stability limit. The second condition, the one that limits Δt based on diffusion number, is usually the real bottleneck.

There are several ways of phrasing this. Keep in mind that the diffusivity ν is going to be the kinematic viscosity when the independent variable u becomes the velocity field. Now for reasons of accuracy, we want to limit the Courant number to $O(1)$. This is true irrespective of the order of the method, the physical basis being the argument given in connection with Figure 7.2. If the Courant condition must be satisfied anyway for accuracy reason, it is not essential to have implicit time-stepping for the advection term. Recall the relation $D = \nu\Delta t/\Delta x^2 = C/\mathrm{Re}_\Delta$, where Re_Δ is the cell Reynolds number. From this it is easy to see that for large Re_Δ the diffusion limit on D will be more constraining than the Courant condition. In which case, for diffusion stability, you may be forced to choose a time step much smaller than what is needed, based either on accuracy or advection stability. There are many occasions where one may want to use a large cell Reynolds number, in particular away from the boundaries.

Such an occasion is demonstrated most clearly in a turbulent flow, where the grid spacing is typically chosen to be of the order of the Kolmogorov length scale (i.e., $\Delta x \approx \eta$). From turbulence theory it is known that the Reynolds number of the Kolmogorov scale eddies is $O(1)$. This can be written as $\eta v_\eta/\nu = 1$, where v_η is the velocity of the Kolmogorov scale eddies. However, the advection velocity that appears in the cell Reynolds number is the local fluid velocity u, which includes contribution from all the scales. In general, $u \gg v_\eta$ and as a result we can expect $\mathrm{Re}_\Delta \gg 1$ and thus diffusion limit on time step will be restrictive in turbulent flows.

Furthermore, in the context of Navier–Stokes equations, the advection term is nonlinear and therefore it is not easy to implement an implicit method for it. Fortunately, in a constant viscosity incompressible flow, the diffusion term is linear and as a result it is possible to implement an implicit scheme for the

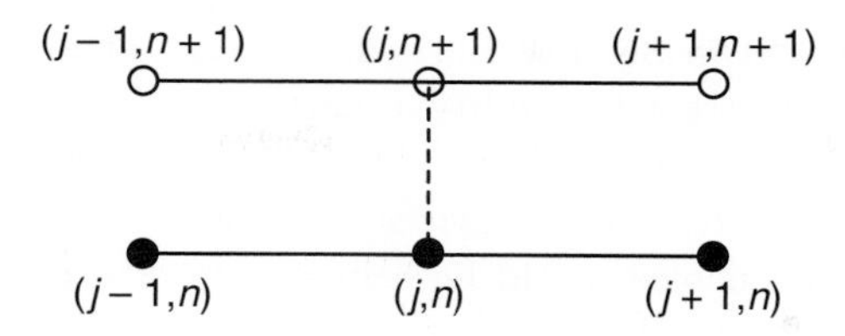

Figure 7.8 Space-time stencil for the Crank–Nicolson method.

diffusion term. Such mixed implicit–explicit schemes are often used in CFD, and we will discuss them in Chapter 8.

7.8.2 The Crank–Nicolson Method

Let us consider an approach that can be thought of as a combination of FTCS, (7.2.23), and FI, the fully implicit method, (7.8.1). Now consider the family of methods symbolically written as

$$M(w) = w(\text{FTCS}) + (1 - w)(\text{FI}), \tag{7.8.2}$$

where w is some weight parameter $0 \le w \le 1$. The notation in (7.8.2) is supposed to denote that the method $M(w)$ arises by weighting FTCS by w and adding it to FI weighted by $1 - w$.

Problem 7.8.2 Using von Neumann analysis obtain the stability condition for the above method as

$$D(2w - 1) \le \frac{1}{2}; \qquad C^2(2w - 1) \le 2D. \tag{7.8.3}$$

The stability conditions given in (7.8.3) reduce to (7.4.13) for $w = 1$, and are identically satisfied for $w = 0$ in accordance with our earlier findings. We note the remarkable fact that for $w = 1/2$ conditions (7.8.3) are satisfied regardless of the values of C and $D > 0$. (The qualification $D > 0$, which we keep running into, means that the term in the advection–diffusion equation with coefficient ν acts as a true, physical diffusion. If this is not the case, the continuum equation itself is not well-posed, and it would be unreasonable to ask any numerical method to have stable solution!) The method that arises for $w = 1/2$ is known as the **Crank–Nicolson method**. Although from desirable stability properties we could choose any value of w in the range 0 to 1/2, as we saw in Chapter 3, only the Crank–Nicolson scheme with the choice $w = 1/2$ is second-order accurate. As a result this scheme is often used in CFD. Figure 7.8 shows the space-time stencil for the Crank–Nicolson method.

The advection–diffusion equation is linear and therefore in (7.8.2) we could easily consider an implicit method for both the advection and diffusion terms.

As remarked in Section 7.8.1, in the case of nonlinear advection, it might be beneficial to treat the diffusive part by an implicit method, while retaining an explicit method for the advective part. In which case we can set $w = 1/2$ for the diffusion term and leave $w = 0$ for the nonlinear advection term (a more appropriate choice would be a higher-order accurate Runge–Kutta or Adams–Bashforth method for the advection term).

7.8.3 Second-order Accuracy

The possibility of higher-order accuracy in both space and time is desired. The possibility of such higher-order methods is important in its own right. We mention two classical examples, the **Lax–Wendroff method** and the **leap-frog DuFort–Frankel method**.

We show the space-time diagram for the leap-frog DuFort–Frankel method in Figure 7.9. In the Lax–Wendroff method three grid values, $j-1$, j and $j+1$ are used to produce two half-step values $(j-1/2, n+1/2)$ and $(j+1/2, n+1/2)$ via Lax's method. These two half-step points along with the original point (j, n) are then used in a leap-frog method to produce the value at $(j, n+1)$. Both the DuFort–Frankel and the Lax–Wendroff methods are three level methods. The decoupling in the leap-frog method leads to a decorrelation of odd and even time level solution values, a phenomenon known as **mesh drifting**. The Lax–Wendroff method cures this problem.

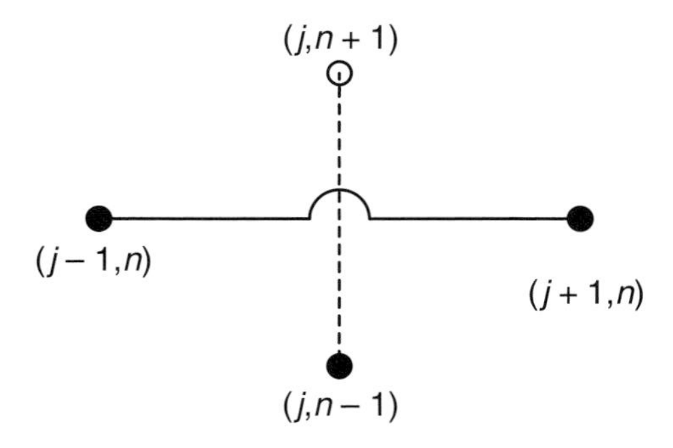

Figure 7.9 Space-time stencil for the leap-frog DuFort–Frankel method.

It turns out that Godunov's theorem can be circumvented and monotonicity can be preserved using a second-order method but one must use a time-stepping procedure that is nonlinear. The discussion of this in Roe (1986) is particularly useful.

7.9 Direct Solution with Matrix Representation

Every one of the space-time discretizations considered so far reduces to a mapping for the unknown vector $\vec{V}_{n+1}$, which contains the value of the function to be evaluated at the grid points at the new time, to the vector of function values $\vec{V}_n$ and possibly $\vec{V}_{n-1}$ that have already been computed in the previous time levels. Here we will illustrate this in the context of the advection–diffusion equation. Following our earlier recommendation we will use an explicit scheme (second-order Adams–Bashforth) for the advection term and an implicit scheme (Crank–Nicolson) for the diffusion term. The resulting equation, where only time has been discretized, can be written as

$$\left(\frac{V_{n+1}-V_n}{\Delta t}\right) = -\frac{3c}{2}\left(\frac{\mathrm{d}V}{\mathrm{d}x}\right)_n + \frac{c}{2}\left(\frac{\mathrm{d}V}{\mathrm{d}x}\right)_{n-1} + \frac{\nu}{2}\left(\frac{\mathrm{d}^2V}{\mathrm{d}x^2}\right)_{n+1} + \frac{\nu}{2}\left(\frac{\mathrm{d}^2V}{\mathrm{d}x^2}\right)_n . \tag{7.9.1}$$

In this semi-discretized notation V_n, V_{n+1}, etc., are functions of x.[2] We have used the second-order Adams–Bashforth (AB2) scheme for illustration purposes only. By now you must know that there can potentially be problems with the above scheme. Recall from Figure 3.5 that the stability diagram of the AB2 scheme does not include any portion of the imaginary axis. Thus AB2 is not a stable scheme for a purely advection problem. Here, the presence of additional physical diffusion can stabilize the scheme. However, we strongly recommend the use of a third-order Adams–Bashforth or third-order Runge–Kutta scheme for the advection operator. Nevertheless, for simplicity of illustration, we will continue to use the AB2 scheme.

Equation (7.9.1) can be rearranged to bring all the unknown quantities to the left-hand side

$$\left[1-\frac{\nu\,\Delta t}{2}\frac{\mathrm{d}^2}{\mathrm{d}x^2}\right]V_{n+1} = \left[1-\frac{3c\,\Delta t}{2}\frac{\mathrm{d}}{\mathrm{d}x}+\frac{\nu\,\Delta t}{2}\frac{\mathrm{d}^2}{\mathrm{d}x^2}\right]V_n + \left[\frac{c\,\Delta t}{2}\frac{\mathrm{d}}{\mathrm{d}x}\right]V_{n-1} . \tag{7.9.2}$$

The above equation is in a convenient form for our later discussion. However, a more transparent way of expressing the left-hand side is to multiply it by a constant $-1/\alpha$, where $\alpha = \nu\Delta t/2$, to obtain

$$\frac{\mathrm{d}^2V_{n+1}}{\mathrm{d}x^2} - \frac{1}{\alpha}V_{n+1} = \mathrm{RHS} . \tag{7.9.3}$$

It can be seen that the effect of semi-discretization in time is to reduce the advection–diffusion equation to a **Helmholtz equation** for V_{n+1}. This simplifies our task to numerically solving the 1D Helmholtz equation at each time-step.

[2] Contrast this against the semi-discretized equation given in (7.2.8), where x was discretized into grid points, whereas time remained continuous. Therefore in (7.2.8) the vector $\vec{V}$ of function values at the grid points depended on t.

We now proceed with spatial discretization. The domain is divided into finite number of grid points and we introduce $\vec{V}_{n+1}$ as the column vector of the function values at these grid points. Furthermore, we choose any one of the finite difference, or compact difference, or spectral approximation to the first and second derivatives. Let $\mathbf{D}_1$ and $\mathbf{D}_2$ be the first- and second-derivative matrices. For example, $\mathbf{D}_1$ and $\mathbf{D}_2$ of the three-point central difference approximation are given in (4.2.2) and (4.2.4). The fully discretized advection–diffusion equation can be expressed in matrix form as

$$\mathbf{L}\,\vec{V}_{n+1} = \mathbf{R}_n\,\vec{V}_n + \mathbf{R}_{n-1}\,\vec{V}_{n-1} = \vec{r}\,, \tag{7.9.4}$$

where the left- and right-hand side matrix operators are

$$\mathbf{L} = \left[\mathbf{I} - \frac{\nu\,\Delta t}{2}\mathbf{D}_2\right], \quad \mathbf{R}_n = \left[\mathbf{I} - \frac{3c\,\Delta t}{2}\mathbf{D}_1 + \frac{\nu\,\Delta t}{2}\mathbf{D}_2\right],$$
$$\mathbf{R}_{n-1} = \left[\frac{c\,\Delta t}{2}\mathbf{D}_1\right]. \tag{7.9.5}$$

The matrix operators $\mathbf{L}$, $\mathbf{R}_n$ and $\mathbf{R}_{n-1}$ can be precomputed. From the known vectors $\vec{V}_n$ and $\vec{V}_{n-1}$ at each time-step, we can calculate the vector of function values at the next time step $\vec{V}_{n+1}$.

The above description is for a periodic problem and therefore no special boundary treatment is necessary. Care must be taken to choose $\mathbf{D}_1$ and $\mathbf{D}_2$ to be circulant matrices that incorporate the periodic boundary condition. Now we will elaborate on the matrix form of the advection–diffusion equation when non-periodic boundary conditions are involved. Consider the following Dirichlet boundary conditions at the two ends

$$u(x=0,t) = U_{\mathrm{L}}(t) \quad \text{and} \quad u(x=2\pi,t) = U_{\mathrm{R}}(t)\,. \tag{7.9.6}$$

The Dirichlet boundary conditions can be easily incorporated into (7.9.4) by redefining the matrix operator on the left-hand side as

$$\mathbf{L}_* = \begin{bmatrix} 1 \;\; 0 \;\; 0 & \cdots & \\ & \text{Rows 2 to } N \text{ same as } \mathbf{L} & \\ & \cdots & 0 \;\; 0 \;\; 1 \end{bmatrix}. \tag{7.9.7}$$

The right-hand side of (7.9.4) is the column vector $\vec{r}$ and the Dirichlet boundary conditions are enforced by modifying it as

$$\vec{r}_* = \begin{bmatrix} U_{\mathrm{L}} \\ \text{Elements 2 to } N \text{ same as } \vec{r} \\ U_{\mathrm{R}} \end{bmatrix}. \tag{7.9.8}$$

The resulting linear system, $\mathbf{L}_*\,\vec{V}_{n+1} = \vec{r}_*$ can be readily solved.

Although the above approach is quite adequate and works well in 1D, in order to develop an efficient direct solver for the Helmholtz equation in two and three dimensions, we now introduce a slightly modified approach as an alternative. We make two observations. First, the two end elements of the vector $\vec{V}_{n+1}$ are already known from the Dirichlet boundary conditions. Second, the discretized equation (7.9.4) needs to be solved only at the interior points. We now refer you back to Section 4.2.1 where we defined the interior second-derivative matrix $\mathbf{D}_{2\mathrm{I}}$. We can make use of it here, since it is constructed in such a manner that it operates only on the interior values of $\vec{V}_{n+1}$ to yield the second-derivative values at the interior points. As can be seen in (4.2.25), the known boundary values of $\vec{V}_{n+1}$ will contribute to the boundary-correction vectors. We now define the interior portion of the vector of unknowns being solved and the interior portion of the known right-hand side as

$$\vec{V}_{\mathrm{I},n+1} = \begin{bmatrix} V_{1,n+1} \\ V_{2,n+1} \\ \vdots \\ \\ V_{N-1,n+1} \end{bmatrix} \quad \text{and} \quad \vec{r}_{\mathrm{I}} = \begin{bmatrix} r_1 \\ r_2 \\ \vdots \\ \\ r_{N-1} \end{bmatrix} . \tag{7.9.9}$$

The discretized Helmholtz equation can now be written in terms of the interior second-derivative matrix $\mathbf{D}_{2\mathrm{I}}$ and the boundary-correction vectors as

$$\mathbf{L}_{\mathrm{I}}\, \vec{V}_{\mathrm{I},n+1} = \vec{r}_{\mathrm{I}} + \frac{\nu\Delta t}{2}\left[U_{\mathrm{L}}(t_{n+1})\vec{C}_{\mathrm{LI}} + U_{\mathrm{R}}(t_{n+1})\vec{C}_{\mathrm{RI}}\right] , \tag{7.9.10}$$

where

$$\mathbf{L}_{\mathrm{I}} = \left[\mathbf{I} - \frac{\nu\,\Delta t}{2}\mathbf{D}_{2\mathrm{I}}\right] . \tag{7.9.11}$$

Note that the above matrix equation is a system of $N-1$ equations for the interior values of $\vec{V}_{n+1}$, while (7.9.4) is a system of $N+1$ equations for all values of $\vec{V}_{n+1}$. Correspondingly, in (7.9.11) and (7.9.6), the identity matrices $\mathbf{I}$ are of size $(N-1)\times(N-1)$ and $(N+1)\times(N+1)$ respectively.

As before, the matrix operators $\mathbf{L}_{\mathrm{I}}$, $\mathbf{R}_n$ and $\mathbf{R}_{n-1}$ can be precomputed. At each time level the steps involved in obtaining $\vec{V}_{I,n+1}$ can be summarized as:

(1) Use previous time-level solutions $\vec{V}_n$ and $\vec{V}_{n-1}$ in (7.9.4) to obtain the right-hand side vector $\vec{r}$.
(2) Strip the end elements of $\vec{r}$ to form $\vec{r}_{\mathrm{I}}$ and add the scaled boundary-correction vector to form the right-hand side of (7.9.10).
(3) Use your favorite linear system solver to solve (7.9.10) for $\vec{V}_{\mathrm{I},n+1}$.
(4) Pad the known boundary values to obtain $\vec{V}_{n+1}$.

The overall computational efficiency depends on how well step (3) is implemented and executed. You must exploit the fact that $\mathbf{L}_{\mathrm{I}}$ is independent of time. So, in addition to precomputing you may want to pre-factorize it as well. This will greatly speed-up the linear system solver.

We are now ready to tackle the viscous Burgers equation given in (7.1.10). The main difference is the advection velocity u varies both in x and t. The above four-step solution procedure must be augmented with an additional step. Note that the right-hand side matrix operators $\mathbf{R}_n$ and $\mathbf{R}_{n-1}$, as defined in (7.9.5), will now vary in time and thus cannot be precomputed. Furthermore, since advection velocity is a function of x in the Burgers equation, they must be defined as

$$\mathbf{R}_n = \left[\mathbf{I} - \frac{3\,\Delta t}{2}\mathbf{V}_n\mathbf{D}_1 + \frac{\nu\,\Delta t}{2}\mathbf{D}_2\right] \quad \text{and} \quad \mathbf{R}_{n-1} = \left[\frac{\Delta t}{2}\mathbf{V}_{n-1}\mathbf{D}_1\right], \tag{7.9.12}$$

where

$$\mathbf{V}_n = \begin{bmatrix} V_{0,n} & & & & \\ & V_{1,n} & & & \\ & & \ddots & & \\ & & & V_{N-1,n} & \\ & & & & V_{N,n} \end{bmatrix} \tag{7.9.13}$$

and similarly $\mathbf{V}_{n-1}$ is defined as a diagonal matrix of size $(N+1) \times (N+1)$ containing the $(N-1)$th time level velocity. Thus, at each time level as the "zeroth" step, one must compute the operators $\mathbf{R}_n$ and $\mathbf{R}_{n-1}$, before proceeding to step 1 of the four-step solution procedure.

The above solution procedure is clearly independent of the method used to approximate the first and second derivatives. The solution procedure equally applies to other boundary conditions as well. Instead of Dirichlet boundary conditions at both ends, if the problem involved Neumann boundary conditions at one end and Dirichlet at the other end, then the results of Problem 4.2.7 could be used to define the appropriate interior second-derivative matrix $\mathbf{D}_{2\mathrm{I}}$ and the left and right correction vectors $\vec{C}_{\mathrm{LI}}$ and $\vec{C}_{\mathrm{RI}}$. Similarly for Neumann–Neumann boundary conditions, the results of Problem 4.2.8 could be used.

We close this section with an important comment regarding implementation. The expressions for $\mathbf{R}_n$ and $\mathbf{R}_{n-1}$ when substituted into (7.9.4) will result in products such as $\mathbf{V}_n\mathbf{D}_1\vec{V}_n$. The most efficient way to calculate this product is to first calculate the matrix–vector product $\mathbf{D}_1\vec{V}_n$, and then multiply by the velocity at every grid point. Clearly this product is the discrete version of the nonlinear advection term, $u(\partial u/\partial x)$. In the context of spectral collocation method, special attention is required in evaluating this quadratic product between u and $\partial u/\partial x$. As discussed in Section 6.4.6 this nonlinear term will result in aliasing error, which can be avoided only by evaluating this product using a de-aliasing technique.

Computer Exercise 7.9.1 In this computer exercise we will numerically solve the advection–diffusion equation (7.1.9) with $c = 1$ and $\nu = 0.1$ in a periodic domain of $0 \le x \le 2\pi$. We will take the initial condition to be

$$u(x, t = 0) = \cos(kx) \quad \text{with} \quad k = 5\,, \tag{7.9.14}$$

and the boundary condition is that the solution is periodic. The exact solution of this problem is

$$u(x,t) = \exp(-\nu k^2 t)\cos(k(x-ct))\,. \tag{7.9.15}$$

Consider a uniform grid of N points discretizing the periodic domain. We will use the Crank–Nicolson scheme for the diffusion terms and for the advection term we use the third-order Adams–Bashforth scheme because its stability diagram includes a portion of the imaginary axis. You may instead choose to use the third-order Runge–Kutta or any other equivalent time integration scheme. In this exercise we will employ direct solution using the spectral collocation method. Since the problem is linear we do not need to use the de-aliasing technique. You should start by constructing the Fourier first- and second-derivative matrices ($\mathbf{D}_1$ and $\mathbf{D}_2$) and testing them. Use these operators to numerically solve the advection–diffusion equation.

We saw in Chapter 6 that the spectral methodology for space discretization will yield exact first- and second-derivatives provided $N > 2k$. Here for $k = 5$ we will try $N = 12, 24$ and 48. The only error will be due to time integration. Also, for stability, Δt must be chosen such that the Courant number is kept within the stability limit. Try different values of Δt so that the numerical solution is not just stable but accurate. The exact solution translates to the right at unit velocity and decays over time. The numerical solution must capture this behavior very accurately for sufficiently small Δt.

Computer Exercise 7.9.2 Repeat the above exercise with (a) the FTCS method, (b) Lax's method, and (c) the Dufort–Frankel method.

Computer Exercise 7.9.3 In this computer exercise we will numerically solve the advection–diffusion equation (7.1.9) with $c = 1$ and $\nu = 0.1$ in a bounded domain of $0 \le x \le 1$. We will take the initial condition to be a Gaussian pulse

$$u(x, t=0) = \exp(-\gamma(x-x_0)^2) \quad \text{with} \quad \gamma = 1381.6\,, \quad x_0 = 0.2\,. \tag{7.9.16}$$

Thus the Gaussian is of unit amplitude and is centered initially about $x_0 = 0.2$ and decays to 10^{-6} within a distance of 0.1. The problem is not periodic in this exercise and the boundary conditions at the two boundaries will be chosen as Dirichlet: $u(x=0,t) = 0$ and $u(x=1,t) = 0$. We will again use the Crank–Nicolson scheme for the diffusion terms and the third-order Adams–Bashforth or third-order Runge–Kutta scheme for the advection term. In this exercise we will employ direct solution using the Chebyshev collocation method. We begin by defining $N + 1$ Chebyshev–Gauss–Lobatto points spanning the domain, then constructing the first- and second-derivative matrices ($\mathbf{D}_1$ and $\mathbf{D}_2$) and testing them. Then, following the steps of Problem 4.2.6, construct the interior second-derivative matrix ($\mathbf{D}_{2\mathrm{I}}$) and the correction vectors $\vec{C}_{\mathrm{LI}}$ and $\vec{C}_{\mathrm{RI}}$. Use these operators to numerically solve the advection–diffusion equation.

The pulse moves to the right at unit velocity and the width of the Gaussian increases due to diffusion. The problem is simple enough that you can solve it analytically, if you ignore the boundary effects. This exact solution is good as long as its value at both the boundaries is smaller than say 10^{-6}. Try different values of N such as 50, 100 and 200. The Gaussian expanded as a polynomial does not terminate, although it does decay quickly, so there will be error due to both spatial discretization and time integration. Try different values of Δt, but make sure it is chosen so that the CFL number is within the stability limit. For sufficiently small Δt the numerical solution must capture the exact behavior very accurately for short times before the Gaussian pulse reaches the right boundary at $x = 1$. When integrated over longer times, the boundary conditions will play a role. Compute and see what happens.

Computer Exercise 7.9.4 Repeat the above exercise with other finite/compact difference space discretization and other time-integration schemes.

8 Multi-dimensional Partial Differential Equations

8.1 Multi-dimensions

We are now ready to move on to multi-dimensional PDEs. As an example, let us consider the advection–diffusion equation of a scalar field, such as temperature, in 2D:

$$\frac{\partial T}{\partial t} = -c_x \frac{\partial T}{\partial x} - c_y \frac{\partial T}{\partial y} + \nu \left[\frac{\partial^2 T}{\partial x^2} + \frac{\partial^2 T}{\partial y^2} \right] . \tag{8.1.1}$$

We begin our discussion with constant advection velocities c_x and c_y in the x- and y-directions. Later in this section we will investigate the effect of replacing them with velocity components u and v. The viscous Burgers equation for velocity in 2D will result in two coupled PDEs for u and v; we will postpone discussion of such coupled multi-dimensional PDEs to a later section.

Again we will use an explicit scheme (second-order Adams–Bashforth) for the advection terms and an implicit scheme (Crank–Nicolson) for the diffusion terms. We first discretize time. The resulting semi-discretized equation is analogous to (7.9.2) and can be written as

$$\left[1 - \alpha \left(\frac{\partial^2}{\partial x^2} + \frac{\partial^2}{\partial y^2} \right) \right] T_{n+1} = r , \tag{8.1.2}$$

where $\alpha = \nu \, \Delta t/2$ and the right-hand side

$$\begin{aligned} r = &-\frac{3\Delta t}{2} \left[c_x \frac{\partial}{\partial x} + c_y \frac{\partial}{\partial y} \right] T_n \\ &+ \frac{\Delta t}{2} \left[c_x \frac{\partial}{\partial x} + c_y \frac{\partial}{\partial y} \right] T_{n-1} + \left[1 + \alpha \left(\frac{\partial^2}{\partial x^2} + \frac{\partial^2}{\partial y^2} \right) \right] T_n , \end{aligned} \tag{8.1.3}$$

is a known function since it depends only on temperature from previous time levels. The above is a 2D Helmholtz equation and here we will discuss numerical methods for solving the multi-dimensional Helmholtz equation. Let us discretize the above equation in space into a two-dimensional grid of equi-spaced points given by

$$\begin{aligned} x_j &= j\Delta x \quad \text{for} \quad j = 0, 1, \ldots, N_x , \\ y_l &= l\Delta y \quad \text{for} \quad l = 0, 1, \ldots, N_y . \end{aligned} \tag{8.1.4}$$

For simplicity we will assume the grid spacing to be the same in both x- and y-directions, i.e., $\Delta x = \Delta y = \Delta$; applying the three-point central difference along both directions will transform (8.1.2) to

$$\left(1 + \frac{4\alpha}{\Delta^2}\right) T_{j,l} - \frac{\alpha}{\Delta^2}\left(T_{j-1,l} + T_{j+1,l} + T_{j,l-1} + T_{j,l+1}\right) = r_{j,l}\,, \tag{8.1.5}$$

where $T_{j,l}$ on the left-hand side is in fact $T_{j,l,n+1}$ and is the temperature at (x_j, y_l) at time level $n+1$. (In the subscript we have suppressed the time level $n+1$. Henceforth, we will continue to suppress such subscripts whenever their omission causes no confusion.)

We collect all the grid point temperature values into a vector

$$\vec{T} = \left[\{T_{0,0}, \ldots, T_{N_x,0}\}, \{T_{0,1}, \ldots, T_{N_x,1}\} \ldots \{T_{0,N_y}, \ldots T_{N_x,N_y}\}\right]^{\mathrm{T}}\,; \tag{8.1.6}$$

it is of length $(N_x+1) \times (N_y+1)$. Similarly the right-hand side can be evaluated at all the grid points and assembled into a long column vector $\vec{r}$. Then (8.1.5) can be expressed as the following linear system:

$$\mathbf{L}\,\vec{T}_{n+1} = \vec{r}\,, \tag{8.1.7}$$

where the matrix operator is a large square matrix of $(N_x+1)\times(N_y+1)$ rows and columns. If we simplify by putting $N_x = N_y = N$, then the memory requirement of the matrix operator goes as N^4 and the number of floating point operations required for its inversion goes as N^6. But the solution of (8.1.7) is not as hard as this scaling suggests. The stencil of the three-point scheme is shown in Figure 8.1, where it can be observed that at the new time level the temperature of five points are coupled. This is the consequence of the implicit treatment of the diffusion term. Without implicitness $\mathbf{L}$ would be diagonal and the solving (8.1.7) would be trivial. As a result of the five coupled points at the $(n+1)$th time level, every row of the matrix operator $\mathbf{L}$ has only five non-zero elements. In other words, $\mathbf{L}$ is very sparse. This sparsity can be exploited to greatly reduce both the memory requirement and the operation count of the solution.

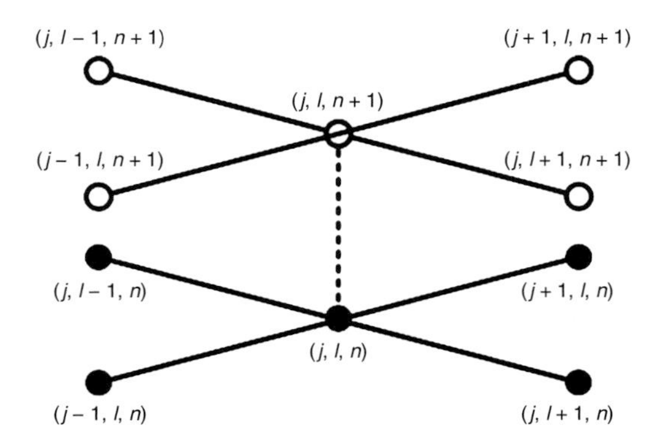

Figure 8.1 Space-time stencil of the three-point implicit scheme in 2D.

Let us consider the following Dirichlet boundary conditions for the 2D advection–diffusion equation:

$$
\begin{aligned}
T(x=0,y) &= T_{\mathrm{L}}(y)\,, \qquad T(x=L_x,y) = T_{\mathrm{R}}(y)\,, \\
T(x,y=0) &= T_{\mathrm{B}}(x) \quad \text{and} \quad T(x,y=L_y) = T_{\mathrm{T}}(x)\,.
\end{aligned} \tag{8.1.8}
$$

These boundary conditions can be included in the matrix equations similar to what was done in the 1D problem in (7.9.7) and (7.9.8). Note that each row of the matrix equation (8.1.7) corresponds to the Helmholtz equation (8.1.2) being applied at a specific grid point. If the point happens to be a boundary point, then we must instead apply the boundary condition appropriate for that boundary point. This can be accomplished by replacing that row of the matrix **L** with all zeros, except for a 1.0 at the diagonal element. Simultaneously the corresponding row of the right-hand side column vector is set to the boundary temperature at that point as given by the boundary condition (8.1.8).

The resulting linear system, with boundary conditions incorporated, is

$$
\mathbf{L}_* \vec{T}_{n+1} = \vec{r}_*\,. \tag{8.1.9}
$$

It can be solved in a number of different ways. Direct solution through inversion, Gaussian elimination or factorization of $\mathbf{L}_*$ are not efficient, since they do not make use of the sparsity of the matrix operator. There are a number of iterative solution methodologies starting from the simplest **point Jacobi method**, where the iteration updates one point at a time. The convergence of the point Jacobi method, if it converges at all, is generally slow but there are more sophisticated iterative methods available. The convergence rate of the **Gauss–Seidel method** is twice that of the point Jacobi method. To further accelerate convergence, you can consider **successive over-relaxation** where an additional numerical parameter is introduced that can be optimized for rapid convergence. Even more sophisticated methods include **preconditioning** of the matrix operator $\mathbf{L}_*$ with a suitable preconditioner, and **multigrid** techniques. In this book we cannot do justice to the large body of literature that exists on solving linear systems. We refer the reader to several excellent textbooks that specialize in iterative solutions to linear systems, for example Varga (1962), Young (2014), Axelsson (1996) and Saad (2003).

For very large linear systems that arise from three-dimensional problems, the recent developments in **Krylov subspace methods** are of particular importance. An attractive feature of this class of iterative methods is that during iteration you need not assemble the matrix operator $\mathbf{L}_*$, which can be a great advantage if $\mathbf{L}_*$ is very large. Only the result of the matrix–vector multiplication $\mathbf{L}_*\vec{V}$ is needed in this iteration. From the definition of $\mathbf{L}_*$ as the discretized version of the left-hand side of (8.1.2), it can be seen that $\mathbf{L}_*\vec{V}$ can easily be calculated in terms of the second derivatives of the two-dimensional field represented by the vector $\vec{V}$ along the x- and y-directions. Since $\mathbf{L}_*$ incorporates Dirichlet boundary conditions, the boundary values of $\vec{V}$ will remain unchanged. Krylov subspace methods such as the Generalized Minimum Residual (**GMRES**) method will iteratively improve upon the solution by repeatedly operating $\mathbf{L}_*$ on the previous level solution. For further details see Saad (2003).

8.1.1 Accuracy and Stability

The space-time discretization (8.1.5) for the 2D advection–diffusion equation is second-order accurate in space and time. In other words, the overall leading order error is $O(\Delta t)^2$, $O(\Delta x)^2$ and $O(\Delta y)^2$. This can be readily verified from the following line of argument. The three-point central difference approximation to the diffusion operator is second-order accurate. Thus, the overall spatial accuracy is $O(\Delta x)^2$ and $O(\Delta y)^2$. For time integration both the Adams–Bashforth (AB2) and Crank–Nicolson schemes are second-order accurate. The orders of accuracy of these time-integration and finite difference schemes were established in Chapters 3 and 4 and they apply even when they are present together as in the 2D advection–diffusion equation. Although a bit tedious, this can be verified by expanding (8.1.5) in a Taylor series about the grid point (x_j, y_l) and time level t_n.

We can get a quick estimate of stability by looking at the advection and diffusion operators individually. The stability region of the Crank–Nicolson scheme is the entire left half of the complex plane, while the eigenvalues of the three-point central difference approximation to $\partial^2/\partial x^2$ and $\partial^2/\partial y^2$ are all real and negative. Thus, the space-time discretization of the diffusion effect is unconditionally stable, whereas the stability region of the AB2 scheme does not include the imaginary axis. On the other hand, the eigenvalues of the three-point central difference approximation to $\partial/\partial x$ and $\partial/\partial y$ are all pure imaginary. Thus, the space-time discretization of the advection effect is unconditionally unstable. This is the reason we in general recommend using the third-order Adams–Bashforth or Runge–Kutta schemes. Nevertheless, it is possible to stabilize the AB2 three-point central scheme for the advection term due to the presence of the added physical diffusion. Here we will explore this possibility with a stability analysis of the combined scheme.

Our analysis will start with (8.1.7), but the right-hand side will be written in terms of $\vec{T}_n$ and $\vec{T}_{n-1}$ as

$$\mathbf{L}\,\vec{T}_{n+1} = \mathbf{R}_1\vec{T}_n + \mathbf{R}_2\vec{T}_{n-1}\,. \tag{8.1.10}$$

The matrix operators $\mathbf{R}_n$ and $\mathbf{R}_{n-1}$ can be easily assembled from the definition of the right-hand side in (8.1.3) with three-point central difference approximation to the first and second spatial derivatives. We now use the matrix method to investigate stability. Since the AB2 is a two-step method, for stability analysis we consider another equation similar to (8.1.10) for time level $(n+2)$ and together write the system as

$$\begin{bmatrix} 0 & \mathbf{L} \\ \mathbf{L} & -\mathbf{R}_1 \end{bmatrix} \begin{bmatrix} \vec{T}_{n+2} \\ \vec{T}_{n+1} \end{bmatrix} = \begin{bmatrix} \mathbf{R}_1 & \mathbf{R}_2 \\ \mathbf{R}_2 & 0 \end{bmatrix} \begin{bmatrix} \vec{T}_n \\ \vec{T}_{n-1} \end{bmatrix}. \tag{8.1.11}$$

The above equation can be explicitly written as a two-time level mapping by taking the block matrix on the left to the right

$$\begin{bmatrix} \vec{T}_{n+2} \\ \vec{T}_{n+1} \end{bmatrix} = \mathbf{A} \begin{bmatrix} \vec{T}_{n} \\ \vec{T}_{n-1} \end{bmatrix} \quad \text{where} \quad \mathbf{A} = \begin{bmatrix} 0 & \mathbf{L} \\ \mathbf{L} & -\mathbf{R}_1 \end{bmatrix}^{-1} \begin{bmatrix} \mathbf{R}_1 & \mathbf{R}_2 \\ \mathbf{R}_2 & 0 \end{bmatrix}. \tag{8.1.12}$$

As discussed in Section 7.6, for stability of the overall scheme we required the spectral radius of $\mathbf{A}$, or its largest eigenvalue in magnitude, to be less than unity.

Several comments are now in order. As pointed out in Section 7.6.1, in order to identify the convective instabilities of the numerical method, it is appropriate to consider the periodic problem in the stability analysis, even when the real problem involves non-periodic boundary conditions. Thus, the matrix operators $\mathbf{L}$, $\mathbf{R}_1$ and $\mathbf{R}_2$ can be assembled for periodic boundary conditions along x and y. With periodic boundary conditions we recall that $\mathbf{L}$ is a large square matrix of size $(N_x \times N_y) \times (N_x \times N_y)$. The matrices $\mathbf{R}_1$ and $\mathbf{R}_2$ are of the same size. Correspondingly the number of rows and columns of $\mathbf{A}$ are twice as large. We recommend Krylov subspace methods for efficient calculation of the spectral radius. Ready-made routines are available in MATLAB, IMSL and other libraries to do your bidding. Finally we point out that even though the numerical solution of the advection–diffusion equation may require a large value of N_x and N_y for accuracy, you may perform the stability analysis with much smaller value of them (and of course with periodic boundary conditions). The spectral radius primarily depends only on the value of Δx and Δy. So keep their values the same as in the numerical method.

Computer Exercise 8.1.1 Here we will use the matrix method to investigate the stability of the three-point central scheme for the advection–diffusion equation as written in (8.1.11). Here we will consider $\Delta x = \Delta y = \Delta = 1$ and $\Delta t = 1$, which is the same as choosing Δ to be the length scale and Δt to be the time scale and working with non-dimensional variables. We will also choose the advection velocity to be the same along both the directions (i.e., $c_x = c_y = c$). Consider a small 2D grid of 10×10 points with periodic boundary conditions. The two parameters of the problem are the Courant number ($C = c\Delta t/\nu$) and the diffusion number $D = \nu \Delta t/\Delta^2$. Vary these parameter values in the range $-5 \le C \le 5$ and $0 \le D < 5$. For each combination of C and D construct the matrix $\mathbf{A}$ and calculate the largest eigenvalue in magnitude (or the spectral radius). Then plot the contour of spectral radius $= 1$ on a plane with $-D$ as the x-axis and C as the y-axis. This is the region of stability and for all combinations of values of the Courant and diffusion numbers within this region the method is stable. Repeat the analysis for a larger grid of 20×20.

8.1.2 Operator Factoring

In going from 1D to 2D the computational cost increased due to the coupling between the x and y grid points at the $(n+1)$th time level. In this section we will explore an approximation that still remains $O(\Delta t)^2$ and $O(\Delta x)^2$ accurate in time and space, but will decouple the solution process along the x- and y-directions and thereby provide improved computational efficiency.

The coupling between the x and y grid points on the left-hand side of (8.1.5) arises from the simultaneous implicit discretization of the x and y derivatives in the semi-discrete equation (8.1.2). The idea is to approximately factor the left-hand side of (8.1.2) to obtain the following semi-discrete equation:

$$\left[1 - \alpha \frac{\partial^2}{\partial x^2}\right] \left[1 - \alpha \frac{\partial^2}{\partial y^2}\right] T_{n+1} = r\,. \tag{8.1.13}$$

The difference between the left-hand side of the above and that of (8.1.2) is $\alpha^2(\partial^4 T_{n+1}/\partial x^2 \partial y^2)$. From the definition of α it is clear that this difference is $O(\Delta t)^2$ and thus is of the order of accuracy of the time-integration scheme. In fact we can do better. In the expression for r in (8.1.3) rewrite the third term on the right-hand side as

$$\left[1 + \alpha \frac{\partial^2}{\partial x^2}\right] \left[1 + \alpha \frac{\partial^2}{\partial y^2}\right] T_n \tag{8.1.14}$$

and let the modified right-hand side be r'. The error in this replacement is again $\alpha^2(\partial^4 T_n/\partial x^2 y^2)$. The two errors on the left- and right-hand sides differ only in the time level of the temperature, i.e., T_{n+1} versus T_n. Thus, the error in replacing (8.1.2) with (8.1.13) can be reduced to $O(\Delta t)^3$ if r is simultaneously replaced by r'.

As we will show below, the real advantage is that the numerical solution of (8.1.13) is easier, since we can proceed one direction at a time. To see this, first define

$$\Theta_{n+1} = \left[1 - \alpha \frac{\partial^2}{\partial y^2}\right] T_{n+1}\,. \tag{8.1.15}$$

With this definition (8.1.13) becomes

$$\left[1 - \alpha \frac{\partial^2}{\partial x^2}\right] \Theta_{n+1} = r'\,. \tag{8.1.16}$$

The numerical solution can now proceed in two steps:

(1) First solve the fully discretized version of (8.1.16). With three-point central difference, the discretized equation can be written as

$$\left(1 + \frac{2\alpha}{\Delta x^2}\right) \Theta_{j,l} - \frac{\alpha}{\Delta x^2} \left(\Theta_{j-1,l} + \Theta_{j+1,l}\right) = {r'}_{j,l}\,. \tag{8.1.17}$$

This step is the same as solving a 1D Helmholtz equation in x. The solutions at all the x points are coupled and must be solved together. But the different

y points are decoupled and can be solved for one l at a time. The cost of a direct solution (or inversion) is only $O((N_x+1)^2 \times (N_y+1))$. Of course for a three-point scheme the matrix operator on the left is tri-diagonal and thus Θ can be obtained much more efficiently with a computational cost of only $O((N_x+1) \times (N_y+1))$ floating point operations.

(2) Next, solve the following fully discretized version of (8.1.15):

$$\left(1+\frac{2\alpha}{\Delta y^2}\right) T_{j,l} - \frac{\alpha}{\Delta y^2}\left(T_{j,l-1}+T_{j,l+1}\right) = \Theta_{j,l}\,. \tag{8.1.18}$$

This step is the same as solving a 1D Helmholtz equation in y. The solutions at all the y points are now coupled, but the different x points are decoupled and can be solved for one j at a time. Again for the three-point central scheme, the tri-diagonal system can be solved very efficiently with $O((N_x+1) \times (N_y+1))$ operations.

If the boundary conditions along x and y are periodic, then the corresponding x and y second-derivative operators will be tri-diagonal circulant matrices (see Section 4.2). If the boundary conditions along x and y are Dirichlet, as given in (8.1.8), additional care is required. In the second step we must apply the boundary conditions

$$T_{j,0} = T_{\mathrm{B}}(x_j) \quad \text{and} \quad T_{j,N_y} = T_{\mathrm{T}}(x_j)\,. \tag{8.1.19}$$

In Section 7.9 we saw how to solve a 1D Helmholtz equation with Dirichlet boundary conditions. We now turn to boundary condition for step (1). From the definition of Θ, the Dirichlet boundary conditions for (8.1.17) become

$$\Theta_{0,l} = \left[T_{\mathrm{L}} - \alpha\frac{\mathrm{d}^2 T_{\mathrm{L}}}{\mathrm{d}y^2}\right]_{y=y_l} \quad \text{and} \quad \Theta_{N_x,l} = \left[T_{\mathrm{R}} - \alpha\frac{\mathrm{d}^2 T_{\mathrm{R}}}{\mathrm{d}y^2}\right]_{y=y_l}\,. \tag{8.1.20}$$

Thus, operator splitting provides an efficient way of solving multi-dimensional problems by attacking one dimension at a time, without any loss of formal order of accuracy. The above splitting approach can easily be extended to the three-dimensional Helmholtz problem by sequentially solving in the x-direction, followed by the y-direction, followed by the z-direction. By rearranging the operators on the left-hand side of (8.1.13) we can change the order of x, y and z.

8.1.3 Alternating Direction Implicit

The idea of operator splitting was originally proposed by Peaceman and Rachford (1955) in a slightly different form than what was presented in the previous section. Their method goes by the name **Alternating Direction Implicit** (ADI). Although they introduced their method for elliptic and parabolic equations, here we will continue to discuss it in the context of the advection–diffusion equation.

The central idea of ADI is to take a half time-step, implicit only in x, followed by a second half time-step, implicit only in y. Similar to (8.1.2) and (8.1.3) the first half-step can be written as

$$\left[1 - \alpha \frac{\mathrm{d}^2}{\mathrm{d}x^2}\right] T_{n+1/2} = r_n \,, \tag{8.1.21}$$

where $\alpha = \nu\, \Delta t/2$, and r_n is given by

$$\begin{aligned} r_n = &-\frac{3\Delta t}{4}\left[c_x \frac{\mathrm{d}}{\mathrm{d}x} + c_y \frac{\mathrm{d}}{\mathrm{d}y}\right] T_n \\ &+\frac{\Delta t}{4}\left[c_x \frac{\mathrm{d}}{\mathrm{d}x} + c_y \frac{\mathrm{d}}{\mathrm{d}y}\right] T_{n-1/2} + \left[1 + \alpha \frac{\mathrm{d}^2}{\mathrm{d}y^2}\right] T_n \,. \end{aligned} \tag{8.1.22}$$

The corresponding semi-discretized equations for the second half-step are

$$\left[1 - \alpha \frac{\mathrm{d}^2}{\mathrm{d}y^2}\right] T_{n+1/2} = r_{n+1/2} \,, \tag{8.1.23}$$

$$\begin{aligned} r_{n+1/2} = &-\frac{3\Delta t}{4}\left[c_x \frac{\mathrm{d}}{\mathrm{d}x} + c_y \frac{\mathrm{d}}{\mathrm{d}y}\right] T_{n+1/2} \\ &+\frac{\Delta t}{4}\left[c_x \frac{\mathrm{d}}{\mathrm{d}x} + c_y \frac{\mathrm{d}}{\mathrm{d}y}\right] T_n + \left[1 + \alpha \frac{\mathrm{d}^2}{\mathrm{d}x^2}\right] T_{n+1/2} \,. \end{aligned} \tag{8.1.24}$$

Thus, the first half-step uses an implicit, respectively explicit, Euler scheme for the x-component, respectively y-component, of diffusion, while the second half-step uses an explicit, respectively implicit, Euler scheme for the x-component, respectively y-component, of diffusion. This scheme is therefore only first-order accurate in time, which is adequate for illustrative purposes. Higher-order time-accurate versions of ADI can easily be designed. But the important feature of ADI is that the first half-step is a 1D Helmholtz equation in x, which must be solved at each y grid point. The second half-step is a 1D Helmholtz equation in y, which must be solved at each x grid point. These half-steps are similar to the two steps outlined in (8.1.16) and (8.1.15). The x boundary conditions are applied in the first half-step, while the y boundary conditions are applied in the second.

Problem 8.1.2 Consider the ADI method for the diffusion equation by ignoring the advection terms in equations (8.1.22) and (8.1.24) (or set $c_x = c_y = 0$). In this case show that the ADI method is identical to the operator factoring method presented in Section 8.1.2.

Problem 8.1.3 Consider the operator factoring method for the diffusion equation. We can investigate the stability of the multi-dimensional method with a von Neumann analysis. Consider a discrete two-dimensional test function of the form

$$T_{j,l}(t_n) = \xi_n \exp\left[iK_x j\right] \exp\left[iK_y l\right] , \qquad (8.1.25)$$

where K_x and K_y are scaled wavenumbers in x and y. The above is a two-dimensional extension of the test function introduced in (4.1.13). Let $\widetilde{\widetilde{K}}_x$ and $\widetilde{\widetilde{K}}_y$ be the modified wavenumbers – see Section 4.1.2 for the definition of modified wavenumber for the numerical approximation to a second derivative. By substituting the test function into (8.1.13) obtain the result

$$\frac{\xi_{n+1}}{\xi_n} = \frac{\left(1 - \alpha\widetilde{\widetilde{K}}_x^2\right)\left(1 - \alpha\widetilde{\widetilde{K}}_y^2\right)}{\left(1 + \alpha\widetilde{\widetilde{K}}_x^2\right)\left(1 + \alpha\widetilde{\widetilde{K}}_y^2\right)} . \qquad (8.1.26)$$

The magnitude of the above ratio is always less than unity. Therefore, we see that the operator factoring method for the diffusion equation is unconditionally stable. From the result of Problem 8.1.2 this result applies for the ADI method as well.

8.1.4 Direct Solution of the Helmholtz Equation

We now present an elegant methodology that exploits eigendecomposition of the Helmholtz operator as the final method for solving a multi-dimensional Helmholtz equation. In this method we solve the discretized version of (8.1.2) directly, without iteration and without operator splitting or factoring. We first rewrite (8.1.2) in the form of a standard Helmholtz equation and discretize along the x- and y-directions. This resulting discretized equation in index notation can be expressed as

$$\left[D_{2x}\right]_{j,p} T_{p,l} + \left[D_{2y}\right]_{l,q} T_{j,q} - \frac{1}{\alpha} T_{j,l} = r''_{j,l} \quad \text{where} \quad r''_{j,l} = -\frac{1}{\alpha} r_{j,l} . \qquad (8.1.27)$$

Here $\mathbf{D}_{2x}$ and $\mathbf{D}_{2y}$ are second-derivative operators along the x- and y-directions. Summation over $p = 0, 1, \ldots, N_x$ and $q = 0, 1, \ldots, N_y$ is assumed and furthermore the above is a linear system of $(N_x + 1) \times (N_y + 1)$ equations for $j = 0, 1, \ldots, N_x$ and $l = 0, 1, \ldots, N_y$. For each pair (j, l), the above equation is the discretized Helmholtz equation applied at the point (x_j, y_l). Note that the grid point temperature values on the left-hand side are at the new time-level $(n + 1)$ and therefore are being solved for, while the right-hand side is known from previous time-level solution.

In the case of periodic boundary conditions, the equation need not be solved at the right boundary ($j = N_x$), and at the top boundary ($l = N_y$) and the size of the linear system reduces to $N_x \times N_y$. In the case of non-periodic boundary conditions it is most convenient to proceed in terms of the interior matrix operator introduced in Section 4.2.1 and discussed in Section 7.9 in the context of 1D Helmholtz equation.

In the multi-dimensional context, along the x-direction we define $\mathbf{D}_{2\text{I}x}$ as the interior second derivative matrix of size $(N_x - 1) \times (N_x - 1)$ and the corresponding boundary correction vectors $\vec{C}_{\text{LI}x}$ and $\vec{C}_{\text{RI}x}$ are of length $(N_x - 1)$. Similarly, along the y-direction we define $\mathbf{D}_{2\text{I}y}$ as the interior second-derivative matrix of size $(N_y - 1) \times (N_y - 1)$ and the corresponding boundary correction vectors $\vec{C}_{\text{LI}y}$ and $\vec{C}_{\text{RI}y}$ are of length $(N_y - 1)$. Refer back to Problems 4.2.6 and 4.2.7 for the definition of the interior operators and correction vectors for Dirichlet and Neumann boundary conditions.

In terms of the interior matrix operators (8.1.27) can be rewritten as

$$[D_{2\text{I}x}]_{j,p}\, T_{p,l} + [D_{2\text{I}y}]_{l,q}\, T_{j,q} - \frac{1}{\alpha} T_{j,l} = [r_\text{I}]_{j,l}\,, \tag{8.1.28}$$

where the boundary-corrected interior right-hand side is

$$[r_\text{I}]_{j,l} = r''_{j,l} - [C_{\text{LI}x}]_j\, T_\text{L}(y_l) - [C_{\text{RI}x}]_j\, T_\text{R}(y_l) - [C_{\text{LI}y}]_l\, T_\text{B}(x_j) - [C_{\text{RI}y}]_l\, T_\text{T}(x_j)\,, \tag{8.1.29}$$

where T_L, T_R, T_B and T_T are, respectively, the left, right, bottom and top boundary conditions specified in (8.1.8). In (8.1.28) summation is over $p = 1, 2 \ldots, (N_x - 1)$ and $q = 1, \ldots, (N_y - 1)$. The discretized 2D Helmholtz equation is now being solved only at the interior grid points, i.e., for $j = 1, 2, \ldots, (N_x - 1)$ and $l = 1, 2, \ldots, (N_y - 1)$. As a result the linear system is of size $(N_x - 1) \times (N_y - 1)$.

We now eigendecompose the interior derivative matrices as

$$\mathbf{D}_{2\text{I}x} = \mathbf{E}_x\, \boldsymbol{\lambda}_x\, \mathbf{E}_x^{-1} \quad \text{and} \quad \mathbf{D}_{2\text{I}y} = \mathbf{E}_y\, \boldsymbol{\lambda}_y\, \mathbf{E}_y^{-1}\,, \tag{8.1.30}$$

where the matrices of eigenvectors $\mathbf{E}_x$ and $\mathbf{E}_y$, and their inverses are of size $(N_x - 1) \times (N_x - 1)$ and $(N_y - 1) \times (N_y - 1)$, respectively. The eigenvalue matrices $\boldsymbol{\lambda}_x$ and $\boldsymbol{\lambda}_y$ are diagonal. Substituting this decomposition into (8.1.28) we get

$$\mathbf{E}_x\, \boldsymbol{\lambda}_x\, \mathbf{E}_x^{-1} \mathbf{T}_\text{I} + \mathbf{T}_\text{I} \mathbf{E}_y^{-\text{T}}\, \boldsymbol{\lambda}_y\, \mathbf{E}_y^\text{T} - \frac{1}{\alpha} \mathbf{T}_\text{I} = \mathbf{r}_\text{I}\,, \tag{8.1.31}$$

where superscript T indicates transpose. The above equation can also be written in index notation as

$$[E_x]_{j,\gamma}\, [\lambda_x]_{\gamma,\beta}\, \left[E_x^{-1}\right]_{\beta,p} T_{p,l} + [E_y]_{l,\delta}\, [\lambda_y]_{\delta,\epsilon}\, \left[E_y^{-1}\right]_{\epsilon,q} T_{j,q} - \frac{1}{\alpha} T_{j,l} = [r_\text{I}]_{j,l}\,, \tag{8.1.32}$$

where summing over $p, \beta, \gamma = 1, 2, \ldots, (N_x - 1)$ and over $\delta, \epsilon, q = 1, 2, \ldots, (N_y - 1)$ is implied. The matrix notation is convenient in the case of a two-dimensional problem since each row and column of $\mathbf{T}$ corresponds to the different x and y grid points. As we will see below in Problem 8.1.4 the matrix notation cannot be easily used in 3D, while the index notation can be readily extended to consider 3D and higher-dimensional Helmholtz equations.

We now transform the grid point values of temperature to the eigenspace with the linear transformation

$$\mathbf{E}_x^{-1}\,\mathbf{T}_\mathrm{I}\,\mathbf{E}_y^{-\mathrm{T}} = \boldsymbol{\psi}\,. \tag{8.1.33}$$

The inverse transform that takes the eigencoefficients to the grid point space can also be defined:

$$\mathbf{E}_x\,\boldsymbol{\psi}\,\mathbf{E}_y^{\mathrm{T}} = \mathbf{T}_\mathrm{I}\,. \tag{8.1.34}$$

Substituting (8.1.34) into (8.1.31) we change the solution variable from temperature to $\boldsymbol{\psi}$ and obtain

$$\mathbf{E}_x\,\boldsymbol{\lambda}_x\,\boldsymbol{\psi}\,\mathbf{E}_y^{\mathrm{T}} + \mathbf{E}_x\,\boldsymbol{\psi}\,\boldsymbol{\lambda}_y\,\mathbf{E}_y^{\mathrm{T}} - \frac{1}{\alpha}\mathbf{E}_x\,\boldsymbol{\psi}\,\mathbf{E}_y^{\mathrm{T}} = \mathbf{r}_\mathrm{I}\,. \tag{8.1.35}$$

The above can be written in index notation as

$$\begin{aligned} &[E_x]_{j,\beta}\,[\lambda_x]_{\beta,\gamma}\,\psi_{\gamma,\epsilon}\,[E_y]_{l,\epsilon} \\ &\quad + [E_x]_{j,\beta}\,\psi_{\beta,\delta}\,[\lambda_y]_{\epsilon,\delta}\,[E_y]_{l,\epsilon} - \frac{1}{\alpha}\,[E_x]_{j,\beta}\,\psi_{\beta,\epsilon}\,[E_y]_{l,\epsilon} = [r_I]_{j,l}\,. \end{aligned} \tag{8.1.36}$$

In the above summation over repeated indices is implied. These repeated indices are dummy indices and have been suitably chosen. Only j and l are the free indices and they go $j = 1, 2, \ldots, (N_x - 1)$ and $l = 1, 2, \ldots, (N_y - 1)$, respectively.

As the final step we also project the right-hand side onto the eigenspace with the linear transformation

$$\mathbf{E}_x\,\widehat{\mathbf{r}}_\mathrm{I}\,\mathbf{E}_y^{\mathrm{T}} = \mathbf{r}_\mathrm{I}\,, \tag{8.1.37}$$

whose inverse is given by

$$\mathbf{E}_x^{-1}\,\mathbf{r}_\mathrm{I}\,\mathbf{E}_y^{-\mathrm{T}} = \widehat{\mathbf{r}}_\mathrm{I}\,. \tag{8.1.38}$$

Substituting (8.1.37) for the right-hand side of (8.1.35), we identify $\mathbf{E}_x$ to be the common left operator and $\mathbf{E}_y^{\mathrm{T}}$ as the common right operator. Canceling these we arrive at

$$\boldsymbol{\lambda}_x\,\boldsymbol{\psi} + \boldsymbol{\psi}\,\boldsymbol{\lambda}_y - \frac{1}{\alpha}\boldsymbol{\psi} = \widehat{\mathbf{r}}_\mathrm{I}\,. \tag{8.1.39}$$

The above in index notation becomes

$$[\lambda_x]_{p,\epsilon}\,\psi_{\epsilon,q} + [\lambda_y]_{q,\epsilon}\,\psi_{p,\epsilon} - \frac{1}{\alpha}\psi_{p,q} = [\widehat{r}_\mathrm{I}]_{p,q}\,. \tag{8.1.40}$$

Furthermore, we recall the lambda matrices are diagonal. With this observation we obtain the final solution in the eigenspace as

$$\psi_{p,q} = \frac{[\widehat{r}_\mathrm{I}]_{p,q}}{(\lambda_x)_p + (\lambda_y)_q - (1/\alpha)}\,. \tag{8.1.41}$$

In the above there is no sum over either p or q. Here $(\lambda_x)_p$ is the pth eigenvalue of $\mathbf{D}_{2\mathrm{I}x}$ and $(\lambda_y)_q$ is the qth eigenvalue of $\mathbf{D}_{2\mathrm{I}y}$. It is remarkable that in the eigenspace the solution fully decouples and $\psi_{p,q}$ can be solved for one value of (p, q) at a time. By looping over $p = 1, 2, \ldots, (N_x - 1)$ and $q = 1, 2, \ldots, (N_y - 1)$ all the values of $\psi_{p,q}$ can be obtained.

Although this derivation was somewhat long, the step-by-step implementation of this multi-dimensional direct method is straightforward. The method involves preprocessing steps which are listed below:

(P1) Choose the number of gridpoints N_x and N_y and their location along the x- and y-directions. With the choice of numerical approximation for the x and y second derivatives, construct the matrix operators $\mathbf{D}_{2x}$ and $\mathbf{D}_{2y}$.
(P2) From the boundary conditions (e.g., Dirichlet–Dirichlet or Neumann–Neumann, etc.) compute the interior second-derivative operators, $\mathbf{D}_{2\mathrm{I}x}$ and $\mathbf{D}_{2\mathrm{I}y}$, and the associated correction vectors. The definitions of the interior second-derivative matrices are given in Section 4.2.
(P3) Eigendecompose the interior second-derivative matrices to compute and store the eigenvalues, eigenvector matrices and their inverse: see (8.1.30).

The above preprocessing steps need to be performed only once at the beginning. At each time-level, the solution of the Helmholtz equation can then be accomplished with the following steps:

(S1) Calculate the right-hand side at all the grid points (i.e., calculate $r_{j,l}$) as defined in (8.1.3). This step involves calculation of the first and second x and y derivatives of the temperature at the past time level. These grid point derivative values are needed to compute the different terms of (8.1.3).
(S2) Using the definitions in (8.1.27) and (8.1.29) calculate the boundary-corrected right-hand side $[r_\mathrm{I}]_{j,l}$ at all the interior points.
(S3) The boundary-corrected right-hand side in the eigenspace can be computed using (8.1.38). Since the inverse matrices have been precomputed, this step involves two matrix–matrix multiplications. The cost of this operation goes as $O(N_x \times N_y \times (N_x + N_y))$.
(S4) Obtain the solution $\psi_{p,q}$ in the eigenspace using (8.1.41). Since (8.1.41) can be computed independently for each (p, q) the cost of this operation is only $O(N_x \times N_y)$.
(S5) Once $\psi_{p,q}$ is known, the corresponding solution in the grid point space can be readily obtained from (8.1.34). This step also involves two matrix–matrix multiplications and the computational cost goes as $O(N_x \times N_y \times (N_x + N_y))$.
(S6) From the above step we now have the temperature at time-level $(n + 1)$ at all the interior nodes. We can append the known Dirichlet boundary temperature values to complete the solution.

We will now discuss a few additional aspects. In general, the advection velocities c_x and c_y will not be constants and they must be replaced by the x- and y-components of fluid velocity $u(x, y, t)$ and $v(x, y, t)$, which are functions of space and time. With this change the advection–diffusion equation (8.1.1) becomes variable coefficient. This affects only step (S1), in the calculation of the right-hand side $r_{j,l}$. In (8.1.3) after the calculation of $\partial T_n/\partial x$ and $\partial T_n/\partial y$ they must be multiplied by u and v at each grid point and similarly for T_{n-1}. In the case of the spectral collocation method, this quadratic product must be computed without aliasing error. Due to the explicit treatment of the advection term, any change in the advection velocity does not alter the Helmholtz operator or its solution procedure.

As the final comment we point out that the above direct solver can be easily extended to the 3D Helmholtz equation by defining the interior derivative matrix in the third direction and eigendecomposing it. Steps (S1) to (S6) of the above algorithm remain essentially the same, but must include all three directions.

Problem 8.1.4 In this problem you will extend the above direct solution methodology to 3D. The solution procedure starts with a three-dimensional discretization of the rectangular computational domain. The interior grid point values of temperature will now be a three-dimensional array denoted by $[T_{\rm I}]_{p,q,l}$ and the corresponding boundary corrected right-hand side that is analogous to (8.1.29) will be denoted by $[r_{\rm I}]_{p,q,l}$. These two quantities are projected onto the eigenspace with the following transformation:

$$\begin{aligned} \psi_{\beta,\gamma,\delta} &= [E_x]_{\beta,p}\,[E_y]_{\gamma,q}\,[E_x]_{\delta,l}\,[T_{\rm I}]_{p,q,l}\,, \\ [\widehat{r}_{\rm I}]_{\beta,\gamma,\delta} &= [E_x]_{\beta,p}\,[E_y]_{\gamma,q}\,[E_x]_{\delta,l}\,[r_{\rm I}]_{p,q,l}\,, \end{aligned} \tag{8.1.42}$$

where sum over p, q and l is assumed. Show that in the eigenspace the solution to the 3D Helmholtz equation reduces to

$$\psi_{p,q,l} = \frac{[\widehat{r}_{\rm I}]_{p,q,l}}{(\lambda_x)_p + (\lambda_y)_q + (\lambda_z)_l - (1/\alpha)}\,. \tag{8.1.43}$$

The eigensolution can then be transformed to the grid points.

8.1.5 Multi-dimensional Poisson Equation

In the case of three-point central difference the operator factoring and ADI methods presented in Sections 8.1.2 and 8.1.3 require the solution of $(N_x-1)+(N_y-1)$ tri-diagonal systems at each time level. Therefore, these methods are computationally more efficient than the direct solver that uses eigendecomposition. However, if the finite difference stencil becomes broader than three points, or when the second-derivative matrices become dense, as in the case of compact difference or spectral methods, the direct method may become computationally comparable to the operator factoring and ADI methods.

The real advantage of the direct method is in the solution of the related Poisson equation. As we will see below in the context of time-splitting techniques for the Navier–Stokes equations, a Poisson equation for pressure must be solved, in order to satisfy the incompressibility condition. Since the solution of the pressure Poisson equation is related to mass conservation, typically it must be satisfied to very high accuracy at each grid point. In this sense, the operator factoring and ADI methods are less than perfect for the pressure Poisson equation, since the discretized equations will satisfy mass conservation to only $O(\Delta t)^3$ accuracy or so. The consequence of not satisfying mass conservation can be severe.

The pressure Poisson equation in 2D can be expressed as

$$\left[\frac{\partial^2}{\partial x^2} + \frac{\partial^2}{\partial y^2}\right] P = r\,, \tag{8.1.44}$$

where r is the inhomogeneous term on the right-hand side. The Poisson equation is solved with homogeneous Neumann boundary conditions on all four sides:

$$\frac{\partial P}{\partial x}(x=0,y) = \frac{\partial P}{\partial x}(x=L_x,y) = \frac{\partial P}{\partial y}(x,y=0) = \frac{\partial P}{\partial y}(x,y=L_y) = 0\,. \tag{8.1.45}$$

The solution procedure follows the steps outlined in the previous section for the 2D Helmholtz equation, with two important differences.

First, the definition of the interior derivative matrix and the correction vectors are now for Neumann–Neumann boundary condition: see Problem 4.2.8, where you derived expressions for $\mathbf{D}_{2\mathrm{I}}$, $\vec{C}_{\mathrm{LI}}$ and $\vec{C}_{\mathrm{RI}}$. You need to use these definitions along both the x- and y-directions and perform eigendecomposition as part of the preprocessing steps.

Note that $\alpha = 0$ for the pressure Poisson equation. Therefore instead of (8.1.41), the final solution in the eigenspace is given by

$$\psi_{p,q} = \frac{[\widehat{r}_{\mathrm{I}}]_{p,q}}{(\lambda_x)_p + (\lambda_y)_q}\,. \tag{8.1.46}$$

The second important difference is in solving the above algebraic equation. Recall our discussion of the null mode in Section 4.2.1 for the case of Neumann–Neumann boundary conditions. For the Poisson equation one of the eigenvalues of λ_x and one of λ_y is zero. Without loss of generality let us say $(\lambda_x)_1 = 0 = (\lambda_y)_1$. Then in (8.1.46) there is a divide-by-zero for the element $\psi_{1,1}$. In Section 5.1.3 we saw in detail that the reason for this divide-by-zero was the non-uniqueness of the solution of the Poisson equation with Neumann boundary conditions. The solution to (8.1.44) with (8.1.45) as boundary conditions is arbitrary to an additive constant. To any solution you could add a constant and it will still be a solution.

In a well-posed numerical method, the right-hand side of the Poisson equation must be such that its projection on the null eigenmode, which in the present discussion is $[\widehat{r}_{\mathrm{I}}]_{1,1}$, must also be zero. In which case, we end up with zero divided by zero for $\psi_{1,1}$. We can then set the mean pressure to zero by setting $\psi_{1,1} = 0$. This special treatment is needed only for the null mode. All other eigenmodes can be solved through (8.1.46).

We now briefly go back to the Helmholtz equation and show that there will be no such difficulty in computing (8.1.41) for any combination of p and q. This is because all the eigenvalues λ_x and λ_y of the second-derivative operator with Dirichlet–Dirichlet boundary conditions are strictly negative. Since α is positive definite, there is no chance of divide-by-zero. It is well known that the solution of the Helmholtz equation is far easier and converges much faster than that of the Poisson equation. Part of the problem arises from the null mode. With the direct method, and with the null mode properly taken care of, the solution of the Poisson equation becomes nearly as efficient as that of the Helmholtz equation.

Computer Exercise 8.1.5 In this computer exercise we will numerically solve the Helmholtz equation

$$\frac{\partial^2 T}{\partial x^2} + \frac{\partial^2 T}{\partial y^2} - \frac{1}{\alpha} T = r''(x, y), \tag{8.1.47}$$

where the right-hand side is given by

$$r'' = \left(4\gamma^2((x - x_0)^2 + (y - y_0)^2) - 4\gamma - \frac{1}{\alpha} \right) \exp(-\gamma(x - x_0)^2) \exp(-\gamma(y - y_0)^2) \tag{8.1.48}$$

with $\alpha = 1$, $\gamma = 1381.6$ and $(x_0, y_0) = (0.6, 0.6)$ in a bounded domain of $0 \le x \le 1$, $0 \le y \le 1$. The solution is a 2D Gaussian pulse

$$T(x, y) = \exp(-\gamma(x - x_0)^2) \exp(-\gamma(y - y_0)^2). \tag{8.1.49}$$

In fact, this problem was written with a known solution in mind, and it's a great way to debug and test your code. This approach to testing code is known as the **method of manufactured solution**. Code that converges to the desired analytic solution with increasing resolution is said to be *verified*, at least for this class of problem. Since the Gaussian pulse decays away rapidly from its center at $(x_0, y_0) = (0.6, 0.6)$, we can assume homogeneous Dirichlet boundary conditions along the four boundaries. For the numerical solution make the following choices:

(1) Use the direct solution approach.
(2) A uniform grid of $(N + 1) \times (N + 1)$ points in the x–y-plane.
(3) A 3-point central difference scheme in both x and y.

Try different values of $N = 50, 100$ and 200. Compare the numerical solution with the analytic solution. Evaluate the maximum error among all the grid points; this is known as the L_∞ error. Also calculate the root mean square of error at all the grid points; this is known as the L_2 error. Plot these errors as a function of N and explain the rate of convergence.

Computer Exercise 8.1.6 Repeat the above exercise by replacing any of the three choices made in the previous exercise. For example, use direct solution approach with Chebyshev–Gauss–Lobatto collocation.

Computer Exercise 8.1.7 Repeat the above exercises with ADI.

Computer Exercise 8.1.8 In this computer exercise we will write some code for numerically solving the Poisson equation

$$\frac{\partial^2 P}{\partial x^2} + \frac{\partial^2 P}{\partial y^2} = r''(x,y)\,. \tag{8.1.50}$$

We will use the method of manufactured solution to test the code. We will take the right-hand side to be

$$r'' = \left(4\gamma^2((x-x_0)^2 + (y-y_0)^2) - 4\gamma\right)\exp(-\gamma(x-x_0)^2)\exp(-\gamma(y-y_0)^2) \tag{8.1.51}$$

with $\gamma = 10$ and $(x_0, y_0) = (0.6, 0.6)$, in a bounded domain of $0 \leq x \leq 1$, $0 \leq y \leq 1$.

The exact solution we are seeking is a 2D Gaussian pulse:

$$P(x,y) = \exp(-\gamma(x-x_0)^2)\exp(-\gamma(y-y_0)^2)\,. \tag{8.1.52}$$

However, due to the smaller value of γ the Gaussian does not decay sufficiently at the domain boundary. We use the known solution to obtain the Neumann boundary conditions at the four boundaries as

$$\begin{aligned}
\left.\frac{\partial P}{\partial x}\right|_{(x=0,y)} &= 2\gamma(x_0)\exp(-\gamma(x_0)^2)\exp(-\gamma(y-y_0)^2)\,,\\
\left.\frac{\partial P}{\partial x}\right|_{(x=1,y)} &= 2\gamma(x_0-1)\exp(-\gamma(1-x_0)^2)\exp(-\gamma(y-y_0)^2)\,,\\
\left.\frac{\partial P}{\partial y}\right|_{(x,y=0)} &= 2\gamma(y_0)\exp(-\gamma(x-x_0)^2)\exp(-\gamma(y_0)^2)\,,\\
\left.\frac{\partial P}{\partial y}\right|_{(x,y=1)} &= 2\gamma(y_0-1)\exp(-\gamma(x-x_0)^2)\exp(-\gamma(1-y_0)^2)\,.
\end{aligned} \tag{8.1.53}$$

For the numerical solution make the following choices:

(1) Use the direct solution approach.
(2) A uniform grid of $(N+1) \times (N+1)$ points in the x–y-plane.
(3) A 3-point central difference scheme in both x and y.

Try different values of $N = 50, 100$ and 200. Compare the numerical solution with the analytic solution. Evaluate the maximum and the root mean square error. Plot these errors as a function of N and explain the rate of convergence. Also repeat the problem with other spatial discretizations and solution methods.

Computer Exercise 8.1.9 Numerically solve the 2D advection–diffusion equation

$$\frac{\partial T}{\partial t} + u\frac{\partial T}{\partial x} + v\frac{\partial T}{\partial y} = \nu\left(\frac{\partial^2 T}{\partial x^2} + \frac{\partial^2 T}{\partial y^2}\right) \tag{8.1.54}$$

in the square domain $0 \le x \le 1$ and $0 \le y \le 1$, starting from the initial Gaussian temperature pulse

$$T(x, y, t = 0) = \exp(-\gamma(x - x_0)^2)\exp(-\gamma(y - y_0)^2)\,. \tag{8.1.55}$$

Here the initial pulse is of unit amplitude and centered about $(x_0, y_0) = (0.6, 0.6)$. By choosing $\gamma = 1381.6$ we ensure that the pulse decays sufficiently at the boundaries. Therefore, we enforce homogeneous Dirichlet boundary condition at the four boundaries of the square domain (i.e., $T = 0$ along all four boundaries). Here we choose $\alpha = 0.1$ and a time-independent cellular velocity field

$$u(x, y) = \sin(2\pi x)\cos(2\pi y) \quad \text{and} \quad v(x, y) = -\cos(2\pi x)\sin(2\pi y)\,. \tag{8.1.56}$$

As a result the pulse goes round and round about the center of the cellular velocity field (i.e., about the point $(0.5, 0.5)$) and diffuses over time.

For the numerical solution make the following choices:

(1) Use the direct solution approach.
(2) Use the Crank–Nicolson scheme for the diffusion term and the AB3 or RK3 scheme for the advection term.
(3) A uniform grid of $(N + 1) \times (N + 1)$ points in the x–y-plane.
(3) A 3-point central difference scheme in both x and y.

Build modular code. You should reuse the Helmholtz solver and the AB3/RK3 schemes written for previous computer exercises. Try different values of $N = 50, 100$ and 200.

Solve the problem also using Chebyshev spectral methods. With a large enough grid and a fine enough Δt, the spectral solution can be considered to be the benchmark. The L_∞ and L_2 errors can be calculated with respect to the benchmark. Integrate the advection–diffusion equation for $t = 5$ and compare the numerical and benchmark solutions. Plot the L_∞ and L_2 errors as a function of N and explain the rate of convergence.

8.2 Navier–Stokes Equations

We now have everything we need to finally address numerical methods for the Navier–Stokes equations. We will restrict attention to constant-property incompressible flows. The governing equations to be solved are the incompressibility condition

$$\nabla \cdot \mathbf{u} = \mathbf{0} \tag{8.2.1}$$

and the momentum equation

$$\frac{\partial \mathbf{u}}{\partial t} + \mathbf{u} \cdot \nabla \mathbf{u} = -\nabla \mathbf{p} + \frac{\mathbf{1}}{\mathrm{Re}} \nabla^{\mathbf{2}} \mathbf{u} \,. \tag{8.2.2}$$

The above equation is in the non-dimensional form where the quantities are non-dimensionalized by suitable length scale d, velocity scale U, and pressure scale ρU^2. Accordingly, the Reynolds number is defined as $\mathrm{Re} = Ud/\nu$. In 3D the above governing equations correspond to a system of four PDEs (continuity and three components of the momentum equations) for the four variables $\mathbf{u} = (\mathbf{u}, \mathbf{v}, \mathbf{w})$ and p. Note that only the momentum equations are prognastic. That is they are equations for the time evolution of the three components of velocity. In contrast, the continuity equation is diagnostic as it contains no time derivative. It is a constraint placed on the three components of velocity and one must find a suitable pressure field that when used in the momentum equation will result in an incompressible velocity field. Here we will demonstrate the numerical methodology in the simpler context of a 2D flow, where the above equations reduce to a system of three PDEs for the two velocity components u and v and the pressure p.

The difficulty with the above Navier–Stokes equations is that the system of four equations (three in 2D) are coupled. To solve the x-, y- or z-momentum equations, one needs the pressure field and to obtain the correct pressure, one needs the velocity fields. We will first consider a simple explicit implementation in Section 8.2.1, where the solution at the new time level is not coupled, and thus the velocity components and pressure can be solved one after the other. However, as we will see in the following section, such an explicit implementation has problems. We will then consider semi-implicit schemes where the viscous and the pressure effects are treated implicitly, while the nonlinear advection is treated explicitly. Any straightforward semi-implicit approach will need to solve all four equations simultaneously, which we will consider first in Section 8.2.2. In Chapter 5 we discussed ways to extend numerical methods that have been developed for a single ODE to handle systems of ODEs. Similarly, we will extend the methods discussed in Section 8.1 to systems of multi-dimensional PDEs. However, the Navier–Stokes equations are made harder by the fact that the incompressibility condition is just a constraint. Thus, in Section 8.1.3 we will pursue a simple plan that will reduce the Navier–Stokes equations to Helmholtz equations for the velocity components and a Poisson equation for pressure. Furthermore, the Helmholtz equations are decoupled and can be solved one at a time. By reducing the Navier–Stokes equations to multi-dimensional scalar Helmholtz and Poisson equations we have essentially solved the problem, since we have already developed numerical methods for solving them.

8.2.1 Explicit Method

We start with a very simple explicit implementation of the Navier–Stokes equations that uses the explicit Euler scheme. Of course we can do much better than that in terms of accuracy with higher-order explicit schemes, but for illustration of the main points the explicit Euler scheme will be sufficient. The Navier–Stokes equations discretized only in time can then be written as

$$\frac{\mathbf{u}_{n+1}-\mathbf{u}_n}{\Delta t} = -\nabla p_n - \mathcal{N}_n + \frac{1}{\mathrm{Re}}\nabla^2 \mathbf{u}_n\,, \tag{8.2.3a}$$

$$\nabla\cdot\mathbf{u_{n+1}} = 0\,, \tag{8.2.3b}$$

where $\mathcal{N}$ represents the nonlinear term $\mathbf{u}\cdot\nabla\mathbf{u}$. For the incompressibility condition we demand the velocity at the next time-level to be divergence-free. After rearranging the above we take the divergence to obtain

$$\nabla\cdot\mathbf{u}_{n+1} = \nabla\cdot\mathbf{u}_n + \Delta t\nabla\cdot\left[-\nabla p_n - \mathcal{N}_n + \frac{1}{\mathrm{Re}}\nabla^2\mathbf{u}_n\right]. \tag{8.2.4}$$

We require $\nabla\cdot\mathbf{u}_{n+1} = 0$ and as a result we also set the divergence of the velocity at the previous time-step to be zero (i.e., $\nabla\cdot\mathbf{u}_n = 0$). This leads to the following equation for pressure:

$$\nabla\cdot\nabla p_n = \nabla\cdot\left[-\mathcal{N}_n + \frac{1}{\mathrm{Re}}\nabla^2\mathbf{u}_n\right]. \tag{8.2.5}$$

On the left-hand side the divergence of the gradient operator becomes the Laplace operator. This analytical equivalence applies only in the continuous sense, before we proceed to discretize space into grid points. In the discrete sense, the discrete divergence operator operating on the discrete gradient operator will not necessarily lead to the discrete Laplace operator.[1] As a result we will retain the left-hand side as it is.

The solution of (8.2.5) requires boundary conditions for pressure which must be obtained from the known velocity boundary conditions. As we will see in the subsequent sections the homogeneous Neumann boundary condition, $\mathbf{n}\cdot\nabla p = 0$, provides a very good approximation, where $\mathbf{n}$ is the outward unit normal at the boundary. We can also derive the pressure boundary condition by evaluating the normal component of the explicit scheme (8.2.3a, 8.2.3b) at the boundary as

$$\mathbf{n}\cdot\nabla p_n = \mathbf{n}\cdot\left[-\mathcal{N}_n + \frac{1}{\mathrm{Re}}\nabla^2\mathbf{u}_n + \frac{1}{\Delta t}(\mathbf{u}_n - \mathbf{u}_{n+1})\right]. \tag{8.2.6}$$

All the terms on the right-hand side are either known from the previous time-velocity or from the velocity boundary condition at the $(n+1)$th time-level. For example, in the case of no-slip and no-penetration boundary conditions, $\mathbf{u}_n$, $\mathbf{u}_{n+1}$ and $\mathcal{N}_n$ are identically zero at the boundary. Also the tangential terms of the Laplacian are zero. This yields the final boundary condition for the pressure gradient as

$$\frac{\partial p_n}{\partial n} = \frac{1}{\mathrm{Re}}\frac{\partial^2 u_n}{\partial n^2}\,, \tag{8.2.7}$$

where $\partial/\partial n$ denotes the normal derivative.

[1] A simple illustration in 1D is as follows. Although analytically $\mathrm{d}^2/\mathrm{d}x^2 = \mathrm{d}/\mathrm{d}x(\mathrm{d}/\mathrm{d}x)$, with the three-point finite difference approximation the second-derivative operator is not equal to the product of the first-derivative matrices (i.e., $\mathbf{D}_2 \neq \mathbf{D}_1\mathbf{D}_1$).

The explicit method for solving the Navier–Stokes equations can then proceed along the following steps:

(1) Starting from the known velocity field $\mathbf{u}_n$ at time-level n, calculate the right-hand side of (8.2.6) and the pressure gradient boundary condition (8.2.7).
(2) Solve (8.2.6) for p_n with the Neumann pressure boundary condition given in (8.2.7).
(3) Use (8.2.3a) at the interior grid points to obtain $\mathbf{u}_{n+1}$.
(4) Apply the velocity boundary condition for the boundary values of $\mathbf{u}_{n+1}$.

Let us now proceed to discretize the spatial coordinates. We will consider a uniform Cartesian **collocated grid** given by $x_j = j\,L_x/N_x$ for $j = 0, 1, \ldots, N_x$, and similarly for y and z. In the collocated grid all three components of velocity and pressure are defined at the same grid points. As we will see the collocated grid leads to difficulties in the pressure calculation. In the context of the three-point central scheme there are two options for representing the left-hand side of (8.2.5). The first is to directly discretize the Laplacian by combining the divergence and the gradient operators. In 2D this yields (dropping the time-level subscript n)

$$\begin{aligned}\frac{p_{i+1,j} - 2p_{i,j} + p_{i-1,j}}{2(\Delta x)^2} &+ \frac{p_{i,j+1} - 2p_{i,j} + p_{i,j-1}}{2(\Delta y)^2} \\ &= \frac{Rx_{i+1,j} - Rx_{i-1,j}}{2\Delta x} + \frac{Ry_{i,j+1} - Rx_{i,j-1}}{2\Delta y}\,, \end{aligned} \tag{8.2.8}$$

where (Rx, Ry) is the term within the square parentheses in (8.2.5). The above leads to a compact five-point stencil (seven-point stencil in 3D) for the solution of p_n. As mentioned in the footnote, the primary difficulty with this solution is that the left-hand side of the above equation is not the discrete representation of $\nabla \cdot \nabla$. As a result of this approximation the final velocity $\mathbf{u}_{n+1}$ will not be divergence-free and violation of mass balance can lead to serious problem.

The alternative approach is to discretize $\nabla \cdot \nabla$ as a combination of a discrete divergence and a discrete gradient. Along the x-direction this becomes

$$\begin{aligned}\frac{(\partial p/\partial x)_{i+1/2,j} - (\partial p/\partial x)_{i-1/2,j}}{\Delta x} &= \frac{1}{2\Delta x}\left[(\partial p/\partial x)_{i,j} + (\partial p/\partial x)_{i+1,j}\right] \\ &\quad - \frac{1}{2\Delta x}\left[(\partial p/\partial x)_{i-1,j} + (\partial p/\partial x)_{i,j}\right] \\ &= \frac{(\partial p/\partial x)_{i+1,j} - (\partial p/\partial x)_{i-1,j}}{2\Delta x} \\ &= \frac{p_{i+2,j} - 2p_{i,j} + p_{i-2,j}}{4(\Delta x)^2}\,. \end{aligned} \tag{8.2.9}$$

With this approximation (8.2.5) becomes

$$\begin{aligned}\frac{p_{i+2,j} - 2p_{i,j} + p_{i-2,j}}{4(\Delta x)^2} &+ \frac{p_{i,j+2} - 2p_{i,j} + p_{i,j-2}}{4(\Delta y)^2} \\ &= \frac{Rx_{i+1,j} - Rx_{i-1,j}}{2\Delta x} + \frac{Ry_{i,j+1} - Rx_{i,j-1}}{2\Delta y}\,. \end{aligned} \tag{8.2.10}$$

The advantage of this revised formulation is that the resulting $\mathbf{u}_n$ will satisfy discrete divergence. However, there are difficulties. First, the left-hand side of (8.2.10) is the same as (8.2.8), but with twice the grid spacing. As a result, the approximation in (8.2.10) is less accurate than that in (8.2.8). But the more serious problem is **odd–even decoupling**. In 2D, (8.2.10) splits into four unconnected problems for pressure: (i) i, j even; (ii) i even, j odd; (iii) i odd, j even; (iv) i, j odd. This leads to a checkerboard pattern for pressure. This pressure oscillation will not affect the velocity field.

The odd–even decoupling problem can be avoided with the use of a staggered grid. We leave it to the reader as an exercise to formulate the explicit method on a staggered grid. However, the explicit method, even with such a grid, will be restricted to a very small time-step due to the more stringent diffusion limit. It is important to consider an implicit scheme for the diffusion term in order to avoid the serious time-step limitation. In the following sections we will discuss methods for the implicit treatment of the diffusion term. We will consider both staggered and collocated grid implementations.

8.2.2 Coupled Method

As the name suggest, in the coupled method the continuity and momentum equations are solved together as required by their coupling. There are many options for discretizing (8.2.1) and (8.2.2) in time, and here, for illustrative purposes, we choose the second-order Adams–Bashforth method for the nonlinear advection, the second-order semi-implicit Crank–Nicolson for the diffusion, and the fully-implicit Euler method for the pressure gradient. The resulting semi-discrete equations are

$$\begin{aligned} \frac{\mathbf{u}_{n+1}-\mathbf{u}_n}{\Delta t} &= -\nabla p_{n+1} - \frac{3}{2}\mathcal{N}_n + \frac{1}{2}\mathcal{N}_{n-1} + \frac{1}{2\,\mathrm{Re}}\left[\nabla^2\mathbf{u}_{n+1} + \nabla^2\mathbf{u}_n\right] \\ \nabla\cdot\mathbf{u_{n+1}} &= 0\,. \end{aligned} \tag{8.2.11}$$

Now we proceed with spatial discretization. To simplify our discussion let us restrict attention to a rectangular two-dimensional (2D) domain and consider a uniform distribution of (N_x+1) points along the x-direction and (N_y+1) points along the y-direction. These grid points form a simple 2D rectangular lattice and are denoted by the index pair (i,j), where $i=0,1,\ldots,N_x$ and $j=0,1,\ldots,N_y$. It is natural to discretize the velocity components, (u,v), and pressure, (p), at these regular (i,j) gridpoints and solve the governing equations there. This arrangement is known as the **collocated grid**. While simpler to implement, this arrangement faces two important difficulties. First, boundary conditions are needed on the four (top, bottom, left and right) faces of the computational domain. The boundary conditions for the velocity components are known from the physical problem at hand, such as no-slip, or no-penetration or inflow velocity boundary conditions, whereas pressure boundary conditions are typically not known from the problem statement. The pressure on the boundaries comes out as a solution to the incompressible Navier–Stokes equations. Note that the incompressibility condition is only a constraint and pressure can be thought of as a Lagrange multiplier. Thus, it is sufficient to only specify the velocity boundary conditions. However, discretization of pressure with a collocated grid requires a numerical pressure boundary condition, which needs to be approximated. The second problem is more technical and perhaps more serious. Due to lack of strong pressure–velocity coupling, the collocated arrangement leads to pressure oscillation similar to the two-delta waves seen in Section 7.6.2. Therefore the collocated arrangement is generally not pursued for the coupled method, although ingenious fixes have been designed to address both the aforementioned problems.

In the applied mathematical literature it is well established that in a numerical discretization of the incompressible Navier–Stokes equations, pressure must be discretized in a space smaller than that of the velocity field. This can be accomplished with a **staggered grid**. The traditional approach is to use a **fully staggered grid** where the pressure is defined at the cell centers (i.e., at $(i+1/2, j+1/2)$) and the velocities are defined on the cell faces (u at $(i+1/2, j)$ points and v at $(i, j+1/2)$ points). To reduce the algorithmic complexity of keeping track of the different cell faces and cell centers, other ways of staggering have been proposed. In the **partially staggered grid** arrangement pressure is defined at the cell centers (i.e., at $(i+1/2, j+1/2)$) and both the velocities are defined at the regular (i, j) grid points. Here we will discuss the coupled method in the context of the partially staggered grid. We will use

$$\vec{u} = \left[\{u_{0,0}, u_{1,0}, \ldots, u_{N_x,0}\}, \{u_{0,1}, \ldots, u_{N_x,1}\}, \ldots, \{u_{0,N_y}, \ldots, u_{N_x,N_y}\}\right]^{\mathrm{T}}$$

to denote the column vector of x-velocity components at the (i, j) grid points. Similarly, we will use $\vec{v}$ to denote the column vector of y-velocity, and $\vec{p}_{\mathrm{c}}$ to denote pressure discretized at the cell centers.

We now discretize (8.2.11) at the partially staggered grid points. The resulting linear system can be expressed by the following block matrix equation:

$$\left[\begin{array}{c|c|c} \mathbf{I} - \alpha(\mathbb{D}_{2x} + \mathbb{D}_{2y}) & 0 & \Delta t\, \mathbb{D}_{1x}\mathbb{I}_{C\to R} \\ \hline 0 & \mathbf{I} - \alpha(\mathbb{D}_{2y} + \mathbb{D}_{2x}) & \Delta t\, \mathbb{D}_{1y}\mathbb{I}_{C\to R} \\ \hline \mathbb{I}_{C\to R}\mathbb{D}_{1x} & \mathbb{I}_{C\to R}\mathbb{D}_{1y} & 0 \end{array}\right] \left[\begin{array}{c} \vec{u}_{n+1} \\ \hline \vec{v}_{n+1} \\ \hline \vec{p}_{c,n+1} \end{array}\right] = \left[\begin{array}{c} \vec{r}_x \\ \hline \vec{r}_y \\ \hline 0 \end{array}\right], \tag{8.2.12}$$

where $\alpha = \Delta t/(2\,\mathrm{Re})$ and the right-hand side vectors that depend on the previous time-level solution are given by

$$\begin{aligned} \vec{r}_x &= -\frac{3\Delta t}{2}\vec{\mathcal{N}}_{x,n} + \frac{\Delta t}{2}\vec{\mathcal{N}}_{x,n-1} + \left[\mathbf{I} + \alpha(\mathbb{D}_{2x} + \mathbb{D}_{2y})\right]\vec{u}_n \\ \vec{r}_y &= -\frac{3\Delta t}{2}\vec{\mathcal{N}}_{y,n} + \frac{\Delta t}{2}\vec{\mathcal{N}}_{y,n-1} + \left[\mathbf{I} + \alpha(\mathbb{D}_{2x} + \mathbb{D}_{2y})\right]\vec{v}_n\,. \end{aligned} \tag{8.2.13}$$

In the above equations $\vec{\mathcal{N}}_{x,n}$ and $\vec{\mathcal{N}}_{y,n}$ are vectors of x- and y-components of the nonlinear term at time-level n, and $\vec{\mathcal{N}}_{x,n-1}$ and $\vec{\mathcal{N}}_{y,n-1}$ are the corresponding definitions for time-level $n-1$. The operators $\mathbb{D}_{1x}$ and $\mathbb{D}_{2x}$ are large square matrices of $(N_x+1)(N_y+1) \times (N_x+1)(N_y+1)$ rows and $(N_x+1)(N_y+1) \times (N_x+1)(N_y+1)$ columns. When they multiply a vector $\vec{u}$ we obtain its first and second x-derivative values, respectively. Similarly, the operators $\mathbb{D}_{1y}$ and $\mathbb{D}_{2y}$ are large square matrices of $(N_x+1)(N_y+1) \times (N_x+1)(N_y+1)$ rows and $(N_x+1)(N_y+1) \times (N_x+1)(N_y+1)$ columns. When they multiply a vector $\vec{u}$ they generate a vector of first and second y-derivative values, respectively.

2D derivative operators

Both $\mathbb{D}_{1x}$ and $\mathbb{D}_{2x}$ are sparse matrices. For example, the non-zero elements of $\mathbb{D}_{2x}$ are from the 1D second-derivative matrix in x (i.e., from $\mathbf{D}_{2x}$). In block matrix form, $\mathbb{D}_{2x}$ consists of $(N_y+1) \times (N_y+1)$ blocks of smaller $(N_x+1) \times (N_x+1)$ matrices and can be expressed as

$$\mathbb{D}_{2x} = \left[\begin{array}{c|c|c|c} \mathbf{D}_{2x} & 0 & 0 & \cdots \\ \hline 0 & \mathbf{D}_{2x} & 0 & \cdots \\ \hline 0 & 0 & \ddots & 0 \\ \hline 0 & \ddots & 0 & \mathbf{D}_{2x} \end{array}\right]. \tag{8.2.14}$$

A similar definition applies for $\mathbb{D}_{1x}$. It can be seen that when the above matrix operates on a vector $\vec{u}$, the operation will yield a vector of second x-derivative values at the grid points.

The y-direction operators $\mathbb{D}_{1y}$ and $\mathbb{D}_{2y}$ are also large matrices of $(N_x+1) \times (N_y+1)$ rows and columns and they compute the first and second derivatives in y. They are also sparse, but their structure is different. For example, in block matrix form $\mathbb{D}_{2y}$ can be written out as $(N_y+1) \times (N_y+1)$ blocks, where each block is now an $(N_x+1) \times (N_x+1)$ diagonal matrix

$$\mathbb{D}_{2y} = \left[\begin{array}{c|c|c} \nwarrow \quad [D_{2y}]_{0,0} \quad \searrow & \cdots & \nwarrow \quad [D_{2y}]_{0,N_y} \quad \searrow \\ \hline \nwarrow \quad \ddots \quad \searrow & \cdots & \nwarrow \quad \ddots \quad \searrow \\ \hline \nwarrow \quad [D_{2y}]_{N_y,0} \quad \searrow & \cdots & \nwarrow \quad [D_{2y}]_{N_y,N_y} \quad \searrow \end{array}\right]. \tag{8.2.15}$$

Note that the vectors $\vec{u}$ and $\vec{v}$ are ordered such that the x grid points are contiguous, while the y grid points are $(N_x + 1)$ elements apart. As a result in $\mathbb{D}_{2x}$ the elements of $\mathbf{D}_{2x}$ appear together along the diagonal blocks, whereas in $\mathbb{D}_{2y}$ the elements of $\mathbf{D}_{2y}$ are dispersed and appear along the diagonal of all the blocks. The structures of the 2D first-derivative operators $\mathbb{D}_{1x}$ and $\mathbb{D}_{1y}$ are similar to those of the corresponding second-derivative operators.

2D interpolation operators

Recall that the pressure $\vec{p}_{c,n+1}$ is defined at the cell centers while in the x- and y-momentum equations (represented by the first two block rows of (8.2.12)) the pressure gradients are required at the regular gridpoints. Therefore before taking the x- or y-derivative, first $\vec{p}_{c,n+1}$ must be interpolated onto the regular grid. This 2D interpolation, represented by the operator $\mathbb{I}_{C\to R}$, can be carried out in two steps. First, $\vec{p}_{c,n+1}$ is interpolated along x from the cell centers to face centers of the left and right faces of each cell. This can be accomplished with a 2D x-interpolation operator that is constructed with the elements of the 1D center-to-regular interpolation operator, $\mathbf{I}_{x,C\to R}$, defined in Section 4.5.2. Here x in the subscript indicates the interpolation is in the x-direction. This construction is identical to how the 2D operator $\mathbb{D}_{2x}$ was constructed from the elements of $\mathbf{D}_{2x}$. The resulting 2D center-to-regular x-interpolation operator can be expressed as

$$\mathbb{I}_{x,C\to R} = \left[\begin{array}{c|c|c|c} \mathbf{I}_{x,C\to R} & 0 & 0 & \cdots \\ \hline 0 & \mathbf{I}_{x,C\to R} & 0 & \cdots \\ \hline 0 & 0 & \ddots & 0 \\ \hline 0 & \ddots & 0 & \mathbf{I}_{x,C\to R} \end{array}\right]. \tag{8.2.16}$$

Note that $\mathbb{I}_{x,C\to R}$ is a rectangular matrix. It consists of $N_y \times N_y$ blocks. But each block is a rectangular matrix of size $(N_x + 1) \times (N_x)$, since along x we are interpolating from cell centers to grid points.

The second step is to interpolate the outcome of the first step from the left and right face centers to the regular grid points. This involves interpolation in y from the center points to the regular points. We can again use the 1D center-to-regular interpolation operator, $\mathbf{I}_{y,C\to R}$, and construct

$$
\mathbb{I}_{y,C\to R} = \left[\begin{array}{c|c|c}
\nwarrow & & \nwarrow \\
[\mathbf{I}_{y,C\to R}]_{0,0} & \cdots & [\mathbf{I}_{y,C\to R}]_{0,N_y-1} \\
\searrow & & \searrow \\
\hline
\nwarrow & & \nwarrow \\
\ddots & \cdots & \ddots \\
\searrow & & \searrow \\
\hline
\nwarrow & & \nwarrow \\
[\mathbf{I}_{y,C\to R}]_{N_y,0} & \cdots & [\mathbf{I}_{y,C\to R}]_{N_y,N_y-1} \\
\searrow & & \searrow
\end{array}\right] . \tag{8.2.17}
$$

In this second step, $\mathbb{I}_{y,C\to R}$ operates on a vector that has already been interpolated to the regular grid points in the x-direction. Thus, each block is a square matrix of size $(N_x+1)\times(N_x+1)$. However, in the y-direction the vector is being interpolated from the center to the regular point and thus there are $(N_y+1)\times N_y$ blocks.

Together the two steps can be applied, one after the other, and accordingly we have the following expression for the 2D cell center-to-regular grid interpolation operator:

$$
\mathbb{I}_{C\to R} = \mathbb{I}_{y,C\to R}\,\mathbb{I}_{x,C\to R} \,. \tag{8.2.18}
$$

From the size of the x and y interpolation operators it can be seen that $\mathbb{I}_{C\to R}$, when operated on $\vec{p}_{c,n+1}$, which is a vector of pressure values at the cell centers, will yield a vector of pressure values at the $(N_x+1)\times(N_y+1)$ regular grid points.

The last block row of (8.2.12) enforces the discrete form of the divergence-free condition at the cell centers, while $\vec{u}_{n+1}$ and $\vec{v}_{n+1}$ are defined at the regular grid points. Thus $\mathbb{I}_{R\to C}$ is the 2D interpolation operator that will take the vector of values from the regular grid points to the cell centers, after the x- and y-derivatives of $\vec{u}$ and $\vec{v}$ have been computed at the regular grid points. We leave the definition of the operator $\mathbb{I}_{R\to C}$ to the reader as an exercise.

Problem 8.2.1 Similar to (8.2.18) we can write

$$
\mathbb{I}_{R\to C} = \mathbb{I}_{y,R\to C}\,\mathbb{I}_{x,R\to C} \,, \tag{8.2.19}
$$

and thereby accomplish the task of interpolating a vector of grid point values to a vector of cell-center values in two steps. The 2D matrix operators $\mathbb{I}_{y,R\to C}$ and $\mathbb{I}_{x,R\to C}$ can be expressed in terms of the elements of the 1D interpolation operator $\mathbf{I}_{R\to C}$ defined in Section 4.5.2. Write out the x and y grid-to-cell-center

2D matrix operators as in (8.2.16) and (8.2.17). Discuss the size of these block matrices.

Problem 8.2.2 In this exercise you will follow the steps outlined above for the partially-staggered grid and formulate a coupled method for the 2D incompressible Navier–Stokes equations using a fully staggered grid. Here there are three grids: (i) the cell centers where pressure is defined and the incompressibility condition is solved; (ii) the left–right face centers where u is defined and the x-momentum is solved; and (iii) the top–bottom face centers where v is defined and the y-momentum is solved. Your first step will be to define 2D interpolation matrices that will take pressure from the cell centers to the two different face centers. Similarly define the interpolation matrices that will take u and v velocity components from their respective face centers to the cell centers. Next define the 2D x and y derivative matrices that will compute derivatives on the two different face-center grids. Now discretize (8.2.11) on the fully staggered grid. Express the resulting discretized equation in a block matrix form that is similar to (8.2.12), using the different interpolation and derivative matrices that you defined in this exercise.

Boundary conditions
So far our discussion pertained only to the discrete implementation of the continuity and momentum equations. Boundary condition must now be included into the linear system (8.2.12). We must identify the rows of the linear system that correspond to the left, bottom, right and top boundaries and these rows must be replaced by the appropriate boundary conditions:

$$\begin{array}{lll} \text{Left:} & u(x=0,y)=U_{\mathrm{L}}(y)\,, & v(x=0,y)=V_{\mathrm{L}}(y)\,, \\ \text{Bottom:} & u(x,y=0)=U_{\mathrm{B}}(x)\,, & v(x,y=0)=V_{\mathrm{B}}(x)\,, \\ \text{Right:} & u(x=L_x,y)=U_{\mathrm{R}}(y)\,, & v(x=L_x,y)=V_{\mathrm{R}}(y)\,, \\ \text{Top:} & u(x,y=L_y)=U_{\mathrm{T}}(x)\,, & v(x,y=L_y)=V_{\mathrm{T}}(x)\,. \end{array} \tag{8.2.20}$$

We identify the first block row of (8.2.12) to be the x momentum equation and it consists of (N_y+1) blocks of (N_x+1) rows and thus a total of $(N_x+1)\times(N_y+1)$ rows. Some of these rows must be modified in order to implement the u velocity boundary conditions. The second block row of (8.2.12) corresponds to the y momentum equation and it also consists of $(N_x+1)\times(N_y+1)$ rows within it. Some of these rows must be modified in order to implement the v velocity boundary conditions. We will refer to these first and second block rows of (8.2.12) as ⓧ and ⓨ, respectively. The boundary conditions can be incorporated in the following eight steps:

(1) The first (N_x+1) rows of ⓧ correspond to the bottom boundary and these

rows in the left-hand matrix operator must be replaced by

$$[\mathbf{I}\ \mathbf{0}\ \ldots|\ldots\ \mathbf{0}\ \ldots|\ldots\ \mathbf{0}\ \ldots]\ ,$$

where each entry is either an identity or a zero matrix of size $(N_x + 1) \times (N_x + 1)$. Correspondingly the first $(N_x + 1)$ elements of the right-hand side vector $\vec{r}_x$ must be replaced by $U_{\mathrm{B}}(x_j)$.

(2) The first $(N_x + 1)$ rows of ⓨ correspond to the bottom boundary in the y-momentum equation and therefore must be replaced by

$$[\ldots\ \mathbf{0}\ \ldots|\mathbf{I}\ \mathbf{0}\ \ldots|\ldots\ \mathbf{0}\ \ldots]\ .$$

The first $(N_x + 1)$ elements of the right-hand side vector $\vec{r}_y$ must be replaced by $V_{\mathrm{B}}(x_j)$.

(3) The last $(N_x + 1)$ rows of ⓧ correspond to the top boundary and therefore must be replaced by

$$[\ldots\ \mathbf{0}\ \mathbf{I}|\ldots\ \mathbf{0}\ \ldots|\ldots\ \mathbf{0}\ \ldots]\ ,$$

and the last $(N_x + 1)$ elements of the right-hand side vector $\vec{r}_x$ must be replaced by $U_{\mathrm{T}}(x_j)$.

(4) The last $(N_x + 1)$ rows of ⓨ must be replaced by

$$[\ldots\ \mathbf{0}\ \ldots|\ldots\ \mathbf{0}\ \mathbf{I}|\ldots\ \mathbf{0}\ \ldots]\ ,$$

and the last $(N_x + 1)$ elements of the right-hand side vector $\vec{r}_y$ must be replaced by $V_{\mathrm{T}}(x_j)$.

(5) The first row of each of the $(N_y + 1)$ blocks of ⓧ correspond to the left boundary. Each of these rows on the left must be replaced by

$$[0\ \ldots\ 1\ 0\ \ldots|\ldots\ 0\ \ldots|\ldots\ 0\ \ldots]\ ,$$

where the only non-zero entry is when the column and row numbers are the same. The corresponding element in the right-hand side vector $\vec{r}_x$ must be replaced by $U_{\mathrm{L}}(y_l)$.

(6) The first row of each of the (N_y+1) blocks of ⓨ on the left must be replaced by

$$[\ldots\ 0\ \ldots|0\ \ldots\ 1\ 0\ \ldots|\ldots\ 0\ \ldots]\ .$$

Again, the above replaces a single row and the only non-zero entry is where the column and row numbers are the same. The corresponding element in the right-hand side vector $\vec{r}_y$ must be replaced by $V_{\mathrm{L}}(y_l)$.

(7) The last row of each of the $(N_y + 1)$ blocks of ⓧ correspond to the right boundary and each of these rows on the left must be replaced by

$$[0\ \ldots\ 1\ 0\ \ldots|\ldots\ 0\ \ldots|\ldots\ 0\ \ldots]\ .$$

The only non-zero entry is where the column and row numbers are the same. The corresponding element in the right-hand side vector $\vec{r}_x$ must be replaced by $U_{\mathrm{R}}(y_l)$.

(8) The last row of each of the (N_y+1) blocks of ⓨ on the left must be replaced by

$$[\dots\ 0\ \dots|0\ \dots\ 1\ 0\ \dots|\dots\ 0\ \dots]\ .$$

The corresponding element in the right-hand side vector $\vec{r}_y$ must be replaced by $V_{\mathrm{R}}(y_l)$.

The above boundary corrections modify (8.2.12) and the resulting linear system can be expressed as

$$\mathbb{L}_* \begin{bmatrix} \vec{u}_{n+1} \\ \vec{v}_{n+1} \\ \vec{p}_{c,n+1} \end{bmatrix} = \begin{bmatrix} \vec{r}_{*x} \\ \vec{r}_{*y} \\ 0 \end{bmatrix}, \qquad (8.2.21)$$

where $\mathbb{L}_*$ is the left-hand side operator modified according to the boundary condition. It is a large square matrix of $(2(N_x+1)\times(N_y+1))+N_xN_y$ rows and $(2(N_x+1)\times(N_y+1))+N_xN_y$ columns. On the right, $\vec{r}_{*x}$ and $\vec{r}_{*y}$ are the boundary-modified right-hand side vectors.

Solution steps

We are now ready to solve complicated flow problems in two directions. This requires a preprocessing step where you will compute the operators $\mathbb{D}_{1x}$, $\mathbb{D}_{2x}$, $\mathbb{D}_{1y}$, $\mathbb{D}_{2y}$, $\mathbb{I}$ and $\mathbb{I}_{R\to C}$, incorporate the boundary corrections and form the left-hand side operator $\mathbb{L}_*$. Then the process of going from the flow field at time-level n to the flow field at time-level $n+1$ (i.e., from $\vec{u}_n$, $\vec{v}_n$ and $\vec{p}_n$ to their new values $\vec{u}_{n+1}$, $\vec{v}_{n+1}$ and $\vec{p}_{n+1}$) is accomplished through the following steps:

(1) From $\vec{u}_n$ and $\vec{v}_n$ compute the nonlinear terms $\vec{\mathcal{N}}_{x,n}$ and $\vec{\mathcal{N}}_{y,n}$.
(2) Compute the right-hand sides $\vec{r}_x$ and $\vec{r}_y$ using (8.2.13).
(3) Incorporate the boundary conditions to obtain $\vec{r}_{*x}$ and $\vec{r}_{*y}$.
(4) Solve the linear system (8.2.21) for $\vec{u}_{n+1}$, $\vec{v}_{n+1}$ and $\vec{p}_{n+1}$.

In the computer exercises below you will solve the classic driven-cavity problem in 2D. Before that let us discuss a few facts that will help you with this task. In 2D, if we consider spatial discretization with $N_x = N_y = 100$ grid points along the two directions, then $\mathbb{L}_*$ is a large matrix of 30,402 rows and columns. If you store the entire matrix, or its factors, the memory requirement is about a billion double precision numbers: yes, you should use 64-bit precision. You should make use of the sparsity of the operator $\mathbb{L}_*$, as only few of the elements are non-zero. Even then, explicit sparse construction of the various matrix operators and the solution can be computationally expensive. Iterative solvers, such as GMRES, offer a computationally attractive alternative. Most importantly, in iterative solvers, only the action of $\mathbb{L}_*$ on a vector $[\vec{u}, \vec{v}, \vec{P}_c]^{\mathrm{T}}$ is needed. We trace back to (8.2.12) and can identify the action of each term of the left-hand block matrix operator. For example, the action of $\mathbf{I} - \alpha(\mathbb{D}_{2x} + \mathbb{D}_{2y})$ on a vector $\vec{u}$ is nothing but calculating $u - \alpha(\partial^2 u/\partial^2 x + \partial^2 u/\partial^2 y)$ at the regular grid points. Similarly, the action of $\mathbb{D}_{1x}\mathbb{I}_{C\to R}$ on a vector $\vec{p}_c$ corresponds to first interpolating

p from the cell center to the regular grid, followed by calculation of $(\partial p/\partial x)$ at the regular grid points. The boundary corrections simply leave the values of u and v on the boundaries unaltered and equal to the boundary condition. Thus, the action of $\mathbb{L}_*$ can be easily computed through elementary derivative and interpolation operations along the x- and y-directions. This is a computationally very efficient strategy for solving the linear system, which is the last of the four steps outlined above.

We will now address some restrictions that apply regarding boundary conditions (8.2.20). In an incompressible flow the net flow into or out of the domain must be zero. In other words, the x-integral of the v-velocity at the bottom and top boundaries, plus the y-integral of the u-velocity at the left and right boundaries must add to zero. This condition must be ensured in the discrete sense as well. In the driven-cavity problem to be considered below, the normal velocities on all four boundaries are identically zero and thus there is no net flow into the cavity. Also, the values of U_l and U_b along the left and bottom boundaries must agree at the bottom-left corner. The same applies to the other three corners and for the v velocity as well. In the case of the driven-cavity problem, the u-velocities along the left, right and bottom walls are zero, while the u-velocity of the top wall is a non-zero constant. Thus, there is a mismatch in the u-velocities at the top-left and top-right corners; this issue has been much addressed in the driven-cavity literature. In the case of a partially staggered grid, the u-velocity is discretized at the regular grid and thus there is confusion as to what should be the numerical u-velocities at the top-right and top-left corner points. This dilemma is avoided with the use of a fully staggered grid, since u needs to be defined only at the left- and right-face center points, i.e., not at the corner points.

Computer Exercise 8.2.3 In this computer exercise we will consider the classic 2D **driven-cavity** problem in a square cavity of unit non-dimensional length along the x- and y-directions. The flow within the cavity is driven by a unit non-dimensional velocity along the top boundary and the boundary conditions on all four sides of the cavity are given by

$$\begin{array}{lll} \text{Left:} & u(x=0,y)=0\,, & v(x=0,y)=0\,, \\ \text{Bottom:} & u(x,y=0)=0\,, & v(x,y=0)=0\,, \\ \text{Right:} & u(x=1,y)=0\,, & v(x=1,y)=0\,, \\ \text{Top:} & u(x,y=1)=1\,, & v(x,y=1)=0\,. \end{array} \tag{8.2.22}$$

Consider a uniform grid of $N_x = N_y = N$ points along the x- and y-directions. Employ the fully staggered grid arrangement and solve the governing incompressible Navier–Stokes equations using the coupled method that you developed in Problem 8.2.2. The driven-cavity problem is a benchmark for which there are many near-exact solutions, for example the pioneering work by Ghia *et al.* (1982). For low enough Reynolds number the flow is steady. Choose $\mathrm{Re} = 100$ and $N = 51$ and time-advance from an initial velocity field of your choice to a steady state. Compare your results with those of Ghia *et al.* Increase grid resolution to $N = 101, 201$, etc., and investigate convergence. You can consider higher Reynolds numbers where the flow will be time dependent. For time-accurate solutions you may need more grid points and smaller time-steps. Consider second-order accurate approximations for both the x- and y-derivatives and choose the corresponding interpolation operators.

Computer Exercise 8.2.4 Repeat the above 2D driven-cavity problem on a partially staggered grid.

Computer Exercise 8.2.5 Repeat the 2D driven-cavity problem using a compact difference scheme for both derivative and interpolation operators on a uniform grid with fully staggered arrangement. Note that the advantage of the above presented framework is that you can essentially use the same code for time integration. The only difference is that all the finite-difference derivative and interpolation operators have to be replaced by the corresponding compact difference operators.

Computer Exercise 8.2.6 Repeat the 2D driven-cavity problem using the Chebyshev spectral method on a fully staggered arrangement. Here the regular grid will be the Chebyshev–Gauss–Lobatto grid, while the cell-center grid will be the Chebyshev–Gauss grid. The 1D derivative operators that you will need in the exercise have been already defined in Chapter 6. You will first need to develop the corresponding Gauss-to-Gauss–Lobatto and Gauss–Lobatto-to-Gauss interpolation operators.

8.2.3 Time-splitting Method

Here we seek a simpler alternative to the coupled method from the previous section. In 2D that method generally requires the iterative solution of a very large linear system, which becomes even larger in 3D. Here we pursue an approach where the governing equations of u, v and p decouple and reduce to the familiar Helmholtz and Poisson equations. The basic idea of the **time-splitting method** is to advance the velocity field over a time-step by splitting the effects of advection (the second term on the left-hand side of (8.2.2)), diffusion (the second term on the right-hand side of (8.2.2)) and pressure gradient separately. The effect of pressure is taken into account in such a way to make the final velocity field incompressible. This divide-and-conquer strategy as an efficient method for solving incompressible Navier–Stokes equations was originally introduced by Chorin (1968) and Teman (1969). In the fluid mechanics literature this method also goes by the names **fractional-step method** and **projection method**. In what follows we will use these terms interchangeably.

Over recent decades there have been many variants of the fractional-step method proposed and analyzed for their accuracy and stability. They can be broadly classified into two groups. In the **pressure-correction approach** the advection and diffusion effects are first taken into account to compute an intermediate velocity, which then is corrected for pressure to get a final divergence-free velocity field. The original method proposed by Chorin and Temam was of this type and other prominent contributions were by Kim and Moin (1985), van Kan (1986) and Brown *et al.* (2001). The alternative is the **velocity-correction approach** introduced by Orszag *et al.* (1986) and Karniadakis *et al.* (1991), where the pressure and the advection effects are taken into account first and then the velocity is corrected with an implicit treatment of the viscous effect. An excellent comparison of the different approaches was given by Guermond *et al.* (2006). The general conclusion is that there is no fundamental difference between the two approaches and the choice is just a matter of taste.

Here we will discuss only the pressure-correction approach. In this approach the velocity at the nth time level, $\mathbf{u}_n$, is first advanced to an intermediate velocity, $\hat{\mathbf{u}}$, taking into account only the advection and diffusion effects. As an example, we will demonstrate this first step with the explicit second-order Adams–Bashforth method for the advection term and the semi-implicit Crank–Nicolson method for the diffusion term. The resulting semi-discrete equation can be expressed as

$$\frac{\hat{\mathbf{u}} - \mathbf{u}_n}{\Delta t} = -\frac{3}{2}\mathcal{N}_n + \frac{1}{2}\mathcal{N}_{n-1} + \frac{1}{2\,\mathrm{Re}}\left[\nabla^2\hat{\mathbf{u}} + \nabla^2\mathbf{u}_n\right]. \tag{8.2.23}$$

Since the pressure effect is not taken into account in the first step, the intermediate velocity field $\hat{\mathbf{u}}$ will not be divergence free in general. Thus, in the second step the velocity field is corrected to obtain a divergence-free velocity field $\mathbf{u}_{n+1}$ as

$$\frac{\mathbf{u}_{n+1} - \hat{\mathbf{u}}}{\Delta t} = -\nabla\phi_{n+1}\,, \tag{8.2.24}$$

where the right-hand side plays the role of pressure gradient and we will relate pressure to the variable ϕ. We obtain the following equation for ϕ by taking the divergence of the above equation and requiring that $\nabla \cdot \mathbf{u}_{n+1} = 0$:

$$\nabla \cdot \nabla\phi_{n+1} = \frac{1}{\Delta t}\nabla \cdot \hat{\mathbf{u}}\,. \tag{8.2.25}$$

After the first step, once $\hat{\mathbf{u}}$ is known, the above Poisson equation is solved to obtain ϕ_{n+1}, which then can be used in (8.2.24) to obtain the final divergence-free velocity field. It is *very important* that the left-hand side of the above equation is discretized precisely as divergence of the gradient operator. Furthermore, numerical discretization of the gradient operator must be the same as that used for calculating $\nabla\phi_{n+1}$ in (8.2.24) and the numerical discretization of the divergence operator must be the same as that used for calculating the right-hand side $\nabla \cdot \hat{\mathbf{u}}$ in (8.2.25). Only then will the resulting pressure-corrected velocity field $\mathbf{u}_{n+1}$ be divergence free. In the continuous sense the divergence of the gradient is identical to the Laplace operator. However replacing the left-hand side of (8.2.25) with $\nabla^2\phi_{n+1}$ and its numerical discretization is not appropriate, since it will not lead to a divergence-free velocity field. The reason for this is in the definition of the discretized first- and second-derivative operators. While $\partial^2/\partial x^2 = \partial/\partial x(\partial/\partial x)$ is guaranteed from basic rules of calculus, the discrete second-derivative operator, in general, will not be equal to product of two first-derivative operators (i.e., $\mathbf{D}_{2x} \neq \mathbf{D}_{1x}\mathbf{D}_{1x}$, and the same is true along y- and z-directions). Thus, while using the method outlined in Section 8.1.5 we will carefully express the left-hand side of (8.2.25) in 2D as

$$\left[\frac{\partial}{\partial x}\frac{\partial}{\partial x} + \frac{\partial}{\partial y}\frac{\partial}{\partial y}\right]\phi\,. \tag{8.2.26}$$

In 2D (8.2.23) can be rewritten as

$$\left[1 - \alpha\left(\frac{\partial^2}{\partial x^2} + \frac{\partial^2}{\partial y^2}\right)\right]\hat{\mathbf{u}} = \mathbf{r}\,, \tag{8.2.27}$$

where $\alpha = \Delta t/(2\,\mathrm{Re})$ and the right-hand side is given by

$$\mathbf{r} = -\frac{3\Delta t}{2}\mathcal{N}_n + \frac{\Delta t}{2}\mathcal{N}_{n-1} + \left[1 + \alpha\left(\frac{\partial}{\partial x^2} + \frac{\partial}{\partial y^2}\right)\right]\mathbf{u}_n\,. \tag{8.2.28}$$

We identify (8.2.27) and (8.2.28) with the semi-discrete advection–diffusion equation discussed in (8.1.2) and (8.1.3), except that the advection velocity is now $\mathbf{u} = (u, v)$ instead of the constant velocity $\mathbf{c} = (c_x, c_y)$. An important point to observe is that the above Helmholtz equations for the intermediate velocity components $\hat{u}$ and $\hat{v}$ are decoupled and can be solved one at a time. The extension to 3D is straightforward. Then, (8.2.27) and (8.2.28) will become three-dimensional Helmholtz equations for the three components of the intermediate velocity, which can be solved independently.

By substituting (8.2.24) into (8.2.23) we can eliminate the intermediate velocity to obtain an equation for the velocity at the $(n+1)$th time-level:

$$\begin{aligned}\frac{\mathbf{u}_{n+1} - \mathbf{u}_n}{\Delta t} &= -\left(\nabla\phi_{n+1} - \frac{\Delta t}{2\,\mathrm{Re}}\nabla^2\nabla\phi_{n+1}\right) - \frac{3}{2}\mathcal{N}_n + \frac{1}{2}\mathcal{N}_{n-1} \\ &\quad + \frac{1}{2\,\mathrm{Re}}\left[\nabla^2\mathbf{u}_{n+1} + \nabla^2\mathbf{u}_n\right].\end{aligned}$$

The first term on the right is an approximation of the pressure gradient. If we take $p_{n+1} = \phi_{n+1}$, we identify $(\Delta t/2\,\mathrm{Re})\nabla^2\nabla\phi_{n+1}$ as the error. Thus, even though the advection and diffusion terms have used second-order methods for temporal discretization, the overall method appears to be only first-order accurate. In fact, Guermond *et al.* in their overview paper show that the above method is less than first-order accurate. In the past it has been argued that higher-order accuracy can be achieved by redefining pressure as

$$\nabla p_{n+1} = \nabla \phi_{n+1} - \frac{\Delta t}{2\,\mathrm{Re}} \nabla^2 \nabla \phi_{n+1}\,, \tag{8.2.29}$$

which, when substituted in (8.2.29), is in agreement with the Navier–Stokes equation. However, as discussed by Perot (1993), this argument is valid only when the right-hand side can indeed be expressed as a gradient of a field. In a continuous sense the Laplacian and the gradient operators commute and the second term of the right-hand side can be expressed as $(\Delta t/2\,\mathrm{Re})\nabla\nabla^2\phi_{n+1}$. However, in general such commutativity will not apply when discretized in space. As a result, even with the approximation (8.2.29) the above fractional-step method will remain lower-order accurate in time. The only exception is when both directions are periodic, in which case with Fourier derivative operators, discrete Laplacian and gradient operators can be made to commute.

Boundary conditions for the intermediate velocity are needed when multi-dimensional Helmholtz equations are solved for $\hat{u}$ and $\hat{v}$. The discretized version of the boundary conditions (8.2.20) is

$$\begin{array}{lll}
\text{Left:} & \hat{u}(x_0, y_l) = U_\mathrm{L}(y_l)\,, & \hat{v}(x_0, y_l) = V_\mathrm{L}(y_l)\,, \\
\text{Bottom:} & \hat{u}(x_j, y_0) = U_\mathrm{B}(x_j)\,, & \hat{v}(x_j, y_0) = V_\mathrm{B}(x_j)\,, \\
\text{Right:} & \hat{u}(x_{N_x}, y_l) = U_\mathrm{R}(y_l)\,, & \hat{v}(x_{N_x}, y_l) = V_\mathrm{R}(y_l)\,, \\
\text{Top:} & \hat{u}(x_j, y_{N_y}) = U_\mathrm{T}(x_j)\,, & \hat{v}(x_j, y_{N_y}) = V_\mathrm{T}(x_j)\,.
\end{array} \tag{8.2.30}$$

It is to be noted that the above boundary values U_L, U_B, etc., are at time-level $(n+1)$ and have been applied for the intermediate velocity. We will have to see if the pressure correction step preserves these boundary conditions, so that they remain unchanged and apply correctly for the final velocity u_{n+1} and v_{n+1}.

With the fractional-step method the numerical solution of the pressure Poisson equation (8.2.25) requires boundary conditions. But there is no natural pressure boundary condition. In fact, in the coupled method, by defining pressure at the cell centers we were able to avoid the need for an artificial pressure boundary condition. The fractional-step method allowed us to decouple the momentum and continuity equations, but the price we pay is to come up with a suitable pressure boundary condition. The criterion we use for the pressure boundary condition is that it must be consistent and transmit the exact velocity boundary conditions imposed on the intermediate velocity in (8.2.30) to the final $(n+1)$th-level velocity. The appropriate pressure boundary conditions are

$$\text{Left \& Right: } \left[\frac{\partial \phi}{\partial x}\right]_{n+1} = 0\,, \quad \text{Top \& Bottom: } \left[\frac{\partial \phi}{\partial y}\right]_{n+1} = 0\,. \tag{8.2.31}$$

More generally, the pressure boundary condition can be written as $\mathbf{n} \cdot \nabla \phi_{n+1} = 0$, where $\mathbf{n}$ is the unit outward normal to the boundary. In other words, the homogeneous Neumann boundary condition is the proper one for the pressure Poisson equation. From the pressure-correction step, the normal and tangential components of velocity at the boundary points can be expressed as

$$\begin{aligned} \mathbf{n} \cdot \mathbf{u}_{n+1} &= \mathbf{n} \cdot \hat{\mathbf{u}} - \Delta t(\mathbf{n} \cdot \nabla \phi_{n+1}) \\ \boldsymbol{\tau} \cdot \mathbf{u}_{n+1} &= \boldsymbol{\tau} \cdot \hat{\mathbf{u}} - \Delta t(\boldsymbol{\tau} \cdot \nabla \phi_{n+1}), \end{aligned} \tag{8.2.32}$$

where $\boldsymbol{\tau}$ is the unit vector along the tangential direction. From the pressure boundary condition it can readily be seen that the normal component of the final velocity is the same as the normal component of the intermediate velocity. In other words, the normal velocity boundary conditions in (8.2.30) are correctly transmitted to the normal velocities at the $(n+1)$th time-level. Since we solve the pressure Poisson equation with the homogeneous Neumann boundary condition, the boundary value of the pressure cannot be simultaneously specified or controlled. In general, the tangential derivative of the boundary pressure will be non-zero (i.e., $\boldsymbol{\tau} \cdot \nabla \phi_{n+1} \neq 0$). As a result, the tangential velocity boundary conditions imposed in (8.2.30) are not accurately transmitted to the $(n+1)$th time-level boundary conditions. This error in the tangential velocity boundary condition is $O(\Delta t)$.

Higher-order accuracy can be achieved by anticipating the tangential pressure gradient and appropriately correcting the boundary condition of the intermediate tangential velocity. For example, in (8.2.30) the tangential velocity boundary condition on the left boundary can be modified as

$$\hat{v}(x_0, y_l) = V_{\mathrm{L}}(y_l) + \Delta t \left(2 \left[\frac{\partial \phi}{\partial y} \right]_{x_0 y_{l n}} - \left[\frac{\partial \phi}{\partial y} \right]_{x_0 y_{l n}} \right), \tag{8.2.33}$$

with similar corrections to the other tangential velocities. In the above, the last term on the right extrapolates to approximate $\boldsymbol{\tau} \cdot \nabla \phi_{n+1}$. With this correction, the error in the $(n+1)$th tangential velocity along the boundaries reduces to $O(\Delta t)^3$. Unfortunately however, this correction does not address the $O(\Delta t)$ splitting error, which remains irreducible. Again the splitting error arises from the fact that the exact pressure Poisson equation obtained by taking the divergence of the Navier–Stokes equation is $\nabla^2 p_{n+1} = -\nabla \cdot (\mathbf{u} \cdot \nabla \mathbf{u})$. Instead, in the fractional-step method, we solve (8.2.25) and the difference gives rise to $O(\Delta t)$ splitting error.

For an excellent discussion of the history and overview of the different fractional-step methods refer to Guermond *et al.* (2006). In this paper the authors explain the stubbornness of the splitting error and the challenge of extending to higher-order accuracy. It is also worth noting that the fractional-step method for incompressible flows ought properly be termed as the **projection method**, since the role of pressure is to project the intermediate velocity onto the space of divergence-free functions. It should also be noted that the final velocity $\mathbf{u}_{n+1}$ is divergence free only in the interior of the computational domain. The divergence at the boundary points will in general be $O(1)$. This is because of the fact the Poisson equation (8.2.25) is satisfied only at the interior grid points. At the boundary points the Neumann boundary conditions are satisfied. As a result, by taking the divergence of (8.2.24) it can be seen that $\nabla \cdot \mathbf{u}_{n+1} \neq 0$ along the boundaries of the domain. This is again a consequence of the fractional-step method.

Computer Exercise 8.2.7 Repeat the 2D driven-cavity problem with the time-splitting method. The advantage now is that you only need to solve Helmholtz and Poisson equations in two dimensions. You should reuse the codes you have written in Computer Exercises 8.1.5, 8.1.6 and 8.1.8. For time-integration, use the Crank–Nicolson scheme for the diffusion terms and AB3 or RK3 for the advection terms. For space discretization, use either the three-point central scheme with a uniform grid or Chebyshev spectral collocation with Gauss–Lobatto grid points. In either case it is sufficient to use a non-staggered grid. Compare your results with those of Ghia *et al.* (1982).

8.2.4 Higher-order Projection Methods

The simplicity of the projection method is its greatest attraction. It is by far the most common methodology that has found widespread usage in CFD. But, as seen in the previous section, the primary difficulty with the method is the splitting error which restricts the overall order of accuracy. Developing higher-order projection methods is currently an important area of active research. We believe many new exciting developments and their implementations will happen in the near future. As an example, we present here an adaptation of the higher-order projection method discussed in Guermond *et al.* (2006) under the title **rotational pressure-correction projection method**. Following their work we will use a second-order backward difference formula to approximate the time derivative. The semi-discretized first step is written as

$$\frac{3\hat{\mathbf{u}} - 4\mathbf{u}_n + \mathbf{u}_{n-1}}{2\Delta t} = -\mathcal{N}_n - \nabla p_n + \frac{1}{\mathrm{Re}}\nabla^2\hat{\mathbf{u}}\,. \tag{8.2.34}$$

The above step accounts for all three: nonlinear, pressure-gradient and viscous effects. However, since only the gradient of the previous time level pressure is used, the resulting intermediate velocity will not be divergence free. The above equation can be rearranged to form a multi-dimensional Helmholtz equation for $\hat{\mathbf{u}}$, which can be numerically solved using any of the methods of Section 8.1. The appropriate boundary conditions for the intermediate velocity at the $(n+1)$th time-level are the exact boundary velocities given in (8.2.30).

The second step is the pressure-correction step given by

$$\frac{3\mathbf{u}_{n+1} - 3\hat{\mathbf{u}}}{2\Delta t} = -\nabla\phi_{n+1}\,, \tag{8.2.35}$$

which, along with the requirement $\nabla \cdot \mathbf{u}_{n+1} = 0$, yields the pressure Poisson equation

$$\nabla \cdot \nabla\phi_{n+1} = \frac{3}{2\Delta t}\nabla \cdot \hat{\mathbf{u}}\,. \tag{8.2.36}$$

The above Poisson equation is solved with homogeneous Neumann boundary conditions given in (8.2.31). The main ingredient of this method is in the proper definition of pressure as

$$p_{n+1} = p_n + \phi_{n+1} - \frac{1}{\mathrm{Re}}\nabla \cdot \hat{\mathbf{u}}\,. \tag{8.2.37}$$

The effect of this modification can be investigated by adding the pressure correction, (8.2.35), to (8.2.34) and replacing ϕ_{n+1} with (8.2.37) to obtain

$$\frac{3\mathbf{u}_{n+1} - 4\mathbf{u}_n + \mathbf{u}_{n-1}}{2\Delta t} = -\mathcal{N}_n - \nabla p_{n+1} + \frac{1}{\mathrm{Re}}\left[\nabla^2 \hat{\mathbf{u}} - \nabla(\nabla \cdot \hat{\mathbf{u}})\right]. \tag{8.2.38}$$

We now use the identity $\nabla \times \nabla \times \hat{\mathbf{u}} = \nabla(\nabla \cdot \hat{\mathbf{u}}) - \nabla^2 \hat{\mathbf{u}}$ and also note from the pressure-correction equation that $\nabla \times \nabla \times \hat{\mathbf{u}} = \nabla \times \nabla \times \mathbf{u}_{n+1}$ to obtain the following overall composite equation that is implied by the two steps

$$\frac{3\mathbf{u}_{n+1} - 4\mathbf{u}_n + \mathbf{u}_{n-1}}{2\Delta t} = -\mathcal{N}_n - \nabla p_{n+1} - \frac{1}{\mathrm{Re}}\nabla \times \nabla \times \mathbf{u}_{n+1}\,. \tag{8.2.39}$$

Finally, since $\nabla \cdot \mathbf{u}_{n+1} = 0$, we recognize the last term on the right to be the diffusion term (i.e., the last term is $\nabla^2 \mathbf{u}_{n+1}/\mathrm{Re}$).

Although homogeneous Neumann boundary conditions are applied in the solution of the Poisson equation for ϕ_{n+1}, the effective boundary condition for pressure can be obtained as

$$\mathbf{n} \cdot \nabla p_{n+1} = -\mathbf{n} \cdot \mathcal{N}_n - \frac{1}{\mathrm{Re}}\mathbf{n} \cdot (\nabla \times \nabla \times \mathbf{u}_{n+1})\,. \tag{8.2.40}$$

Thus, it can be seen that the effective pressure boundary condition that results from the above two-step projection method is consistent. As a result of this consistent boundary treatment, Guermond *et al.* (2006) show that the above method is second-order accurate in velocity. The main error is in the tangential velocity. As discussed in the previous section, since the tangential gradient of ϕ_{n+1} is not controlled in the solution of the Poisson equation, the tangential velocity boundary conditions applied for the intermediate velocity in the first step are altered in the pressure-correction step. Although there is no rigorous analysis yet, one could try a modification such as (8.2.33) to reduce the boundary error in the tangential velocity.

Computer Exercise 8.2.8 Repeat the 2D driven-cavity problem with the higher-order time-splitting method. For the same spatial and temporal resolution compare your results with those of the lower-order time-splitting method.

8.3 Navier–Stokes Equations in Spectral Form

To round out our presentation of numerical methods for Navier–Stokes equations, we show how the governing equation in a periodic box can be written in terms of Fourier coefficients. In the terminology of Chapter 6 on spectral methods, the method to be discussed will be a three-dimensional Fourier–Galerkin approach. This methodology is widely used in the simulation of isotropic turbulence and some homogeneous turbulence problems. Although its applicability is limited to these canonical turbulent flows, we hope to learn much from writing the Navier–Stokes equations in spectral form. Also, there are many fluid-mechanical problems where periodic boundary conditions are appropriate in one or two directions. In which case a Fourier spectral method can be used along those periodic directions in combination with finite difference, or spectral discretization along the non-periodic direction. The discussion to follow below is relevant for how periodic directions can be handled in such mixed problems.

Let the velocity field be expanded in Fourier series along all three directions:

$$\mathbf{u}(\mathbf{x},t) = \sum_{\mathbf{k}} \hat{\mathbf{u}}(\mathbf{k},t)\exp(i\mathbf{k}\cdot\mathbf{x})\,. \tag{8.3.1}$$

Here $\mathbf{k} = (2\pi/L)(k_1,k_2,k_3)$ is the wavenumber vector, where k_1, k_2 and k_3 are integers, for a cubic box of side L. The summation in the above equation is in fact the product of three nested summations over integer values of k_1, k_2 and k_3. Here the 1D truncated Fourier expansion given in (6.2.9) has been extended to 3D. Here we have listed $\mathbf{k}$ as an argument of the Fourier coefficient $\hat{\mathbf{u}}$ rather than a subscript for later notational convenience. Then the condition that the velocity is divergence free becomes

$$\nabla\cdot\mathbf{u} = \sum_{\mathbf{k}} i\mathbf{k}\cdot\hat{\mathbf{u}}(\mathbf{k},t)\exp(i\mathbf{k}\cdot\mathbf{x}) = 0\,. \tag{8.3.2}$$

Thus, for each $\mathbf{k}$ we must have

$$\mathbf{k}\cdot\hat{\mathbf{u}}(\mathbf{k},t) = 0\,. \tag{8.3.3}$$

This is the incompressibility condition written in Fourier space.

In the Navier–Stokes equation (8.2.2) the first term on the left and the second term on the right are easy to write in terms of Fourier coefficients:

$$\frac{\partial\mathbf{u}}{\partial t} = \sum_{\mathbf{k}} \frac{\mathrm{d}\hat{\mathbf{u}}}{\mathrm{d}t}\exp(i\mathbf{k}\cdot\mathbf{x})\,, \tag{8.3.4}$$

and

$$\frac{1}{\mathrm{Re}}\nabla^2\mathbf{u} = \frac{1}{\mathrm{Re}}\sum_{\mathbf{k}}(-k^2)\hat{\mathbf{u}}\exp(i\mathbf{k}\cdot\mathbf{x})\,, \tag{8.3.5}$$

where $k^2 = \mathbf{k}\cdot\mathbf{k}$ is the square of the magnitude of the wavenumber vector.

The αth component of the nonlinear term leads to

$$(\mathbf{u}\cdot\nabla)u_\alpha = u_\beta\frac{\partial u_\alpha}{\partial x_\beta} = \sum_{\mathbf{q}} \hat{u}_\beta(\mathbf{q},t)\exp(i\mathbf{q}\cdot\mathbf{x})\sum_{\mathbf{q}'}(iq'_\beta)\hat{u}_\alpha(\mathbf{q}',t)\exp(i\mathbf{q}'\cdot\mathbf{x})\,. \tag{8.3.6}$$

We are using the convention that repeated indices are summed over $1,2,3$. We regroup the terms in the sum (8.3.6) by introducing $\mathbf{k} = \mathbf{q}' + \mathbf{q}$ as a new summation variable. This gives

$$
\begin{aligned}
u_\beta \frac{\partial u_\alpha}{\partial x_\beta} &= \sum_{\mathbf{k}} \exp(i\mathbf{k}\cdot\mathbf{x}) \sum_{\mathbf{q}} i\,(k_\beta - q_\beta)\hat{u}_\beta(\mathbf{q},t)\hat{u}_\alpha(\mathbf{k}-\mathbf{q},t) \\
&= \sum_{\mathbf{k}} \exp(i\mathbf{k}\cdot\mathbf{x}) \left[i\, m_\beta \sum_{\mathbf{q}+\mathbf{m}=\mathbf{k}} \hat{u}_\beta(\mathbf{q},t)\hat{u}_\alpha(\mathbf{m},t) \right] \\
&= \sum_{\mathbf{k}} \exp(i\mathbf{k}\cdot\mathbf{x}) \left[i\, k_\beta \sum_{\mathbf{q}+\mathbf{m}=\mathbf{k}} \hat{u}_\beta(\mathbf{q},t)\hat{u}_\alpha(\mathbf{m},t) \right] .
\end{aligned}
\tag{8.3.7}
$$

The last expression was obtained by using (8.3.3) and the last sum in the final two expressions is over all pairs of wave vectors $\mathbf{m}$ and $\mathbf{q}$ that add to $\mathbf{k}$. Hence, the term within the square parentheses is the Fourier coefficient of the nonlinear term.

Finally, there is the pressure term. By taking the divergence of the Navier–Stokes equation we obtain the pressure Poisson equation

$$
\nabla^2 p = -\nabla\cdot((\mathbf{u}\cdot\nabla)\mathbf{u}) = -\frac{\partial}{\partial x_\alpha}\left(u_\beta \frac{\partial u_\alpha}{\partial x_\beta} \right) . \tag{8.3.8}
$$

Now we can substitute the Fourier expansion of the nonlinear term from (8.3.7). Furthermore, the operator $-\frac{\partial}{\partial x_\alpha}$ becomes the multiplicative factor $-i\,k_\alpha$ in the Fourier space. Similarly, the Laplacian operator on the left-hand side becomes $-k^2$. Substituting we get

$$
\sum_{\mathbf{k}} -k^2 \hat{p} \exp(i\mathbf{k}\cdot\mathbf{x}) = \sum_{\mathbf{k}} \exp(i\mathbf{k}\cdot\mathbf{x}) \left[k_\alpha k_\beta \sum_{\mathbf{q}+\mathbf{m}=\mathbf{k}} \hat{u}_\beta(\mathbf{q},t)\hat{u}_\alpha(\mathbf{m},t) \right] . \tag{8.3.9}
$$

From the above we obtain the Fourier coefficient of the jth component of the pressure gradient term in the Navier–Stokes equation as

$$
i \frac{k_j k_\alpha k_\beta}{k^2} \sum_{\mathbf{q}+\mathbf{m}=\mathbf{k}} \hat{u}_\beta(\mathbf{q},t)\hat{u}_\alpha(\mathbf{m},t) . \tag{8.3.10}
$$

We now have all the pieces needed in order to write the Navier–Stokes equation in spectral form. We take the jth component of (8.3.4) and (8.3.5), express (8.3.7) for the jth component (instead of α), and include the pressure gradient term to obtain

$$
\frac{\mathrm{d}\hat{u}_j}{\mathrm{d}t} + \frac{k^2}{\mathrm{Re}}\hat{u}_j = -\left(\delta_{j\beta} - \frac{k_j k_\beta}{k^2} \right) \left[i k_\alpha \sum_{\mathbf{q}+\mathbf{m}=\mathbf{k}} \hat{u}_\beta(\mathbf{q},t)\hat{u}_\alpha(\mathbf{m},t) \right] . \tag{8.3.11}
$$

The above is the evolution equation for the Fourier coefficients of velocity. This is the spectral form of the Navier–Stokes equation in a periodic box.

The above equation presents a simple interpretation. The term within the square parentheses is the Fourier transform of the nonlinear term, which we can represent as $\hat{\mathcal{N}}_\beta(\mathbf{k},t)$. The quadratic nature of this term appears as the convolution sum in the spectral representation. The quantity

$$\mathcal{P}_{j,\beta}(\mathbf{k}) = \left(\delta_{j\beta} - \frac{k_j k_\beta}{k^2}\right) \tag{8.3.12}$$

is a projection operator that takes any vector field that it operates on to the divergence-free space. In spectral space it projects the vector field onto the plane perpendicular to $\mathbf{k}$. In other words, for any vector field whose Fourier coefficients are represented by $\hat{V}_\beta(\mathbf{k},t)$, the operation $\mathcal{P}_{j,\beta}(\mathbf{k})\hat{V}_\beta(\mathbf{k},t)$ will yield the divergence-free part of the vector field $\mathbf{V}$, since

$$k_j\, \mathcal{P}_{j,\beta}(\mathbf{k})\hat{V}_\beta(\mathbf{k},t) = 0\,. \tag{8.3.13}$$

Here the projection operator projects the nonlinear effect onto divergence-free space. In the context of a periodic box, with all periodic boundaries, this is sufficient to ensure that the resulting flow field will remain divergence free. Thus, together with the viscous effect represented by the second term on the left-hand side, all three effects are taken into account to advance the flow in time. If we use AB2 method for the advection term and Crank–Nicolson for the viscous term, the time integration of the Fourier coefficients can be expressed as

$$\frac{\hat{u}_{j,n+1} - \hat{u}_{j,n}}{\Delta t} = -\frac{k^2}{2\,\mathrm{Re}}\left[\hat{u}_{j,n+1} + \hat{u}_{j,n}\right] - \frac{3}{2}\mathcal{P}_{j,\beta}\hat{\mathcal{N}}_{\beta,n} + \frac{1}{2}\mathcal{P}_{j,\beta}\hat{\mathcal{N}}_{\beta,n-1}\,. \tag{8.3.14}$$

In the above equation the $\mathbf{k}$-dependence of the Fourier coefficients is suppressed.

A simple step-by-step process for time advancing the flow field from time-level n to time-level $(n+1)$ can now be prescribed:

(1) Given the velocity Fourier coefficients $\hat{\mathbf{u}}_n(\mathbf{k})$, first compute the convolution, represented as the term within the square parentheses on the right-hand side of (8.3.11), to obtain the Fourier coefficients of the nonlinear term, $\hat{\mathcal{N}}_{\beta,n}$ for all values of $\mathbf{k}$. For practical purposes the range of wavenumbers must be limited by considering the values of k_1, k_2 and k_3 to be between $-N/2$ and $N/2$.

(2) Project the nonlinear term to obtain the divergence-free part. That is compute $\mathcal{P}_{j,\beta}\hat{\mathcal{N}}_{\beta,n}$.

(3) Combine with the previous time-level value of the projected nonlinear term to form the last two terms of the right-hand side of (8.3.14).

(4) Solve (8.3.14) to obtain $\hat{\mathbf{u}}_{n+1}(\mathbf{k})$.

Note that the above time-advancement scheme is entirely in the spectral space. In this strategy, if you are given a velocity field $\mathbf{u}(\mathbf{x})$ within the 3D periodic box at $t = 0$, you will need to first perform a three-dimensional Fourier transform to obtain the Fourier coefficients $\hat{\mathbf{u}}(\mathbf{k})$ at $t = 0$. As mentioned in step (1) it will be necessary to truncate the number of Fourier modes. The Fourier coefficients can also be obtained from the velocity being specified on a uniform grid of $N \times N \times N$ points within the periodic box. Then the above step-by-step process can be followed to integrate the Fourier coefficients over time. At any later time when the velocity field $\mathbf{u}(\mathbf{x})$ is needed, the Fourier coefficients can be inverse Fourier transformed to obtain the velocity at the $N \times N \times N$ grid points.

The convolution sum in step (1) is the most complicated and computationally expensive part of the above algorithm. Note that the number of Fourier modes in each direction is N, and therefore the cost of this convolution sum scales as N^6. We recall that nonlinear multiplication is far easier and cheaper to perform in the real space. However, as discussed in Section 6.4.6, this process of performing nonlinear multiplication at the grid points involves aliasing error. In order to compute the nonlinear multiplication in a de-aliased manner, it must be performed on a finer grid of M points along all three directions, and M must be chosen greater than $3N/2$. So step (1) of the above algorithm can be replaced by the following sub-steps:

(1a) Given the velocity Fourier coefficients $\hat{\mathbf{u}}_n(\mathbf{k})$ for $-N/2 \leq (k_1, k_2, k_3) \leq N/2$, expand the Fourier coefficients to a larger range of $-M/2 \leq (k_1, k_2, k_3) \leq M/2$ by setting the additional Fourier coefficients to zero. This substep is called "padding" with zero.

(1b) From the expanded Fourier coefficients compute the corresponding velocity at the M^3 uniformly spaced grid points within the periodic box. This step involves a 3D inverse Fourier transform of the padded $\hat{\mathbf{u}}_n(\mathbf{k})$, and can be accomplished very efficiently using a fast Fourier transform (FFT) which is included with many software packages.

(1c) Derivatives are easier to compute in spectral space. For example, Fourier coefficients of $\partial u/\partial x$ are simply $-i\, k_1\, \hat{\mathbf{u}}_n(\mathbf{k})$. Once computed in spectral space it can be padded with zeros and inverse Fourier transformed to obtain the values of $\partial u/\partial x$ at the M^3 uniformly spaced grid points. This step can be repeated to calculate all other components of the velocity gradient tensor at the M^3 grid points.

(1d) Nonlinear products such as $u(\partial u/\partial x)$ can be readily performed by multiplying these quantities at every grid point. This process is repeated to calculate the three components of the nonlinear term $\mathbf{u} \cdot \nabla \mathbf{u}$ at the M^3 uniformly spaced grid points.

(1e) The three components of the nonlinear term at the M^3 gridpoints are Fourier transformed to obtain the corresponding M^3 Fourier coefficients for $-M/2 \leq (k_1, k_2, k_3) \leq M/2$.

(1f) Truncate the M^3 Fourier coefficients to retain only the N^3 modes corresponding to $-N/2 \leq (k_1, k_2, k_3) \leq N/2$. This "stripping" step is the opposite of padding. These truncated modes will be used in steps (2) to (4).

This strategy of going back and forth between Fourier modes and grid points is called the pseudo-spectral method and it is computationally very efficient. The forward and backward FFTs are the most computationally intensive parts of the above pseudo-spectral algorithm. The operation count of a FFT, on the fine grid, scales as $O(M^3 \log M)$, which is orders of magnitude lower than the $O(N^6)$ operation count of the convolution sum. In comparison to the FFT, the projection operation, step (2), can be done for each $\mathbf{k}$ one mode at a time. Thus, the operation counts of steps (2), (3) and (4) scale only as $O(N^3)$.

8.3.1 Sample Applications

We conclude this section with a couple of examples of calculations using spectral methods in a periodic box. The pioneering work of Brachet *et al.* (1983) investigated fully developed 3D turbulence starting from the so-called Taylor–Green vortex. They studied a range of Reynolds numbers up to Re = 3000. You will recall the limitations on simulating very high Re. This particular flow has been of continuing interest because the Fourier coefficients have several symmetries that allow it to be run at higher Reynolds numbers, based on the available grid size, than would be possible for a flow without such symmetries. It has also been subjected to comprehensive theoretical analyses. Another example is the spectral methods calculation of three-dimensional vortex reconnection in a periodic box, for example by Zabusky and Melander (1989). They noticed the teasing out of the initial filaments and the formation of fingers or bridges as the vortex reconnection progresses.

Numerical simulation of isotropic turbulence in a periodic box using spectral methods has a very rich history starting from the pioneering work of Orszag and Patterson (1972). Also of particular significance is the work of Eswaran and Pope (1988), who introduced a rigorous approach to spectrally forcing the flow to obtain a stationary state for isotropic turbulence. From turbulence theory we know that a simulation of the Navier–Stokes equations, as given in (8.2.2), starting from an initial isotropic turbulence in a periodic box will continue to decay over time, since there is no turbulence production. Eswaran and Pope (1988) introduced added external forcing to the Navier–Stokes equations that maintains a statistically stationary state, without interfering with the turbulence cascade and dissipative process.

Very large simulations of isotropic turbulence in a periodic box are now routine and they are being used to investigate a wide range of fluid mechanical problems. For example, Matsuda *et al.* (2014) have used simulations of isotropic turbulence in a periodic box to investigate the influence of turbulent clustering of cloud droplets at the microscale on radar reflectivity and cloud observation. They have used a fine grid of up to 2000 uniformly spaced points that has enabled them to reach a reasonably large Reynolds number of 5595 (this Reynolds number is based on an integral length scale). In addition to solving the fluid flow they also tracked the motion of cloud droplets of varying size by integrating their equations of motion. (We saw methods for tracking such Lagrangian particles in earlier chapters.) Up to 50 million droplets were tracked within the periodic box. Figure 8.2, taken from their paper, shows spatial distributions of droplets obtained from the simulation for different sized droplets, characterized by their Stokes number: a larger number indicates a larger droplet. Strange clustering of particles can be observed even in a turbulent flow. This phenomenon is called the **preferential accumulation** of particles. It is due to the inertial behavior

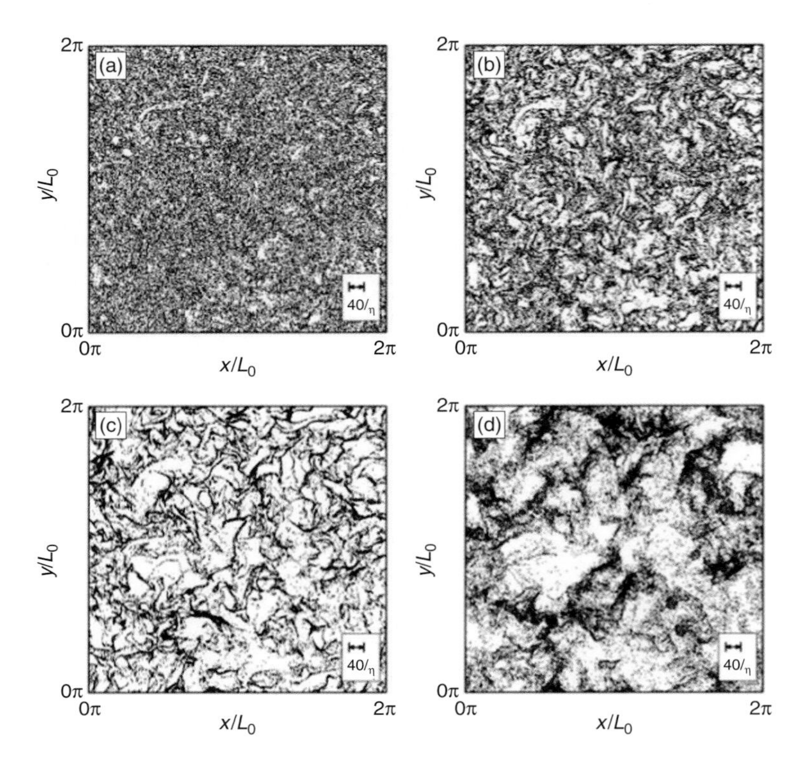

Figure 8.2 Spatial distributions of droplets in a periodic box of isotropic turbulence. The four panels show droplets of four different Stokes numbers equal to (a) 0.05, (b) 0.2, (c) 1.0 and (d) 5.0. Figure from Matsuda *et al.* (2014), © American Meteorological Society, used with permission.

of the particles and it is most amplified when the Stokes number, which is the ratio of particle to flow time scale, is of order unity.

Computer Exercise 8.3.1 Write a computer program to solve the 2D or 3D Navier–Stokes equations in a periodic box using the pseudo-spectral method. Use available packages for FFTs, and implement de-aliasing using the 3/2-rule. Use a third-order time-stepping method for the advection term. It is beneficial to do the advective terms explicitly and the dissipative terms implicitly. Investigate initial conditions such as randomly excited low modes as in Eswaran and Pope (1988), or a shear layer, or a couple of concentrated vortex pairs.

8.4 Finite Volume Formulation

The focus of this book is not on finite volume methods, as there are many excellent texts on this subject. However, here we will present a simple implementation of the time-splitting method in the context of a second-order-accurate finite volume formulation. Later in Section 8.6 that we will build upon this formulation to describe how the influence of the embedded solid boundaries can be accounted in the computation using the sharp interface method. In the finite volume methodology to be described below, we will consider a fully collocated approach, where the velocity components and pressure are defined at the cell centers. Staggered finite volume implementations, although not discussed below, can follow along similar lines. In the collocated finite volume formulation in addition to the cell-center velocity, $\mathbf{u}$, the corresponding face-center velocity, $\mathbf{U}$, will also be defined and used appropriately (see Figure 8.3).

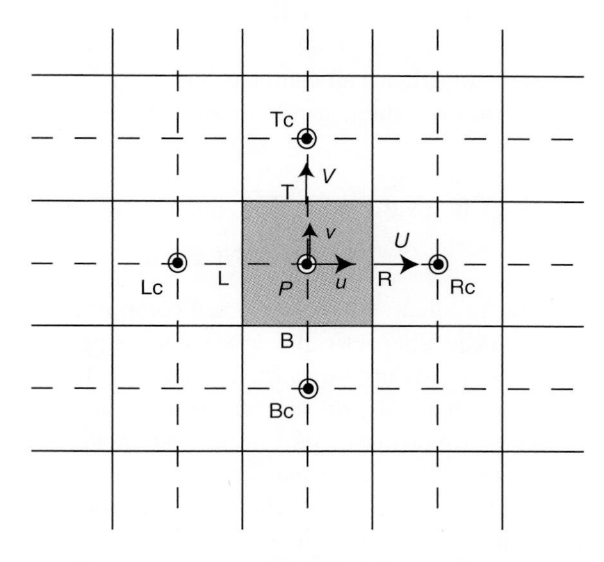

Figure 8.3 Schematic of the underlying Cartesian mesh in two dimensions showing clearly the four neighbors of the cell marked P. Also marked are the face centers and the face-center velocities. This figure is from Ye *et al.* (1999), with permission.

As the first step of the time-split methodology an intermediate cell-center velocity is computed by accounting for only the advection and diffusion terms. The finite volume analog of (8.2.23) can be written as

$$\int_{\mathrm{cv}} \frac{\hat{\mathbf{u}} - \mathbf{u}_n}{\Delta t} \mathrm{d}V = -\frac{3}{2} \int_{\mathrm{cs}} \mathbf{u}_n (\mathbf{U}_n \cdot \mathbf{n}) \mathrm{d}S + \frac{1}{2} \int_{\mathrm{cs}} \mathbf{u}_{n-1} (\mathbf{U}_{n-1} \cdot \mathbf{n}) \mathrm{d}S$$
$$+ \frac{1}{2\,\mathrm{Re}} \int_{\mathrm{cs}} (\nabla \hat{\mathbf{u}} + \nabla \mathbf{u}_n) \cdot \mathbf{n}\, \mathrm{d}S\,, \tag{8.4.1}$$

where cv and cs represent the control volume of a cell and its corresponding control surface. The left-hand side will be approximated as

$$\frac{[\hat{\mathbf{u}}]_P - [\mathbf{u}_n]_P}{\Delta t} V_P\,, \tag{8.4.2}$$

where $[.]_P$ indicates evaluation of the function within square parentheses at the Pth cell center and V_P is the volume of the Pth cell. Note that the fluxes on the control surface are computed appropriately using the nth and $(n-1)$th level face-center velocities. For example the integral in the first term on the right-hand side can be evaluated as the sum of fluxes on the four faces as (for the 2D geometry shown in Figure 8.3)

$$\int_{\text{cs}} \mathbf{u}_n(\mathbf{U}_n \cdot \mathbf{n}) \mathrm{d}S = [\mathbf{u}_n(\mathbf{U}_n \cdot \mathbf{n})]_{\mathrm{R}}\, A_{\mathrm{R}} + [\mathbf{u}_n(\mathbf{U}_n \cdot \mathbf{n})]_{\mathrm{B}}\, A_{\mathrm{B}} + [\mathbf{u}_n(\mathbf{U}_n \cdot \mathbf{n})]_{\mathrm{L}}\, A_{\mathrm{L}} + [\mathbf{u}_n(\mathbf{U}_n \cdot \mathbf{n})]_{\mathrm{T}}\, A_{\mathrm{T}} \,. \quad (8.4.3)$$

In the above A_{R}, A_{B}, A_{L} and A_{T} are the areas of the right, bottom, left and top faces of the cell and all velocities are from the nth time-level. The face-center velocities are readily available at the face centers, while $[\mathbf{u}_n]_{\mathrm{R}}$, etc., must be interpolated from their cell-center values.

The integral over the control surface for the diffusive fluxes can be similarly expressed as

$$\int_{\text{cs}} (\nabla \hat{\mathbf{u}} + \nabla \mathbf{u}_n)\, \mathrm{d}S = [\nabla \hat{\mathbf{u}} + \nabla \mathbf{u}_n]_{\mathrm{R}}\, A_{\mathrm{R}} + [\nabla \hat{\mathbf{u}} + \nabla \mathbf{u}_n]_{\mathrm{B}}\, A_{\mathrm{B}} + [\nabla \hat{\mathbf{u}} + \nabla \mathbf{u}_n]_{\mathrm{L}}\, A_{\mathrm{L}} + [\nabla \hat{\mathbf{u}} + \nabla \mathbf{u}_n]_{\mathrm{T}}\, A_{\mathrm{T}} \,. \quad (8.4.4)$$

The velocity gradients at the faces are computed using standard central difference between the adjacent cell-center values. Approximations (8.4.2), (8.4.3) and (8.4.4) can be substituted into (8.4.1), and all the terms involving $\hat{\mathbf{u}}$ can be collected on the left-hand side with the rest of the terms moved to the right-hand side. This resulting equation is a linear system which must be solved. The boundary conditions for the intermediate velocity to be applied at the boundaries of the computational domain are the same as those for the final pressure-corrected velocity. Slip velocity error on the boundaries can be reduced by estimating and correcting for the tangential pressure gradient as given in (8.2.33). Before solving the boundary conditions must be incorporated into the linear system.

Once the intermediate cell-center velocities are obtained the corresponding intermediate face-center velocities, $\hat{\mathbf{U}}$, are obtained by interpolation as

$$(\hat{U})_{\mathrm{L}} = \frac{((\hat{u})_P + (\hat{u})_{\mathrm{R}})}{2} \quad (8.4.5)$$

and

$$(\hat{V})_{\mathrm{T}} = \frac{((\hat{v})_P + (\hat{v})_{\mathrm{T}})}{2} \,. \quad (8.4.6)$$

The finite volume version of the projection operation onto the divergence-free space can be accomplished by first computing ϕ_{n+1} at the cell centers as

$$\int_{\text{cs}} (\nabla \phi_{n+1}) \cdot \mathbf{n}\, \mathrm{d}S = \frac{1}{\Delta t} \int_{\text{cs}} \left(\hat{\mathbf{U}} \cdot \mathbf{n} \right) \mathrm{d}S \,, \quad (8.4.7)$$

which is the integral version of the Poisson equation. The boundary condition to be applied at all the boundaries is homogeneous Neumann: $\nabla \phi_{n+1} \cdot \mathbf{n} = 0$. The corrected final cell-center and face-center velocities are given by

$$\mathbf{u}_{n+1} = \hat{\mathbf{u}} - \Delta t\, (\nabla \phi_{n+1})_{\text{cc}} \quad (8.4.8)$$

and

$$\mathbf{U}_{n+1} = \hat{\mathbf{U}} - \Delta t \left(\nabla \phi_{n+1}\right)_{\text{fc}} , \tag{8.4.9}$$

where with reference to Figure 8.3 the gradients are computed as

$$\left(\partial \phi / \partial x\right)_P = \frac{(\phi_{\text{R}} - \phi_{\text{L}})}{2\Delta x} \tag{8.4.10}$$

and

$$\left(\partial \phi / \partial x\right)_{\text{R}} = \frac{(\phi_{\text{R}} - \phi_P)}{\Delta x} . \tag{8.4.11}$$

A similar formulation is followed for defining gradients along the other directions. Cell-center pressure at the intermediate time can be obtained from ϕ_{n+1}, as given in (8.2.29).

It must be pointed out that $\mathbf{U}_{n+1}$ evaluated as given above is not a simple interpolation of the corresponding cell-center velocity, $\mathbf{u}_{n+1}$. The above formulation is necessary in order to ensure a truly divergence-free velocity field at the end of the time-step. As shown by Ye *et al.* (1999) the above fully collocated finite-volume formulation avoids the grid-to-grid pressure oscillation that is present in certain other collocated formulations. The primary advantage of the fully collocated formulation is that it is simpler to implement than a staggered formulation as both the momentum and pressure equations are solved at the same location (cell centers). This simplicity becomes important, especially in terms of how the embedded boundaries are handled, as will be addressed below.

8.5 CFD for Complex Geometries

So far our discussion of computational fluid dynamics has been limited to solving the Navier–Stokes equations in simple rectangular Cartesian boxes. But often problems of interest involve flow through complex geometries. There are two distinct possibilities for handling the geometric complexities. The first approach is to precisely define the complex boundaries and use a **body-fitted grid**. Here again one can pursue a **structured grid**, or an **unstructured grid**. Typically one starts with a surface mesh that defines the complex boundary and then extends the surface mesh into the flow domain to define a curvilinear mesh. The curvilinear grid that conforms to the boundaries, greatly simplifies specification of boundary conditions. Furthermore, numerical methods on curvilinear grids are well developed and a desired level of accuracy can be maintained both within the flow and along the boundaries. The main difficulty with this approach is grid generation. For example, for the case of flow in a pipe with several hundred cylinders or spheres embedded within the pipe, obtaining an appropriate body-fitted grid around all the cylinders or spheres is a non-trivial task. There are a few unstructured and structured grid generation packages that are readily available in the market. Nevertheless, the computational cost of grid generation can increase rapidly as geometric complexity increases. This can be severely restrictive, particularly, in cases where the boundaries are in time-dependent motion. The cost of grid generation can overwhelm the cost of computing the flow.

The alternative to body-fitted grid, that is increasingly becoming popular, are the **Cartesian grid methods**. Here, irrespective of the complex nature of the internal and external boundaries, only a regular Cartesian mesh is employed and the governing equations are solved on this simple mesh. The domain boundaries are treated as embedded interfaces within the regular Cartesian mesh. The two key issues to consider are: (1) how the embedded interfaces are defined and advected on the Cartesian mesh and (2) how the boundary conditions are enforced at the embedded interfaces. Based on how the interface is defined we have two different classes of methods: the **sharp interface method** and the **immersed boundary method**.

In the sharp interface Cartesian grid approach the sharpness of the boundary is recognized both in its advection through the Cartesian mesh and also in accounting for its influence on the fluid motion (LeVeque and Li, 1994; Pember *et al.*, 1995; Almgren *et al.*, 1997; Ye *et al.*, 1999; Calhoun and Leveque, 2000; Udaykumar *et al.*, 1997, 2001; Marella *et al.*, 2005). In the computational cells cut through by the interface the governing equations are solved by carefully computing the various local fluxes (or gradients). The above references have shown that by carefully modifying the computational stencil at the interface cells a formal second-order accuracy can be maintained.

In the immersed boundary method the embedded interface is defined by a large number of Lagrangian marker points. The influence of the boundary on the fluid is enforced as forcing through source terms in the governing equations and in this process the boundary is typically diffused over a few mesh cells. Different variants of feedback and direct forcing have been proposed for accurate implementation of the interface (Peskin, 1977; Goldstein *et al.*, 1993; Saiki and Biringen, 1996; Fadlun *et al.*, 2000; Kim *et al.*, 2001; Mittal and Iaccarino, 2005). Thus the jump conditions across the interface are enforced only in an integral sense. The advantage of the immersed boundary method over the sharp interface approach is that the former is simpler to implement in a code than the latter. Furthermore, as the interface moves across the fixed Cartesian grid, the sharp interface approach requires recomputation of the discrete operators close to the interface. Accomplishing this in three dimensions in an efficient manner can be a challenge.

In this section the body-fitted mesh approach will be briefly introduced first in the context of finite difference and finite volume implementations. Then, as another example, we will discuss a special implementation of spectral collocation methodology on a structured body-fitted grid in spherical coordinates for spectrally accurate simulation of complex flows around an isolated sphere. There is a large body of literature on fully resolved fluid flow simulations in complex geometries and it is growing rapidly. The discussion presented below closely follows that presented in Chapter 4 of the book by Prosperetti and Tryggvason (2009). The rest of this chapter is intended to just give a flavor and as a result the coverage will not be exhaustive. The reader should consult the other chapters in the above-mentioned book for an excellent discussion on other approaches to solving Navier–Stokes equations in complex geometries.

8.5.1 Structured Curvilinear Body-fitted Grid

In this subsection we will address computational techniques for solving Navier–Stokes equations in complex geometries using a structured body-fitted grid. The boundaries are first accurately defined with a two-dimensional structured discretization, which is then extended into the fluid domain with a structured three-dimensional mesh.

Consider the simple case of a spherical particle located within a rectangular channel flow. A structured mesh that covers the fluid region outside the sphere and extends over the entire Cartesian geometry of the channel is required. There are different approaches to constructing such a structured grid: algebraic methods, methods based on solving partial differential equations, variational approach, etc.

For problems involving even more complicated fluid–solid interfaces, one might need to resort to a block structured approach. Here the overall computational domain is divided into blocks and within each block a three-dimensional structured mesh is defined and the adjacent blocks communicate at the block interface. The simplest implementation is one where the discretization within the blocks are conforming, i.e., at the block boundaries the adjacent blocks share common nodal points. A more complex implementation involves non-conforming blocks. This requires interpolation between the different meshes at the block boundaries. The non-conforming blocks provide greater flexibility in handling more complex geometries. Nevertheless within each block the grid can be considered to provide a local curvilinear coordinate.

Within each block the physical domain is mapped onto a computational domain, which is usually defined to be a unit square in two dimensions or a unit cube in three dimensions. A transformation from the physical coordinate (x, y, z) to the computational coordinate (ξ, η, ζ) can then be defined:

$$\xi = \xi(x, y, z), \qquad \eta = \eta(x, y, z) \qquad \text{and} \qquad \zeta = \zeta(x, y, z). \tag{8.5.1}$$

The above transformation can be inverted to obtain

$$x = x(\xi, \eta, \zeta) \qquad y = y(\xi, \eta, \zeta) \qquad \text{and} \qquad z = z(\xi, \eta, \zeta). \tag{8.5.2}$$

From the above coordinate transformations, transformation of gradients can also be defined by chain rule as

$$\begin{bmatrix} \dfrac{\partial}{\partial x} \\ \dfrac{\partial}{\partial y} \\ \dfrac{\partial}{\partial z} \end{bmatrix} = \begin{bmatrix} \xi_x & \eta_x & \zeta_x \\ \xi_y & \eta_y & \zeta_y \\ \xi_z & \eta_z & \zeta_z \end{bmatrix} \begin{bmatrix} \dfrac{\partial}{\partial \xi} \\ \dfrac{\partial}{\partial \eta} \\ \dfrac{\partial}{\partial \zeta} \end{bmatrix}, \tag{8.5.3}$$

and

$$\begin{bmatrix} \dfrac{\partial}{\partial \xi} \\ \dfrac{\partial}{\partial \eta} \\ \dfrac{\partial}{\partial \zeta} \end{bmatrix} = \begin{bmatrix} x_\xi & y_\xi & z_\xi \\ x_\eta & y_\eta & z_\eta \\ x_\zeta & y_\zeta & z_\zeta \end{bmatrix} \begin{bmatrix} \dfrac{\partial}{\partial x} \\ \dfrac{\partial}{\partial y} \\ \dfrac{\partial}{\partial z} \end{bmatrix}. \tag{8.5.4}$$

Here the forward and backward transformation matrices are related as

$$\begin{bmatrix} \xi_x & \xi_y & \xi_z \\ \eta_x & \eta_y & \eta_z \\ \zeta_x & \zeta_y & \zeta_z \end{bmatrix} = \begin{bmatrix} x_\xi & x_\eta & x_\zeta \\ y_\xi & y_\eta & y_\zeta \\ z_\xi & z_\eta & z_\zeta \end{bmatrix}^{-1}. \tag{8.5.5}$$

The **Jacobian** of the transformation is given by

$$J = \frac{\partial(\xi, \eta, \zeta)}{\partial(x, y, z)} = \det \begin{bmatrix} \xi_x & \xi_y & \xi_z \\ \eta_x & \eta_y & \eta_z \\ \zeta_x & \zeta_y & \zeta_z \end{bmatrix}. \tag{8.5.6}$$

In the above the derivatives ξ_x, ξ_y, ... are called the **metric of the geometric transformation**.

The computational coordinate can be easily discretized along the ξ- and η-directions independently, since the computational domain is a unit square. For example, ξ and η can be discretized with $(N+1)$ uniformly distributed grid points. The mapping (8.5.2) can then be used to compute the corresponding $(N+1) \times (N+1)$ grid points in the physical domain. Furthermore, the metrics of the transformation and the Jacobian can be calculated at all the grid points analytically using the above relations. They can also be computed using numerical differentiation.

The above transformations allow for the governing equations to be written in terms of the computational coordinates. We first introduce the contravariant velocities as

$$\left.\begin{aligned} U &= u\xi_x + v\xi_y + w\xi_z\,, \\ V &= u\eta_x + v\eta_y + w\eta_z, \\ W &= u\zeta_x + v\zeta_y + w\zeta_z\,. \end{aligned}\right\} \tag{8.5.7}$$

In terms of the contravariant velocities the incompressibility equation becomes

$$\frac{\partial U}{\partial \xi} + \frac{\partial V}{\partial \eta} + \frac{\partial W}{\partial \zeta} = 0\,. \tag{8.5.8}$$

To transform the momentum equations we first write in the conservative form as

$$\frac{\partial \mathbf{u}}{\partial t} + \frac{\partial \mathbf{F}_{ix}}{\partial x} + \frac{\partial \mathbf{F}_{iy}}{\partial y} + \frac{\partial \mathbf{F}_{iz}}{\partial z} = \frac{\partial \mathbf{F}_{vx}}{\partial x} + \frac{\partial \mathbf{F}_{vy}}{\partial y} + \frac{\partial \mathbf{F}_{vz}}{\partial z} \tag{8.5.9}$$

where $\mathbf{u} = [u, v, w]^{\mathrm{T}}$ and the inviscid fluxes are given by

$$\mathbf{F}_{ix} = \begin{bmatrix} u^2 + p \\ uv \\ uw \end{bmatrix}, \quad \mathbf{F}_{iy} = \begin{bmatrix} vu \\ v^2 + p \\ vw \end{bmatrix}, \quad \mathbf{F}_{iz} = \begin{bmatrix} wu \\ wv \\ w^2 + p \end{bmatrix}, \tag{8.5.10}$$

and the viscous fluxes are given by

$$\mathbf{F}_{vx} = \begin{bmatrix} \tau_{xx} \\ \tau_{xy} \\ \tau_{xz} \end{bmatrix}, \quad \mathbf{F}_{vy} = \begin{bmatrix} \tau_{yx} \\ \tau_{yy} \\ \tau_{yz} \end{bmatrix}, \quad \mathbf{F}_{vz} = \begin{bmatrix} \tau_{zx} \\ \tau_{zy} \\ \tau_{zz} \end{bmatrix}. \tag{8.5.11}$$

The Navier–Stokes equations when transformed to the computational coordinates can also be written in conservative form as follows:

$$\frac{\partial \hat{\mathbf{u}}}{\partial t} + \frac{\partial \hat{\mathbf{F}}_{ix}}{\partial \xi} + \frac{\partial \hat{\mathbf{F}}_{iy}}{\partial \eta} + \frac{\partial \hat{\mathbf{F}}_{iz}}{\partial \zeta} = \frac{\partial \hat{\mathbf{F}}_{vx}}{\partial \xi} + \frac{\partial \hat{\mathbf{F}}_{vy}}{\partial \eta} + \frac{\partial \hat{\mathbf{F}}_{vz}}{\partial \zeta} \tag{8.5.12}$$

where $\hat{\mathbf{u}} = (1/J)[u, v, w]^{\mathrm{T}}$ and the inviscid fluxes are given in terms of contravariant velocity by

$$\hat{\mathbf{F}}_{ix} = \frac{1}{J}\begin{bmatrix} uU + p\xi_x \\ vU + p\xi_y \\ wU + p\xi_z \end{bmatrix}, \quad \hat{\mathbf{F}}_{iy} = \frac{1}{J}\begin{bmatrix} uV + p\eta_x \\ vV + p\eta_y \\ wv + p\eta_z \end{bmatrix}, \quad \hat{\mathbf{F}}_{iz} = \frac{1}{J}\begin{bmatrix} uW + p\zeta_x \\ vW + p\zeta_y \\ wW + p\zeta_z \end{bmatrix}, \tag{8.5.13}$$

and similarly the viscous fluxes are given by

$$\begin{aligned}
\hat{\mathbf{F}}_{vx} &= \frac{1}{J}\left(\xi_x \mathbf{F}_{vx} + \xi_y \mathbf{F}_{vy} + \xi_z \mathbf{F}_{vz}\right), \\
\hat{\mathbf{F}}_{vy} &= \frac{1}{J}\left(\eta_x \mathbf{F}_{vx} + \eta_y \mathbf{F}_{vy} + \eta_z \mathbf{F}_{vz}\right), \\
\hat{\mathbf{F}}_{vz} &= \frac{1}{J}\left(\zeta_x \mathbf{F}_{vx} + \zeta_y \mathbf{F}_{vy} + \zeta_z \mathbf{F}_{vz}\right).
\end{aligned} \tag{8.5.14}$$

Problem 8.5.1 Apply the coordinate transformation and obtain (8.5.12) starting from the conservative form of the Navier–Stokes equations in Cartesian coordinates.

In the context of finite difference approaches, the coupled methods presented in Section 8.2.2 and the time-splitting methods presented in Section 8.2.3 can now be applied to the transformed Navier–Stokes equations given in (8.5.12) in much the same way as they were applied to the original Navier–Stokes equations. However, computational complexity has substantially increased, since the governing equations are far more complex now than in their original form. First and foremost, the metrics ξ_x, ξ_y, ... and the Jacobian J must be computed at all the grid points, since they are needed in the calculation of the contravariant velocities and the inviscid and the viscous fluxes. If the physical grid remains the same over time, the computation of the metrics and Jacobian can be done once at the beginning and stored for repeated use. In the more complex case where the geometry of the flow domain changes over time, not only a suitable curvilinear grid must be generated at each time, but the metrics and Jacobian must be recomputed at each time. Often the mappings (8.5.1) and (8.5.2) are not known analytically and the metrics must also be computed using finite difference. For example, the first component of the metric at the (i, j, k)th point can be approximated as

$$\left.\frac{\partial \xi}{\partial x}\right|_{(i,j,k)} = \frac{\xi|_{(i+1,j,k)} - \xi|_{(i-1,j,k)}}{x|_{(i+1,j,k)} - x|_{(i-1,j,k)}} . \tag{8.5.15}$$

Other components of the metric can be defined similarly and from them the Jacobian can be evaluated.

The inviscid fluxes $\hat{\mathbf{F}}_{ix}$, $\hat{\mathbf{F}}_{iy}$ and $\hat{\mathbf{F}}_{iz}$ and their ξ, η and ζ derivatives, respectively, are required in equation (8.5.12). From the value of the velocity field (u, v, w) at all the grid points at the nth time level, these terms can be calculated by following their definitions step-by-step. Thus, an explicit treatment of the inviscid flux (or the nonlinear term of the Navier–Stokes equations) is straightforward.

So far we have pursued implicit treatment of the viscous term in order for the time step Δt to be not constrained by the more restrictive diffusion number. In the context of the time-splitting approach for the standard Navier–Stokes equations in Cartesian coordinates, the implicit treatment of the viscous term resulted in a 3D Helhmoltz equation for each component of velocity. In Sections 8.1.2 to 8.1.4 we presented operator factoring, ADI and eigendecomposition as possible solution techniques for the 3D Helmholtz equation. Implicit treatment of the viscous term in equation (8.5.12) is not as easy. By substituting (8.5.11) into (8.5.14), and then into (8.5.12) it can seen that the viscous term does not reduce to a simple Laplace operator.

First, there are cross derivative terms, such as $\partial^2/\partial\xi\partial\eta$. The numerical implementation of these cross derivative terms is complicated, and therefore they are typically treated explicitly. In the case of an orthogonal curvilinear grid, where lines of constant ξ, η and ζ are curved but orthogonal to each other, it can be shown that these cross derivative terms vanish. Therefore, to the extent possible we must choose the curvilinear grid to be as close to orthogonal as possible.

Second, the viscous term will involve (ξ, η, ζ)-dependent variable coefficients due to the metric terms. Therefore, even if the cross derivative terms are either avoided with an orthogonal curvilinear grid, or treated explicitly, the remaining viscous term of the Navier–Stokes equations will involve both the first- and second-derivative terms, with ξ, η and ζ-dependent variable coefficients. The advection–diffusion step of the time-splitting method will not lead to the standard Helmholtz equation for the three velocity components. The resulting equation for the intermediate velocity $\hat{\mathbf{u}}$ can be rearranged to take the following form, instead of (8.2.27),

$$\left[1 - \alpha\left(\frac{\partial}{\partial \xi}\left[h_1 \frac{\partial}{\partial \xi}\right] + \frac{\partial}{\partial \eta}\left[h_2 \frac{\partial}{\partial \eta}\right] + \frac{\partial}{\partial \zeta}\left[h_3 \frac{\partial}{\partial \zeta}\right]\right)\right] \hat{\mathbf{u}} = \mathbf{r}\,, \tag{8.5.16}$$

where the right-hand side $\mathbf{r}$ now includes the cross derivative and other viscous terms that do not fit the form of the left-hand side, and as a result are treated explicitly. The coefficients (h_1, h_2, h_3) are dependent on the metrics of the transformation and are functions of ξ, η and ζ. Because of the variable coefficients, the above equation cannot be readily eigendecomposed along the ξ-, η- and ζ-directions separately, and as a result the direct solution approach is not possible. The operator factoring and ADI techniques can be implemented, since by construction they consider only one direction implicit at a time. Even then, the 1D Helmholtz equation along the three directions will involve (ξ, η, ζ)-dependent variable coefficients, which must be treated appropriately.

The resulting intermediate velocity $\hat{\mathbf{u}}$ can be transformed to the contravariant velocity using (8.5.7). The resulting $\hat{\mathbf{U}}$ will not satisfy the incompressibility condition (8.5.8). The following pressure correction step can be carried out for the contravariant velocity to enforce the incompressibility condition:

$$\left.\begin{aligned} U_{n+1} &= \hat{U} - \Delta t\, \frac{\partial \phi_{n+1}}{\partial \xi} \\ V_{n+1} &= \hat{V} - \Delta t\, \frac{\partial \phi_{n+1}}{\partial \eta} \\ W_{n+1} &= \hat{W} - \Delta t\, \frac{\partial \phi_{n+1}}{\partial \zeta}\,. \end{aligned}\right\} \tag{8.5.17}$$

From the incompressibility condition (8.5.8) we obtain the following Poisson equation in the computational coordinates for the pressure-like variable ϕ_{n+1}:

$$\frac{\partial^2 \phi_{n+1}}{\partial \xi^2} + \frac{\partial^2 \phi_{n+1}}{\partial \eta^2} + \frac{\partial^2 \phi_{n+1}}{\partial \zeta^2} = 0\,. \tag{8.5.18}$$

Using (8.5.7), the contravariant velocity at the next time can be inverted to get $\mathbf{u}_{n+1}$. The boundary conditions for the intermediate velocity are those for the pressure-corrected final velocity. If higher-order accuracy is desired, in order to contain the slip velocity errors, one could enforce a tangential velocity correction similar to (8.2.33). The boundary conditions for the pressure Poisson equation are the homogeneous Neumann boundary conditions.

The above outlined methodology is only one possible approach to solving the Navier–Stokes equations in curvilinear coordinates. It closely follows the steps taken for the Cartesian geometry. However, many other methodologies have been developed to handle curvilinear coordinates that are often better suited than the one presented above. Our purpose has been to demonstrate that the basic concepts presented for the Cartesian geometry, with appropriate modifications, can be extended to more complex geometries and to reassure readers that they are not too far away from doing complex CFD. The Navier–Stokes equations in curvilinear coordinates are often solved with finite volume technique. The governing equations (8.5.12) are written in conservative form and therefore are ideally suited for the implementation of the finite volume technique. The convective and diffusive fluxes through the four bounding surfaces of each computational cell (six surfaces in three-dimensions), as well as the requisite gradients, can be computed in terms of differences between the function values at the adjacent cell volumes, closely following the procedure outlined in Section 8.4. These standard steps are well addressed in textbooks on finite volume and finite difference methods (e.g., Pletcher *et al.*, 2012) and therefore will not be covered here in any more detail.

The key point to note here is the fluid–solid interface does not require any special treatment, since it is naturally one of the surfaces bounding the boundary cells. From the specified boundary condition, fluxes through this surface and requisite gradients can be computed. Thus, the governing equations can be solved in all the cells in a uniform manner, without any special treatment. Finite volume and related finite difference methods have been widely used in complex flow simulations. For example, in the context of flow over a single spherical particle, Dandy and Dwyer (1990), Magnaudet *et al.* (1995) have successfully employed the finite volume technique and Johnson and Patel (1999), Kim *et al.* (1998), Kurose and Komori (1999), Kim and Choi (2002), Mei and Adrian (1992) and Shirayama (1992) have employed the finite difference technique.

8.5.2 Spectral Methods with Body-fitted Grid

Spectral methods are global in nature and therefore provide the highest possible accuracy. In finite volume and finite difference techniques, gradients and fluxes are computed in terms of only the few neighboring cells that form part of the stencil. In contrast in a fully spectral methodology gradients are effectively computed based on the broadest possible stencil and thus it provides spectral accuracy. As we saw in Chapter 6, with spectral methods, as the grid resolution N increases, error decreases exponentially as e^{-N} and not algebraically. This best possible accuracy, however, comes at a cost. Due to its global nature, spectral methodology is limited to only simple geometries, which have been generally limited to Cartesian, cylindrical and spherical coordinates. Furthermore, all the boundaries of the computational domain, which include solid surfaces and other inflow and outflow surfaces, must conform to the chosen coordinate system. However, in problems where it can be applied, spectral methodology provides the best possible option in terms of accuracy and computational efficiency. For instance, a good application for the use of spectral methodology has been in the investigation of particle–turbulence interaction (Bagchi and Balachandar, 2003, 2004). Here the geometric complexity has been simplified to a single sphere subjected to an isotropic turbulent cross flow. By considering such a simplified geometric situation all the computational resources have been devoted to accurately addressing the details of turbulence and its interaction with the sphere.

An important effort in adapting spectral methodology to more complex geometries is the development of spectral element methodology (Deville *et al.*, 2002; Karniadakis and Sherwin, 2013). Here the computational domain is divided into elements and within each element a spectral representation is used with an orthogonal polynomial basis. Across the interface between the elements continuity of the function and its derivatives is demanded. Thus, with increasing number of polynomial bases, spectral convergence is obtained within each element; and, with increasing number of elements, higher-order convergence is obtained. Again as an example, in the context of flow over a spherical particle, spectral (Chang and Maxey, 1994; Mittal, 1999; Fornberg, 1988; Bagchi and Balachandar, 2002b,c), spectral element (Fischer *et al.*, 2002) and mixed spectral element/Fourier expansion (Tomboulides and Orszag, 2000) have been employed.

In the following discussion we will showcase the fully spectral methodology for the solution of complex flow over a single rigid spherical particle. Consider the general case of a spherical particle freely translating and rotating in response to hydrodynamic and external forces and torque acting on it. We will formulate the problem in a frame of reference translating with the particle so that in this moving frame of reference the particle and the computational domain centered on the particle will remain fixed. Computing the governing equations in a suitably chosen frame of reference that is different from the fixed laboratory frame of reference is a powerful tool and it can greatly simplify the computation. We will also take this opportunity to introduce an important aspect of simulation that pertains to inflow and outflow specification.

Governing equations

We consider the particle to translate at velocity $\mathbf{V}(t)$ and rotate at angular velocity $\Omega_{\mathrm{p}}(t)$. The ambient flow that approaches the sphere in the laboratory frame of reference is $\mathbf{U}(\mathbf{X}, t)$. The ambient flow can be a uniform flow, or a linear shear flow or a turbulent flow. Accordingly, the ambient flow will be considered to be both time and space varying. The ambient pressure P is related to $\mathbf{U}$ by

$$-\frac{1}{\rho_f}\nabla P = \frac{\partial \mathbf{U}}{\partial t} + \mathbf{U}\cdot\nabla\mathbf{U} - \nu\nabla^2\mathbf{U}\,. \tag{8.5.19}$$

For simple ambient flows such as linear shear, or vortical flow the corresponding ambient pressure P can be readily obtained. The presence of the particle and its relative motion with respect to the ambient flow introduces a disturbance field $\mathbf{U}'(\mathbf{X}, t)$. The resultant total velocity field $\mathbf{U} + \mathbf{U}'$ and the corresponding total pressure $P + p'$ satisfy the Navier–Stokes equations. By subtracting the equation for the undisturbed ambient flow, the governing equations for the perturbation flow in dimensional terms can be written as

$$\begin{gathered}\nabla \cdot \mathbf{U}' = 0\,, \\ \frac{\partial \mathbf{U}'}{\partial t} + \mathbf{U}' \cdot \nabla \mathbf{U}' + \mathbf{U}' \cdot \nabla \mathbf{U} + \mathbf{U} \cdot \nabla \mathbf{U}' = -\frac{1}{\rho_f} \nabla p' + \nu \nabla^2 \mathbf{U}' \,. \end{gathered} \tag{8.5.20}$$

At large distances from the sphere the disturbance field decays to zero, while on the surface of the sphere, the no-slip and no-penetration conditions require

$$\mathbf{U}'(\mathbf{X}, t) = \mathbf{V}(t) + \Omega_{\mathrm{p}}(t) \times (\mathbf{X} - \mathbf{X}_{\mathrm{p}}) - \mathbf{U}(\mathbf{X}), \tag{8.5.21}$$

where $\mathbf{X}_{\mathrm{p}}$ is the current location of the center of the particle in the laboratory frame of reference.

It is convenient to attach the reference frame to the moving sphere and the non-inertial frame can be expressed as $\mathbf{x} = \mathbf{X} - \mathbf{X}_{\mathrm{p}}$. In the non-inertial reference frame, the perturbation velocity transforms as

$$\mathbf{u}(\mathbf{x}, t) = \mathbf{U}'(\mathbf{X}, t) - \mathbf{V}(t)\,, \tag{8.5.22}$$

and the governing equations can be rewritten as

$$\left.\begin{aligned} \nabla \cdot \mathbf{u} &= 0\,, \\ \frac{\partial \mathbf{u}}{\partial t} + \frac{\mathrm{d}\mathbf{V}}{\mathrm{d}t} + \mathbf{u} \cdot \nabla \mathbf{u} + \mathbf{U} \cdot \nabla \mathbf{u} + \mathbf{u} \cdot \nabla \mathbf{U} + \mathbf{V} \cdot \nabla \mathbf{U} &= -\frac{1}{\rho_f} \nabla p' + \nu \nabla^2 \mathbf{u}\,. \end{aligned}\right\} \tag{8.5.23}$$

The boundary conditions for $\mathbf{u}$ are as follows: (i) far away from the particle as the undisturbed ambient flow is approached, $\mathbf{u} = -\mathbf{V}$; and (ii) on the surface of the particle, which is now defined by $|\mathbf{x}| = R$, $\mathbf{u} = \Omega_{\mathrm{p}} \times \mathbf{x} - \mathbf{U}(\mathbf{x}, t)$. Note that the term $\mathrm{d}\mathbf{V}/\mathrm{d}t$ on the left-hand side of the above equation accounts for the acceleration of the non-rotating, non-inertial frame of reference. The advantage of the moving frame of reference is that in this frame the geometry of the computational domain remains fixed.

Problem 8.5.2 Transform (8.5.20) to the moving frame of reference and obtain (8.5.23).

Translational and rotational motions of the spherical particle are governed by the following equations of motion:

$$\left.\begin{aligned} m_{\mathrm{p}} \frac{\mathrm{d}\mathbf{V}}{\mathrm{d}t} &= \mathbf{F} + \mathbf{F}_{\mathrm{ext}}\,, \\ \frac{\mathrm{d}\mathbf{X}_{\mathrm{p}}}{\mathrm{d}t} &= \mathbf{V}\,, \\ I_{\mathrm{p}} \frac{\mathrm{d}\boldsymbol{\Omega}_{\mathrm{p}}}{\mathrm{d}t} &= \mathbf{T} + \mathbf{T}_{\mathrm{ext}}\,, \end{aligned}\right\} \tag{8.5.24}$$

where m_{p} is the mass of the particle and I_{p} is the moment of inertia of the particle. The net hydrodynamic force ($\mathbf{F}$) and torque ($\mathbf{T}$) acting on the particle can be obtained by integrating the pressure, and the tangential stresses on the surface of the particle:

$$\mathbf{F} = \int_S \left[-(p' + P + \tau_{rr})\, \mathbf{e}_r + \tau_{r\theta}\, \mathbf{e}_\theta + \tau_{r\phi}\, \mathbf{e}_\phi \right] \mathrm{d}S, \tag{8.5.25}$$

$$\mathbf{T} = R \int_S \mathbf{e}_r \times (\tau_{r\theta}\mathbf{e}_\theta + \tau_{r\phi}\mathbf{e}_\phi)\, \mathrm{d}S\,, \tag{8.5.26}$$

where the integral is taken over the surface of the sphere of radius R and the stress components, $\tau_{r\theta}$ and $\tau_{r\phi}$, are computed on the surface from the total velocity field. In the above $\mathbf{e_r}$, $\mathbf{e}_\theta$ and $\mathbf{e}_\phi$ are the unit vectors along the spherical coordinate with its origin at the center of the particle. $\mathbf{F}_{\mathrm{ext}}$ and $\mathbf{T}_{\mathrm{ext}}$ are the externally applied force and torque, respectively. When $\mathbf{F}_{\mathrm{ext}} = -\mathbf{F}$ the particle is forced to be at a fixed location, and when $\mathbf{F}_{\mathrm{ext}} = \mathbf{0}$ or $m_{\mathrm{p}}\mathbf{g}$ the particle is allowed to move freely either without or with gravitational effect. Similarly, when $\mathbf{T}_{\mathrm{ext}} = -\mathbf{T}$ the angular acceleration of the particle is suppressed, and when $\mathbf{T}_{\mathrm{ext}} = \mathbf{0}$ the particle is allowed to spin freely.

The governing equations (8.5.23) are non-dimensionalized with the diameter of the particle $d = 2R$ as the length scale and the relative velocity $|\mathbf{u}_{\mathrm{r}}| = |\mathbf{U}(\mathbf{X}_{\mathrm{p}}) - V|$ as the velocity scale. The resulting non-dimensional equations are the same as in (8.5.23), except that the fluid density is set to unity and the fluid viscosity is replaced by the inverse of the Reynolds number $\mathrm{Re} = d|\mathbf{u}_{\mathrm{r}}|/\nu$. Of interest are the non-dimensional force and torque coefficients, which are defined as follows:

$$\mathbf{C}_{\mathrm{F}} = \frac{\mathbf{F}}{\frac{1}{2}\,\rho_f\, |\mathbf{u}_{\mathrm{r}}|^2\, \pi (d/2)^2} \tag{8.5.27}$$

$$\mathbf{C}_{\mathrm{T}} = \frac{\mathbf{T}}{\frac{1}{2}\rho_f |\mathbf{u}_{\mathrm{r}}|^2 \pi (d/2)^3}\,. \tag{8.5.28}$$

The force coefficient in the flow direction (here the x-direction) is called the drag coefficient and the force coefficient in the perpendicular direction is called the lift coefficient:

$$C_{\mathrm{D}} = \mathbf{C}_{\mathrm{F}} \cdot \mathbf{e}_x \quad \text{and} \quad C_{\mathrm{L}y} = \mathbf{C}_{\mathrm{F}} \cdot \mathbf{e}_y\,. \tag{8.5.29}$$

Temporal discretization

The time-splitting technique of Section 8.2.3 can be applied to the above governing equations just as efficiently as to the original Navier–Stokes equations. In the evaluation of the intermediate velocity in (8.2.23), the nonlinear term will now be given by

$$\mathcal{N} = \frac{\mathrm{d}\mathbf{V}}{\mathrm{d}t} + \mathbf{u} \cdot \nabla \mathbf{u} + \mathbf{U} \cdot \nabla \mathbf{u} + \mathbf{u} \cdot \nabla \mathbf{U} + \mathbf{V} \cdot \nabla \mathbf{U}\,. \tag{8.5.30}$$

Although $\mathbf{u} \cdot \nabla \mathbf{u}$ is the only true nonlinear term, the entire $\mathcal{N}$ will be treated with an explicit time-advancement scheme. If needed, the $\mathrm{d}\mathbf{V}/\mathrm{d}t$ and $\mathbf{V} \cdot \nabla \mathbf{U}$ terms can be treated implicitly. After solving the Helmholtz equations for the intermediate velocity, the Poisson equation for ϕ is solved. In the calculation of (8.5.26) the pressure and surface stresses must be evaluated at the same time-level. The equations of particle motion and rotation (8.5.24) must also be advanced in time with a consistent scheme. A simple approach will be to use the same explicit scheme as the nonlinear terms of the Navier–Stokes equation.

Computational domain
Although the mathematical problem is usually posed in terms of an unbounded domain, typically the computational domain is truncated to a large but finite outer boundary on which the far-field boundary condition is imposed: see Figure 8.4(a). The placement of the outer boundary is dictated by a few competing requirements:

(i) The computational domain must be small enough that the flow can be adequately resolved with fewer grid points, thus reducing the computational cost.
(ii) The inflow and free-stream sections of the outer boundary must be sufficiently far removed from the sphere. This will allow the perturbation velocity $\mathbf{U}'$ to have sufficiently decayed at the outer boundary, so that $\mathbf{U} - \mathbf{V}$ can be applied as the far-field boundary condition with high accuracy.
(iii) The outflow section of the outer boundary must also be sufficiently downstream of the sphere that any unavoidable approximation in the outflow boundary condition is isolated and prevented from propagating upstream.

A convenient option is to choose the outer domain to be a large sphere of radius R_{o}, which can be chosen to be about 15 times the particle diameter, based on the requirements stated above (Kurose and Komori, 1999; Johnson and Patel, 1999; Mittal, 1999; Bagchi and Balachandar, 2002b). This choice of a large spherical outer domain places the inflow and outflow boundaries sufficiently away from the particle. Also in the plane normal to the flow direction the particle covers only about 0.1% of the total area of cross-section of the computational domain. Thus the blockage effect due to the particle is small.

The influence of the placement of the outer boundary was investigated (Bagchi and Balachandar, 2002b,c) by comparing the results from an outer domain of $15\times$ the particle diameter to those from an even larger outer domain of $30\times$ the particle diameter. The degree of resolution was maintained the same in both cases by proportionately increasing the number of grid points in the larger domain. The drag and lift coefficients obtained from the two different domains for a linear shear flow past a spherical particle at $\mathrm{Re} = 100$ are 1.1183, -0.0123 and 1.1181, -0.0123, respectively, clearly implying the domain independence of the results. With the choice of a spherical outer boundary, a spherical coordinate system can be conveniently adopted. Such geometric simplicity is essential for the implementation of the spectral methodology.

An alternate choice for the computational outer domain has been a large cylinder-like domain (Legendre and Magnaudet, 1998; Tomboulides and Orszag, 2000; Kim and Choi, 2002) surrounding the particle (see Figure 8.4(b)). The

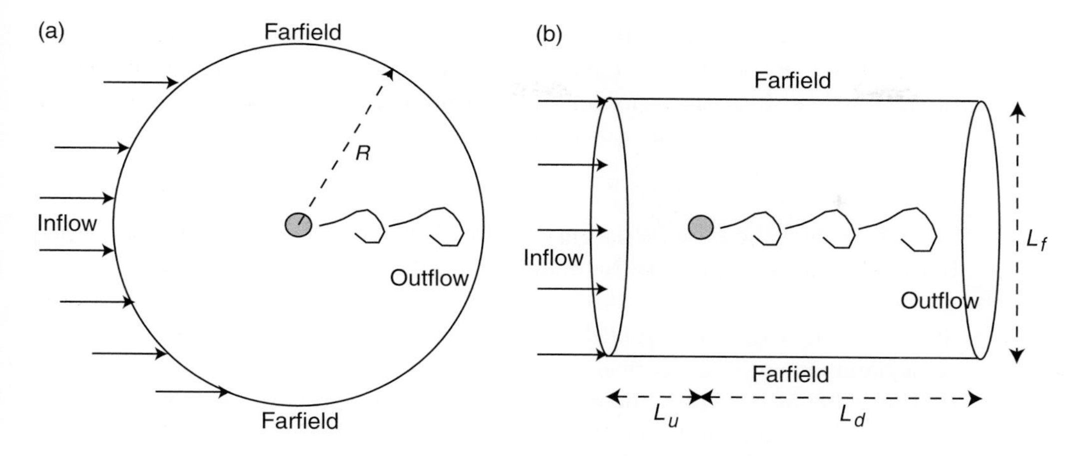

Figure 8.4 Schematic of the computational domain. (a) Shows a spherical outer domain, which includes inflow, outflow and far-field boundaries. (b) Shows an alternative where the outer computational domain is a cylinder. With this option the placement of upstream, downstream and far-field boundaries can be independently controlled.

advantage of this choice is that the distance from the particle to the inflow, L_u, outflow, L_d, and far-field boundaries, $L_f/2$, can be independently controlled. The geometric transition from the inner spherical boundary to the cylindrical outer boundary can be easily handled with an unstructured mesh (Tomboulides and Orszag, 2000) or with a structured mesh using generalized curvilinear coordinates (Magnaudet *et al.*, 1995; Legendre and Magnaudet, 1998).

Boundary conditions

Appropriate specification of boundary conditions, especially at the outflow section of the outer boundary plays an important role. Only carefully derived boundary conditions will produce stable and accurate results. In the context of the time-splitting technique, boundary conditions are required for the intermediate velocity, $\hat{\mathbf{u}}$, in the solution of the Helmholtz equations for the velocity components, and for the pseudo-pressure, ϕ_{n+1}, in the solution of the Poisson equation. As explained in Section 8.2.3, the homogeneous Neumann boundary condition for pressure,

$$\mathbf{n} \cdot \nabla \phi_{n+1} = 0, \tag{8.5.31}$$

is applied both at the outer boundary and on the surface of the particle, where $\mathbf{n}$ indicates the outward normal to the surface. Note that the total pressure does not satisfy a Neumann condition, since depending on the nature of the imposed ambient flow the gradient of the background ambient pressure $\partial P/\partial n$ may not be zero on the boundary. $\partial P/\partial n$ will account for the pressure gradient force arising from the ambient flow. The Neumann boundary condition for the perturbation pressure provides a good representation to capture the viscous effects (Bagchi and Balachandar, 2002b, 2003).

On the surface of the particle, no-slip and no-penetration conditions are imposed. These conditions are equivalent to $\mathbf{u}_{n+1} = \Omega_{\mathrm{p}} \times \mathbf{x} - \mathbf{U}$ on the surface of the particle. In anticipation of the pressure correction step, the appropriate boundary condition for the intermediate velocity is then

$$\hat{\mathbf{u}} = \Omega_{\mathrm{p}} \times \mathbf{x} - \mathbf{U} + \Delta t(2\nabla\phi'_n - \nabla\phi'_{n-1})\,. \tag{8.5.32}$$

The last term is the same as that in (8.2.33). The above boundary condition combined with the homogeneous Neumann boundary condition for pressure (8.5.31), guarantees zero penetration through the surface of the sphere. The no-slip condition is satisfied to order $O(\Delta t^3)$, since the pressure gradient terms on the right-hand side of (8.5.32) provide a higher-order approximation for $\nabla {p'}_{n+1}$. A typical value of the non-dimensional time-step Δt is about two orders of magnitude smaller than unity and thus the slip error arising from the time-splitting scheme is usually quite small and negligible.

At the inflow section of the outer boundary, a Dirichlet boundary condition specifying the undisturbed ambient flow is enforced. For the intermediate velocity this translates to the following boundary condition:

$$\text{Inflow:} \quad \hat{\mathbf{u}} = -\mathbf{V} + \Delta t(2\nabla\phi'_n - \nabla\phi'_{n-1})\,. \tag{8.5.33}$$

The elliptic nature of the Helmholtz equation requires a boundary condition even at the outflow section of the outer boundary. Since no natural velocity condition exists at the outflow, its specification can be a challenge. In the context of finite difference and finite volume computations, approximate boundary conditions such as (Kim and Choi, 2002; Magnaudet *et al.*, 1995; Kim *et al.*, 1998; Shirayama, 1992)

$$\begin{aligned}
\textbf{convective:} \quad & \frac{\partial \hat{\mathbf{u}}}{\partial t} + c\frac{\partial \hat{\mathbf{u}}}{\partial n} = 0, \\
\textbf{parabolic:} \quad & \frac{\partial^2 \hat{u}_n}{\partial n^2} = 0, \quad \frac{\partial \hat{u}_\tau}{\partial n} = 0, \quad \frac{\partial^2 p_{n+1}}{\partial n \partial \tau} = 0, \\
\textbf{zero gradient:} \quad & \frac{\partial \hat{\mathbf{u}}}{\partial n} = 0, \\
\textbf{zero second gradient:} \quad & \frac{\partial^2 \hat{\mathbf{u}}}{\partial n^2} = 0,
\end{aligned} \tag{8.5.34}$$

have been used in the past. In the above, n and τ indicate directions normal and tangential to the outer boundary. However, owing to their global nature, spectral simulations place a more stringent non-reflection requirement at the outflow boundary. A buffer domain or viscous sponge technique is often used to implement a non-reflecting outflow boundary condition (Streett and Macaraeg, 1989; Karniadakis and Triantafyllou, 1992; Mittal and Balachandar, 1996). The idea is to parabolize the governing equations smoothly by multiplying the diffusion term in the Helmholtz equation by a filter function as

$$f(r,\theta)\left(\frac{\partial \hat{\mathbf{u}}}{\partial r^2} + \frac{\partial \hat{\mathbf{u}}}{\partial \theta^2} + \frac{\partial \hat{\mathbf{u}}}{\partial \phi^2}\right) - \frac{2\,\mathrm{Re}}{\Delta t}\hat{\mathbf{u}} = \mathbf{RHS}\,. \tag{8.5.35}$$

The filter function $f(r, \theta)$ is defined such that $f \to 1$ over most of the computational domain, and as the outflow section of the outflow boundary is approached it smoothly approaches $f \to 0$. Thus, the diffusion term remains unaltered over the bulk of the flow, while at the outflow boundary the equation for $\hat{\mathbf{u}}$ is parabolized. Hence the method does not require any outflow boundary condition, and (8.5.35) can be solved to obtain the velocity there. Instead of filtering the entire viscous term one could just filter the radial component of the viscous term:

$$f(r, \theta) \frac{\partial \hat{\mathbf{u}}}{\partial r^2} + \frac{\partial \hat{\mathbf{u}}}{\partial \theta^2} + \frac{\partial \hat{\mathbf{u}}}{\partial \phi^2} - \frac{2\,\mathrm{Re}}{\Delta t} \hat{\mathbf{u}} = \mathbf{RHS} \,, \tag{8.5.36}$$

in which case a 2D Helmholtz equation in θ and ϕ must be solved just at the outer boundary, along with the inflow condition, in order to obtain the outflow boundary velocity. The filter function can be taken to be axisymmetric and dependent only on r and θ. For example, a wake filter function of the following form was employed (Bagchi and Balachandar, 2002b):

$$f(r, \theta) = 1 - \exp\left[-\nu_1 \left| \frac{r - R_\mathrm{o}}{1/2 - R_\mathrm{o}} \right|^{\gamma_1} \right] \exp\left[-\nu_2 \left| \frac{\theta}{\pi} \right|^{\gamma_2} \right] , \tag{8.5.37}$$

where R_o is the radius of the outer spherical boundary. Large values for ν_1 and ν_2 of the order of 40 and values for γ_1 and γ_2 of the order of 4 were used to localize the filtered region close to the outer boundary in the outflow region.

Spectral discretization

The requirement of geometric simplicity demands that the outer boundary of the computational domain is a sphere of large non-dimensional radius, R_o, concentric with the particle. This defines the computational domain to be

$$1/2 \le r \le R_\mathrm{o} \,, \quad 0 \le \theta \le \pi \quad \text{and} \quad 0 \le \phi \le 2\pi \,.$$

A Chebyshev expansion is used along the inhomogeneous radial direction. The grid points along this direction are Gauss–Lobatto collocation points, which are first defined on the interval $[-1, 1]$ as

$$\xi_i = -\cos\left[\frac{\pi(i-1)}{N_r - 1} \right] , \tag{8.5.38}$$

for $i = 1, 2, \ldots, N_\mathrm{r}$, where N_r is the number of radial grid points. Then an algebraic mapping is used to map the grid points from $[-1, 1]$ to $[1/2, R_\mathrm{o}]$. The mapping can also be used to redistribute the points radially in order to obtain enhanced resolution near the surface of the sphere, by clustering points close to the sphere. For example, Bagchi and Balachandar (2002b) used the following radial stretching in their simulations of flow over a sphere:

$$\hat{\xi} = C_0 + C_1 \xi - C_0 \xi^2 + C_3 \xi^3 \,, \tag{8.5.39}$$

$$C_1 = \frac{1}{2}(-\gamma_1 + 2C_0 + 3), \quad C_3 = \frac{1}{2}(\gamma_1 - 2C_0 - 1) \,. \tag{8.5.40}$$

The parameters C_0 and γ_1 control the amount of stretching. The computational points in physical space are obtained using the mapping

$$r_i = \frac{1}{2}\hat{\xi}_i\left(\frac{1}{2} - R_o\right) + \frac{1}{2}\left(\frac{1}{2} + R_o\right). \tag{8.5.41}$$

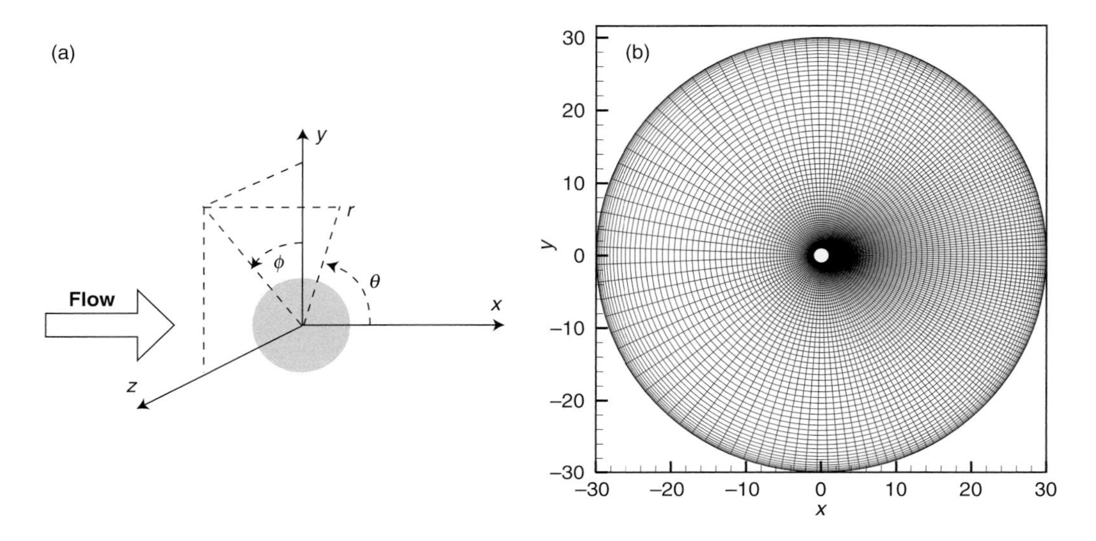

Figure 8.5 Computational grid. (a) Schematic of the spherical coordinate. (b) A ϕ-projection of the computational grid.

The azimuthal direction ϕ is periodic over 2π, and a Fourier expansion is used along this direction. The collocation points in ϕ are

$$\phi_k = \frac{2\pi(k-1)}{N_\phi}, \tag{8.5.42}$$

for $k = 1, 2, \ldots, N_\phi$, where N_ϕ is the number of grid points in ϕ. In the tangential direction θ, it is sufficient to define the variables over the interval $0 \le \theta \le \pi$; however, the variables are periodic over 2π, and not over π. Thus a Fourier collocation in θ can be used only with some symmetry restrictions. One may note that a scalar, the radial component of a vector and the radial derivative of a scalar are continuous over the poles ($\theta = 0$ and π), but the tangential and the azimuthal components of a vector change sign across the poles. The tangential and azimuthal derivatives of a scalar also change sign. This is the so-called "parity" problem in spherical geometry, and has been discussed by Merilees (1973), Orszag (1974) and Yee (1981). The problem of pole parity does not arise if surface harmonics are used (Fornberg, 1998). However, spectral methods using surface harmonics require O(N) operations per mode, while those based on Fourier series require only O($\log N$) operations. Bagchi and Balachandar (2002b) used a Fourier expansion in θ.

A typical term in the expansion can be written as

$$\begin{Bmatrix} c \\ u_r \\ u_\theta \end{Bmatrix} = \begin{Bmatrix} \alpha \\ \beta \\ \gamma \end{Bmatrix} r^p \exp(im\theta)\exp(ik\phi)\,, \tag{8.5.43}$$

where c represents a scalar. The requirement of analyticity at the poles results in the following acceptable expansions:

$$c = \begin{cases} \sum \alpha_{pmk} T_p(r)\cos(m\theta)\exp(ik\phi) & \text{even} \quad k \\ \sum \alpha_{pmk} T_p(r)\sin(m\theta)\exp(ik\phi) & \text{odd} \quad k \end{cases}, \tag{8.5.44}$$

$$u_r = \begin{cases} \sum \beta_{pmk} T_p(r)\cos(m\theta)\exp(ik\phi) & \text{even} \quad k \\ \sum \beta_{pmk} T_p(r)\sin(m\theta)\exp(ik\phi) & \text{odd} \quad k \end{cases} \tag{8.5.45}$$

and

$$u_\theta = \begin{cases} \sum \gamma_{pmk} T_p(r)\sin(m\theta)\exp(ik\phi) & \text{even} \quad k \\ \sum \gamma_{pmk} T_p(r)\cos(m\theta)\exp(ik\phi) & \text{odd} \quad k \end{cases}, \tag{8.5.46}$$

where T_p represents the pth Chebyshev polynomial and m and k are the wavenumbers in the θ and ϕ directions. Here α, β and γ are the coefficients in the expansions and are functions of p, m and k. The expansion for u_ϕ follows that of u_θ.

The collocation points in θ are normally equi-spaced as

$$\hat{\theta}_j = \frac{\pi}{N_\theta}\left[j - \frac{1}{2}\right], \tag{8.5.47}$$

for $j = 1, 2, \ldots, N_\theta$, where N_θ is the number of grid points in θ. However, for the problem of flow over a sphere, the complexity of the flow, and therefore the resolution requirement, is greater in the wake of the sphere than on the windward side. A grid stretching that will cluster points in the wake region of the particle will make better use of the grid resolution. One requirement, however, is that the grid stretching must preserve the periodic nature of the θ direction. Such a grid stretching along the periodic direction has been suggested by Augenbaum (1989) and employed in the simulation by Bagchi and Balachandar (2002b,c), and it can be defined as

$$\theta_j = \tan^{-1}\left[\frac{\sin(\hat{\theta}_j)(1-\hbar^2)}{\cos(\hat{\theta}_j)(1+\hbar^2) - 2\hbar}\right], \tag{8.5.48}$$

where $\hbar$ is the parameter that controls the degree of stretching. A value of $\hbar = -0.35$ has been shown to provide sufficient resolution in the particle wake. A $\phi = 0$ projection of a typical mesh with enhanced resolution in the wake through grid stretching is shown in Figure 8.5(b). It must be pointed out that enhanced resolution of the wake through grid stretching has been widely employed in most simulations of flow over a sphere.

Due to the topology of the grid, the azimuthal (ϕ) resolution is spatially non-uniform. The resolution is much higher near the poles compared to the equator. Furthermore, the azimuthal grid spacing linearly increases with the radial location from the surface of the particle to the outer boundary. The viscous stability constraint due to such non-uniform resolution is avoided by the implicit treatment of the radial and azimuthal diffusion terms. However, the time-step size is still restricted by the convective stability (CFL) condition.

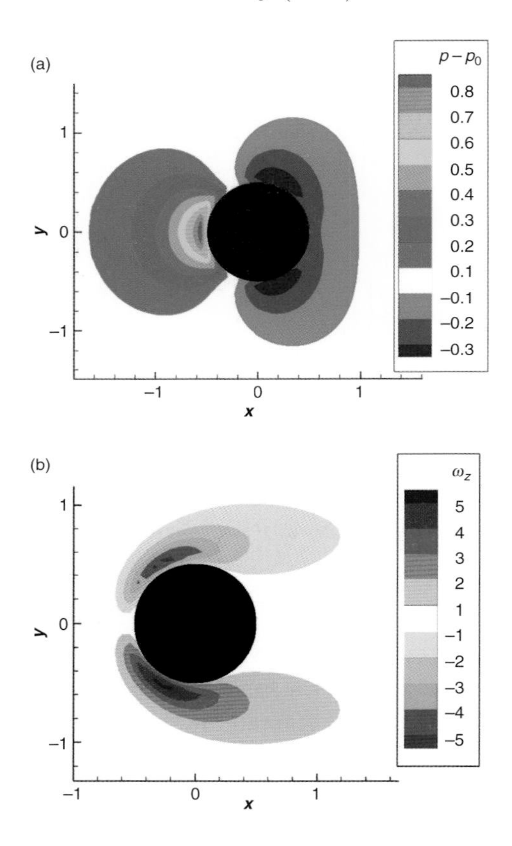

Figure 8.6 (a) Contours of pressure for a uniform flow over a sphere at $\mathrm{Re} = 16.7$. (b) Corresponding vorticity contours. The flow is axisymmetric.

8.5.3 Sample Applications

Here we will first present some sample results obtained for a fully spectral simulation of a uniform flow over a spherical particle. Figure 8.6 shows contours of pressure and vorticity around a stationary sphere for a uniform far-field flow at a particle Reynolds number of 16.7. Uniform flow over an isolated sphere at finite

Reynolds number is a canonical problem and a high-resolution spectral simulation with a body-fitted grid can serve as a benchmark against which results from other numerical methods can be compared. For Reynolds numbers below about 210 the flow remains axisymmetric and thus a 2D simulation with $m = 0$ will be sufficient. Above this critical Reynolds number the flow undergoes a bifurcation and becomes non-axisymmetric, although it remains steady for a small range of Re above 210. With further increase in Reynolds number the flow becomes unsteady and starts shedding wake vortices and eventually the wake flow becomes chaotic and turbulent.

As the next example of the body-fitted spectral methodology we will present the results from a more complex problem of interaction of isotropic turbulence with a rigid spherical particle. The isotropic turbulent field was generated separately using the pseudo-spectral methodology outlined in Section 8.3 using a 256^3 grid in a $(2\pi)^3$ box. The turbulent field is periodic in all three directions. The Kolmogorov length (η) and velocity (v_k) are chosen as the reference scales. The isotropic turbulence is characterized by the microscale Reynolds number of $\mathrm{Re}_\lambda = \lambda U_{\mathrm{rms}}/\nu = 164$, where the non-dimensional rms turbulent velocity fluctuation (U_{rms}/v_k) is 6.5, the box size (L/η) is 757.0, and the Taylor microscale (λ/η) is 25.2.

The sphere was subjected to a uniform cross flow of velocity, V, upon which an instantaneous realization of isotropic turbulence is superposed. Figure 8.7 shows a plane of the isotropic turbulence with a particle of diameter $d/\eta = 10$ drawn to scale placed within it. The larger circle represents the computational outer sphere around the particle within which the details of the flow around the particle were simulated. In a frame attached to the particle, as the box of isotropic turbulence advects past the particle (say from left to right), the turbulent field is interpolated onto the left half of the computational outer boundary as inflow. With this interpolated velocity as the inflow condition, the details of the flow over the particle were simulated.

The grid resolution was chosen to satisfy two criteria: first, the size of the largest grid in the spherical domain is less than that of the grid used to simulate isotropic turbulence, in order to guarantee resolution of the free-stream turbulence. Second, the grid is adequate to resolve the thin shear layers and the wake structures generated by the particle. Typical grids used in the simulations have 141 points in the radial direction, 160 in the θ-direction and 128 in the ϕ-direction.

An important outcome of their study was a direct measurement of the effect of free-stream turbulence on particle drag force. The mean drag coefficient obtained as a function of Re from a variety of experimental sources is shown in Figure 8.8. Also plotted in the figure for reference is the standard drag correlation applicable for the case of a stationary particle in a steady uniform ambient flow. The scatter in the experimental data clearly illustrates the degree of disagreement as to the effect of turbulence. For example, in the moderate Reynolds number regime, the measurements of Uhlherr and Sinclair (1970), Zarin and Nicholls (1971), and

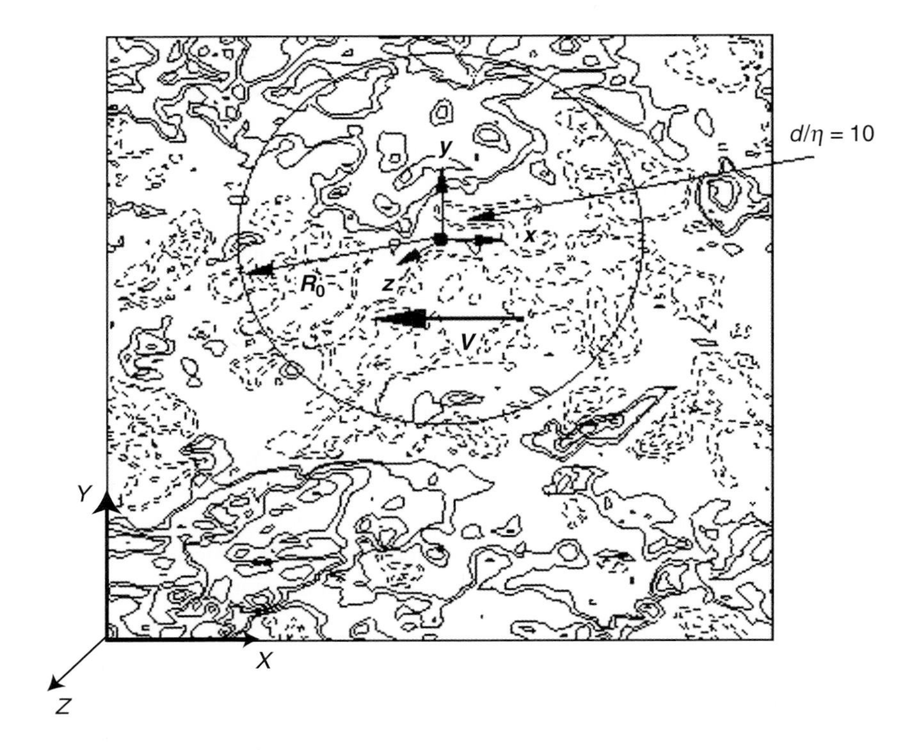

Figure 8.7 Schematic of the particle flow configuration. Drawn to scale, a particle of $d/\eta = 10$ is shown here. The larger circle surrounding the particle represents the outer boundary of the spectral computational domain attached to the particle. The outer box represents the $(2\pi)^3$ box in which the isotropic turbulence was generated.

Brucato *et al.* (1998) indicated a substantial increase in the drag coefficient in a turbulent flow. On the other hand, the results of Rudoff and Bachalo (1988) tend to suggest a reduction in the drag coefficient due to ambient turbulence. The results of the spectral simulation are also shown in the figure. There is substantial scatter in the experimental results. Nevertheless, the simulation results are in reasonable agreement with the experimental data. It can be concluded that free-stream turbulence has little systematic effect on the mean drag. However, as can be expected turbulence has substantial effect on fluctuating drag and lift forces (Bagchi and Balachandar, 2003).

8.6 Sharp Interface Cartesian Grid Method

In this section we will discuss the sharp interface Cartesian grid approach. Consider the case of a fluid flow around a complex distribution of rigid solid spheres. For the initial discussion also consider the solid spheres (particles) to be stationary, so that the interface between the fluid and solid phases is stationary.

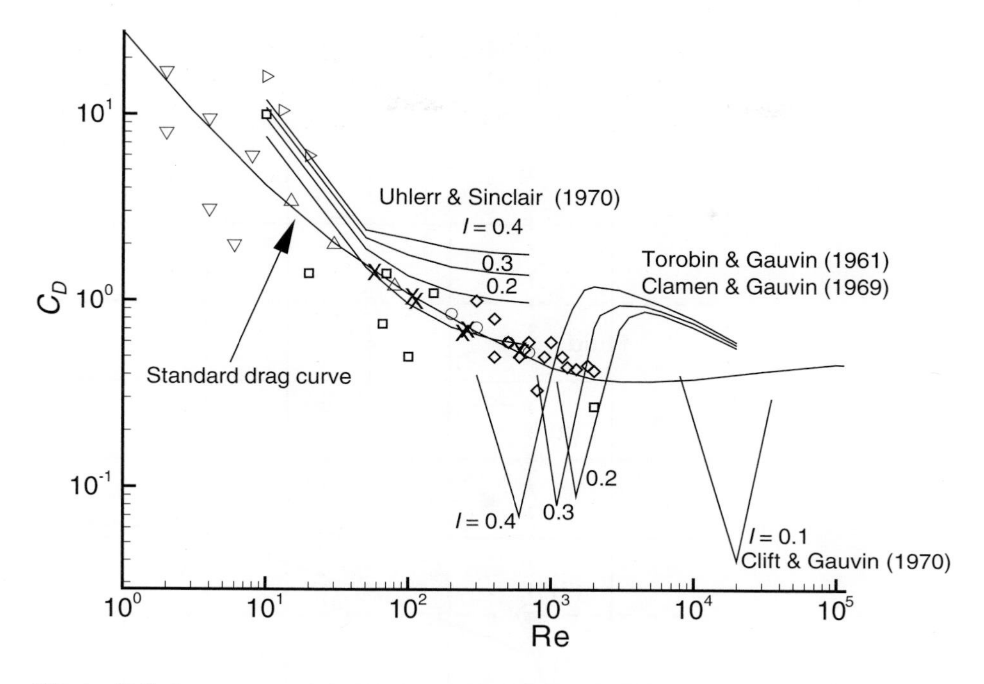

Figure 8.8 A summary of the results on the effect of turbulence on the drag coefficient. $\times$ Present results; $\square$ (Gore and Crowe, 1990); $\diamond$ (Sankagiri and Ruff, 1997); $\circ$ (Zarin and Nicholls, 1971); Δ (Warnica *et al.*, 1994); ∇ (Rudoff and Bachalo, 1988); $\triangleright$ (Brucato *et al.*, 1998). The standard drag curve is obtained using the Schiller–Neumann formula $C_D = (24/\,\mathrm{Re})(1 + 0.15\,\mathrm{Re}^{0.687})$ (see Clift *et al.*, 1978). The parameter I is the ratio of the rms velocity of the free-stream turbulence to the mean relative velocity between the particle and the fluid.

We start with a regular Cartesian grid and the grid may be clustered in order to achieve enhanced resolution in the neighborhood of the embedded particles. Other than this resolution requirement, the choice of the grid can be quite independent of the actual shape, size or location of the particles. The sharp interface methodology to be described below is based on Ye *et al.* (1999) to which the reader is referred for complete details.

In the sharp interface approach the interface is represented using marker points which are connected by piecewise linear or quadratic curves. These boundaries overlay on the background Cartesian mesh and here we will describe how to treat the Cartesian cells that are cut through by the interfacial boundaries.

The first step is to go through a *reshaping* procedure, where the shape of the cells adjacent to the interfaces is redefined and the surfaces of adjacent cells are redefined accordingly. As outlined in Ye *et al.* (1999) the cells, which are cut by the interface, whose cell-center lies within the fluid, are reshaped by discarding the part of the cell that lies within the solid. On the other hand, if the center of the cut cell lies within the solid, then it is not treated independently, but is attached to a neighboring cell to form a larger cell. Thus the resulting control volumes are of trapezoidal shape as shown in Figure 8.9. The details of this reshaping procedure can be found in Udaykumar *et al.* (1997, 1999).

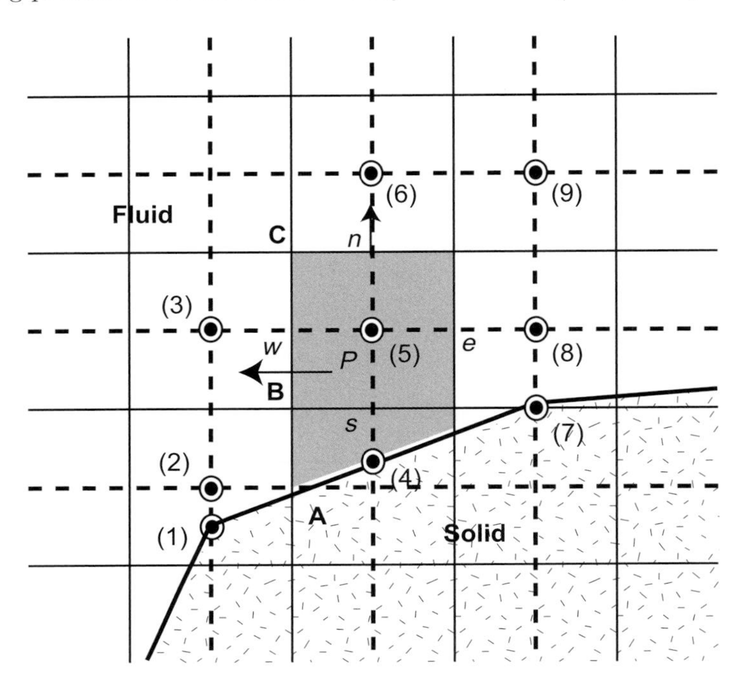

Figure 8.9 Schematic of the fluid–solid interface cutting through the underlying Cartesian mesh. The figure shows the reshaping of the shaded cell located adjacent to the interface. Also marked are the nine points that form the stencil in two dimensions, which are required to compute the necessary fluxes and gradients for the cell marked P. Three of the nine points are interface points and the other five are cell centers. This figure is taken from Ye *et al.* (1999) with permission.

The sharp interface method is typically built on top of an underlying Cartesian grid finite volume approach, such as the one described in Section 8.4. The main issue is to define fluxes and gradients across the faces of the trapezoidal cell in terms of interface information and available cell-center values in the neighborhood. Particular attention must be paid to maintain second-order accuracy. A multi-dimensional polynomial interpolation using the neighboring cell centers and interface points is constructed, which then is used to calculate appropriate fluxes and gradient information. For example, for calculating the fluxes through the left face of the trapezoidal cell (marked AC in Figure 8.9) the following interpolation function can be used:

$$u(x, y) = c_1 xy^2 + c_2 y^2 + c_3 xy + c_4 y + c_5 x + c_6 \,, \tag{8.6.1}$$

where the coefficients of the polynomial are obtained by interpolating through the six points marked 1 to 6 in Figure 8.9. These six points are carefully chosen to surround the left face and they include four cell centers and two points along the interface. The six coefficients, c_1–c_6, are evaluated in terms of the value of u at these six points. The function value and the gradient at the midpoint of the face (marked B) can be evaluated from (8.6.1) by substituting $x = x_{\mathrm{B}}$ and $y = y_{\mathrm{B}}$. The result can be rearranged to express velocity and its x-gradient as a weighted sum of the velocity at the six points:

$$u_w = \sum_{j=1}^{6} \alpha_j u_j \qquad \text{and} \qquad \left(\frac{\partial u}{\partial x}\right)_w = \sum_{j=1}^{6} \beta_j u_j \,. \tag{8.6.2}$$

Here the coefficients α_j and β_j depend only on the cell geometry. Provided the interface remains stationary over time, these coefficients can be computed once at the beginning and stored for repeated use. The above polynomial is clearly asymmetric in x and y; however, it can be shown that it is the minimal polynomial needed to obtain second-order accuracy for the x-flux and x-gradient.

The wall-normal gradient on the slanted face of the trapezoid is decomposed into its x- and y-components. The y-gradient is obtained to second-order accuracy by fitting a quadratic polynomial in y through the three points marked (4), (5) and (6) in Figure 8.9. Owing to the orientation of the cell the calculation of the x-gradient is a bit more involved. Here, for second-order accuracy, a polynomial collocating at points marked (1), (2), (3), (7), (8) and (9) must be constructed for computing the x-gradient at the midpoint of the slanted face. The fluxes and the gradients on the other faces can be defined similarly (for details see Ye *et al.*, 1999). When all the different terms are plugged into the finite volume version of the governing equations, the resulting compact stencil involves nine points for the trapezoidal cell, which are shown in Figure 8.9. Three of the nine points are boundary points while the other six are cell centers. Note that the nine-point stencil is larger than the regular five-point stencil used for the interior cells (see Figure 8.3). The intersection of the fluid–solid interface with the Cartesian mesh will result in trapezoidal cells of other types as well and the stencil for such cells can be similarly formed.

The above finite volume discretization of (8.4.1) results in a linear system of the following form for the x-component of velocity:

$$\sum_{k=1}^{M} \chi_P^k u_x^k = R_{\mathrm{p}} \tag{8.6.3}$$

where P denotes the cell in which the control volume momentum balance is applied and it runs over all the cells. The coefficients χ correspond to the weights for the M nodes of the stencil around the point P. In 2D, the stencil of interior cells is of width $M = 5$ and for the interface cells the stencil is wider and $M = 9$. The banded linear operator on the left-hand side corresponds to the finite volume discretization of the Helmholtz operator and the right-hand side is the net effect of the explicit terms in (8.4.1). A similar linear system results from the finite-volume discretization of the Poisson equation (8.4.7). These linear systems are generally very large and as a result a direct solution can be very expensive. Several options exist for their iterative solution. Bi-CGSTAB has been recommended for the solution of the Helmholtz equation for the intermediate velocity and multigrid methods are preferred for the Poisson equation (Udaykumar *et al.*, 2001). GMRES is an attractive alternative for the iterative solution of the above large linear system (Saad, 2003).

The above discretization has been explained for a two-dimensional geometry. Extension of the formalism to three-dimensional geometries is in principle straightforward; a three-dimensional implementation can, however, be complicated. Udaykumar *et al.* (2001) presented an extension of the above formalism to account for a moving fluid–solid interface over a fixed Cartesian grid. An interesting aspect of the moving interface is that new cells can appear in the fluid domain, with no prior velocity or pressure information in them, since they were previously part of the solid. Strategies for handling such situations were discussed in Udaykumar *et al.* (2001).

Other approaches have been proposed for handling complex interfaces maintaining their sharpness within a Cartesian discretization. In particular, the approach proposed by Calhoun and Leveque (2000) is quite interesting. A capacity function is introduced to account for the fact that some of the Cartesian cells are only partly in the fluid region. This allows the partial cells to be treated in essentially the same way as the other interior full cells. Calhoun and Leveque demonstrated the second-order accuracy of the sharp interface methodology.

8.7 Immersed Boundary Method

The immersed boundary method (IBM) was pioneered by Peskin and his group at the Courant Institute (Peskin, 1977; McQueen and Peskin, 1989). Currently, it is the most popular method for solving Navier–Stokes equations in complex geometries. Its attraction is its simplicity. The IBM presents a relatively simple technique to simulate immersed bodies, such as flow over an airfoil or flow around an array of cylinders or spheres. It is a very powerful methodology and can be easily used to simulate moving immersed bodies, as in the case of a multiphase flow involving a large array of freely moving particles in a flow. The IBM was originally developed to investigate flow around heart valves and therefore can very efficiently handle deformable fluid–solid interfaces as well. As a result it is well suited to simulate complex problems such as instability and atomization of a liquid jet into a droplet spray. Its appeal lies in its ability to use a fixed Eulerian grid, and it does not require a body-fitted mesh. This is an immense advantage in the case of moving and deforming interface.

Here we will describe the method in the simpler context of a fluid flow around a complex distribution of rigid solid particles. The fluid–solid interface is represented by Lagrangian markers that move with the particles. A forcing term (or appropriate momentum and sometimes mass sources) at these markers is added to the governing equations to account for the boundary conditions on the interface. The differences between the different implementations of IBM lie in the method by which the forcing term is applied. Our discussion will again be brief and interested readers are referred to Mittal and Iaccarino (2005), Uhlmann (2005) and Tseng and Ferziger (2003).

The interface between the fluid and the solid is defined by the surface, S. A body-force, $\mathbf{f}$, is introduced into the Navier–Stokes equations as

$$\frac{\partial \mathbf{u}}{\partial t} + \mathbf{u} \cdot \nabla \mathbf{u} = -\nabla p + \frac{1}{Re} \nabla^2 \mathbf{u} + \mathbf{f}\,, \tag{8.7.1}$$

and the forcing is chosen such that it enforces the desired velocity boundary condition at the interface, S. Ideally we desire the forcing to be identically zero in the entire fluid region, so that the standard Navier–Stokes equation is faithfully solved there. The forcing must be localized along the interface and perhaps extended into the solid. The force distribution will be chosen appropriately so that the computed velocity field will satisfy the requisite boundary condition. Clearly there is no explicit restriction to the shape of the interface or the boundary condition that can be satisfied along the interface. Irrespective of these details the flow can be solved on a fixed Eulerian grid that is suitably chosen to provide adequate resolution everywhere needed.

The key aspect of IBM is the discretization of the interface S with a distribution of *Lagrangian markers* within the fixed Eulerian grid used for solving the Navier–Stokes equation. As an example, Figure 8.10(a) shows the fixed Eulerian grid used for the simulation of flow within a rectangular channel, while (b) on the right shows the distribution of Lagrangian markers used to define a spherical particle embedded within the rectangular channel with the center of the sphere at the center of the channel. In most IBM simulations the fixed Eulerian grid is typically uniformly spaced and as a consequence the Lagrangian markers are uniformly distributed over the interface S. In the case of a sphere, a perfectly uniform distribution of Lagrangian markers on the surface of the sphere still remains an open mathematical question, but there are algorithms for obtaining a nearly uniform distribution (Saff and Kuijlaars, 1997). For a uniform Eulerian grid, Uhlmann (2005) showed that the Lagrangian markers should be spaced such that the surface area associated with each Lagrangian marker must be approximately equal to the square of the Eulerian grid spacing. Any further increase in the number of Lagrangian markers does not contribute to finer resolution of the immersed interface, since interface resolution will be controlled by Eulerian grid spacing. However, a lower number of Lagrangian markers will contribute to under-resolution of the immersed boundary as seen by the Eulerian grid on which the flow is resolved.

Implementation of IBM is easier in the context of a uniformly spaced Eulerian grid, since the corresponding Lagrangian markers need to be nearly uniformly distributed on the interface, with the surface area associated with each Lagrangian marker of the order of the surface area of the uniformly sized Eulerian cell. The distribution of Lagrangian markers becomes more complicated in the

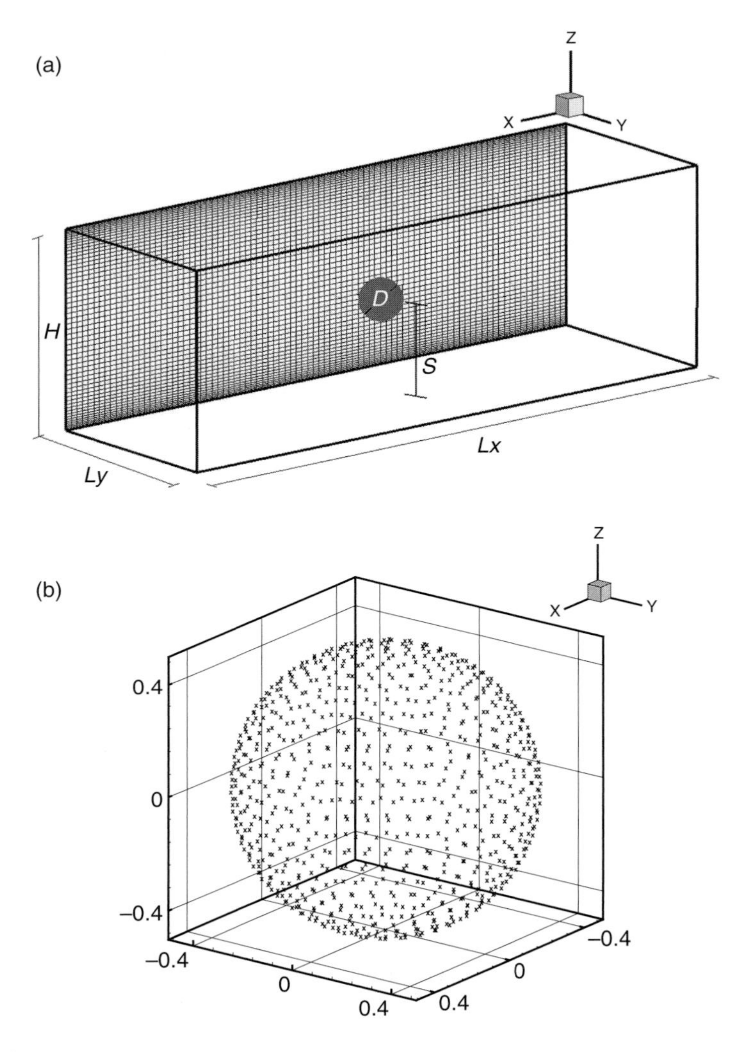

Figure 8.10 (a) The fixed Eulerian grid used for the simulation of flow within a rectangular channel. (b) The distribution of Lagrangian markers used to define a spherical particle embedded within the rectangular channel with the center of the sphere at the center of the channel.

case of a non-uniform Eulerian mesh. A uniform Lagrangian marker distribution can still be used based on the smallest Eulerian grid that overlaps the interface S. However, this can result in substantial over-resolution of the interface in regions where the Eulerian mesh is much coarser. An algorithm for non-uniformly distributing Lagrangian markers was considered by Akiki and Balachandar (2016), but their focus was limited to immersed spherical particles. An approach to getting around this difficulty is to use a fine uniform mesh in the region that embeds the immersed boundary and a stretched non-uniform grid away from that vicinity (Lee *et al.*, 2011). For more complex problems one can pursue the idea of reproducing kernel particle method (RKPM) used in the implementation of immersed finite element method (Zhang *et al.*, 2004) and it works for both uniform and non-uniform Eulerian meshes.

8.7.1 Direct Forcing Method

There are different variants of the immersed boundary technique and they differ primarily in the manner in which the force field is computed. Some of the forcing techniques that have been advanced are *feedback forcing* (Goldstein *et al.*, 1993), *direct forcing* (Fadlun *et al.*, 2000; Uhlmann, 2005) and *discrete mass and momentum forcing* (Kim *et al.*, 2001). Here we will discuss the direct forcing method. (Mohd-Yusof, 1997) developed a very simple direct procedure for forcing the required velocity condition at the interface, which was later adapted and tested in (Fadlun *et al.*, 2000). In this approach the forcing to be applied at each time is determined directly without solving an additional integro-differential term for the force. The initial implementations faced some difficulty in the case of moving interfaces as the interface moved across grid cells. Here we will describe the direct forcing method introduced by Uhlmann (2005), whose implementation is quite easy and avoids force oscillations. We will adopt the time-splitting methodology and consider the forcing to be enforced during the advection–diffusion stage. Thus, the forcing will be applied to give the appropriate intermediate velocity, $\hat{\mathbf{u}}$, at the interface, which in turn will make the final divergence-free velocity satisfy the specified interface condition. The pressure Poisson equation and the pressure-correction step remain unaffected by the added forcing.

For simplicity we have so far only employed the second-order Adams–Bashforth method for the advection term, with the advice that the reader should in fact use a third-order AB or RK method for ensuring stability. Here we will describe the direct forcing IBM with a third-order RK method for the advection term and the Crank–Nicolson method for the diffusion term. The implementation of IBM is along the following steps:

(1) The first step is to get an explicit estimate of the velocity without the forcing, so that the estimate can be used to evaluate the force that is required to nullify the velocity at the interface Lagrangian markers to obtain the desired interface velocity. The estimated velocity is given by

$$\tilde{\mathbf{u}}^{(m)} = \mathbf{u}^{(m-1)} + c_{2m}\mathbf{J}^{(m-1)} + \frac{2c_{dm}}{\mathrm{Re}}\nabla^2\mathbf{u}^{(m-1)} - 2c_{dm}\nabla p^{(m-1)}\,, \tag{8.7.2}$$

where the nonlinear term is cumulated over the RK stages and is given by

$$\mathbf{J}^{(m-1)} = \Delta t\left[(\mathbf{u}\cdot\nabla)\mathbf{u}\right]^{(m-1)} - c_{1m}\mathbf{J}^{(m-2)}\,. \tag{8.7.3}$$

In the above the three stages of the RK method are denoted by superscripts

$m = 1, 2$ and 3. The coefficients of the three RK stages are given by

$$\left.\begin{aligned} c_{1m} &= \left[0, -\frac{5}{9}, -\frac{153}{128}\right], \\ c_{2m} &= \left[\frac{1}{3}, \frac{15}{16}, \frac{8}{15}\right], \\ c_{dm} &= \left[\frac{\Delta t}{6}, \frac{5\Delta t}{24}, \frac{\Delta t}{8}\right]. \end{aligned}\right\} \tag{8.7.4}$$

It can be verified that the three stages respectively correspond to (1/3)rd, (5/12)th and (1/4)th of a time-step and together contribute to a full time-step Δt. Note that the first RK stage will start with the solution at time t_n (i.e., $\mathbf{u}^0 = \mathbf{u}_n$) and, at the end of the third RK stage, the solution will correspond to t_{n+1} (i.e., $\mathbf{u}^3 = \mathbf{u}_{n+1}$). Note that the above velocity estimate includes all three advection, diffusion and pressure-gradient terms. The semi-discretization is fully explicit and as a result $\tilde{\mathbf{u}}^{(m)}$ will not be divergence free. Nor will it satisfy the boundary conditions. As we will see below this is not a problem since $\tilde{\mathbf{u}}^{(m)}$ will be used only to evaluate the force distribution.

(2) We define the Lagrangian marker points to be $\mathbf{X}_l$ for $l = 1, 2, \ldots, N_L$, and in general these points will not coincide with the Eulerian grid points ($\mathbf{x}$). Thus, the next step is to interpolate the estimated velocity to the Lagrangian markers as

$$\tilde{\mathbf{U}}(\mathbf{X}_l) = \sum_{\forall \mathbf{x}} \tilde{\mathbf{u}}(\mathbf{x})\, \delta(\mathbf{x} - \mathbf{X}_l)\, dv(\mathbf{x}), \tag{8.7.5}$$

where $dv(\mathbf{x})$ is the volume associated with the Eulerian grid point. Although the above summation is over all the Eulerian grid points, the value of $\tilde{\mathbf{U}}$ at the Lagrangian marker $\mathbf{X}_l$ is influenced only by those Eulerian grid points in the immediate neighborhood. This is ensured by the following definition of the discrete delta function, whose compact support extends over only a few Eulerian grid points around each Lagrangian marker:

$$\delta(\mathbf{x} - \mathbf{X}_l) = \delta_x(x - x_l)\, \delta_y(y - y_l)\, \delta_z(z - z_l), \tag{8.7.6}$$

where

$$\left.\begin{aligned} \delta_x(x - X_l) &= \frac{1}{h_x}\, \psi\left(\frac{x - X_l}{h_x}\right), \\ \delta_y(y - Y_l) &= \frac{1}{h_y}\, \psi\left(\frac{y - Y_l}{h_y}\right), \\ \delta_z(z - Z_l) &= \frac{1}{h_z}\, \psi\left(\frac{z - Z_l}{h_z}\right). \end{aligned}\right\} \tag{8.7.7}$$

The dilation parameters h_x, h_y and h_z define the characteristic width of the discrete delta function and on a uniform grid these dilation parameters are chosen to be the grid spacings Δx, Δy and Δz, respectively. Following Roma *et al.* (1999), the function ψ is chosen to be

$$\psi(r) = \begin{cases} \frac{1}{6}(5 - 3|r| - \sqrt{-3(1-|r|)^2 + 1}), & 0.5 \le |r| \le 1.5, \\ \frac{1}{3}(1 + \sqrt{-3r^2 + 1}), & |r| \le 0.5, \\ 0, & \text{otherwise}. \end{cases} \tag{8.7.8}$$

The above *discrete delta function* extends over three Eulerian grid cells around the Lagrangian marker and satisfies the following conservation of the zeroth and first moments:

$$\sum_{\text{for all } \mathbf{x}} \delta(\mathbf{x} - \mathbf{X}_l)\, dv(\mathbf{x}) = 1 \quad \text{and} \quad \sum_{\text{for all } \mathbf{x}} (\mathbf{x} - \mathbf{X}_l)\, \delta(\mathbf{x} - \mathbf{X}_l)\, dv(\mathbf{x}) = 0\,. \tag{8.7.9}$$

(3) The Lagrangian force needed to drive the velocity at the Lagrangian marker towards the desired target velocity $\mathbf{V}(\mathbf{X}_l)$ is calculated as

$$\mathbf{F}(\mathbf{X}_l) = \frac{\mathbf{V}(\mathbf{X}_l) - \tilde{\mathbf{U}}(\mathbf{X}_l)}{\Delta t}\,. \tag{8.7.10}$$

When applied at the Lagrangian marker points this force will give the desired velocity.

(4) However, the force must be applied at the Eulerian grid points. Therefore, the next step is to spread the Lagrangian force from the Lagrangian markers to the neighboring Eulerian grid points. This projection step also uses the discrete delta function as

$$\mathbf{f}(\mathbf{x}) = \sum_{l=1}^{N_L} \mathbf{F}(\mathbf{X}_l)\, \delta(\mathbf{x} - \mathbf{X}_l)\, \Delta V_l\,, \tag{8.7.11}$$

where ΔV_l is the weight associated with the Lagrangian marker $\mathbf{X}_l$. For a uniform distribution of Lagrangian markers these weights are a constant equal to the cell volume $\Delta x \Delta y \Delta z$. The conservation properties of the discrete delta function given in (8.7.9) play an important role in ensuring conservation of force and torque upon mapping from the Lagrangian markers to the Eulerian grid points.

(5) Now that the forces to be applied at the Eulerian grid points near the Lagrangian markers are known, we are ready to perform the advection–diffusion step, including the forcing term. This time we use the semi-implicit algorithm, where the RK3 scheme is used for the advection term and the Crank–Nicolson scheme for the diffusion term. The resulting Helmholtz equation for the intermediate velocity can be expressed as

$$\nabla^2 \hat{\mathbf{u}}^{(m)} - \frac{\text{Re}}{c_{dm}} \hat{\mathbf{u}}^{(m)} = -\frac{\text{Re}}{c_{dm}} \Big\{ \mathbf{u}^{(m-1)} + c_{2m}\, \mathbf{J}^{(m-1)} + \frac{c_{dm}}{\text{Re}} \nabla^2 \mathbf{u}^{(m-1)} + \Delta t\, \mathbf{f}^{(m)} \Big\}\,. \tag{8.7.12}$$

The nonlinear term is the same as computed in step (1) and therefore can be reused. The intermediate velocity is not divergence free. So the next two steps are to project the velocity onto the divergence-free space.

(6) In the pressure-correction process, the first step is to solve the following Poisson equation for the pseudo-pressure:

$$\nabla^2 \phi^{(m)} = \frac{1}{2c_{dm}} \nabla \cdot \hat{\mathbf{u}}^{(m)} . \tag{8.7.13}$$

(7) The final pressure-correction step can be carried out as

$$\mathbf{u}^{(m)} = \hat{\mathbf{u}}^{(m)} - 2c_{dm} \nabla \phi^{(m)} . \tag{8.7.14}$$

Also the final pressure can be expressed in terms of the pseudo-pressure as

$$p^{(m)} = p^{(m-1)} + \phi^{(m)} - \frac{c_{dm}}{\mathrm{Re}} \nabla^2 \phi^{(m)} . \tag{8.7.15}$$

Variants of the above immersed boundary method have been advanced over the past decade. One of the challenges with the above implementation is the determination of the Lagrangian weights in the case of non-uniform distribution of Lagrangian markers. Pinelli *et al.* (2010) and Akiki and Balachandar (2016) have presented ways to calculate non-uniform weights based on the distribution of Lagrangian markers. Also, since forcing is being applied to control the interfacial velocity, the resulting flow field will include a nonphysical velocity distribution in the region covered by the solid. Goldstein *et al.* (1993) used this solid phase velocity to smoothen the discontinuity in velocity derivative that might otherwise arise at the fluid–solid interface. However, in their application Saiki and Biringen (1996) observed the above approach sometimes leads to an overall nonphysical solution. This erroneous behavior was rectified by imposing an extended forcing that will enforce the solid–phase velocity not only at the interface, but also inside the solid region. This difficulty is greatly mitigated in other applications, such as at a liquid–liquid interface, when both the phases undergo deformation. Tennetti *et al.* (2011) proposed the particle-resolved uncontaminated fluid reconcilable immersed boundary method (PUReIBM), whose advantages are that the Lagrangian weights are unambiguously determined and also it does not apply the forcing in the fluid phase.

8.7.2 Feedback Forcing and Other Immersed Boundary Methods

Goldstein *et al.* (1993) proposed a forcing of the following form for enforcing a Dirichlet velocity along the interface S:

$$\mathbf{f}(\mathbf{X}_l, t) = \alpha \int_0^t [\hat{\mathbf{u}}(\mathbf{X}_l, t') - \mathbf{V}(\mathbf{X}_l, t')] \, \mathrm{d}t' + \beta \, [\hat{\mathbf{u}}(\mathbf{X}_l, t) - \mathbf{V}(\mathbf{X}_l, t)] . \tag{8.7.16}$$

Here $\mathbf{V}$ is the velocity boundary condition to be satisfied at the point $(\mathbf{X}_l)$ on the fluid–solid interface. The above forcing attempts to drive the intermediate velocity $\hat{\mathbf{u}}$ towards the target $\mathbf{V}$ as a damped oscillator. The constants α and β determine the frequency and damping of the oscillator as $(1/2\pi)\sqrt{|\alpha|}$ and

$-\beta/(2\sqrt{|\alpha|})$. We desire the frequency of damping to be much larger than all relevant inverse time-scales of fluid motion and the damping coefficient to be adequately large, in order for the velocity condition at the interface to be satisfied rapidly. This requires the constants α and β to be chosen as large positive constants.

Goldstein *et al.* applied the feedback forcing in the context of a pseudo-spectral numerical methodology and difficulties with localizing the forcing within the global nature of pseudo-spectral methodology were addressed. Saiki and Biringen (1996) showed that the feedback forcing is far more effective in the context of finite difference schemes. The feedback force evaluated with the above equation replaces the step (3) of the direct forcing method described in the previous section. The rest of the steps can be the same as those of the direct forcing methodology.

One of the primary difficulties with the feedback forcing has been the severe restriction it poses on the computational time step. For an explicit treatment of forcing using the Adams–Bashforth scheme, the time-step limitation has been shown to be (Goldstein *et al.*, 1993)

$$\Delta t \leq \frac{-\beta - \sqrt{(\beta^2 - 2\alpha k)}}{\alpha}, \tag{8.7.17}$$

where k is an $O(1)$ flow-dependent constant. The above time-step limitation generally results in time-steps that are more than an order of magnitude smaller than that required by the CFL condition. Such very small values for the time-step arises from the large values of α and β chosen in (8.7.16) and the associated very small time-scale of the oscillator. It was shown in Fadlun *et al.* (2000) that if an implicit scheme is used for the forcing, instead of the explicit scheme, considerably larger time steps can be employed. Nevertheless, even with the implicit treatment, the time step that satisfies stability requirements is still small compared to CFL condition. One of the primary advantages of direct forcing over feedback forcing is that the time-step limitation is not so severe. In fact, it has been demonstrated that with direct forcing, time-steps as large as the CFL limit can be taken without any numerical instability.

Kim *et al.* (2001) pointed out that the direct forcing employed by Fadlun *et al.* (2000) applies the forcing inside the fluid, i.e., even outside the solid region. Kim *et al.* proposed a modification to the direct forcing strategy that strictly applies the momentum forcing only on the fluid–solid interface and inside the solid region. Thus the Navier–Stokes equations are faithfully solved in the entire region occupied by the fluid. Their reformulation, however, requires that a mass source be also included along the interface. In other words, the continuity equation is modified to

$$\nabla \cdot \mathbf{u} = q, \tag{8.7.18}$$

where q is strictly zero inside the fluid, but can take non-zero values along the interface and possibly in the solid region (where the velocity field is of no interest). In this approach discrete-time mass forcing is used in addition to momentum forcing in order to enforce the interface velocity.

8.7.3 Sample Applications

There is now a very vast body of literature on the application of IBM for solving the Navier–Stokes equations using direct forcing and other methods. Two simple examples will be presented here. In the first, a sphere, located close to a flat wall, is subjected to a linear shear flow – see Figure 8.11(a) – and the flow around the sphere is simulated with IBM (Lee *et al.*, 2011; Lee and Balachandar, 2010). The spherical particle is of diameter $\tilde{d}$, rotates at an angular velocity $\tilde{\Omega}$ about the z-axis and translates at a velocity $\tilde{V}_{\mathrm{p}}$ parallel to the flat plate along the flow direction (x-axis). The sphere's vertical location is fixed at a distance $\tilde{L}$ from the bottom wall and when $\tilde{L} = \tilde{d}/2$ the sphere is rolling/sliding on the wall. Note tilde indicates dimensional quantities. The ambient flow approaching the particle is a linear shear flow, whose only non-zero velocity component is along the x-axis parallel to the wall. The shear flow can be written as $\tilde{U} = \tilde{G}(\tilde{y} + \tilde{L})$, where $\tilde{G}$ is the dimensional shear-rate of the ambient flow. For a stationary particle in a shear flow, $\tilde{G} \neq 0$, $\tilde{V}_{\mathrm{p}} = 0$, and for a translating particle in a quiescent medium, $\tilde{G} = 0$, $\tilde{V}_{\mathrm{p}} \neq 0$.

They chose the particle diameter to be the length scale and the translational velocity of the particle to be the velocity scale. The time and pressure scales are defined accordingly. This results in $\mathrm{Re}_{\mathrm{t}} = \tilde{d}\tilde{V}_{\mathrm{p}}/\tilde{\nu}$ as the important non-dimensional parameter. They employed the direct forcing IBM technique of Uhlmann (2005) to impose the appropriate velocity boundary conditions on the surface of the translating/rotating spherical particle, as it moves through a fixed Cartesian grid. The problem is solved in a frame of reference fixed on the translating sphere and the non-dimensional governing equations are those presented in Section 8.5.2. In this moving frame of reference the computational grid around the particle becomes time independent, while the ambient shear flow becomes $\tilde{G}(\tilde{y} + \tilde{L}) - \tilde{V}_{\mathrm{p}}$.

A regular Cartesian grid is used to simulate the flow and the projection of the Cartesian grid on the boundaries of the rectangular domain within which the Navier–Stokes equations are solved is shown in Figure 8.11(b). The Eulerian grid points are clustered close to the wall along the wall-normal direction, and clustered around the sphere along the spanwise and streamwise directions. The clustering of points is important in order to achieve a high degree of resolution around the sphere. In the immersed boundary region that embeds the sphere a uniform Cartesian mesh of equal resolution along all three directions is used ($\Delta x = \Delta y = \Delta z$). This high-resolution region extends over $[-L, 1] \times [-1, 1]$ along the wall-normal and spanwise directions. In the frame centered on the sphere the region of high resolution is limited along the streamwise direction over the domain $[-1, 1]$. Outside this region of uniform grid, a geometric grid stretching is used. Each successive grid spacing is progressively increased by a multiplicative factor.

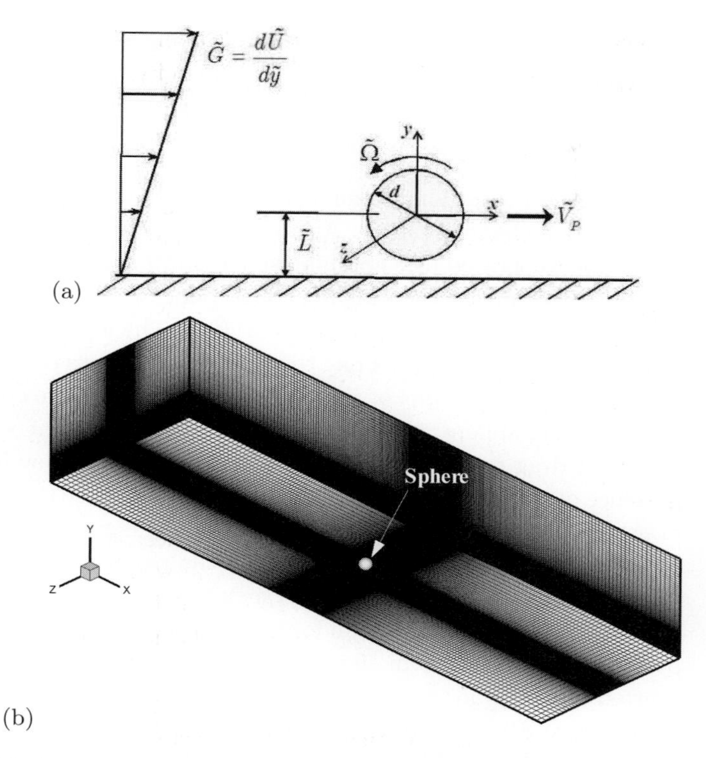

Figure 8.11 (a) Schematic of a wall-bounded linear shear flow over a sphere located at a distance $\tilde{L}$ from the flat wall. The sphere translates at velocity $\tilde{V}_{\mathrm{p}}$ parallel to the wall and rotates at angular velocity $\tilde{\Omega}$. (b) The non-uniform Cartesian grid used for solving the Navier–Stokes equations. A uniform fine grid is used around the sphere along the streamwise, spanwise and wall-normal directions. Away from the high-resolution region a stretched grid is used. Taken from Lee *et al.* (2011), with permission.

The overall grid resolution used in the simulations was $371 \times 301 \times 151$ points along the streamwise, wall-normal and spanwise directions. The computational domain extends over $[-25, 25] \times [-L, 8] \times [-7, 7]$ non-dimensional units along the streamwise, wall-normal and spanwise directions. The size of the domain and the grid resolution was sufficient to obtain grid- or domain-independent converged solutions (Zeng *et al.*, 2009). The simulations typically involved 40 grid points resolving one particle diameter in the high-resolution region. Correspondingly the number of Lagrangian markers was chosen to be of the order of $(40)^2$. The Lagrangian markers were uniformly distributed on the surface of the sphere using the algorithm of Saff and Kuijlaars (1997).

The above simulation framework has been used to study a variety of problems:

(i) wall-bounded linear shear flow over a stationary sphere;

(ii) a spherical particle translating parallel to a wall in otherwise quiescent flow;

(iii) a spinning spherical particle located close to a wall in otherwise quiescent flow;
(iv) a particle rolling or sliding on a flat wall in response to an ambient wall-bounded linear shear flow.

For details on these simulations and their results refer to Lee *et al.* (2011) and Lee and Balachandar (2010). As an example, Figure 8.12 shows the streamlines of the flow generated by a spinning sphere located close to a flat wall. The axis of rotation is parallel to the wall and counterclockwise. The Reynolds number based on particle angular velocity is $\mathrm{Re}_\Omega = \tilde{\Omega}\tilde{d}^2/\tilde{\nu}$. The left-hand frames correspond to a Reynolds number of 10, while the right-hand frames correspond to $\mathrm{Re}_\Omega = 100$. Three different locations of the sphere from the wall are considered.

A spinning sphere in an unbounded fluid will experience only a torque and not a net drag or lift force. The axisymmetry of the flow field is broken by the presence of the wall. Therefore, a spinning sphere close to a wall will experience a drag (parallel to the wall) and a lift (perpendicular to the wall) force. With IBM methodology the total force exerted by the particle on the surrounding fluid is given by the Lagrangian force $\mathbf{F}(\mathbf{X}_l)$ summed over all the Lagrangian markers, or equivalently by the Eulerian force field $\mathbf{f}(\mathbf{x})$ summed over all the Eulerian grid points. The hydrodynamic force on the particle is just the opposite of this force exerted on the fluid.

In Figure 8.13 we present the drag and lift forces on a sphere located very close to a flat wall. The gap between the bottom of the sphere and the wall is only 0.5% of the sphere diameter. Thus, the sphere is practically sitting on the wall. Three different cases are considered:

(i) a stationary sphere subjected to a uniform linear shear flow;
(ii) a sphere translating parallel to the wall at a constant velocity in an otherwise quiescent fluid;
(iii) a sphere spinning at a constant angular velocity in an otherwise quiescent fluid.

The relevant parameters in the latter two cases have been already defined to be Re_t and Re_Ω. For the case of linear shear the appropriate shear Reynolds number is $\mathrm{Re}_\mathrm{s} = \tilde{G}\tilde{L}\tilde{d}/\tilde{\nu}$. The drag and lift coefficients for the three different cases are defined as

$$\left.\begin{aligned}
C_{\mathrm{Ds}} &= \frac{\tilde{F}_x}{\frac{\pi}{8}\tilde{\rho}\tilde{G}|\tilde{G}|\tilde{L}^2\tilde{d}^2} &\quad \text{and} \quad C_{\mathrm{Ls}} &= \frac{\tilde{F}_y}{\frac{\pi}{8}\tilde{\rho}\tilde{G}^2\tilde{L}^2\tilde{d}^2} \\
C_{\mathrm{Dt}} &= \frac{\tilde{F}_x}{\frac{\pi}{8}\tilde{\rho}\tilde{V}_\mathrm{p}|\tilde{V}_\mathrm{p}|\tilde{d}^2} &\quad \text{and} \quad C_{\mathrm{Lt}} &= \frac{\tilde{F}_y}{\frac{\pi}{8}\tilde{\rho}\tilde{V}_\mathrm{p}^2\tilde{d}^2} \\
C_{\mathrm{D}\Omega} &= -\frac{\tilde{F}_x}{\frac{\pi}{32}\tilde{\rho}\tilde{\Omega}|\tilde{\Omega}|\tilde{d}^4} &\quad \text{and} \quad C_{\mathrm{L}\Omega} &= \frac{\tilde{F}_y}{\frac{\pi}{32}\tilde{\rho}\tilde{\Omega}^2\tilde{d}^4}\,,
\end{aligned}\right\} \tag{8.7.19}$$

where $\tilde{F}_x$ and $\tilde{F}_y$ are the dimensional streamwise and wall-normal forces on the particle. From the figure it is clear that the drag and lift forces due to sphere

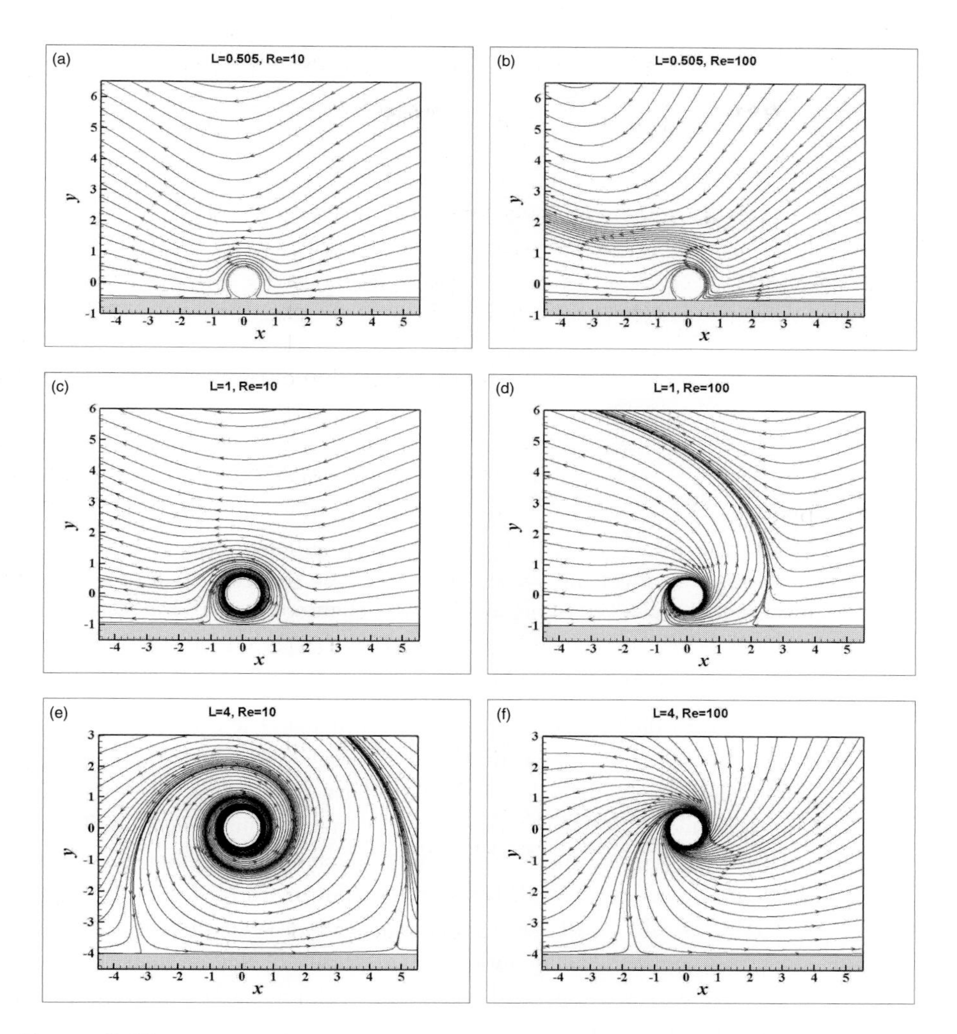

Figure 8.12 Streamlines of the flow generated by a spinning sphere located close to a flat wall. The axis of rotation is parallel to the wall and counterclockwise. (a) $L = 0.505$, $\mathrm{Re}_\Omega = 10$; (b) $L = 0.505$, $\mathrm{Re}_\Omega = 100$; (c) $L = 1$, $\mathrm{Re}_\Omega = 10$; (d) $L = 1$, $\mathrm{Re}_\Omega = 100$; (e) $L = 4$, $\mathrm{Re}_\Omega = 10$; (f) $L = 4$, $\mathrm{Re}_\Omega = 100$; Taken from Lee *et al.* (2011), with permission.

rotation are smaller than those from shear and translation. At low to moderate Reynolds numbers translational drag is higher, whereas at larger Reynolds numbers the drag coefficient from translation and shear are about equal. In all cases the effect of lift force is to push the particle away from the wall. Clearly, shear-induced lift force is much stronger than the other two contributions.

You may consider the problem of a single sphere in a rectangular box to be geometrically simple enough that a Navier–Stokes simulation with a body-fitted grid must be possible. Indeed such simulations have been performed. For example, Zeng *et al.* (2009) considered the same problem of a linear shear flow

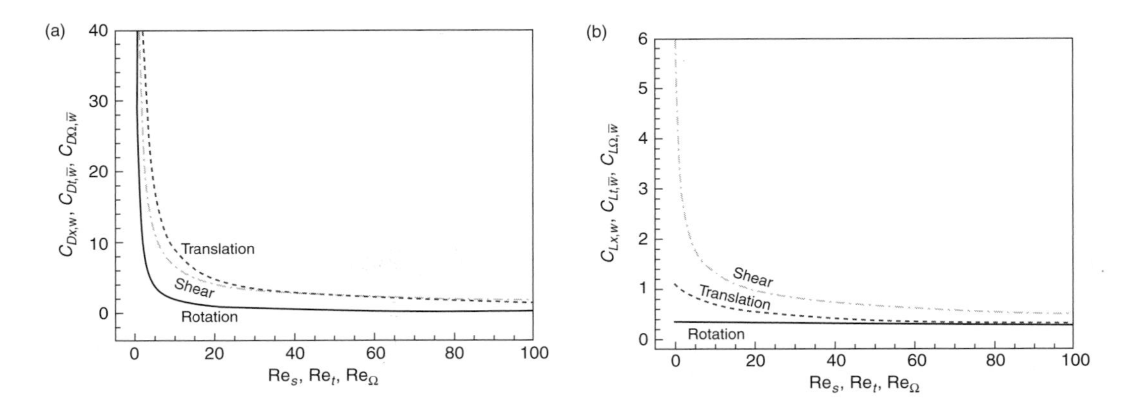

Figure 8.13 Drag and lift coefficients plotted as a function of Reynolds number for sphere located very close to a flat wall. The gap between the bottom of the sphere and the wall is only 0.5% of the sphere diameter. Three different cases are considered: (i) a stationary sphere subjected to a uniform linear shear flow; (ii) a sphere translating parallel to the wall at a constant velocity in an otherwise quiescent fluid; (iii) a sphere spinning at a constant angular velocity in an otherwise quiescent fluid. Taken from Lee *et al.* (2011) with permission.

past a stationary sphere near a flat wall using a body-fitted spectral element grid. The body-fitted grid provided a precise resolution of the spherical particle, while the finite width of the discrete delta function in IBM necessarily diffused the interface between the particle and the surrounding fluid. Furthermore, the spectral element methodology provided spectral or higher-order accuracy, while the formal accuracy of IBM is second order or lower. Lee *et al.* (2011) compared the IBM results with those of spectral element simulation and demonstrated that drag and lift coefficients can be reliably obtained with IBM with the same level of accuracy as spectral element methodology. However, to obtain the same accuracy, a much higher overall resolution is required with IBM.

This comparison also demonstrates the need for *benchmark* solutions in CFD against which new computational methods and codes can be compared to establish their performance and order of accuracy. The results that will serve as the benchmark solution must establish a very high standard and demonstrate higher order accuracy and complete convergence. In other words, the benchmark must serve as a "numerically obtained analytic solution."

From such comparison the resolution requirement of IBM has been established for this class of problems involving flows over spherical particles. The real power of IBM is realized in much more geometrically complex problems. For example, consider the case of a spherical particle sitting on a rough wall made of hemispherical bumps, subjected to a linear shear flow (see Figure 8.14(a)). Another example is the problem of turbulence generated by the gravitational settling of a random array of finite-sized particles in a periodic channel considered by Uhlmann and Doychev (2014). A snapshot of the particle distribution

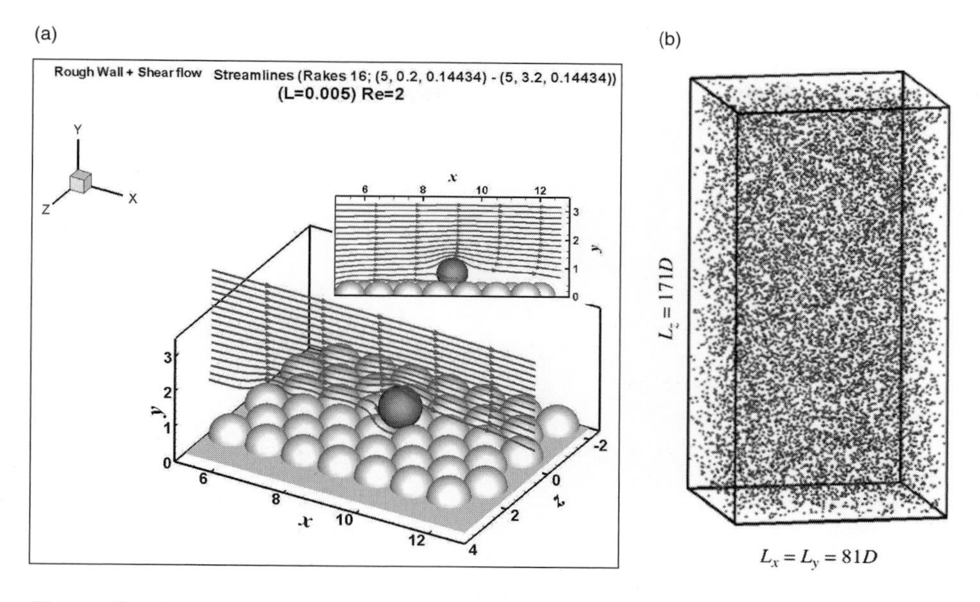

Figure 8.14 Two complex examples that demonstrate the power of IBM. (a) Flow around a spherical particle sitting on a rough wall made of hemispherical bumps, subjected to a linear shear flow (taken from Lee and Balachandar, 2017). The shear Reynolds number of the flow is $\mathrm{Re}_s = 2$. (b) Snapshot of a random array of 11867 spheres gravitationally settling in a periodic channel considered by Uhlmann and Doychev (2014).

at one time instant is shown in Figure 8.14(b). This computationally demanding simulation employed a uniform Eulerian grid of $2048 \times 2048 \times 4096$ points and immersed within the flow are 11867 spheres, each of which was resolved with $O(24)^2$ Lagrangian markers. Such simulations are clearly possible only with IBM methodology. In this case of falling particles, the relative position of the particles is ever changing and generating a body-fitted grid in an efficient manner is out of current ability.

References

Abramowitz, M. and Stegun, I. (eds.) (1965). *Handbook of Mathematical Functions.* Dover Publications.

Akiki, G. and Balachandar, S. (2016). Immersed boundary method with non-uniform distribution of Lagrangian markers for a non-uniform Eulerian mesh. *J. Comp. Phys.* **307**, 34–59.

Almgren, A.S., Bell, J.B., Colella, P. and Marthaler, T. (1997). A Cartesian grid projection method for the incompressible Euler equations in complex geometries. *SIAM J. Sci. Comput.* **18**, 1289.

Anderson, S.L. (1990). Random number generation on vector supercomputers and other advanced architectures. *SIAM Rev.* **32**, 221–51.

Aref, H. (1979). Motion of three vortices. *Phys. Fluids* **22**, 393–400.

Aref, H. (1982). Point vortex motions with a center of symmetry. *Phys. Fluids* **25**, 2183–7.

Aref, H. (1983). Integrable, chaotic and turbulent vortex motion in two-dimensional flows. *Ann. Rev. Fluid Mech.* **15**, 345–89.

Aref, H. (1984). Stirring by chaotic advection. *J. Fluid Mech.* **143**, 1–21.

Aref, H. (2002). The development of chaotic advection. *Phys. Fluids* **14**, 1315–25.

Aref, H. and Balachandar, S. (1986). Chaotic advection in a Stokes flow. *Phys. Fluids* **29**, 3515–21.

Aref, H. and Jones, S.W. (1993). Chaotic motion of a solid through ideal fluid. *Phys. Fluids* **5**, 3026–8.

Aref, H., Jones, S.W., Mofina, S. and Zawadzki, I. (1989). Vortices, kinematics and chaos. *Physica D* **37**, 423–40.

Aref, H., Rott, N. and Thomann, H. (1992). Gröbli's solution of the three-vortex problem. *Ann. Rev. Fluid Mech.* **24**, 1–20.

Arnold, V.I. (1978). *Mathematical Methods of Classical Mechanics.* Springer.

Arnold, V.I. and Avez, A. (1968). *Ergodic Problems of Classical Mechanics.* W.A. Benjamin, Inc.

Augenbaum, J.M. (1989). An adaptive pseudospectral method for discontinuous problems. *Appl. Numer. Math.* **5**, 459–80.

Axelsson, O. (1996). *Iterative Solution Methods.* Cambridge University Press.

Bagchi, P. and Balachandar, S. (2002a). Shear versus vortex-induced lift force on a rigid sphere at moderate Re. *J. Fluid Mech.* **473**, 379–88.

Bagchi, P. and Balachandar, S. (2002b). Effect of free rotation on the motion of a solid sphere in linear shear flow at moderate Re. *Phys. Fluids* **14**, 2719–37.

Bagchi, P. and Balachandar, S. (2002c). Steady planar straining flow past a rigid sphere at moderate Reynolds number. *J. Fluid Mech.* **466**, 365–407.

Bagchi, P. and Balachandar, S. (2003). Effect of turbulence on the drag and lift of a particle. *Phys. Fluids* **15**(11), 3496–513.

Bagchi, P. and Balachandar, S. (2004). Response of the wake of an isolated particle to an isotropic turbulent flow. *J. Fluid Mech.* **518**, 95–123.

Bai-Lin, H. (1984). *Chaos.* World Scientific Publishing Co.

Balachandar, S. and Eaton, J.K. (2010). Turbulent dispersed multiphase flow. *Ann. Rev. Fluid Mech.* **42**, 111–33.

Balachandar, S. and Maxey, M.R. (1989). Methods for evaluating fluid velocities in spectral simulations of turbulence. *J. Comp. Phys.* **83**(1), 96–125.

Ballal, B.Y. and Rivlin, R.S. (1976). Flow of a Newtonian fluid between eccentric rotating cylinders: inertial effects. *Arch. Rational Mech. Anal.* **62**(3), 237–94.

Barenblatt, G.I. (1996). *Scaling, Self-similarity, and Intermediate Asymptotics: Dimensional Analysis and Intermediate Asymptotics.* Cambridge University Press.

Bashforth, F. and Adams, J.C. (1883). *An Attempt to Test the Theories of Capillary Action: By Comparing the Theoretical and Measured Forms of Drops of Fluid.* University Press.

Basset, A.B. (1888). *A Treatise on Hydrodynamics.* Deighton, Bell and Company.

Batchelor, G.K. (1967). *An Introduction to Fluid Dynamics.* Cambridge University Press.

Ben-Jacob, E. and Garik, P. (1990). The formation of patterns in non-equilibrium growth. *Nature* **343**, 523–30.

Berry, M.V., Percival, I.C. and Weiss, N.O. (1987). *Dynamical Chaos.* The Royal Society, London. (First published as *Proc. R. Soc. London A* **413**, 1–199.)

Birkhoff, G. and Fisher, J. (1959). Do vortex sheets roll up? *Rend. Circ. Mat. Palermo* **8**, 77–90.

Boris, J.P. (1989). New directions in computational fluid dynamics. *Ann. Rev. Fluid Mech.* **21**, 345–85.

Boussinesq, J. (1885). Sur la résistance qu'oppose un liquide indéfini au repos au mouvement varié d'une sphére solide. *C. R. Acad. Sci. Paris* **100**, 935–7.

Boyd, J.P. (2001). *Chebyshev and Fourier Spectral Methods.* Courier Dover Publications.

Brachet, M.E., Meiron, D.I., Orszag, S.A., Nickel, B.G., Morf, R.H. and Frisch, U. (1983). Small scale structure of the Taylor–Green vortex. *J. Fluid Mech.* **130**, 411–52.

Brown, D.L., Cortez, R. and Minion, M.L. (2001). Accurate projection methods for the incompressible Navier–Stokes equations. *J. Comput. Phys.* **168**(2), 464–99.

Brucato, A., Grisafi, F. and Montante, G. (1998). Particle drag coefficients in turbulent fluids. *Chem. Eng. Sci.* **53**, 3295–314.

Calhoun, D. and Leveque, R.J. (2000). A Cartesian grid finite-volume method for the advection–diffusion equation in irregular geometries. *J. Comput. Phys.* **157**, 143–80.

Canuto, C., Hussaini, M.Y., Quarteroni, A. and Zang, T.A. (2006). *Spectral Methods in Fluid Dynamics.* Springer.

Carnahan, B., Luther, H.A. and Wilkes, J.O. (1969). *Applied Numerical Methods.* Wiley.

Chaiken, J., Chevray, R., Tabor, M. and Tan, Q.M. (1986). Experimental study of Lagrangian turbulence in a Stokes flow. *Proc. R. Soc. London A* **408**, 165–74.

Chaiken, J., Chu, C.K., Tabor, M. and Tan, Q.M. (1987). Lagrangian turbulence and spatial complexity in a Stokes flow. *Phys. Fluids* **30**, 687–94.

Chandrasekhar, S. (1961). *Hydrodynamic and Hydromagnetic Stability.* Oxford University Press.

Chang, E.J. and Maxey, M.R. (1994). Unsteady flow about a sphere at low to moderate Reynolds number. Part 1: Oscillatory motion. *J. Fluid Mech.* **277**, 347–79.

Chapra, S.C. (2002). *Numerical Methods for Engineers*, 4th edn. McGraw–Hill.

Chorin, A.J. (1968). Numerical solution of the Navier–Stokes equations. *Math. Comput.* **22**, 745–62.

Chorin, A.J. (1976). Random choice solution of hyperbolic systems. *J. Comp. Phys.* **22**(4), 517–33.

Chung, T.J. (2010). *Computational Fluid Dynamics.* Cambridge University Press.

Clift, R., Grace, J.R. and Weber, M.E. (1978). *Bubbles, Drops and Particles.* Academic Press.

Cochran, W.G. (1934). The flow due to a rotating disk. *Proc. Camb. Phil. Soc.* **30**, 365–75.

Collatz, L. (1960). *The Numerical Treatment of Differential Equations.* Springer.

Cossu, C. and Loiseleux, T. (1998). On the convective and absolute nature of instabilities in finite difference numerical simulations of open flows. *J. Comp. Phys.* **144**(1), 98–108.

Criminale, W.O., Jackson, T.L. and Joslin, R.D. (2003). *Theory and Computation of Hydrodynamic Stability.* Cambridge University Press.

Crutchfield, J.P., Farmer, J.D., Packard, N.H. and Shaw, R.S. (1986). Chaos. *Sci. Amer.* **255**, 46–57.

Curle, N. (1957). On hydrodynamic stability in unlimited fields of viscous flow. *Proc. Royal Soc. London A* **238**, 489–501.

Curry, J.H., Garnett, L. and Sullivan, D. (1983). On the iteration of a rational function: computer experiments with Newton's method. *Comm. Math. Phys.* **91**, 267–77.

Curry, J.H., Herring, J.R., Loncaric, J. and Orszag, S.A. (1984). Order and disorder in two- and three-dimensional Benard convection. *J. Fluid Mech.* **147**, 1–38.

Dandy, D.S. and Dwyer, H.A. (1990). A sphere in shear flow at finite Reynolds number: effect of shear on particle lift, drag and heat transfer. *J. Fluid Mech.* **216**, 381–410.

Devaney, R.L. (1989). *An Introduction to Chaotic Dynamical Systems*, 2nd edn. Addison–Wesley.

Deville, M.O., Fischer, P.F. and Mund, E.H. (2002). *High-order Methods for Incompressible Fluid Flow.* Cambridge University Press.

Dombre, T., Frisch, U., Greene, J.M., Hénon, M., Mehr, A. and Soward A.M. (1986). Chaotic streamlines and Lagrangian turbulence: the ABC flows. *J. Fluid Mech.* **167**, 353–91.

Dongarra, J.J., Croz, J.D., Hammarling, S. and Duff, I.S. (1990). A set of level 3 basic linear algebra subprograms. *ACM Trans. Math. Software (TOMS)* **16**(1), 1–17.

Donnelly, R.J. (1991). Taylor–Couette flow: the early days. *Phys. Today*, November, 32–9.

Drazin, P.G. and Reid, W.H. (2004). *Hydrodynamic Stability*, 2nd edn. Cambridge University Press.

Eckhardt, B. and Aref, H. (1988). Integrable and chaotic motions of four vortices II. Collision dynamics of vortex pairs. *Phil. Trans. Royal Soc. London A* **326**, 655–96.

Edelsbrunner, H. (2001). *Geometry and Topology for Mesh Generation.* Cambridge University Press.

Eiseman, P.R. (1985). Grid generation for fluid mechanics computations. *Ann. Rev. Fluid Mech.* **17**, 4875–22.

Emmons, H.W. (1970). Critique of numerical modelling of fluid mechanics phenomena. *Ann. Rev. Fluid Mech.* **2**, 15–37.

Eswaran, V. and Pope, S.B. (1988). An examination of forcing in direct numerical simulations of turbulence. *Comp. Fluid.* **16**(3), 257–78.

Evans, G., Blackledge, J. and Yardley, P. (2012). *Numerical Methods for Partial Differential Equations.* Springer.

Fadlun, E.A., Verzicco, R., Orlandi, P. and Mohd-Yusof, J. (2000). Combined immersed-boundary finite-difference methods for three-dimensional complex flow simulations. *J. Comput. Phys.* **161**, 35–60.

Fatou, P. (1919). Sur les equations fontionelles. *Bull. Soc. Math. France* **47**, 161–271; also **48**, 33–94; 208–314.

Feigenbaum, M.J. (1978). Quantitative universality for a class of non-linear transformations. *J. Stat. Phys.* **19**, 25–52.

Feigenbaum, M.J. (1980). The metric universal properties of period doubling bifurcations and the spectrum for a route to turbulence. *Ann. N.Y. Acad. Sci.* **357**, 330–6.

Ferziger, J.H. and Peric, M. (2012). *Computational Methods for Fluid Dynamics.* Springer.

Fink, P.T. and Soh, W.K. (1978). A new approach to roll-up calculations of vortex sheets. *Proc. Royal Soc. London A* **362**, 195–209.

Finlayson, B.A. and Scriven, L.E. (1966). The method of weighted residuals: a review. *Appl. Mech. Rev.* **19**(9), 735–48.

Finn, R. (1986). *Equilibrium Capillary Surfaces.* Springer.

Fischer, P.F., Leaf, G.K. and Restrepo, J.M. (2002). Forces on particles in oscillatory boundary layers. *J. Fluid Mech.* **468**, 327–47.

Fjørtoft, R. (1950). *Application of Integral Theorems in Deriving Criteria of Stability for Laminar Flows and for the Baroclinic Circular Vortex.* Geofysiske Publikasjoner, Norske Videnskaps-Akademi i Oslo.

Fletcher, C.A.J. (1991). *Computational Techniques for Fluid Dynamics*, Vol I and II. Springer.

Fornberg, B. (1988). Steady viscous flow past a sphere at high Reynolds numbers. *J. Fluid Mech.* **190**, 471.

Fornberg, B. (1998). *A Practical Guide to Pseudospectral Methods.* Cambridge University Press.

Fox, R.O. (2012). Large-eddy-simulation tools for multiphase flows. *Ann. Rev. Fluid Mech.* **44**, 47–76.

Franceschini, V. and Tebaldi, C. (1979). Sequences of infinite bifurcations and turbulence in a five-mode truncation of the Navier–Stokes equations. *J. Stat. Phys.* **21**(6), 707–26.

Funaro, D. (1992). *Polynomial Approximation of Differential Equations.* Springer.

Gatignol, R. (1983). The Faxén formulas for a rigid particle in an unsteady non-uniform Stokes-flow. *J. Mécanique Théorique Appliquée* **2**(2), 143–60.

Gear, C.W. (1971). *Numerical Initial Value Problems in Ordinary Differential Equations.* Prentice–Hall.

Ghia, U.K.N.G., Ghia, K.N. and Shin, C.T. (1982). High-Re solutions for incompressible flow using the Navier–Stokes equations and a multigrid method. *J. Comp. Phys.* **48**(3), 387–411.

Glendinning, P. (1994). *Stability, Instability and Chaos: an Introduction to the Theory of Nonlinear Differential Equations.* Cambridge University Press.

Glowinski, R. and Pironneau, O. (1992). Finite element methods for Navier–Stokes equations. *Ann. Rev. Fluid Mech.* **24**, 167–204.

Godunov, S.K. (1959). A finite difference method for the numerical computation of discontinuous solutions of the equations of fluid dynamics. *Math. Sbornik* **47**, 271–306. Translated as US Joint Publ. Res. Service, JPRS 7226 (1969).

Goldstein, D., Handler, R. and Sirovich, L. (1993). Modeling a no-slip flow boundary with an external force field. *J. Comput. Phys.* **105**, 354–66.

Goldstine, H.H. (1972). *The Computer from Pascal to Von Neumann.* Princeton University Press.

Gollub, J.P. and Benson, S.V. (1980). Many routes to turbulent convection. *J. Fluid Mech.* **100**, 449–70.

Golub, G.H. (ed.) (1984). *Studies in Numerical Analysis.* The Mathematical Association of America.

Golub, G.H. and Kautsky, J. (1983). Calculation of Gauss quadratures with multiple free and fixed knots. *Numerische Mathematik* **41**(2), 147–63.

Golub, G.H. and van Loan, C.F. (2012). *Matrix Computations.* Johns Hopkins University Press.

Gore, R.A. and Crowe, C.T. (1990). Discussion of particle drag in a dilute turbulent two-phase suspension flow. *Int. J. Multiphase Flow* **16**, 359–61.

Gottlieb, D. and Orszag, S.A. (1983). *Numerical Analysis of Spectral Methods: Theory and Applications.* SIAM.

Guermond, J.L., Minev, P. and Shen, J. (2006). An overview of projection methods for incompressible flows. *Comp. Meth. Appl. Mech. Eng.* **195**(44), 6011–45.

Gustafsson, B., Kreiss, H.-O. and Sundstrøm, A. (1972). Stability theory of difference approximations for mixed initial boundary value problems, II. *Math. Comput.* **26**, 649–86.

Hamming, R.W. (1973). *Numerical Methods for Scientists: No-slip Engineers*, 2nd edn. McGraw–Hill. (Republished by Dover Publications, 1986.)

Harrison, W.J. (1908). The influence of viscosity on the oscillations of superposed fluids. *Proc. London Math. Soc.* **6**, 396–405.

Hartland, S. and Hartley, R.W. (1976). *Axisymmetric Fluid–Liquid Interfaces.* Elsevier Scientific.

Hartree, D.R. (1937). On an equation occurring in Falkner and Skan's approximate treatment of the equations of the boundary layer. *Proc. Camb. Phil. Soc.* **33**, 223–39.

Hénon, M. (1969). Numerical study of quadratic area-preserving mappings. *Q. Appl. Math.* **27**, 291–312.

Hénon, M. (1982). On the numerical computation of Poincaré maps. *Physica D* **5**, 412–4.

Higdon, J.J.L. (1985). Stokes flow in arbitrary two-dimensional domains: shear flow over ridges and cavities. *J. Fluid Mech.* **159**, 195–226.

Hildebrand, F.B. (1974). *Introduction to Numerical Analysis*, 2nd edn. McGraw–Hill. (Republished by Dover Publications).

Hirsch, C. (2007). *Numerical Computation of Internal and External Flows: the Fundamentals of Computational Fluid Dynamics*, 2nd edn. Butterworth–Heinemann.

Hirsch, M.W., Smale, S. and Devaney, R.L. (2004). *Differential Equations, Dynamical Systems, and an Introduction to Chaos.* Academic Press.

Holt, M. (1977). *Numerical Methods in Fluid Dynamics.* Springer.

Howard, L.N. (1958). Hydrodynamic stability of a jet. *J. Math. Phys.* **37**(1), 283–98.

Huntley, H.E. (1967). *Dimensional Analysis.* Dover Publications.

Johnson, T.A. and Patel, V.C. (1999). Flow past a sphere up to a Reynolds number of 300. *J. Fluid Mech.* **378**, 19–70.

Julia, G. (1918). Memoire sur l'iteration des fonctions rationelles. *J. Math. Pures Appl.* **4**, 47–245.

Kac, M. (1938). Sur les fonctions $2^n t - [2^n t] - \frac{1}{2}$. *J. London Math. Soc.* **13**, 131–4.

Kamath, V. and Prosperetti, A. (1989). Numerical integration methods in gas–bubble dynamics. *J. Acoust. Soc. Amer.* **85**, 1538–48.

Karniadakis, G.E. and Sherwin, S. (2013). *Spectral/hp Element Methods for Computational Fluid Dynamics.* Oxford University Press.

Karniadakis, G.E. and Triantafyllou, G.E. (1992). Three-dimensional dynamics and transition to turbulence in the wake of bluff objects. *J. Fluid Mech.* **238**, 1–30.

Karniadakis, G.E., Israeli, M. and Orszag, S.A. (1991). High-order splitting methods for the incompressible Navier–Stokes equations. *J. Comput. Phys.* **97**, 414–43.

Keller, H.B. (1968). *Numerical Methods for Two-point Boundary-value Problems.* Dover Publications.

Kida, S. and Takaoka, M. (1987). Bridging in vortex reconnection. *Phys. Fluids* **30**(10), 2911–4.

Kim, D. and Choi, H. (2002). Laminar flow past a sphere rotating in the streamwise direction. *J. Fluid Mech.* **461**, 365–86.

Kim, J. and Moin, P. (1985). Application of a fractional-step method to incompressible Navier–Stokes equations. *J. Comp. Phys.* **59**, 308–23.

Kim, I., Elghobashi, S. and Sirignano, W.A. (1998). Three-dimensional flow over 3-spheres placed side by side. *J. Fluid Mech.* **246**, 465–88.

Kim, J., Kim, D. and Choi, H. (2001). An immersed-boundary finite-volume method for simulations of flow in complex geometries. *J. Comput. Phys.* **171**, 132–50.

Kirchhoff, G.R. (1876). *Vorlesungen Uber Mathematische Physik*, Vol. 1. Teubner.

Knapp, R.T., Daily, J.W. and Hammitt, F.G. (1979). *Cavitation.* Institute for Hydraulic Research.

Knuth, D.E. (1981). *The Art of Computer Programming*, 2nd edn., Vol. 2. Addison–Wesley.

Kopal, Z. (1961). *Numerical Analysis.* Chapman and Hall.

Kozlov, V.V. and Onischenko, D.A. (1982). Nonintegrability of Kirchhoff's equations. *Sov. Math. Dokl.* **26**, 495–8.

Krasny, R. (1986a). A study of singularity formation in a vortex sheet by the point-vortex approximation. *J. Fluid Mech.* **167**, 65–93.

Krasny, R. (1986b). Desingularization of periodic vortex sheet roll-up. *J. Comput. Phys.* **65**, 292–313.

Kreiss, H.-O. and Oliger, J. (1972). Comparison of accurate methods for the integration of hyperbolic equations. *Tellus* **24**(3), 199–215.

Krishnan, G.P. and Leighton Jr, D.T. (1995). Inertial lift on a moving sphere in contact with a plane wall in a shear flow. *Phys. Fluids* **7**(11), 2538–45.

Kurose, R. and Komori, S. (1999). Drag and lift forces on a rotating sphere in a linear shear flow. *J. Fluid Mech.* **384**, 183–206.

Lamb, Sir H. (1932). *Hydrodynamics*, 6th edn. Cambridge University Press.

Lanczos, C. (1938). Trigonometric interpolation of empirical and analytical functions. *J. Math. Phys.* **17**, 123–99.

Landau, L.D. and Lifshitz, E.M. (1987). *Fluid Mechanics*, 2nd edn. Pergamon Press.

Ledbetter, C.S. (1990). A historical perspective of scientific computing in Japan and the United States. *Supercomputing Rev.* **3**(11), 31–7.

Lee, H. and Balachandar, S. (2010). Drag and lift forces on a spherical particle moving on a wall in a shear flow at finite Re. *J. Fluid Mech.* **657**, 89–125.

Lee, H. and Balachandar, S. (2017). Effects of wall roughness on drag and lift forces of a particle at finite Reynolds number. *Int. J. Multiphase Flow*, **88**, 116–132.

Lee, H., Ha, M.Y. and Balachandar, S. (2011). Rolling/sliding of a particle on a flat wall in a linear shear flow at finite Re. *Int. J. Multiphase Flow* **37**(2), 108–124.

Legendre, D. and Magnaudet, J. (1998). The lift force on a spherical bubble in a viscous linear shear flow. *J. Fluid Mech.* **368**, 81–126.

Leighton, D. and Acrivos, A. (1985). The lift on a small sphere touching a plane in the presence of a simple shear flow. *ZAMP* **36**(1), 174–8.

Lele, S.K. (1992). Compact finite difference schemes with spectral-like resolution. *J. Comp. Phys.* **103**(1), 16–42.

Leonard, B.P. (1979). A stable and accurate convective modelling procedure based on quadratic upstream interpolation. *Comp. Meth. Appl. Mech. Eng.* **19**(1), 59–98.

LeVeque, R.J. (2007). *Finite Difference Methods for Ordinary and Partial Differential Equations: Steady-state and Time-dependent Problems.* SIAM.

LeVeque, R.J. and Li, Z. (1994). The immersed interface method for elliptic equations with discontinuous coefficients and singular sources. *SIAM J. Numer. Anal.* **31**, 1019.

Li, T.Y. and Yorke, J.A. (1975). Period three implies chaos. *Amer. Math. Monthly* **82**, 985–92.

Lighthill, M.J. (1978). *Waves in Fluids.* Cambridge University Press.

Lin, C.C. (1943). *On the Motion of Vortices in Two Dimensions.* University of Toronto Press.

Liseikin, V.D. (2009). *Grid Generation Methods.* Springer.

Lorenz, E.N. (1963). Deterministic nonperiodic flow. *J. Atmos. Sci.* **20**(2), 130–41.

Luke, Y.L. (1969). *The Special Functions and their Approximations.* Academic Press.

Lundgren, T.S. and Pointin, Y.B. (1977). Statistical mechanics of two-dimensional vortices. *J. Stat. Phys.* **17**, 323–55.

MacCormack, R.W. (1969). The effect of viscosity in hypervelocity impact cratering. AIAA Paper 69–354.

MacCormack, R.W. and Lomax, H. (1979). Numerical solution of compressible viscous flows. *Ann. Rev. Fluid Mech.* **11**, 289–316.

Mack, L.M. (1976). A numerical study of the temporal eigenvalue spectrum of the Blasius boundary layer. *J. Fluid Mech.* **73**(3), 497–520.

MacKay, R.S. and Meiss, J.D. (1987). *Hamiltonian Dynamical Systems: A Reprint Selection.* Adam Hilger.

Maeder, R.E. (1995). Function iteration and chaos. *Math. J.* **5**, 28–40.

Magnaudet, J., Rivero, M. and Fabre, J. (1995). Accelerated flows past a rigid sphere or a spherical bubble. Part 1: Steady straining flow. *J. Fluid Mech.* **284**, 97–135.

Mandelbrot, B.B. (1980). Fractal aspects of $z \to \Lambda z(1-z)$ for complex Λ and z. *Ann. N.Y. Acad. Sci.* **357**, 249–59.

Mandelbrot, B.B. (1983). *The Fractal Geometry of Nature.* Freeman.

Marella, S., Krishnan, S.L.H.H., Liu, H. and Udaykumar, H.S. (2005). Sharp interface Cartesian grid method I: an easily implemented technique for 3D moving boundary computations. *J. Comp. Phys.* **210**(1), 1–31.

Matsuda, K., Onishi, R., Hirahara, M., Kurose, R., Takahashi, K. and Komori, S. (2014). Influence of microscale turbulent droplet clustering on radar cloud observations. *J. Atmos. Sci.* **71**(10), 3569–82.

Maxey, M.R. and Riley, J.J. (1983). Equation of motion for a small rigid sphere in a non-uniform flow. *Phys. Fluids* **26**(4), 883–9.

May, R.M. (1976). Simple mathematical models with very complicated dynamics. *Nature* **261**, 459–67.

McLaughlin, J.B. (1991). Inertial migration of a small sphere in linear shear flows. *J. Fluid Mech.* **224**, 261–74.

McQueen, D.M. and Peskin, C.S. (1989). A three-dimensional computational method for blood flow in the heart. II. Contractile fibers. *J. Comput. Phys.* **82**, 289.

Mei, R. and Adrian, R.J. (1992). Flow past a sphere with an oscillation in the free-stream and unsteady drag at finite Reynolds number. *J. Fluid Mech.* **237**, 133–74.

Meiron, D.I., Baker, G.R. and Orszag, S.A. (1982). Analytic structure of vortex sheet dynamics 1. Kelvin–Helmholtz instability. *J. Fluid Mech.* **114**, 283–98.

Mercier, B. (1989). *An Introduction to the Numerical Analysis of Spectral Methods.* Springer.

Merilees, P.E. (1973). The pseudospectral approximation applied to the shallow water equations on a sphere. *Atmosphere* **11**(1), 13–20.

Mitchell, A.R. and Griffiths, D.F. (1980). *The Finite Difference Method in Partial Differential Equations.* John Wiley.

Mittal, R. (1999). A Fourier–Chebyshev spectral collocation method for simulating flow past spheres and spheroids. *Int. J. Numer. Meth. Fluids* **30**, 921–37.

Mittal, R. and Balachandar, S. (1996). Direct numerical simulations of flow past elliptic cylinders. *J. Comput. Phys.* **124**, 351–67.

Mittal, R. and Iaccarino, G. (2005). Immersed boundary methods. *Ann. Rev. Fluid Mech.* **37**, 239–61.

Mohd-Yusof, J. (1997). Combined immersed boundaries B-spline methods for simulations of flows in complex geometries. *CTR Annual Research Briefs.* NASA Ames, Stanford University.

Moin, P. (2010a). *Fundamentals of Engineering Numerical Analysis.* Cambridge University Press.

Moin, P. (2010b). *Engineering Numerical Analysis.* Cambridge University Press.

Moin, P. and Krishnan M. (1998). Direct numerical simulation: a tool in turbulence research. *Ann. Rev. Fluid Mech.* **30**(1), 539–78.

Moore, D.W. (1979). The spontaneous appearance of a singularity in the shape of an evolving vortex sheet. *Proc. Royal Soc. London A* **365**, 105–19.

Morton, K.W. (1980). Stability of finite difference approximations to a diffusion–convection equation. *Int. J. Numer. Method. Eng.* **15**(5), 677–83.

Moser, J. (1973). *Stable and Random Motions in Dynamical Systems.* Ann. Math. Studies No.77. Princeton University Press.

Onsager, L. (1949). Statistical hydrodynamics. *Nuovo Cim.* **6** (Suppl.), 279–87.

Orszag, S.A. (1969). Numerical methods for the simulation of turbulence. *Phys. Fluids* **12**(12), Supp. II, 250–7.

Orszag, S.A. (1970). Transform method for the calculation of vector-coupled sums: application to the spectral form of the vorticity equation. *J. Atmos. Sci.* **27**(6), 890–5.

Orszag, S.A. (1974). Fourier series on spheres. *Monthly Weather Rev.* **102**(1), 56–75.

Orszag, S.A. and Israeli, M. (1974). Numerical simulation of viscous incompressible flows. *Ann. Rev. Fluid Mech.* **6**, 281–318.

Orszag, S.A. and McLaughlin, N.N. (1980). Evidence that random behavior is generic for nonlinear differential equations. *Physica D* **1**, 68–79.

Orszag, S.A. and Patterson Jr, G.S. (1972). Numerical simulation of three-dimensional homogeneous isotropic turbulence. *Phys. Rev. Lett.* **28**(2), 76.

Orszag, S.A., Israeli, M. and Deville, M. (1986). Boundary conditions for incompressible flows. *J. Sci. Comput.* **1**, 75–111.

Oseen, C.W. (1927). *Hydrodynamik.* Akademische Verlagsgesellschaft.

Ottino, J.M. (1989). *The Kinematics of Mixing: Stretching, Chaos and Transport.* Cambridge University Press.

Patterson, G. (1978). Prospects for computational fluid mechanics. *Ann. Rev. Fluid Mech.* **10**, 289–300.

Peaceman, D.W. and Rachford, Jr, H.H. (1955). The numerical solution of parabolic and elliptic differential equations. *J. Soc. Indus. Appl. Math.* **3**(1), 28–41.

Peitgen, H.-O. and Richter, P.H. (1986). *The Beauty of Fractals.* Springer.

Peitgen, H.-O. and Saupe, D. (eds.) (1988). *The Science of Fractal Images.* Springer.

Peitgen, H.-O., Saupe, D. and Haessler, F.V. (1984). Cayley's problem and Julia sets. *Math. Intelligencer* **6**, 11–20.

Pember, R.B., Bell, J.B., Colella, P., Crutchfield, W.Y. and Welcome, M.L. (1995). An adaptive Cartesian grid method for unsteady compressible flow in irregular regions. *J. Comput. Phys.* **120**, 278–304.

Perot, J.B. (1993). An analysis of the fraction step method. *J. Comput. Phys.* **108**, 51–8.

Peskin, C.S. (1977). Numerical analysis of blood flow in the heart. *J. Comput. Phys.* **25**, 220–52.

Peyret, R. and Taylor, T.D. (1983). *Computational Methods for Fluid Flow.* Springer.

Pinelli, A., Naqavi, I.Z., Piomelli, U. and Favier, J. (2010). Immersed-boundary methods for general finite-difference and finite-volume Navier–Stokes solvers. *J. Comp. Phys.* **229**(24), 9073–91.

Pirozzoli, S. (2011). Numerical methods for high-speed flows. *Ann. Rev. Fluid Mech.* **43**, 163–94.

Plesset, M.S. and Prosperetti, A. (1977). Bubble dynamics and cavitation. *Ann. Rev. Fluid Mech.* **9**, 145–85.

Pletcher, R.H., Tannehill, J.C. and Anderson, D. (2012). *Computational Fluid Mechanics and Heat Transfer.* CRC Press.

Pozrikidis, C. (2011). *Introduction to Theoretical and Computational Fluid Dynamics.* Oxford University Press.

Pozrikidis, C. and Higdon, J.J.L. (1985). Nonlinear Kelvin–Helmholtz instability of a finite vortex layer. *J. Fluid Mech.* **157**, 225–63.

Press, W.H., Flannery, B.P., Teukolsky, S.A. and Vetterling, W.T. (1986). *Numerical Recipes: The Art of Scientific Programming.* Cambridge University Press.

Prosperetti, A. and Tryggvason, G. (eds.) (2009). *Computational Methods for Multiphase Flow.* Cambridge University Press.

Proudman, I. and Pearson, J.R.A. (1957). Expansions at small Reynolds numbers for the flow past a sphere and a circular cylinder. *J. Fluid Mech.* **2**(3), 237–62.

Rayleigh, Lord (1880). On the stability, or instability, of certain fluid motions. *Proc. London Math. Soc.* **11**, 57–72.

Rice, J.R. (1983). *Numerical Methods, Software, and Analysis*, IMSL Reference Edition. McGraw–Hill.

Richtmyer, R.D., and Morton, K.W. (1994). *Difference Methods for Initial-Value Problems*, 2nd edn. Krieger Publishing Co.

Roache, P.J. (1976). *Computational Fluid Dynamics.* Hermosa.

Roberts, K.V. and Christiansen, J.P. (1972). Topics in computational fluid mechanics. *Comput. Phys. Comm.* **3**(Suppl.), 14–32.

Roe, P.L. (1986). Characteristic-based schemes for the Euler equations. *Ann. Rev. Fluid Mech.* **18**, 337–65.

Roma, A.M., Peskin, C.S. and Berger, M.J. (1999). An adaptive version of the immersed boundary method. *J. Comp. Phys.* **153**(2), 509–34.

Rosenblum, L.J. (1995). Scientific visualization: advances and challenges. *IEEE Comp. Sci. Eng.* **2**(4), 85.

Rosenhead, L. (1931). The formation of vortices from a surface of discontinuity. *Proc. Roy. Soc. London Ser. A* **134**, 170–92.

Rosenhead, L. (ed.) (1963). *Laminar Boundary Layers.* Oxford University Press.

Rudoff, R.C. and Bachalo, W.D. (1988) Measurement of droplet drag coefficients in polydispersed turbulent flow fields. *AIAA Paper*, 88–0235.

Saad, Y. (2003). *Iterative Methods for Sparse Linear Systems.* SIAM.

Saff, E.B. and Kuijlaars, A.B. (1997). Distributing many points on a sphere. *Math. Intelligencer* **19**(1), 5–11.

Saffman, P.G. (1965). The lift on a small sphere in a slow shear flow. *J. Fluid Mech.*, **22**, 385–400.

Saiki, E.M. and Biringen, S. (1996). Numerical simulation of a cylinder in uniform flow: application of a virtual boundary method. *J. Comput. Phys.* **123**, 450–65.

Sankagiri, S. and Ruff, G.A. (1997). Measurement of sphere drag in high turbulent intensity flows. *Proc. ASME FED.* **244**, 277–82.

Sansone, G. (1959). *Orthogonal Functions.* Courier Dover Publications.

Sato, H. and Kuriki, K. (1961). The mechanism of transition in the wake of a thin flat plate placed parallel to a uniform flow. *J. Fluid Mech.* **11**(3), 321–52.

Schlichting, H. (1968). *Boundary-Layer Theory.* McGraw–Hill.

Shampine, L.F. and Gordon, M.K. (1975). *Computer Solution of Ordinary Differential Equations.* W.H. Freeman and Co.

Shirayama, S. (1992). Flow past a sphere: topological transitions of the vorticity field. *AIAA J.* **30**, 349–58.

Shu, S.S. (1952). Note on the collapse of a spherical cavity in a viscous, incompressible fluid. In *Proc. First US Nat. Congr. Appl. Mech. ASME*, pp. 823–5.

Siemieniuch, J.L. and I. Gladwell. (1978). Analysis of explicit difference methods for a diffusion–convection equation. *Int. J. Numer. Method. Eng.* **12**(6), 899–916.

Silcock, G. (1975). *On the Stability of Parallel Stratified Shear Flows.* PhD Dissertation, University of Bristol.

Smereka, P., Birnir, B. and Banerjee, S. (1987). Regular and chaotic bubble oscillations in periodically driven pressure fields. *Phys. Fluids* **30**, 3342–50.

Smith, G.D. (1985). *Numerical Solution of Partial Differential Equations: Finite Difference Methods.* Oxford University Press.

Sommerfeld, A. (1964). *Mechanics of Deformable Bodies.* Academic Press.

Spalart, P.R. (2009). Detached-eddy simulation. *Ann. Rev. Fluid Mech.* **41**, 181–202.

Squire, H.B. (1933). On the stability for three-dimensional disturbances of viscous fluid flow between parallel walls. *Proc. Roy. Soc. London A* **142**, 621–8.

Stewartson, K. (1954). Further solutions of the Falkner–Skan equation. *Math. Proc. Cambridge Phil. Soc.* **50**(3), 454–65.

Strang, G. (2016). *Introduction to Linear Algebra*, 5th edn. Wellesley-Cambridge Publishers.

Streett, C.L. and Macaraeg, M. (1989). Spectral multi-domain for large-scale fluid dynamics simulations. *Appl. Numer. Math.* **6**, 123–39.

Struik, D.J. (1961). *Lectures on Classical Differential Geometry*, 2nd edn. Addison–Wesley. (Reprinted by Dover Publications, 1988.)

Swanson, P.D. and Ottino, J.M. (1990). A comparative computational and experimental study of chaotic mixing of viscous fluids. *J. Fluid Mech.* **213**, 227–49.

Taneda, S. (1963). The stability of two-dimensional laminar wakes at low Reynolds numbers. *J. Phys. Soc. Japan* **18**(2), 288–96.

Temam, R. (1969). Sur l'approximation de la solution des équations de Navier–Stokes par la méthode des pas fractionnaires (II). *Arch. Ration. Mech. Anal.* **33**, 377–85.

Tennetti, S., Garg, R. and Subramaniam, S. (2011). Drag law for monodisperse gas–solid systems using particle-resolved direct numerical simulation of flow past fixed assemblies of spheres. *Int. J. Multiphase Flow* **37**(9), 1072–92.

Theofilis, V. (2011). Global linear instability. *Ann. Rev. Fluid Mech.* **43**, 319–52.

Thomas, J.W. (2013). *Numerical Partial Differential Equations: Finite Difference Methods.* Springer.

Thompson, J.F., Warsi, Z.U.A. and Mastin, C.W. (1982). Boundary-fitted coordinate systems for numerical solution of partial differential equations – a review. *J. Comp. Phys.* **47**, 1–108.

Thompson, J.F., Warsi, Z.U.A. and Mastin, C.W. (1985). *Numerical Grid Generation.* North–Holland.

Tomboulides, A.G. and Orszag, S.A. (2000). Numerical investigation of transitional and weak turbulent flow past a sphere. *J. Fluid Mech.* **416**, 45–73.

Toro, E.F. (1999). *Riemann Solvers and Numerical Methods for Fluid Dynamics.* Springer.

Trefethen, L.N. (1982). Group velocity in finite difference schemes. *SIAM Rev.* **24**(2), 113–36.

Trefethen, L.N. and Embree, M. (2005). *Spectra and Pseudospectra: The Behavior of Non-normal Matrices and Operators.* Princeton University Press.

Tseng, Y.-H. and Ferziger, J.H. (2003). A ghost-cell immersed boundary method for flow in complex geometry. *J. Comp. Phys.* **192**(2), 593–623.

Udaykumar, H.S., Kan, H.-C., Shyy, W. and Tran-son-Tay, R. (1997). Multiphase dynamics in arbitrary geometries on fixed Cartesian grids. *J. Comput. Phys.* **137**, 366–405.

Udaykumar, H.S., Mittal, R. and Shyy, W. (1999). Solid–fluid phase front computations in the sharp interface limit on fixed grids. *J. Comput. Phys.* **153**, 535–74.

Udaykumar, H.S., Mittal, R., Rampunggoon, P. and Khanna, A. (2001). A sharp interface Cartesian grid method for simulating flows with complex moving boundaries. *J. Comput. Phys.* **174**, 345–80.

Uhlherr, P.H.T. and Sinclair, C.G. (1970). The effect of freestream turbulence on the drag coefficients of spheres. *Proc. Chemca.* **1**, 1–12.

Uhlmann, M. (2005). An immersed boundary method with direct forcing for the simulation of particulate flows. *J. Comp. Phys.* **209**(2), 448–76.

Uhlmann, M. and Doychev, T. (2014). Sedimentation of a dilute suspension of rigid spheres at intermediate Galileo numbers: the effect of clustering upon the particle motion. *J. Fluid Mech.* **752**, 310–48.

Ulam, S.M. and von Neumann, J. (1947). On combinations of stochastic and deterministic processes. *Bull. Amer. Math. Soc.* **53**, 1120.

van de Vooren, A.I. (1980). A numerical investigation of the rolling-up of vortex sheets. *Proc. Roy. Soc. London A.* **373**, 67–91.

van Dyke, M. (1964). *Perturbation Methods in Fluid Mechanics.* Academic Press.

van Dyke, M. (1982). *An Album of Fluid Motion.* Parabolic Press.

van Dyke, M. (1994). Computer-extended series. *Ann. Rev. Fluid Mech.* **16**(1), 287–309.

van Kan, J. (1986). A second-order accurate pressure-correction scheme for viscous incompressible flow. *SIAM J. Sci. Stat. Comput.* **7**(3), 870–91.

van Leer, B. (1974). Towards the ultimate conservative difference scheme. II. Monotonicity and conservation combined in a second-order scheme. *J. Comp. Phys.* **14**(4), 361–70.

Varga, R.S. (1962). *Matrix Iterative Methods.* Prentice–Hall Inc.

Versteeg, H.K. and Malalasekera, W. (2007). *An Introduction to Computational Fluid Dynamics: The Finite Volume Method.* Pearson Education.

Vrscay, E.R. (1986). Julia sets and Mandelbrot-like sets associated with higher-order Schröder rational iteration functions: a computer assisted study. *Math. Comp.* **46**, 151–69.

Wang, M., Freund, J.B. and Lele, S.K. (2006). Computational prediction of flow-generated sound. *Ann. Rev. Fluid Mech.* **38**, 483–512.

Warnica, W.D., Renksizbulut, M. and Strong, A.B. (1994). Drag coefficient of spherical liquid droplets. *Exp. Fluids* **18**, 265–70.

Wazzan, A.R., Okamura, T. and Smith, A.M.O. (1968). The stability of water flow over heated and cooled flat plates. *J. Heat Transfer* **90**(1), 109–14.

Wendroff, B. (1969). *First Principles of Numerical Analysis: An Undergraduate Text.* Addison–Wesley.

White, F.M. (1974). *Viscous Flow Theory.* McGraw–Hill.

Whittaker, E.T. (1937). *A Treatise on the Analytical Mechanics of Particles and Rigid Bodies*, 4th edn. Cambridge University Press.

Wilkinson, J.H. (1965). *The Algebraic Eigenvalue Problem.* Clarendon Press.

Yakhot, V., Bayly, B. and Orszag, S.A. (1986). Analogy between hyperscale transport and cellular automaton fluid dynamics. *Phys. Fluids* **29**, 2025–7.

Ye, T., Mittal, R., Udaykumar, H.S. and Shyy, W. (1999). An accurate Cartesian grid method for viscous incompressible flows with complex immersed boundaries. *J. Comput. Phys.* **156**, 209–40.

Yee, S.Y. (1981). Solution of Poisson's equation on a sphere by truncated double Fourier series. *Monthly Weather Rev.* **109**(3), 501–5.

Yeung, P.K. and Pope, S.B. (1988). An algorithm for tracking fluid particles in numerical simulations of homogeneous turbulence. *J. Comp. Phys.* **79**(2), 373–416.

Young, D.M. (2014). *Iterative Solution of Large Linear Systems.* Elsevier.

Young, D.M. and Gregory, R.T. (1988). *A Survey of Numerical Mathematics*, Vols. I and II. Dover Publications.

Zabusky, N.J. (1987). A numerical laboratory. *Phys. Today* **40**, 28–37.

Zabusky, N.J. and Kruskal, M.D. (1965). Interaction of solitons in a collisionless plasma and the recurrence of initial states. *Phys. Rev. Lett.* **15**(6), 240.

Zabusky, N.J. and Melander, M.V. (1989). Three-dimensional vortex tube reconnection: morphology for orthogonally-offset tubes. *Physica D* **37**, 555–62.

Zarin, N.A. and Nicholls, J.A. (1971). Sphere drag in solid rockets – non-continuum and turbulence effects. *Comb. Sci. Tech.* **3**, 273–80.

Zeng, L., Najjar, F., Balachandar, S. and Fischer, P. (2009). Forces on a finite-sized particle located close to a wall in a linear shear flow. *Phys. Fluids* **21**(3), 033302.

Zhang, L., Gerstenberger, A., Wang, X. and Liu, W.K. (2004). Immersed finite element method. *Comp. Meth. Appl. Mech. Eng.* **193**(21), 2051–67.

Index